T0330923

Stochastic Volatility Modeling

CHAPMAN & HALL/CRC
Financial Mathematics Series

Aims and scope:
The field of financial mathematics forms an ever-expanding slice of the financial sector. This series aims to capture new developments and summarize what is known over the whole spectrum of this field. It will include a broad range of textbooks, reference works and handbooks that are meant to appeal to both academics and practitioners. The inclusion of numerical code and concrete real-world examples is highly encouraged.

Series Editors

M.A.H. Dempster
Centre for Financial Research
Department of Pure
Mathematics and Statistics
University of Cambridge

Dilip B. Madan
Robert H. Smith School
of Business
University of Maryland

Rama Cont
Department of Mathematics
Imperial College

Published Titles

American-Style Derivatives; Valuation and Computation, *Jerome Detemple*

Analysis, Geometry, and Modeling in Finance: Advanced Methods in Option
   Pricing, *Pierre Henry-Labordère*

Commodities, *M. A. H. Dempster and Ke Tang*

Computational Methods in Finance, *Ali Hirsa*

Counterparty Risk and Funding: A Tale of Two Puzzles, *Stéphane Crépey and
   Tomasz R. Bielecki, With an Introductory Dialogue by Damiano Brigo*

Credit Risk: Models, Derivatives, and Management, *Niklas Wagner*

Engineering BGM, *Alan Brace*

Financial Mathematics: A Comprehensive Treatment, *Giuseppe Campolieti and
   Roman N. Makarov*

Financial Modelling with Jump Processes, *Rama Cont and Peter Tankov*

Interest Rate Modeling: Theory and Practice, *Lixin Wu*

Introduction to Credit Risk Modeling, Second Edition, *Christian Bluhm,
   Ludger Overbeck, and Christoph Wagner*

An Introduction to Exotic Option Pricing, *Peter Buchen*

Introduction to Risk Parity and Budgeting, *Thierry Roncalli*

Introduction to Stochastic Calculus Applied to Finance, Second Edition,
   Damien Lamberton and Bernard Lapeyre

Monte Carlo Methods and Models in Finance and Insurance, *Ralf Korn, Elke Korn,
   and Gerald Kroisandt*

Monte Carlo Simulation with Applications to Finance, *Hui Wang*

Nonlinear Option Pricing, *Julien Guyon and Pierre Henry-Labordère*

Numerical Methods for Finance, *John A. D. Appleby, David C. Edelman,
   and John J. H. Miller*

Option Valuation: A First Course in Financial Mathematics, *Hugo D. Junghenn*

Portfolio Optimization and Performance Analysis, *Jean-Luc Prigent*

Quantitative Finance: An Object-Oriented Approach in C++, *Erik Schlögl*

Quantitative Fund Management, *M. A. H. Dempster, Georg Pflug,
    and Gautam Mitra*

Risk Analysis in Finance and Insurance, Second Edition, *Alexander Melnikov*

Robust Libor Modelling and Pricing of Derivative Products, *John Schoenmakers*

Stochastic Finance: An Introduction with Market Examples, *Nicolas Privault*

Stochastic Finance: A Numeraire Approach, *Jan Vecer*

Stochastic Financial Models, *Douglas Kennedy*

Stochastic Processes with Applications to Finance, Second Edition,
    Masaaki Kijima

Stochastic Volatility Modeling, *Lorenzo Bergomi*

Structured Credit Portfolio Analysis, Baskets & CDOs, *Christian Bluhm
    and Ludger Overbeck*

Understanding Risk: The Theory and Practice of Financial Risk Management,
    David Murphy

Unravelling the Credit Crunch, *David Murphy*

Proposals for the series should be submitted to one of the series editors above or directly to:
CRC Press, Taylor & Francis Group
3 Park Square, Milton Park
Abingdon, Oxfordshire OX14 4RN
UK

Chapman & Hall/CRC FINANCIAL MATHEMATICS SERIES

Stochastic Volatility Modeling

Lorenzo Bergomi

CRC Press
Taylor & Francis Group
Boca Raton   London   New York

CRC Press is an imprint of the
Taylor & Francis Group, an **informa** business
A CHAPMAN & HALL BOOK

Typeset in Linux Libertine

CRC Press
Taylor & Francis Group
6000 Broken Sound Parkway NW, Suite 300
Boca Raton, FL 33487-2742

© 2016 by Taylor & Francis Group, LLC
CRC Press is an imprint of Taylor & Francis Group, an Informa business

No claim to original U.S. Government works

Printed on acid-free paper
Version Date: 20151105

International Standard Book Number-13: 978-1-4822-4406-9 (Hardback)

Library of Congress Cataloging-in-Publication Data

Names: Bergomi, Lorenzo, author.
Title: Stochastic volatility modeling / Lorenzo Bergomi.
Description: Boca Raton : CRC Press, [2016] | Series: Chapman & Hall/CRC financial mathematics series | Includes bibliographical references and index.
Identifiers: LCCN 2015037873 | ISBN 9781482244069 (alk. paper)
Subjects: LCSH: Finance--Mathematical models. | Securities--Mathematical models. | Stochastic models.
Classification: LCC HG106 .B39 2016 | DDC 332.63/2220151922--dc23
LC record available at http://lccn.loc.gov/2015037873

Visit the Taylor & Francis Web site at
http://www.taylorandfrancis.com

and the CRC Press Web site at
http://www.crcpress.com

Contents

Preface

Pour soulever un poids si lourd
Sisyphe, il faudrait ton courage!
Bien qu'on ait du cœur à l'ouvrage,
L'Art est long et le Temps est court.

Baudelaire, *Les Fleurs du mal*

Tu quid? Quousque sub alio moveris? Impera et dic quod memoriae tradatur,
aliquid et de tuo profer.

Seneca, *Letters to Lucilius, XXXIII*

C'est icy un livre de bonne foy, lecteur.

Montaigne, *Essais*

– This, Reader, is an honest book. It warns you at the outset that it is not a treatise on stochastic volatility.

Nor is it a mathematical finance textbook – there are treatises and textbooks galore on the shelves of bookshops and university libraries.

Rather, my intention has been to explain how stochastic volatility – and which kind of stochastic volatility – can be used to address practical issues arising in the modeling of derivatives.

Modeling in finance is an engineering field: while our task as engineers is to frame problems in mathematical terms and solve them using sophisticated machinery whenever necessary, the problems themselves originate in the form of embarrassingly practical trading questions.

I have been fortunate to have spent my career as a quant in an institution – Société Générale – and in a field – equity derivatives – that the derivatives community has come to associate. I have tried to convey the experience thus gained. My main objective has been to clearly state the motivation for each question I address and to represent the thought process one follows when designing or using a model.

A model's (ir)relevance is measured by its (in)ability to account for all non-linearities of a derivative payoff with respect to hedging instruments, and to adequately reflect the former in the prices it produces.

Thus have I often lamented the propensity of "pricing quants" or "expectation calculators" to envision their task as that of producing (real) numbers auspiciously called prices, and to favor analytical tractability or computational speed at the expense of model relevance. What is the point in ultrafast mispricing?[1]

---

[1]Kind souls will allege it is still preferable to ultraslow mispricing.

This book is intended to be read in sequence. It assumes familiarity with the basic concepts and models of quantitative finance. The motivation for stochastic volatility is the subject of the first chapter; this is followed by a chapter on local volatility, both a special breed of stochastic volatility and a market model, which I survey as such.

I urge you, Reader, to read it fully, as I introduce notions and discuss issues that are referred to repeatedly throughout the book. After a warm-up with forward-start options we embark on stochastic volatility. Jump and Lévy models are briefly dealt with at the end of Chapter 10.

This work was written mostly at night, during hours normally devoted to rest and sleep. Thus, dear Reader, I beg your forgiveness. You will do me a great service by reporting typos, inaccuracies, downright errors, to lb.svbook@gmail.com.

For everything I have learned so far I have to be thankful to more people than my memory can remember. I wish to thank the practitioners and academics that I have met regularly at conferences and seminars, and also fellow workers at Société Générale: quants, traders, structurers.

I am especially indebted to my coworkers in the Quantitative Research team, past and present. They will find here a reflection of the very many discussions we have had over the years.

The generous help of colleagues – quants and traders – in proofreading the manuscript is also gratefully acknowledged, chief among them Florent Bersani and Pierre Henry-Labordère whose eagle eyes have helped clear the text of many imprecisions.

Finally, I would like to express my gratitude to Rena, Elisa and Chiara for their patience.

Thank you Rena for occasionally filling the loneliness of nightly writing sessions with music. This has resulted for the author in some unlikely pairings, e.g. local volatility and the piano part of the Kreutzer sonata, the Heston model and the Chopin Barcarolle, Rachmaninoff's Third and options on variance, etc.

Paris, Summer 2015

Chapter 1

Introduction

Why would a trader use a stochastic volatility model? What for? Which issues does one address by using a stochastic volatility model? Why aren't practitioners content with just delta-hedging their derivative books? These are the questions we address in this introduction.

We begin our analysis by reviewing the Black-Scholes model and how it is used on trading desks. It may come as a surprise to many that, despite the widely publicized inconsistency between the actual dynamics of financial securities, as observed in reality, and the idealized lognormal dynamics that the Black-Scholes model postulates, it is still used daily in banks to risk-manage derivative books.

One may think that a model derives its legitimacy and usefulness from the accuracy with which it captures the historical dynamics of the underlying security – hence the scorn demonstrated by econometricians and econophysicists for the Black-Scholes model and its simplistic assumptions, upon first encounter. With regard to models, many are used to a normative thought process. Given the behavior of securities' prices – as specified for example by an xx-GARCH or xxx-GARCH model – this is what the price of a derivative should be. Models not conforming to such type of specification – or to some canonical set of *stylized facts* – are deemed "wrong".

This would be suitable if (a) the realized dynamics of securities benevolently complied with the model's specification and (b) if practitioners only engaged in delta-hedging. The dynamics of real securities, however, is not regular enough, nor can it be characterized with sufficient accuracy that the normative stance is appropriate. Moreover, such an approach skirts the issue of dynamical trading in options – at market prices – and of marking to market.

Rather than calibrating their favorite model to historical data for the spot process, and, armed with it and trusting its seaworthiness, endeavor to ride out the rough seas of financial markets, derivatives practitioners will be content with barely floating safely and making as few assumptions as possible about future market conditions.

Still, this requires some modeling infrastructure – hence this book: while they do not use the models' predictive power and may have little confidence in the reliability of the models' underlying assumptions, practitioners do need and make use of the models' *pricing equations*. This is an important distinction: while the Black-Scholes *model* is not used on derivatives desks, everybody uses the Black-Scholes *pricing equation*.

Indeed, a pricing equation is essentially an analytical accounting device: rather than predicting anything about the future dynamics of the underlying securities, a model's pricing equation supplies a decomposition of the profit and loss (P&L) experienced on a derivative position as time elapses and securities' prices move about. It allows its user to anticipate the sign and size of the different pieces in his/her P&L. We will illustrate this below with the example of the Black-Scholes equation, which could be motivated by elementary break-even *accounting* criteria.

More sophisticated models enable their users to characterize more precisely their P&L and the conditions under which it vanishes, for example by separating contributions from different effects that may be lumped together in simpler models. Again, the issue, from a practitioner's perspective, is not to be able to *predict* anything, but rather to be able to *differentiate* risks generated by these different contributions to his/her P&L and to ensure that the model offers the capability of *pricing* these different types of risk consistently across the book at levels that can be *individually controlled*.

It is then a trading decision to either hedge away some of these risks, by taking offsetting positions in more liquid – say vanilla – options or by taking offsetting positions in other exotic derivatives, or to keep these risks on the book.

1.1 Characterizing a usable model – the Black-Scholes equation

Imagine we are sitting on a trading desk and are tasked with pricing and risk-managing a short position in an option – say a European option of maturity T whose payoff at $t = T$ is $f(S_T)$, where S is the underlying.

The bank quants have coded up a pricing function: $P(t, S)$ is the option's price in the library model. Assume we don't know anything about what was implemented. How can we assess whether using the black-box pricing function $P(t, S)$ for risk-managing a derivative position is safe, that is whether the library model is usable?

We assume here that the underlying is the only hedging instrument we use. The case of multiple hedging instruments is examined next.

- The first sanity check we perform is set $t = T$ and check that P equals the payoff:

$$P(t = T, S) = f(S), \forall S. \tag{1.1}$$

Provided (1.1) holds, we proceed to consider the P&L of a delta-hedged position. For the purpose of splitting the total P&L incurred over the option's lifetime into pieces that can be ascribed to each time interval in between two successive delta rehedges, we can assume that we sell the option at time t, buy it back at $t + \delta t$ then start over again. δt is typically 1 day. Let Δ be the number of shares we buy at t as delta-hedge.

Our P&L consists of two pieces: the P&L of the option itself, of which we are short, which comprises interest earned on the premium received at t, and the P&L generated by the delta-hedge, which incorporates interest we pay on money we have borrowed to buy Δ shares, as well as money we make by lending shares out during δt:

$$P\&L = -[P(t+\delta t, S+\delta S) - P(t,S)] + rP(t,S)\,\delta t + \Delta(\delta S - rS\delta t + qS\delta t)$$

where δS is the amount by which S moves during δt. r is the interest rate and q the repo rate, inclusive of dividend yield.

How should we choose Δ? We pick $\Delta = \frac{dP}{dS}$ so as to cancel the first-order term in δS in the P&L above.

We now expand the P&L in powers of δS and δt. We would like to stop at the lowest non-trivial orders for δt and δS: order one in δt, and order two in δS, as the order one contribution is canceled by the delta-hedge. What about cross-terms such as $\delta S \delta t$?

In practice, this term, as well as higher order terms in δS, are smaller than δS^2 and δt terms. Indeed, to a good approximation, the variance of returns scales linearly with their time scale, thus $\langle \delta S^2 \rangle$ is of order δt and δS is of order $\sqrt{\delta t}$.[1] The contributions at order one in δt and order two in δS are then both of order δt while the cross-term $\delta S \delta t$ and terms of higher order in δS are of higher order in δt, thus become negligible as $\delta t \to 0$.[2]

We then get the following expression for our carry P&L – the standard denomination for the P&L of a hedged option position:

$$P\&L = -\left(\frac{dP}{dt} - rP + (r-q)S\frac{dP}{dS}\right)\delta t - \frac{1}{2}S^2\frac{d^2P}{dS^2}\left(\frac{\delta S}{S}\right)^2 \quad (1.2)$$

- The first piece – called the theta portion – is deterministic. It is given by the time derivative of the option's price (sometimes theta is used to denote $\frac{dP}{dt}$ only), corrected for the financing cost/gain during δt of the delta hedge and the premium.

---

[1] The property that the variance of returns scales linearly with their time scale is equivalent to the property that returns have no serial correlation. Securities' returns do in fact exhibit some amount of serial correlation at varying time scales, of the order of several days down to shorter time scales and this is manifested in the existence of "statistical arbitrage" desks. Serial correlation in itself is of no consequence for the pricing of derivatives, however the measure of realized volatility will depend on the time scale of returns used for its estimation. As will be clear shortly, for a derivatives book, the relevant time scale is that of the delta-hedging frequency.

[2] How small should δt be so that this is indeed the case? The order of magnitude of δS is $S\sigma\sqrt{\delta t}$ where σ is the volatility of S. It turns out that for equities, volatility levels are such that for $\delta t = 1$ day, higher order terms can usually be ignored. There is nothing special about daily delta rebalancing; for much higher volatility levels, intra-day delta re-hedging would be mandatory.

- The second piece is random and quadratic in δS, as the linear term is cancelled by the delta position. $\frac{d^2 P}{dS^2}$ is called "gamma". We usually prefer to work with the "dollar gamma" $S^2 \frac{d^2 P}{dS^2}$, as it has the same dimension as P.

Our daily P&L reads:

$$P\&L = -A(t,S)\,\delta t - B(t,S)\left(\frac{\delta S}{S}\right)^2 \tag{1.3}$$

where $A = \left(\frac{dP}{dt} - rP + (r-q)S\frac{dP}{dS}\right)$ and $B = \frac{1}{2}S^2\frac{d^2 P}{dS^2}$. Because the second piece in the P&L is random we cannot demand that the P&L vanish altogether.

- What if $A \geq 0$ and $B \geq 0$? We lose money, regardless of the value of δS. This means P cannot be used for risk-managing our option. The initial price $P(t=0, S_0)$ we have charged is too low. We should have charged more so as not to keep losing money as we delta-hedge our option.

- What if $A \leq 0$ and $B \leq 0$? We make "free" money, regardless of δS. While less distressing than persistently losing money, the consequence is identical: P cannot be used for risk-managing our option. The initial price $P(t=0, S_0)$ we have charged is too high.

- The model is thus usable only if the signs of $A(t,S)$ and $B(t,S)$ are different, $\forall t, \forall S$. The values of $\frac{\delta S}{S}$ such that money is neither made nor lost are $\frac{\delta S}{S} = \pm\sqrt{-\frac{A(t,S)}{B(t,S)}}\sqrt{\delta t}$.

This condition is necessary, otherwise the model is unusable. We now introduce a further reasonable requirement.

While daily returns are random, empirically their squares average out over time to their realized variance. Let us call $\hat\sigma$ the (lognormal) historical volatility of S: $\left\langle\left(\frac{\delta S}{S}\right)^2\right\rangle = \hat\sigma^2\delta t$. Requiring that we do not lose or make money on average is a natural risk-management criterion – it reads: $A(t,S) = -\hat\sigma^2 B(t,S), \forall S, \forall t$.

- Replacing A and B with their respective expression yields the following identity that $P_{\hat\sigma}$ ought to obey:

$$\frac{dP_{\hat\sigma}}{dt} - rP_{\hat\sigma} + (r-q)S\frac{dP_{\hat\sigma}}{dS} = -\frac{\hat\sigma^2}{2}S^2\frac{d^2 P_{\hat\sigma}}{dS^2} \tag{1.4}$$

where subscript $\hat\sigma$ keeps track of the dependence of P on the break-even level of volatility $\hat\sigma$.

Plugging now in (1.2) the expression for $\left(\frac{dP}{dt} - rP + (r-q)S\frac{dP}{dS}\right)$ in (1.4) yields:

$$P\&L = -\frac{S^2}{2}\frac{d^2 P_{\hat\sigma}}{dS^2}\left(\frac{\delta S^2}{S^2} - \hat\sigma^2\delta t\right) \tag{1.5}$$

The condition for the two pieces in the P&L to offset each other is then expressed very simply as a condition on the realized variance of S: the P&L will be positive or negative depending upon whether $\frac{\delta S^2}{S^2}$ is larger or smaller than $\widehat{\sigma}^2 \delta t$.

In the absence of a volatility market for S, $\widehat{\sigma}$ should be chosen as our best estimate of future realized volatility, weighted by the option's dollar gamma.[3]

For vanilla options that can be bought or sold at market prices we can define the notion of implied volatility – hence the hat: $\widehat{\sigma}$ is such that $P_{\widehat{\sigma}}$ is equal to the market price of the option considered.

(1.4) is in fact the Black-Scholes equation. Together with condition (1.1) it defines $P_{\widehat{\sigma}}(t, S)$.

Starting from expression (1.2) for our P&L and imposing the basic accounting criterion that the P&L vanish for $\left(\frac{\delta S}{S}\right)^2 = \widehat{\sigma}^2 \delta t$, at order one in δt and two in δS, a (gifted) trader would thus have obtained the Black-Scholes pricing equation (1.4), though he may not have known anything about Brownian motion and may have been reluctant to assume that real securities are lognormal. The Black-Scholes model is typical of the market models considered in this book:

- there exists a well-defined break-even level for $\left(\frac{\delta S}{S}\right)^2$ such that the P&L at order two in δS of a delta-hedged position vanishes,

- this break-even level does not depend on the specific payoff of the option at hand.

This last condition is important: should the gamma of an options portfolio vanish – that is the portfolio is locally riskless – then theta should vanish as well. If break-even levels were payoff-dependent, we could possibly run into one of the two absurd situations considered above, with $B = 0$ and $A \neq 0$, at the portfolio level.

A model not conforming to these criteria is unsuitable for trading purposes.[4]

Multiple hedging instruments

What if our pricing function is a function of several asset values: $P(t, S_1 \ldots S_n)$ where the S_i are market values of our hedge instruments – either different underlyings, or one underlying and its associated vanilla options?

---

[3]This is not exactly true. Equation (1.5) shows that the situation is different depending on whether our position is short gamma $\left(\frac{d^2 P}{dS^2} > 0\right)$ or long gamma $\left(\frac{d^2 P}{dS^2} < 0\right)$. In the short gamma situation, our gain is bounded while our loss is potentially unbounded – the reverse is true in the long gamma situation: our bid/offer levels for $\widehat{\sigma}$ will likely be shifted with respect to an unbiased estimate of future realized volatility.

[4]We may have more complex requirements, for example that our P&L vanishes on average, *inclusive* of P&Ls generated by stress-tests scenarios, or inclusive of a tax levied by the bank on our desk to cover losses generated by these stress test-scenarios. This leads to a different pricing equation than (1.4) – see Appendix A of Chapter 10, page 407.

Exceptions to the rule that break-even levels should not depend on the payoff occur if we explicitly demand that $\widehat{\sigma}$ be an increasing function of $S^2 \frac{d^2 P_{\widehat{\sigma}}}{dS^2}$, to ensure that, for larger gammas, the ratio of theta to gamma is increased, for the sake of conservativeness, with the deliberate consequence that the resulting model is non-linear. One example is the Uncertain Volatility Model, covered in Appendix A of Chapter 2.

Running through the same derivation that led to (1.3), the P&L in the multi-asset case reads:

$$P\&L = -A(t, S)\,\delta t - \frac{1}{2}\Sigma_{ij}\phi_{ij}(t, S)\frac{\delta S_i}{S_i}\frac{\delta S_j}{S_j} \tag{1.6}$$

where $\phi_{ij}(t, S) = S_i S_j \left.\frac{d^2 P}{dS_i dS_j}\right|_{t,S}$ and S denotes the vector of the S_i.

Let us diagonalize ϕ, a real symmetric matrix, and denote by φ_k its eigenvalues and T_k the associated eigenvectors. Also denote by φ the diagonal matrix with the φ_k on its diagonal. We have:

$$\phi = T\varphi T^{\mathsf{T}}$$

The gamma portion of our P&L can be rewritten as:

$$\Sigma_{ij}\phi_{ij}\frac{\delta S_i}{S_i}\frac{\delta S_j}{S_j} = U^{\mathsf{T}}\phi U = U^{\mathsf{T}}T\varphi T^{\mathsf{T}}U = (T^{\mathsf{T}}U)^{\mathsf{T}}\varphi(T^{\mathsf{T}}U) = \Sigma_k\varphi_k\delta z_k{}^2$$

where $U_i = \frac{\delta S_i}{S_i}$ and $\delta z_k = T_k^{\mathsf{T}}U$. Our P&L now reads:

$$P\&L = -A\delta t - \frac{1}{2}\Sigma_k\varphi_k\delta z_k{}^2$$

which is a sum of P&Ls of the type in (1.3).

The δz_k are variations of particular baskets of the hedge instruments S_i. These baskets can be considered our effective hedge instruments, since the T_k form a basis.

$\delta z_k{}^2$ is always positive. As in the mono-asset case, the condition for our model to be usable is that there exist n positive numbers ω_k such that:

$$A = -\frac{1}{2}\Sigma_k\varphi_k\omega_k \tag{1.7}$$

so that our P&L reads:

$$P\&L = -\frac{1}{2}\Sigma_k\varphi_k\left(\delta z_k{}^2 - \omega_k\delta t\right)$$

Let us express A differently, so as to give our P&L in (1.6) a more symmetrical form. Denote by ω the diagonal matrix with the ω_k on the diagonal. We have:

$$A = -\frac{1}{2}\Sigma_k\varphi_k\omega_k = -\frac{1}{2}\mathrm{tr}(\varphi\omega) = -\frac{1}{2}\mathrm{tr}(T^{\mathsf{T}}\phi T\omega) = -\frac{1}{2}\mathrm{tr}\left(\phi T\omega T^{\mathsf{T}}\right) = -\frac{1}{2}\mathrm{tr}\left(\phi C\right)$$
$$= -\frac{1}{2}\Sigma_{ij}\phi_{ij}C_{ij}$$

where $C = T\omega T^{\mathsf{T}}$ is a positive matrix by construction, as the ω_k are positive.

Our P&L then reads:

$$P\&L = -\frac{1}{2}\Sigma_{ij}\phi_{ij}\left(\frac{\delta S_i}{S_i}\frac{\delta S_j}{S_j} - C_{ij}\delta t\right) \tag{1.8}$$

Because C is a positive matrix, it can be interpreted as an (implied) covariance matrix; its elements are implied covariance break-even levels.

We have just shown that on the condition that our model is usable, there exists a positive break-even covariance matrix C such that our P&L reads as in (1.8).

In our construction C is given by: $C = TwT^\intercal$, based on expression (1.7) for A. Is this restrictive, or is P&L (1.8) guaranteed to be nonsensical, for *any* positive matrix C? The answer is yes.[5]

Conclusion

In the general case of multiple hedge instruments, the condition that our model is usable – no situation in which our carry P&L is systematically positive or negative – is that there exists a positive break-even covariance matrix $C(t, S)$, $\forall S, \forall t$, such that the model's theta and cross-gammas are related through:

$$A = -\frac{1}{2}\Sigma_{ij}\phi_{ij}C_{ij}$$

Again, a model not meeting this criterion is unsuitable for trading purposes. In the sequel, suitable models are also called *market models*.

The important thing here is that cross-gammas ϕ_{ij} involve derivatives with respect to values of *actual hedge instruments*, not model-specific state variables.

We will see in Chapter 2 that the local volatility model is a market model, in Chapter 7 that forward variance models are market models, and in Chapter 12 that most local-stochastic volatility models are not.

Specifying a break-even condition for the carry P&L at order 2 in δS leads to pricing equation (1.4). It so happens that the latter – a parabolic equation – has a probabilistic interpretation: the solution can be written as the expectation of the payoff applied to the terminal value of a stochastic process for S_t that is a diffusion: $dS_t = (r - q)S_t dt + \hat{\sigma}S dW_t$.

The argument goes this way and not the other way around – modeling in finance *does not start* with the assumption of a stochastic process for S_t and has little to do with Brownian motion.

Expression (1.5) is a useful accounting tool – and the Black-Scholes equation (1.4) can be used to risk-manage options – despite the fact that real securities are not lognormal and do not exhibit constant volatility.

---

[5] Assume our P&L reads as in (1.8) with C an arbitrary positive matrix. We have:

$$A = -\frac{1}{2}\text{tr}(\phi C) = -\frac{1}{2}\text{tr}\left(T\varphi T^\intercal C\right) = -\frac{1}{2}\text{tr}\left(\varphi T^\intercal CT\right) = -\frac{1}{2}\Sigma_k\varphi_k\left(T_k^\intercal CT_k\right) = -\frac{1}{2}\Sigma_k\varphi_k\alpha_k$$

where $\alpha_k = T_k^\intercal CT_k$ are positive numbers as C is positive. C can thus be any positive matrix.

1.2 How (in)effective is delta hedging?

Expression (1.5) quantifies the P&L of a short delta-hedged option position. The aim of delta-hedging is to reduce uncertainty in our final P&L – it removes the linear term in δS: is this sufficient from a practical point of view? How large is the gamma/theta P&L (1.5)? More precisely, how large is the average and standard deviation of the total P&L incurred over the option's life?

It can be shown – this is the principal result of the Black-Scholes-Merton analysis – that:

- if the underlying security indeed follows a lognormal process with the same volatility σ as that used for pricing and delta-hedging the option; that is, S follows the Black-Scholes *model* with volatility σ

- and if we take the limit of very frequent hedging: $\delta t \to 0$

then the sum of P&Ls (1.5) incurred over the option's life vanishes with probability 1.

In real life delta-hedging occurs discretely in time, typically on a daily basis, and real securities do not follow diffusive lognormal processes. Thus, the sum of P&Ls (1.5) over the option's life will not vanish. Already in the lognormal case, if S follows a lognormal process but with a different volatility – say higher – than the implied volatility $\hat{\sigma}$, the sum of P&Ls (1.5) will not vanish in the limit $\delta t \to 0$.

Obviously, the condition that the final P&L vanishes on average requires that the implied volatility $\hat{\sigma}$ used for pricing and risk-managing the option match on average the future realized volatility weighted by the option's dollar gamma over the option's life:

$$\left\langle \int_0^T e^{-rt} S^2 \frac{d^2 P_{\hat{\sigma}}}{dS^2} \sigma_t^2 dt \right\rangle = \left\langle \int_0^T e^{-rt} S^2 \frac{d^2 P_{\hat{\sigma}}}{dS^2} \hat{\sigma}^2 dt \right\rangle$$

where σ_t is the instantaneous *realized* volatility defined by: $\sigma_t^2 \delta t = \frac{\delta S_t^2}{S_t^2}$ and the discount factor e^{-rt} is used to convert $P\&L$ generated at time t into $P\&L$ at $t = 0$.

Throughout this book, we use $\langle \rangle$ to denote either an average or a quadratic (co)variation – context should dispel any ambiguity as to which is intended.

Let us assume that this condition holds, so that our final P&L is not biased on average and let us concentrate on the dispersion – the standard deviation – of the final P&L. It vanishes in the Black-Scholes case with continuous hedging. How large is it, first in the Black-Scholes case with discrete hedging and then in the case of discrete hedging with real securities?

Assume that the option is delta-hedged daily at times t_i: $\delta t = 1$ day. The total P&L over the option's life, discounted at time $t = 0$, is:

$$P\&L = -\sum_i e^{-rt_i} \frac{S_i^2}{2} \frac{d^2 P_{\hat{\sigma}}}{dS^2} (t_i, S_i) \left(r_i^2 - \hat{\sigma}^2 \delta t\right) \qquad (1.9)$$

where r_i are daily returns, given by $r_i = \frac{S_{i+1} - S_i}{S_i}$. As expression (1.9) shows, at order 2 in δS and order 1 in δt, the total P&L is given by the sum of the differences between *realized* daily quadratic variation $\frac{\delta S_i^2}{S_i^2}$ and the *implied* quadratic variation $\hat{\sigma}^2 \delta t$, weighted by the prefactor $e^{-rt_i} \frac{S_i^2}{2} \frac{d^2 P_{\hat{\sigma}}}{dS^2} (t_i, S_i)$, which is payoff-dependent and involves the gamma of the option. $\hat{\sigma}$ is the implied volatility we are using to risk-manage our option position.

Let us make the approximation that the option's discounted dollar gamma $e^{-rt_i} S^2 \frac{d^2 P_{\hat{\sigma}}}{dS^2}$ is a constant, equal to its initial value $S_0^2 \frac{d^2 P_{\hat{\sigma}}}{dS^2} (t_0, S_0)$ – this removes one source of randomness in the P&L.[6] The standard deviation of the total P&L depends on the variances of individual daily P&Ls as well as on their covariances. Let us write the daily return r_i as:

$$r_i = \sigma_i \sqrt{\delta t} z_i \qquad (1.10)$$

where σ_i is the realized volatility for day i, and z_i is centered and has unit variance: $\langle z_i \rangle = 0$, $\langle z_i^2 \rangle = 1$. Let us assume that the z_i are iid and are independent of the volatilities σ_i.

Because the z_i are independent, returns r_i have no serial correlation but are not independent, as daily volatilities σ_i may be correlated. Expression (1.10) allows separation of the effects of the scale σ_i of return r_i on one hand, and of the distribution of r_i – which up to a rescaling is given by that of z_i – on the other hand. Our total P&L now reads:

$$P\&L = -\frac{S_0^2}{2} \frac{d^2 P_{\hat{\sigma}}}{dS^2} (t_0, S_0) \sum_i (\sigma_i^2 z_i^2 - \hat{\sigma}^2) \delta t$$

Let us assume that the process for the σ_i is time-homogeneous so that, in particular, $\langle \sigma_i^2 \rangle$ does not depend on i and let us take $\hat{\sigma}^2 = \langle \sigma_i^2 \rangle$. The variance of

---

[6]There exists actually a European payoff whose discounted dollar gamma is constant and equal to 1. It is called the log contract and pays at maturity $-2 \ln S$; see Section 3.1.4.

$\sum_i (\sigma_i^2 z_i^2 - \widehat{\sigma}^2)\delta t$ is given by:

$$\left\langle \sum_{ij} (\sigma_i^2 z_i^2 - \widehat{\sigma}^2)\delta t \; (\sigma_j^2 z_j^2 - \widehat{\sigma}^2)\delta t \right\rangle$$

$$= \sum_i \left(\langle \sigma_i^4 z_i^4 \rangle + \widehat{\sigma}^4 - 2\widehat{\sigma}^4 \right) \delta t^2 + \sum_{i \neq j} \langle \sigma_i^2 \sigma_j^2 z_i^2 z_j^2 \rangle + \widehat{\sigma}^4 - 2\widehat{\sigma}^4 \rangle \delta t^2$$

$$= \sum_i (2 + \kappa)\,\widehat{\sigma}^4 \delta t^2 + \sum_{i \neq j} (\langle \sigma_i^2 \sigma_j^2 \rangle - \widehat{\sigma}^4)\delta t^2$$

$$= \sum_i (2 + \kappa)\,\widehat{\sigma}^4 \delta t^2 + (\langle \sigma^4 \rangle - \widehat{\sigma}^4)\sum_{i \neq j} f_{ij}\delta t^2$$

$$= \widehat{\sigma}^4 \left(\sum_i (2 + \kappa)\,\delta t^2 + \Omega \sum_{i \neq j} f_{ij}\delta t^2 \right) \qquad (1.11)$$

where we have introduced the (excess) kurtosis κ of returns r_i and the variance/variance correlation function f defined by:

$$\kappa = \frac{\langle \sigma_i^4 z_i^4 \rangle}{\widehat{\sigma}^4} - 3, \quad f_{ij} = \frac{\langle (\sigma_i^2 - \widehat{\sigma}^2)(\sigma_j^2 - \widehat{\sigma}^2) \rangle}{\sqrt{\langle \sigma_i^4 \rangle - \widehat{\sigma}^4}\sqrt{\langle \sigma_j^4 \rangle - \widehat{\sigma}^4}}$$

and where the dimensionless factor Ω, which quantifies the variance of daily variances σ_i^2 is given by:

$$\Omega = \frac{\langle \sigma^4 \rangle - \widehat{\sigma}^4}{\widehat{\sigma}^4} = \frac{\langle \sigma^4 \rangle - \langle \sigma^2 \rangle^2}{\langle \sigma^2 \rangle^2}$$

We then get:

$$\text{StDev}\,(P\&L) = \left| \frac{S_0^2}{2}\frac{d^2 P_{\widehat{\sigma}}}{dS^2}(t_0, S_0) \right| \sqrt{\widehat{\sigma}^4 \left(\sum_i (2 + \kappa)\,\delta t^2 + \Omega \sum_{i \neq j} f_{ij}\delta t^2 \right)}$$

It is useful to measure the standard deviation of the final P&L in units of the option's vega, the sensitivity of the option's price to the implied volatility $\widehat{\sigma}$. In the Black-Scholes model, for European options the following relationship linking vega and gamma holds:

$$\frac{dP_{\widehat{\sigma}}}{d\widehat{\sigma}} = S^2 \frac{d^2 P_{\widehat{\sigma}}}{dS^2}\widehat{\sigma}T \qquad (1.12)$$

where T is the residual option's maturity – this is derived in Appendix A of Chapter 5, page 181. Using now the vega, the final expression for the standard deviation of the P&L is:

$$\text{StDev}\,(P\&L) = \left| \widehat{\sigma}\frac{dP_{\widehat{\sigma}}}{d\widehat{\sigma}} \right| \frac{1}{2T}\sqrt{\sum_i (2 + \kappa)\,\delta t^2 + \Omega \sum_{i \neq j} f_{ij}\delta t^2} \qquad (1.13)$$

1.2.1 The Black-Scholes case

Let us first assume that S follows the lognormal Black-Scholes dynamics. σ_i is constant, equal to $\hat{\sigma}$, hence $\Omega = 0$. Since $\sigma_i\sqrt{\delta t}$ is small (δt is one day and typically $\sigma_i = 20\%$, so that $\sigma_i\sqrt{\delta t} \simeq 0.01$), daily returns can be considered Gaussian: $\kappa = 0$. $\Sigma_i\,(2+\kappa)\,\delta t^2 = \frac{2T^2}{N}$, where T is the option's maturity and N is the number of delta rehedges: $N\delta t = T$. Expression (1.13) becomes:

$$\text{StDev}\,(P\&L) = \frac{1}{\sqrt{2N}}\left|\hat{\sigma}\frac{dP_{\hat{\sigma}}}{d\hat{\sigma}}\right| \tag{1.14}$$

Thus, provided it is small, the standard deviation of our final P&L is equivalent to the impact on the option's price of a relative perturbation of $\hat{\sigma}$ of size $\frac{1}{\sqrt{2N}}$.

Note that $\frac{\hat{\sigma}}{\sqrt{2N}}$ is approximately the standard deviation of the historical volatility estimator. The standard variance estimator is given by:

$$\overline{\sigma}^2 = \frac{1}{N\delta t}\sum_i\left(\frac{S_{i+1}-S_i}{S_i}\right)^2$$

In the Black-Scholes case, for daily returns, $\frac{S_{i+1}-S_i}{S_i}$ is approximately Gaussian and we have:

$$\overline{\sigma}^2 \simeq \frac{\hat{\sigma}^2}{N}\sum_i z_i^2$$

where z_i are standard normal random variables. The variance of $\overline{\sigma}^2$ is $\frac{2\hat{\sigma}^4}{N}$, thus the *relative* standard deviation $\text{StDev}(\overline{\sigma}^2)\,/\,\langle\overline{\sigma}^2\rangle$ is $\sqrt{\frac{2}{N}}$ and, if it is not too large, the *relative* standard deviation of the *volatility* estimator $\overline{\sigma}$ is approximately half of this, that is $\frac{1}{\sqrt{2N}}$.

The standard deviation observed on our final P&L is then approximately given by the option's vega multiplied by the standard deviation of the volatility estimator built on the same schedule as that of the delta rehedges.

Consider the example of a one-year at-the-money call option, with $\hat{\sigma} = 20\%$, vanishing interest rates, repo and dividends, and $S = 1$. The option's price is then $P = 7.97\%$. There are about 250 trading days in one year, which gives $\frac{1}{\sqrt{2N}} \simeq 0.045$. An at-the-money option has the property that its price is approximately linear in $\hat{\sigma}$ for short maturities: $P_{\hat{\sigma}} \simeq \frac{1}{\sqrt{2\pi}}S\hat{\sigma}\sqrt{T}$, thus $\hat{\sigma}\frac{dP_{\hat{\sigma}}}{d\hat{\sigma}} \simeq P$ (using this approximation yields a price of 7.98%).

We then get for the one-year at-the-money option: $\text{StDev}(P\&L) \simeq 0.045P$: the standard deviation of our final P&L is about 5% of the option's price we charged at inception.

5% of the premium charged for the option – or equivalently 5% of the volatility – may seem a very reasonable risk to take. Alternatively, adjusting the option's price to account for one standard deviation of our final P&L would result in a relative bid/offer spread on the option price of about 10%.

1.2.2   The real case

In real life, in contrast to the Black-Scholes case, the second term in the square root in (1.13) does not vanish. It involves the variance/variance correlation function f_{ij}. We have made the (reasonable) assumption that the process for the σ_i is time-homogeneous: f_{ij} is then a function of the difference $j - i$, actually a function of $|j - i|$.

As δt is small compared to the option's maturity, we convert the sums in (1.11) into integrals:

$$\sum_{ij} f_{ij} \delta t^2 \simeq \int_0^T du \int_0^T dt f\,(t - u) \;=\; 2 \int_0^T (T - \tau) f\,(\tau)\, d\tau$$

We now have from (1.13):

$$\mathrm{StDev}\,(P\&L) \simeq \left| \widehat{\sigma} \frac{dP_{\widehat{\sigma}}}{d\widehat{\sigma}} \right| \frac{1}{2T} \sqrt{(2 + \kappa)\frac{T^2}{N} + 2\Omega \int_0^T (T - \tau) f\,(\tau)\, d\tau}$$

$$= \left| \widehat{\sigma} \frac{dP_{\widehat{\sigma}}}{d\widehat{\sigma}} \right| \sqrt{\frac{2 + \kappa}{4N} + \frac{\Omega}{2T^2} \int_0^T (T - \tau) f\,(\tau)\, d\tau} \qquad (1.15)$$

Let us now examine the two contributions to StDev($P\&L$).

Imagine first that daily variances are constant: Ω vanishes and the first piece alone contributes to the standard deviation of the P&L. Just as in the Black-Scholes case (equation (1.14)), the variance of the final P&L scales like $1/N$, where N is the number of daily rehedges, which is natural as the final P&L is the sum of N identically distributed and independent daily P&Ls.

In contrast to the Black-Scholes case though, in which daily returns are approximately Gaussian, the effect of the tails of the distribution of daily returns appears through the kurtosis κ. By setting $\kappa = 0$ we recover result (1.14).

Consider now the second contribution in (1.15). The prefactor Ω quantifies the dispersion of daily variances while $f\,(\tau)$ quantifies how a fluctuation in daily variance σ_i^2 on day t_i impacts daily variances $\sigma_{i+\tau}^2$ on subsequent days. If f decays slowly, daily variances will be very correlated: in case one daily variance was higher than $\widehat{\sigma}$, daily variances for the following days are likely to be higher as well, resulting in daily gamma/theta P&Ls all having the same sign – thus generating strong correlation among daily P&Ls and increasing the variance of our final P&L.

For example, assume that daily variances are perfectly correlated: $f\,(\tau) = 1$. The second piece in (1.15) is then simply equal to $\frac{\Omega}{4}$. If Ω is small, the contribution of this term is then equivalent to the impact of a relative displacement of $\widehat{\sigma}$ by $\widehat{\sigma}\frac{\sqrt{\Omega}}{2}$, regardless of the number N of daily rehedges.[7]

---

[7]The case $f\,(\tau) = 1$ is unrealistic in that daily variances are random, but are all identical: the underlying security follows a lognormal dynamics with a constant volatility whose value is drawn randomly at inception.

Estimating $f(\tau), \Omega, \kappa$

Consider now the dynamics of daily variances σ_i in the case of real securities. Separating in r_i the contributions from σ_i and z_i is difficult if the only daily data we have are daily returns. In what follows we have estimated daily volatilities σ_i using 5-minute returns: σ_i is given by the square root of the sum of squared 5-minute returns during the exchange's opening hours, plus the square of the close-to-open return. Figure 1.1 shows the autocorrelation function f averaged over a set of European financial stocks, evaluated on a two-year sample: [August 2008, August 2010].[8]

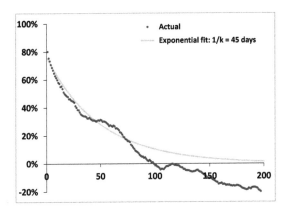

Figure 1.1: Correlation function $f(\tau)$ of daily volatilities evaluated on a basket of financial stocks. τ is in business days.

For $\tau = 0$, $f(\tau) = 1$. As is customary with correlation functions, however, $\lim_{\tau \to 0} f(\tau) \neq 1$, and the discontinuity in $\tau = 0$ quantifies the signal-to-noise ratio of our measurement of daily volatilities. As Figure 1.1 shows, this discontinuity is rather moderate and we get a robust estimation of the autocorrelation of daily volatilities up to time scales $\tau \simeq 100$ days.

For larger τ, Figure 1.1 displays negative autocorrelations: this is unphysical and most likely due to the fact that, over our historical sample (2 years), for $\tau > 100$ days, i and $i + \tau$ fall into two different regimes of respectively low and high volatilities. We have also graphed in Figure 1.1 an exponential fit to f: $f(\tau) = \rho e^{-k\tau}$, with $\rho = 0.78$ and $\frac{1}{k} = 45$ days. The agreement of f with the exponential form is acceptable in the region $\tau < 100$ days, where our measurement is reliable.

Using this form for $f(\tau)$ yields our final expression for the standard deviation of the P&L:

$$\frac{\text{StDev}(P\&L)}{\left| \hat{\sigma} \frac{dP_{\hat{\sigma}}}{d\hat{\sigma}} \right|} \simeq \sqrt{\frac{2+\kappa}{4N} + \frac{\rho\Omega}{2} \frac{kT - 1 + e^{-kT}}{(kT)^2}} \qquad (1.16)$$

---

[8]I am grateful to Benoît Humez for generating these data as well as estimates of Ω.

Ω quantifies the relative variance of daily variances σ_i^2. It varies appreciably, even among stocks of the same sector: a typical range for Ω is $[1.5, 4]$. Let us use the value $\Omega = 2$.

Estimating the unconditional kurtosis κ is also difficult, as the 4th order moment of daily returns converges slowly, so slowly that it is unreasonable to assume that the same regime of kurtosis holds throughout the historical sample used for its estimation: a typical order of magnitude is $\kappa = 5$.

1.2.3   Comparing the real case with the Black-Scholes case

We now use the typical values for Ω, κ, ρ, k estimated above in expression (1.16). Figure 1.2 shows the right-hand side of equation (1.16), that is the relative displacement of $\widehat{\sigma}$ that produces a variation of the option's price P equal to one standard deviation of the P&L. For an at-the-money option, whose price is approximately linear in $\widehat{\sigma}$, this number is also the ratio of one standard deviation of the P&L to the option's price itself.

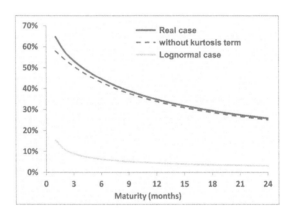

Figure 1.2: Right-hand side in equation 1.16 (darker line), as a function of maturity, compared to the same quantity, but without the kurtosis term (dashed line), and the lognormal case (lighter line).

Figure 1.2 also displays the same quantity, but without the term $\frac{2+\kappa}{4N}$, to remove the effect of the tails of the daily returns, as well as the standard deviation of the P&L in the lognormal, Black-Scholes case (1.14).

We can see that the standard deviation of the final P&L of a delta-hedged option in the real case is much larger than its estimation in the Black-Scholes case.

Consider again the example of a 1-year at-the-money option, with $\widehat{\sigma} = 0.2$, with $P = 7.97\%$. As Figure 1.2 shows, while in the Black-Scholes case, the standard deviation is 4.5% of the option's price, that is 0.35%, in the real case, for a 1-year maturity it is equal to 35% of the option's price, that is 2.8%.

Comparison of the dark and dashed lines in Figure 1.2 shows that, but for very short maturities, the dispersion of the P&L is mostly generated by correlation of daily volatilities rather than the thickness of the tails of daily returns.

Delta-hedging our one-year at-the-money option position exposes us to the risk of making or losing about one third of the option premium[9] – this is an unreasonable risk to take considering that, typically, commercial fees charged by banks on option transactions are much smaller than the option's value.

The conclusion is that, in real life, delta-hedging is not sufficient: while delta-hedging removes the linear term in δS in our daily P&L, the effect of the δS^2 term is still too large: the only way to remove it is to use other options – for example vanilla options – to offset the gamma of the option we are risk-managing.

This was expressed bluntly to the author upon starting his career in finance by Nazim Mahrour, an FX option trader: "options are hedged with options".

1.3   On the way to stochastic volatility

Let us then use other options to offset the gamma of the exotic option we are risk-managing: assume for simplicity that we use a single vanilla option, whose implied volatility is $\widehat{\sigma}_O$. The P&L of a delta-hedged position in the vanilla option O has the same form as in equation (1.5), except it involves the implied volatility $\widehat{\sigma}_O$:

$$P\&L_O = -\frac{S^2}{2}\frac{d^2 O}{dS^2}\left(\frac{\delta S^2}{S^2} - \widehat{\sigma}_O^2 \delta t\right) \tag{1.17}$$

The number λ of vanilla options O we are buying as gamma hedge is :

$$\lambda = \frac{1}{\frac{d^2 O}{dS^2}}\frac{d^2 P}{dS^2} \tag{1.18}$$

The gamma profiles of P and O are unlikely to be homothetic, thus this gamma hedge will be efficient only locally; as time elapses and S moves, we need to readjust the hedge ratio λ.

We could decide to risk-manage each option P and O with its own implied volatility $\widehat{\sigma}$ and $\widehat{\sigma}_O$, but this leads to incongruous carry P&Ls.

Indeed by selecting λ as specified in (1.18) we cancel the gamma of the hedged position. The P&Ls of options O and P are both of the form in (1.17). If $\widehat{\sigma} \neq \widehat{\sigma}_O$, the theta portion of our global P&L does not vanish, even though gamma vanishes, a situation as nonsensical as those encountered in Section 1.1, when A and B have the same sign – see also the discussion in Section 2.8 below.

---

[9]Remember that we have made the unrealistic assumption that we were able to predict the average realized volatility. Uncertainty about the future average level of realized volatility would push the standard deviation of our final P&L even higher.

We must thus choose $\widehat{\sigma} = \widehat{\sigma}_O$. We now have for P a pricing function that explicitly depends on two dynamical variables: S and $\widehat{\sigma}_O$:

$$P(t, S, \widehat{\sigma}_O)$$

which is natural as we are using two instruments as hedges.

This is an elementary instance of calibration: we *decide* to make our exotic option's price a function of other derivatives' prices. It is a trading decision.

In the unhedged case we were free to chose the implied volatility $\widehat{\sigma}$ as our best estimate of future realized volatility and kept it constant throughout: no P&L was generated by the variation of $\widehat{\sigma}$.

Unlike $\widehat{\sigma}$, however, $\widehat{\sigma}_O$ is a market implied volatility and cannot be kept constant. As S moves and time flows we readjust λ, thus buying or selling the vanilla option at prevailing market prices: $\widehat{\sigma}_O$ will move so as to reflect the market price O of the vanilla option. Daily P&Ls for O and P will include extra terms involving $\delta\widehat{\sigma}_O$. At second order in $\delta\widehat{\sigma}_O$:

$$P\&L_O = -\frac{S^2}{2}\frac{d^2O}{dS^2}\left(\frac{\delta S^2}{S^2} - \widehat{\sigma}_O^2\delta t\right) - \frac{dO}{d\widehat{\sigma}_O}\delta\widehat{\sigma}_O - \frac{1}{2}\frac{d^2O}{d\widehat{\sigma}_O^2}\delta\widehat{\sigma}_O^2 - \frac{d^2O}{dSd\widehat{\sigma}_O}\delta S\delta\widehat{\sigma}_O$$

$$(1.19)$$

The expansion of the P&L of the *hedged* position at second order in δS, $\delta\widehat{\sigma}_O$ and order 1 in δt reads:

$$P\&L = -\left(\frac{dP}{d\widehat{\sigma}_O} - \lambda\frac{dO}{d\widehat{\sigma}_O}\right)\delta\widehat{\sigma}_O \qquad (1.20)$$

$$-\frac{1}{2}\left(\frac{d^2P}{d\widehat{\sigma}_O^2} - \lambda\frac{d^2O}{d\widehat{\sigma}_O^2}\right)\delta\widehat{\sigma}_O^2 - \left(\frac{d^2P}{dSd\widehat{\sigma}_O} - \lambda\frac{d^2O}{dSd\widehat{\sigma}_O}\right)\delta S\delta\widehat{\sigma}_O$$

This is an accounting equation: the P&L generated by these three terms is no less real than the usual gamma/theta P&L – it is usually called *mark-to-market* P&L, while the gamma/theta P&L is typically called *carry* P&L.[10]

There is no contribution from δS^2 as $\frac{d^2P}{dS^2} = \lambda\frac{d^2O}{dS^2}$ by construction. Exotic options are typically path-dependent options: their final payoff is a function of values of S observed at discrete dates, specified in the option's term sheet. Between two observation dates, the pricing equation for P in the Black-Scholes framework is the same as that of a European option. Since P and O are given by a Black-Scholes pricing equation with the same implied volatility $\widehat{\sigma}_O$, cancellation of gamma implies cancellation of theta as well: there is no δt term in (1.20).

Consider the last two terms in $\delta\widehat{\sigma}_O^2$ and $\delta S\delta\widehat{\sigma}_O$ in (1.19) and (1.20). While their contributions to $P\&L_O$ and $P\&L$ look similar, they have a different status and have to be treated differently. Expression (1.19) is the P&L of a vanilla option position.

---

[10]The distinction between mark-to-market and carry P&L is somewhat arbitrary. Usually mark-to-market P&L refers to P&L generated by the variation of parameters that were supposed to stay constant in the pricing model: typically, in the Black-Scholes model a change in $\widehat{\sigma}$ generates mark-to-market P&L.

The extra terms that come in addition to the gamma/theta P&L do not warrant any adjustment to the price of the vanilla option: their contribution to the P&L is already priced-in in the market price of the vanilla option.

In expression (1.20), however, what appear as prefactors of $\delta\widehat{\sigma}_O^2$ and $\delta S\delta\widehat{\sigma}_O$ are the second-order sensitivities of the *hedged* position. We then need to adjust the price $P(t, S, \widehat{\sigma}_O)$ of our exotic option for the cost of these two contributions to the P&L.

What matters in the evaluation of extra-model cost is not so much the second-order sensitivities of the *naked* exotic option, but the residual sensitivities of the *hedged* position.

Three observations are in order:

- We now have a vega term in $\delta\widehat{\sigma}_O$. If P is a European option with the same maturity as O, the vega of a gamma-hedged position cancels out, owing to relationship (1.12) linking gamma and vega in the Black-Scholes model. A European payoff is statically hedged with a portfolio of vanilla options of the same maturity; it can hardly be called an exotic derivative.

  The situation we have in mind is that of real exotics that has no static hedge, whose hedge portfolio comprises vanilla options of different maturities: gamma cancellation does not imply vega cancellation. Depending on the relative sizes of the gamma and vega risks we may prefer to gamma-hedge or vega-hedge our exotic option: this is a trading decision. In practice an exotics book is a large caldron where mitigation of the gamma and vega risks of many different exotic and vanilla options takes place: gamma and vega hedging, unachievable on a deal-by-deal basis, can be reasonably achieved at the book level.[11]

- Our P&L does not involve realized volatility anymore. Instead, we have acquired sensitivity to $\widehat{\sigma}_O$. While in the unhedged case we were exposed to *realized* volatility, we are now exposed to the dynamics of the *implied* volatility $\widehat{\sigma}_O$.[12]

- Unlike in the unhedged case for the δS^2 term, no deterministic δt term is now offsetting the $\delta\widehat{\sigma}_O^2$ and $\delta S\delta\widehat{\sigma}_O$ terms: depending on their realized values and the signs of their prefactors, we may systematically make or lose money. This is a serious issue. While in the Black-Scholes pricing equation we had a parameter – the implied volatility – to control how the gamma and theta terms for the spot offset each other, we have no equivalent parameter at our disposal to control break-even levels for gammas on $\widehat{\sigma}_O$: no implied volatility of $\widehat{\sigma}_O$ and no implied correlation of S and $\widehat{\sigma}_O$. P and O should then be given by a different pricing equation than Black-Scholes', that explicitly includes

---

[11]Client preferences, pressure from the salesforce, unwillingness of other counterparties to take on exotic risks, may lead an exotics desk to pile up one-way risk. In normal circumstances, though, as exposure to a particular risk builds up, traders will be willing to quote aggressive prices for payoffs that offset this risk so as to keep the overall risk levels of the book under control.

[12]This is not exactly true – there remains a residual sensitivity to realized volatility in the covariance term $\delta S\delta\widehat{\sigma}_O$.

these new parameters so as to generate additional theta terms in the P&L: this is the general task of stochastic volatility models.[13]

The general conclusion is that by using options as hedges we lower – or cancel – our exposure to realized volatility, but acquire an exposure to the dynamics of implied volatilities. However, while the Black-Scholes pricing equation provides a theta term to offset the gamma term for S, no provision of a theta is made to offset the gamma P&Ls experienced on the variation of implied volatilities of options used as hedges.

This is not surprising as the notion of dynamic implied volatilities is alien to the Black-Scholes framework.

This is where stochastic volatility models are called for: their aim is not to model the dynamics of *realized* volatility, which is hedged away by trading other options, but to model the dynamics of *implied* volatilities, and provide their user with simple break-even accounting conditions for the P&L of a hedged position.

---

[13] The vanna-volga method – see [29] – once used on FX desks for generating FX smiles is a poor man's answer to this issue, with "exotic" option P a European option.

- Rather than using a single vanilla option O we use 3 of them, and find quantities λ_i so that the 3 sensitivities $\frac{d}{d\sigma}$, $\frac{d^2}{dSd\sigma}$, $\frac{d^2}{d\sigma^2}$ of the hedged position $P - \Sigma_{i=1}^3 \lambda_i O_i$ vanish (a) in the Black-Scholes model for an implied volatility $\hat{\sigma}_0$, (b) for current values of t, S. Cancellation of $\frac{d}{d\sigma}$ is equivalent to cancellation of $\frac{d^2}{dS^2}$, owing to the vega/gamma relationship in the Black-Scholes model – see Section A.1 of Chapter 5.

- The hedging options are bought/sold at market prices, at implied volatilities $\hat{\sigma}_i$, thus the difference $O_i^{\mathrm{BS}}(\hat{\sigma}_i) - O_i^{\mathrm{BS}}(\hat{\sigma}_0)$ has to be passed on to the client as a hedging cost. We thus define the "market-adjusted" price P^{Mkt} of option P as:

$$P^{\mathrm{Mkt}} = P^{\mathrm{BS}}(\hat{\sigma}_0) + \Sigma_i \lambda_i \left(O_i^{\mathrm{BS}}(\hat{\sigma}_i) - O_i^{\mathrm{BS}}(\hat{\sigma}_0) \right) \tag{1.21}$$

The hedge portfolio is only effective for current values of t, S. It needs to be readjusted whenever either moves – the corresponding rehedging costs are not factored in P^{Mkt}.

- As observed in [29], the vanna-volga price in (1.21) can be written as:

$$
\begin{aligned}
P^{\mathrm{Mkt}} &= P^{\mathrm{BS}}(\hat{\sigma}_0) + y_\sigma \left.\frac{dP^{\mathrm{BS}}}{d\hat{\sigma}_0}\right|_{S,\hat{\sigma}_0} + y_{\sigma^2} \left.\frac{d^2 P^{\mathrm{BS}}}{d\hat{\sigma}_0^2}\right|_{S,\hat{\sigma}_0} + y_{S\sigma} \left.\frac{d^2 P^{\mathrm{BS}}}{dSd\hat{\sigma}_0}\right|_{S,\hat{\sigma}_0} \\
&= P^{\mathrm{BS}}(\hat{\sigma}_0) + y_{S^2} \left.\frac{d^2 P^{\mathrm{BS}}}{dS^2}\right|_{S,\hat{\sigma}_0} + y_{\sigma^2} \left.\frac{d^2 P^{\mathrm{BS}}}{d\hat{\sigma}_0^2}\right|_{S,\hat{\sigma}_0} + y_{S\sigma} \left.\frac{d^2 P^{\mathrm{BS}}}{dSd\hat{\sigma}_0}\right|_{S,\hat{\sigma}_0}
\end{aligned} \tag{1.22}
$$

where the second line again follows from the vega/gamma relationship in the Black-Scholes model: $y_{S^2} = y_\sigma S^2 \hat{\sigma}_0 T$. The interpretation of (1.22) is: we supplement the Black-Scholes price at implied volatility $\hat{\sigma}_0$ with an estimation of future gamma P&Ls calculated (a) with current values of the gammas and cross-gammas, (b) values for $y_{S^2}, y_{S\sigma}, y_{\sigma^2}$ such that market prices for the three vanilla options O_i are recovered; $y_{S^2}, y_{S\sigma}, y_{\sigma^2}$ only depend on the $\hat{\sigma}_i$, not on P. This underscores how local the vanna-volga adjustment is – it cannot replace a genuine model for pricing volatility-of-volatility risk.

- Historically, the vanna-volga method has been used for interpolating implied volatilities: pick a vanilla option of strike K and use (1.21) to generate the corresponding adjusted "market price" – hence implied volatility. There is obviously no guarantee that the resulting interpolation $\hat{\sigma}^{\mathrm{Mkt}}(K, \hat{\sigma}_0, \hat{\sigma}_i)$ is arbitrage-free.

In practice, for liquid securities such as equity indexes, there are plenty of options available: rather than one implied volatility $\widehat{\sigma}_O$, one needs to model the dynamics of all implied volatilities $\widehat{\sigma}_{KT}$, where K and T are, respectively, the strikes and maturities of vanilla options. The two-dimensional set $\widehat{\sigma}_{KT}$ is known as the *volatility surface*.

While a stochastic volatility model should ideally offer maximum flexibility as to the range of dynamics of the volatility surface it is able to produce, we may not be able to build such a flexible model on one hand, and on the other hand we may not need so much versatility: some classes of exotic options are only sensitive to specific features of the dynamics of the volatility surface.

Before we delve into stochastic volatility models, we present two examples of exotic options whose type of volatility risk can be exactly pinpointed.

1.3.1   Example 1: a barrier option

Consider an option of maturity one year that pays at maturity 1 unless S_t hits the barrier $L = 120$, in which case the option expires worthless. The initial spot value is $S_0 = 100$. The pricing function $F(t, S)$ of this barrier option has to satisfy the terminal condition at maturity: $F(T, S) = 1$, for $S < L$ as well as the boundary condition $F(t, L) = 0$ for all $t \in [0, T]$.

How do we hedge this barrier option with vanilla options? Peter Carr and Andrew Chou show in [22] that, given a barrier option with payoff $f(S)$ and upper barrier L, it is possible to find a European payoff $g(S)$ of maturity T such that in the Black-Scholes model its value $G(t, S)$ exactly equals that of the barrier option, $F(t, S)$ for $S \leq L$, at all times.

The condition that $G(t, S) = F(t, S)$ at $t = T$ implies that $g(S) = f(S)$ for $S < L$. For $S > L$, f is not defined, but we have to find $g(S)$ such that $G(t, S = L)$ vanishes for all $t < T$.

Imagine we are able to find g such that this condition is satisfied. Then we have a European payoff that: (a) has the same final payoff as the barrier option, (b) satisfies the same boundary condition for $S = L$ and (c) solves the same pricing equation over $[0, L]$: this implies that $F(t, S) = G(t, S)$ for all $S \in [0, L]$, $t \in [0, T]$: the barrier option is statically hedged by the European payoff G.

Carr and Chou give the following explicit expression for g, in the Black-Scholes model:

$$S < L \quad g(S) = f(S) \tag{1.23a}$$

$$S > L \quad g(S) = -\left(\frac{L}{S}\right)^{\frac{2r}{\sigma^2} - 1} f\left(\frac{L^2}{S}\right) \tag{1.23b}$$

where r is the interest rate and σ the volatility. Let us assume vanishing interest rates. The replicating European payoff for our barrier options is:

$$S < L \quad g(S) = 1$$

$$S > L \quad g(S) = -\frac{S}{L} = -1 - \frac{1}{L}(S-L)^+$$

This static hedge thus consists of two European digital options struck at L, each of which pays 1 if $S_T < L$ and 0 otherwise, minus (a) one zero-coupon bond that pays 1, $\forall S_T$, and (b) $\frac{1}{L}$ call options of strike L. S_T is the value of S at maturity.

Equations (1.23a), (1.23b) for $g(S)$ show that if $f(L) \neq 0$, g has a discontinuity in $S = L$ whose magnitude is twice that of f. The replicating European payoff includes a digital option whose role is instrumental in replicating the sharp variation of F in the vicinity of L.

Let us consider for simplicity that we are only using the double European digital option: it pays 1 at T if $S_T < L$ and -1 if $S_T > L$. Even though European digitals are not liquid, they can be synthesized just like any European payoff by trading an appropriate set of vanilla options, in our case a very tight put spread, that is the combination of $\frac{1}{2\varepsilon}$ puts struck at $L + \varepsilon$ minus $\frac{1}{2\varepsilon}$ puts struck at $L - \varepsilon$.

The values of the barrier option, F, and of the double European digital option – minus the zero-coupon bond – are shown as a function of S at $t = 0$ on the left-hand side of Figure 1.3 while the right-hand side shows the dollar gamma for both options. We have used $\sigma = 20\%$.

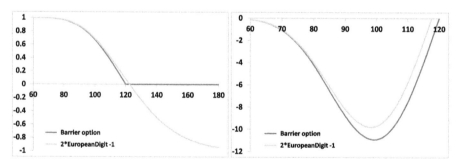

Figure 1.3: Value (left) and dollar gamma (right) of barrier option and double European digital option.

Had we used the exact static European hedge, curves would have overlapped exactly, in both graphs, by construction. Simply using the double European digital option still provides an acceptable hedge. Let us assume that we have sold at $t = 0$ the barrier option and have simultaneously purchased the double European digital as a hedge.

Which price do we quote for the barrier option? We are using as hedge a double European digital option whose market price will likely differ from its Black-Scholes price. The price we charge must thus be equal to the Black-Scholes price of the barrier option augmented by the difference between market and Black-Scholes prices of the double European digital: this extra charge covers the cost of actually purchasing the European hedge.[14]

If we reach maturity without hitting the barrier $L = 120$, the payoffs of the barrier option and the static European hedge exactly match: the hedge is perfect.

What if instead S hits the barrier? When S hits L at time τ, the barrier option expires worthlessly and we need to unwind our static European hedge. By construction, in the Black-Scholes model, its value for $S = L$ approximately vanishes.[15]

How about in reality? In reality, the value of our European static hedge will depend on market implied volatilities at time τ for European options of maturity T and will likely not vanish.

Let us make this dependence more explicit: the value D of the double European digital is given by:

$$D = 2\frac{\mathcal{P}_{L+\varepsilon} - \mathcal{P}_{L-\varepsilon}}{2\varepsilon} - 1 = 2\left.\frac{d\mathcal{P}_K}{dK}\right|_L - 1$$

where $\mathcal{P}_K$ denotes the value of a put option of strike K, which is given by the Black-Scholes formula for put options, using the implied volatility for strike K : $\mathcal{P}_K = \mathcal{P}_K^{BS}(\hat{\sigma} = \hat{\sigma}_K)$. We have:

$$\frac{d\mathcal{P}_K}{dK} = \frac{d\mathcal{P}_K^{BS}(\hat{\sigma}_K)}{dK} = \frac{d\mathcal{P}_K^{BS}}{dK} + \frac{d\mathcal{P}_K^{BS}}{d\hat{\sigma}}\frac{d\hat{\sigma}_K}{dK}$$
$$= \mathcal{D}_K^{BS}(\hat{\sigma}_K) + \frac{d\mathcal{P}_K^{BS}}{d\hat{\sigma}}\frac{d\hat{\sigma}_K}{dK}$$

where $\mathcal{D}^{BS}$ is the value of a (single) European digital option, which pays 1 if $S_T < L$ and 0 otherwise, in the Black-Scholes model. We get the following value for the double European digital:

$$\mathcal{D} = 2\left(\mathcal{D}_L^{BS}(\hat{\sigma}_L) + \left.\frac{d\mathcal{P}_L^{BS}}{d\hat{\sigma}}\frac{d\hat{\sigma}_K}{dK}\right|_L\right) - 1 \qquad (1.24)$$

$\mathcal{D}^{BS}$ is evaluated for $S = L$; as can be checked numerically, $\mathcal{D}^{BS}$ for $S = L$ is almost equal to 50% and has little sensitivity to the implied volatility $\hat{\sigma}_L$. Expression (1.24) shows, though, that the value of the double European digital is very sensitive to $\left.\frac{d\hat{\sigma}_K}{dK}\right|_L$ which is the at-the-money skew at the time S hits L. Take the example of a one-year ATM digital option; while $\mathcal{D}_L^{BS}(\hat{\sigma}_L)$ is about 50%, the size of the correction term $\left.\frac{d\mathcal{P}_L^{BS}}{d\hat{\sigma}}\frac{d\hat{\sigma}_K}{dK}\right|_L$ for an equity index is typically about 8%: this is not a small effect.

---

[14]Black-Scholes prices are computed with the volatility σ that we choose to risk-manage the barrier option.

[15]It would vanish exactly had we used the exact static European hedge.

Thus, as we unwind our static hedge the magnitude of the then-prevailing at-the-money skew will determine whether we make or lose money. The Black-Scholes price of the barrier option has then to be adjusted manually to include an estimation of this gain or loss.

The lesson of this example is that the price of a barrier option is mostly dependent on the dynamics of the at-the-money skew conditional on S hitting the barrier.[16] A stochastic volatility model for barrier options would need to provide a direct handle on this precise feature of the dynamics of the volatility surface so as to appropriately reflect its P&L impact in the option price.

1.3.2   Example 2: a forward-start option

Forward-start options – also called cliquets[17] – involve the ratio of a security's price observed at two different dates – they are considered in detail in Chapter 3. Let T_1 and T_2 be two dates in the future and consider the case of a simple call cliquet whose payoff at T_2 is given by

$$\left(\frac{S_{T_2}}{S_{T_1}} - k \right)^+ \tag{1.25}$$

Let us choose $k = 100\%$ – this is called a forward-start at-the-money call. The price P of this option in the Black-Scholes model, because of homogeneity, does not depend on S and only depends on volatility. Assuming zero interest rates for simplicity, for $k = 100\%$, the Black-Scholes price of our cliquet is approximately given by:

$$P \simeq \frac{1}{\sqrt{2\pi}} \sigma \sqrt{T_2 - T_1} \tag{1.26}$$

The fact that P does not depend on S is worrisome: the only instrument whose dynamics is accounted for in the Black-Scholes model is S, yet S is not appearing in the pricing function.

P is only a function of volatility σ – σ is in fact the real underlying of the cliquet option.

A cliquet is an option on volatility, more precisely on forward implied volatility, that is the future implied volatility observed at T_1 for maturity T_2. At $t = T_1$, the cliquet becomes a vanilla option of maturity T_2, in our case a call option struck at kS_{T_1}. A suitable hedging strategy needs to generate at time T_1 the money needed to purchase a call option of maturity T_2 struck at kS_{T_1}.

While payoff formula (1.25) suggests that the cliquet is an option of maturity T_2 on S observed at T_1 and T_2, it is in fact an option of maturity T_1 whose underlying

---

[16]Besides the forward-skew risk we have just analyzed, the price of the barrier option needs to be adjusted for gap risk. Unwinding the European hedge – or unwinding the delta – cannot be done instantaneously as S crosses L. In our case, the delta of the barrier option we have sold is negative: we will need to buy stocks (or sell the double digital option) at a spot level that is presumably larger than L, thus incurring a loss. We must thus adjust the price charged for the barrier option to cover, on average, this loss.

[17]Ratchet, in French.

is the at-the-money implied volatility for maturity T_2, observed at T_1. This is the quantity whose dynamics a stochastic volatility model ought to provide a handle on.

1.3.3 Conclusion

Running an exotics book entails trading options dynamically to hedge other options. Vanilla options should be considered as hedging instruments in their own right and their dynamics modeled accordingly; as such the task of a stochastic volatility model is to model the joint dynamics of the underlying security and its associated volatility surface.

Chapter's digest

▶ Delta hedging removes the order-one contribution of δS to the P&L of an option position. Specifying a break-even condition for the lowest-order portion – the second order in δS – of the residual P&L leads to the Black-Scholes pricing equation – a parabolic equation. The latter has a probabilistic interpretation: the solution can be expressed as the expectation of the payoff under a density which is generated by a diffusion for S_t.

The argument goes this way and not the other way around – modeling does not start with the assumption of a diffusion for S_t and has little to do with Brownian motion; in this respect we refer the reader to Section 4.2 of [53].[18] For alternative break-even criteria that involve higher-order terms in δS see Chapter 10.

When there are multiple hedge instruments, the suitability of a model depends on the existence of a – possibly state- and time-dependent – break-even covariance matrix for hedge instruments that ensures gamma/theta cancellation.

▶ Delta hedging is not adequate for reducing the standard deviation of the P&L of an option position to reasonable levels. The sources of the dispersion of this P&L are: (a) the tails of returns, (b) the volatility of realized volatility and the correlation of future realized volatilities – see (1.15). Except for very short options, the latter effect prevails, because of the long-ranged nature of volatility/volatility correlations.

▶ Using options for gamma-hedging immunizes us against realized volatility. Dynamical trading of vanilla options, however, exposes us to uncertainty as to future levels of implied volatilities. Stochastic volatility models are thus needed for modeling the dynamics of implied volatilities, rather than that of realized volatility.

▶ Exotic options often depend in a complex way on the dynamics of implied volatilities. Some specific classes of options, such as barrier options, or cliquets, are such that their volatility risk can be pinpointed, enabling an easier assessment of the suitability of a given model.

---

[18]This is not to mean we can write down just *any* pricing equation. It has to comply with the basic requirements that (a) given two payoffs f and g, if $g(S) \geq f(S)\ \forall S$ then g should be more expensive than f – this expresses absence of arbitrage, and for a linear pricing equation implies the existence of a (risk-neutral) density and (b) that it obeys the convex order condition – see Section 2.2.2, page 29.

Chapter 2

Local volatility

This chapter covers the simplest and most widely used stochastic volatility model: the local volatility model. Local volatility [37], [40] was introduced as an extension of the Black-Scholes model that can be exactly calibrated to the whole volatility surface $\hat{\sigma}_{KT}$.

While its proponents did not have stochastic volatility in mind, local volatility is a particular breed of stochastic volatility. It is also the simplest *market model*.

2.1 Introduction – local volatility as a market model

Market models aim at treating vanilla options on the same footing as the underlying itself: vanilla option prices observed at $t = 0$ are initial values of hedge instruments to be used as inputs in the model.[1] A stochastic volatility model should be able to accommodate as initial condition any configuration of these asset values, provided it is not nonsensical: for example, call option prices for a given maturity should be a decreasing function of the option's strike. We will see in Chapter 4 that this basic capability is difficult to achieve – most stochastic volatility models cannot be calibrated to the volatility surface exactly.

The enduring popularity of the local volatility model lies in its ability – as a market model – to take as input an arbitrary volatility surface provided it is free of arbitrage. Because any European option can be synthesized using call and put options of the European option's maturity (see Section 3.1.3, page 106), the local volatility model prices European options exactly.

It is a peculiar market model, however, because calibration on the market smile fully determines the model. Such frugality comes at a price: the dynamics it generates for the volatility surface is fully fixed by the vanilla smile used for calibration, it is not explicit, and must be extracted *a posteriori*. Mathematically, it is a market model that possesses a Markov representation in terms of t, S_t.

---

[1]We can use a subset of vanilla options in our market model or other types of options. The models of Chapter 7 are market models for log-contracts, or for a term structure of vanilla options of an arbitrary moneyness. In Section 4.3 of Chapter 4 we show how power payoffs can be used as well.

Historically, the local volatility model has been published and presented as a variant of the Black-Scholes model such that the instantaneous volatility is a deterministic function of t, S: $\sigma(t, S)$.

The local volatility function $\sigma(t, S)$, however, only serves an ancillary purpose and has no physical significance. It is a by-product of the fact that the model has a one-dimensional Markov representation in terms of t, S_t.

Traditional presentations of the local volatility model make $\sigma(t, S)$ a central object: the local volatility function is calibrated at $t = 0$ on the market smile and kept frozen afterwards. This contravenes the typical trading practice of recalibrating the local volatility function on a daily basis – which then seems to amount to an improper use of the model.

We will see instead in Section 2.7 that:

- this is how the local volatility model should be used,

- the resulting carry P&L has the standard expression in terms of offset-ting spot/volatility gamma/theta contributions with well-defined and payoff-independent break-even levels – the trademark of a market model.

Our aim in the following sections is to characterize the dynamics generated for implied volatilities and then discuss the issue of the delta and the carry P&L. We will first need to establish the relationships linking local and implied volatilities.

2.1.1   SDE of the local volatility model

In the local volatility model, all assets have a one-dimensional representation in terms of t, S. The stochastic differential equation (SDE) for S_t is:

$$dS_t = (r - q) S_t dt + \sigma(t, S_t) S_t dW_t \qquad (2.1)$$

where r is the interest rate and q the repo rate inclusive of the dividend yield. The pricing equation is identical to the Black-Scholes equation (1.4), except $\sigma(t, S)$ now replaces $\hat{\sigma}$:

$$\frac{dP}{dt} + (r - q) S \frac{dP}{dS} + \frac{\sigma(t, S)^2}{2} S^2 \frac{d^2 P}{dS^2} = rP \qquad (2.2)$$

Given a particular local volatility function $\sigma(t, S)$ one can get prices of vanilla options by setting $P(t = T, S)$ equal to the option's payoff and solving equation (2.2) backwards from T to t, to generate $P(t, S)$. Conversely, given a configuration of vanilla options' prices, can we find a function $\sigma(t, S)$ such that, by solving equation (2.2) they are recovered? What is the condition for the existence of a $\sigma(t, S)$?

The expression of $\sigma(t, S)$, which we derive below, was found by Bruno Dupire [40]:

$$\sigma(t, S)^2 = 2 \frac{\frac{dC}{dT} + qC + (r - q) K \frac{dC}{dK}}{K^2 \frac{d^2 C}{dK^2}} \Bigg|_{\substack{K = S \\ T = t}} \qquad (2.3)$$

where $C(K,T)$ is the price of a call option of strike K and maturity T. This equation expresses the fact that the local volatility for spot S and time t is reflected in the differences of option prices with strikes straddling S and maturities straddling t.

2.2 From prices to local volatilities

2.2.1 The Dupire formula

Consider the following diffusive dynamics for S_t:

$$dS_t = (r-q)S_t dt + \sigma_t S_t dZ_t$$

where σ_t is for now an arbitrary process. By only using vanilla option prices, how precisely can we characterize σ_t?

The price of a call option is given by:

$$C(K,T) = e^{-rT} E[(S_T - K)^+]$$

The dynamics of S_t on the interval $[T, T+dT]$ determines how much prices of options of maturities T and $T+dT$ differ. Let us write the Itô expansion for $(S_T - K)^+$ over $[T, T+dT]$:

$$d(S_T - K)^+$$
$$= \frac{d(S_T - K)^+}{dS_T}\left((r-q)S_T dT + \sigma_T S_T dZ_T\right)$$
$$+ \frac{1}{2}\frac{d^2(S_T - K)^+}{dS_T^2}\sigma_T^2 S_T^2 dT$$
$$= \theta(S_T - K)\left((r-q)S_T dT + \sigma_T S_T dZ_T\right) + \frac{1}{2}\delta(S_T - K)\sigma_T^2 S_T^2 dT \quad (2.4)$$

where $\theta(x)$ is the Heaviside function: $\theta(x) = 1$ for $x > 0$, $\theta(x) = 0$ for $x < 0$, and δ is the Dirac delta function.

For simplicity let us switch temporarily to *undiscounted* option prices $\mathcal{C}(K,T)$. Taking derivatives with respect to K of the identity: $\mathcal{C}(K,T) = E[(S_T - K)^+]$ we get:

$$E[\theta(S_T - K)] = -\frac{d\mathcal{C}}{dK}, \quad E[\delta(S_T - K)] = \frac{d^2\mathcal{C}}{dK^2} \quad (2.5)$$

The second equation expresses the well-known property that the second derivative of undiscounted call or put prices with respect to their strike yields the pricing (or risk-neutral) density of S_T.

From the identity

$$\mathcal{C} = E[(S_T - K)^+] = E[(S_T - K)\theta(S_T - K)]$$
$$= E[S_T\theta(S_T - K)] - KE[\theta(S_T - K)]$$

we get:

$$E[S_T\theta(S_T - K)] = C - K\frac{dC}{dK}$$

Now take the expectation of both sides of equation (2.4). In the left-hand side, $E[d(S_T - K)^+] = dE[(S_T - K)^+]$, that is the difference of the undiscounted prices of two call options of strike K expiring at T and $T + dT$: this is equal to $\frac{dC}{dT}dT$.

$$\frac{dC}{dT}dT = (r-q)\left(C - K\frac{dC}{dK}\right)dT + \frac{K^2}{2}E[\sigma_T^2\delta(S_T - K)]dT$$

yields:

$$E[\sigma_T^2\delta(S_T - K)] = \frac{2}{K^2}\left(\frac{dC}{dT} - (r-q)\left(C - K\frac{dC}{dK}\right)\right)$$

Dividing the left-hand side by $E[\delta(S_T - K)]$ and the right-hand side by $\frac{d^2C}{dK^2}$, which are equal, we get:

$$\frac{E[\sigma_T^2\delta(S_T - K)]}{E[\delta(S_T - K)]} = 2\frac{\frac{dC}{dT} - (r-q)\left(C - K\frac{dC}{dK}\right)}{K^2\frac{d^2C}{dK^2}}$$

and reverting back to discounted option prices: $C = e^{-rT}\mathcal{C}$:

$$E[\sigma_T^2|S_T = K] = \frac{E[\sigma_T^2\delta(S_T - K)]}{E[\delta(S_T - K)]} = 2\frac{\frac{d\mathcal{C}}{dT} + q\mathcal{C} + (r-q)K\frac{d\mathcal{C}}{dK}}{K^2\frac{d^2\mathcal{C}}{dK^2}} \qquad (2.6)$$

This identity, known as the Dupire equation, expresses a general relationship linking the expectation of the instantaneous variance conditional on the spot price to the maturity and strike derivatives of vanilla option prices.

It holds in diffusive models for S_t: knowledge of vanilla options prices is not sufficient to pin down the process σ_t, but characterizes the class of diffusive processes that yield the same vanilla option prices. Two processes σ_t, σ_t' generate the same vanilla smile if $E[\sigma_T^2|S_T = K] = E[\sigma_T'^2|S_T = K]$ for all K, T.[2]

A stochastic volatility model aiming to reproduce at time $t = 0$ the market smile has to satisfy this condition. The simplest way of accommodating this constraint is to take for process σ_t a deterministic function of t, S:

$$\sigma_t \equiv \sigma(t, S)$$

The conditional expectation of the instantaneous variance on the left-hand side of (2.6) is then simply $\sigma(t, S)^2$ and we get the Dupire formula (2.3). The Dupire

---

[2]The general result that the marginals of an arbitrary diffusive process with instantaneous volatility σ_t are exactly recovered by using an effective local volatility model whose local volatility is given by $\sigma^2(t, S) = E[\sigma_t^2|S]$ is due to Gyöngy – see [54].

equation can also be used to compute call and put option prices for a known local volatility function $\sigma(t, S)$ – let us rewrite it as:

$$\frac{dC}{dT} + (r - q) K \frac{dC}{dK} - \frac{\sigma^2(t = T, S = K)}{2} K^2 \frac{d^2C}{dK^2} = -qC \qquad (2.7)$$

This is called the forward equation. Unlike the usual pricing equation, which is a backward equation and provides prices of a single call option with given maturity and strike for a range of initial spot prices, the forward equation, with initial condition $C(K, T = 0) = (S_0 - K)^+$, supplies prices for a single value of the spot price S_0, but for all K and T: this makes it attractive in situations when derivatives with respect to S_0 are not needed. Put option prices are obtained by changing the initial condition to $(K - S_0)^+$.

In the derivation above, we have made the assumption of a diffusive process for S_t. Consider now a given market smile – does there exist a local volatility function $\sigma(t, S)$ that is able to reproduce it? Choosing $\sigma(t, S)$ as specified by (2.3) will do the job, but what if the numerator or denominator in the right-hand side of (2.3) are negative? We now prove that this cannot be the case unless vanilla option prices are arbitrageable.

2.2.2 No-arbitrage conditions

Strike arbitrage
The denominator in (2.6) involves the second derivative of the call price with respect to K:

$$\frac{d^2C(K, T)}{dK^2} = \lim_{\varepsilon \to 0} \frac{C(K - \varepsilon, T) - 2C(K, T) + C(K + \varepsilon, T)}{\varepsilon^2}$$

Consider the European payoff consisting of $\frac{1}{\varepsilon^2}$ calls of strike $K - \varepsilon$, $\frac{1}{\varepsilon^2}$ calls of strike $K + \varepsilon$ and $-\frac{2}{\varepsilon^2}$ calls of strike K – this is known as a butterfly spread.

The payout at maturity as a function of S_T has a triangular shape whose surface area is unity: it vanishes for $S_T \leq K - \varepsilon$ and $S_T \geq K + \varepsilon$ and is equal to $\frac{1}{\varepsilon}$ for $S_T = K$. For $\varepsilon \to 0$ it becomes a Dirac delta function. It either vanishes or is strictly positive depending on S_T: its price at inception must be positive.

Options' markets are arbitraged well enough that butterfly spreads do not have negative prices:[3] the denominator in the Dupire formula (2.3) is positive.

In a model, $\frac{d^2C(K,T)}{dK^2}$ is related to the probability density of S_T through:

$$\frac{d^2C(K, T)}{dK^2} = e^{-rT} E[\delta(S_T - K)] \qquad (2.8)$$

---

[3]Bid/offer spreads of options are usually not negligible: arbitrage opportunities may appear more attractive than they really are.

thus is positive by construction. The condition $\frac{d^2 C(K,T)}{dK^2} > 0$ is equivalent to requiring that the market implied density be positive. Violation of the positivity of the denominator of (2.6) is called a strike arbitrage.

Maturity arbitrage

What about the numerator in (2.3)? It can be rewritten as:

$$e^{-qT} \frac{d}{dT}[e^{qT} C(Ke^{(r-q)T}, T)]$$

For it to be positive, e^{qT} times the price of a call option struck at a strike that is a fixed proportion of the forward $F_T = Se^{(r-q)T}$ – that is $K = kF_T$ – must be an increasing function of maturity. For $T_1 \le T_2$:

$$e^{qT_1} C\left(kF_{T_1}, T_1\right) \le e^{qT_2} C\left(kF_{T_2}, T_2\right) \tag{2.9}$$

Imagine that this condition is violated: there exist two maturities $T_1 < T_2$ and k such that:

$$e^{qT_1} C\left(kF_{T_1}, T_1\right) > e^{qT_2} C\left(kF_{T_2}, T_2\right)$$

Set up the following strategy: buy one option of maturity T_2, strike kF_{T_2} and sell $e^{-q(T_2-T_1)}$ options of maturity T_1, strike kF_{T_1}: we pocket a net premium at inception. At T_1 take the following Δ position on S:

$$\text{if } S_{T_1} < kF_{T_1} : \quad \Delta = 0$$
$$\text{if } S_{T_1} > kF_{T_1} : \quad \Delta = -1$$

Our P&L at T_2 comprises the payout of the T_2 option which we receive, the payout of the T_1 option which we pay, capitalized up to T_2, and the P&L generated by the delta position entered at T_1, which we unwind at T_2 – inclusive of financing costs. Its expression is:

$$(S_{T_2} - kF_{T_2})^+ - e^{r(T_2-T_1)} e^{-q(T_2-T_1)} (S_{T_1} - kF_{T_1})^+ + \Delta \left(S_{T_2} - \frac{F_{T_2}}{F_{T_1}} S_{T_1}\right)$$

$$= (S_{T_2} - kF_{T_2})^+ - \left[\frac{F_{T_2}}{F_{T_1}} (S_{T_1} - kF_{T_1})^+ + 1_{S_{T_1} > kF_{T_1}} \left(S_{T_2} - \frac{F_{T_2}}{F_{T_1}} S_{T_1}\right)\right]$$

$$= (S_{T_2} - kF_{T_2})^+ - \left[\left(S_{T_1}^\star - kF_{T_2}\right)^+ + 1_{S_{T_1}^\star > kF_{T_2}} \left(S_{T_2} - S_{T_1}^\star\right)\right]$$

where $S_{T_1}^\star = \frac{F_{T_2}}{F_{T_1}} S_{T_1}$. The last equation reads:

$$f(S_{T_2}) - \left[f\left(S_{T_1}^\star\right) + \frac{df}{dx}\left(S_{T_1}^\star\right)\left(S_{T_2} - S_{T_1}^\star\right)\right]$$

with $f(x) = (x - kF_{T_2})^+$. Since f is convex this is positive. Our strategy not only produces strictly positive P&L at inception; it also generates positive P&L at T_2.

Real markets are sufficiently arbitraged that arbitrage opportunities of this type do not exist: market prices of vanilla options are such that the numerator in the Dupire equation (2.3) is always positive.

In a model, the numerator in (2.3) is positive by construction. Writing the price of an option of maturity T_2 as an expectation and conditioning with respect to S_{T_1} at T_1 we get, using Jensen's inequality:

$$e^{qT_2} C(kF_{T_2}, T_2)$$
$$= e^{qT_2} e^{-rT_2} E[(S_{T_2} - kF_{T_2})^+] = e^{-(r-q)T_2} E[E[(S_{T_2} - kF_{T_2})^+ | S_{T_1}]]$$
$$\geq e^{-(r-q)T_2} E[(\frac{F_{T_2}}{F_{T_1}} S_{T_1} - kF_{T_2})^+] = e^{-(r-q)T_2} \frac{F_{T_2}}{F_{T_1}} E[(S_{T_1} - kF_{T_1})^+]$$
$$\geq e^{qT_1} C(kF_{T_1}, T_1)$$

Violation of the positivity of the numerator of (2.6) is called a maturity arbitrage.

Conclusion

In conclusion, a violation of (2.9) can be arbitraged and the local volatility given by the Dupire equation is well-defined for any arbitrage-free smile.

Contrary to a frequently heard assertion, the steep skews observed for short-maturity equity smiles are no evidence that jumps are needed to generate them – as long as they are non-arbitrageable, local volatility will be happy to oblige.

See also Section 8.7.2 for an example of how a two-factor stochastic volatility model is also able to generate the typical term structures of ATMF skews observed for equity indexes.

2.2.2.1 Convex order condition for implied volatilities

What does the convex order condition (2.9) for prices mean for implied volatilities?

Consider a call option of maturity T for a strike $K = kF_T$, whose implied volatility we denote by $\hat{\sigma}_{kT}$. We have:

$$e^{qT} C_{BS}(kF_T, T, \hat{\sigma}_{kT}) = e^{-(r-q)T} E[(S_T - kF_T)^+]$$
$$= S_0 E[(U_{\tau(T)} - k)^+] = S_0 f(k, \tau) \qquad (2.10)$$

where we have introduced U_τ defined by: $U_\tau = e^{-\frac{\tau}{2} + W_\tau}$, $\tau(T) = \hat{\sigma}_{kT}^2 T$ and f is defined by:

$$f(k, \tau) = E[(U_\tau - k)^+] \qquad (2.11)$$

U_τ is a martingale: for $\tau_1 \leq \tau_2$ $E[U_{\tau_2} | U_{\tau_1}] = U_{\tau_1}$. We could use Jensen's inequality exactly as above: for $\tau_1 \leq \tau_2$: $E[(U_{\tau_2} - k)^+] = E[E[(U_{\tau_2} - k)^+ | U_{\tau_1}]] \geq E[(U_{\tau_1} - k)^+]$ thus

$$\tau_1 \leq \tau_2 \Rightarrow f(k, \tau_1) \leq f(k, \tau_2) \qquad (2.12)$$

What we need, however, is the reverse implication.

From (2.11) $f(\tau, k)$ is the price of call option of strike k in the Black-Scholes model where the underlying U starts from $U_0 = 1$ and has a constant volatility equal to 1. It obeys the following forward PDE:

$$\frac{df}{d\tau} = \frac{1}{2}k^2\frac{d^2 f}{dk^2} \tag{2.13}$$

with initial condition $f(k, \tau = 0) = (1 - k)^+$.

From (2.8) $\frac{d^2 f}{dk^2}$ is proportional to the risk-neutral density of U_τ, which in a lognormal model with constant volatility, is *strictly* positive. Thus, from (2.13) $\frac{df}{d\tau} > 0$.

We now have property (2.12), with a *strict* inequality: $\tau_1 < \tau_2 \Rightarrow f(k, \tau_1) < f(k, \tau_2)$. This, together with (2.12) yields the following equivalence:

$$\tau_1 \leq \tau_2 \Leftrightarrow f(k, \tau_1) \leq f(k, \tau_2)$$

which, using (2.10), translates into:

$$e^{qT_1}C_{BS}(kF_{T_1}, T_1, \widehat{\sigma}_{kT_1}) \leq e^{qT_2}C_{BS}(kF_{T_2}, T_2, \widehat{\sigma}_{kT_2}) \Leftrightarrow T_1\widehat{\sigma}^2_{kT_1} \leq T_2\widehat{\sigma}^2_{kT_2} \tag{2.14}$$

Thus, in an arbitrage-free smile, the integrated variance corresponding to any given moneyness k is an increasing function of maturity:

$$T_1\widehat{\sigma}^2_{kF_{T_1},T_1} \leq T_2\widehat{\sigma}^2_{kF_{T_2},T_2} \tag{2.15}$$

2.2.2.2 Implied volatilities of general convex payoffs

The notion of implied volatility is not a privilege of hockey-stick payoffs. One can show that, in the absence of arbitrage, the notion of (lognormal) implied volatility can be defined for any convex payoff. Moreover, consider a family of European options such that the payoff $f(S_T)$ for maturity T is given by:

$$f(S_T) = h(x) \quad \text{with } x = \frac{S_T}{F_T} \text{ and } h \text{ convex.} \tag{2.16}$$

It is shown in [81] that:

- there exists one single Black-Scholes implied volatility $\widehat{\sigma}_T$ that matches a given market price for payoff f.

- no-arbitrage in market prices for maturities T_1, T_2 implies that the following convex order condition holds:

$$T_2\widehat{\sigma}^2_{T_2} \geq T_1\widehat{\sigma}^2_{T_1} \tag{2.17}$$

Vanilla options are but a particular case of convex payoffs – the payoffs of maturities T_1, T_2 used above to derive (2.15) are of type (2.16), with $h(x) = (x - k)^+$.

We will consider in Section 4.3 the particular class $h(x) = x^p$ and will focus on the special case $p \to 0$.

A note on "arbitrage" arguments

In all fairness, the type of arbitrage strategy we have outlined – entering a position and keeping it until maturity to pocket the (positive) arbitrage profit – is a bit unrealistic as it does not take into account mark-to-market P&L and the discomfort that comes with it, in the case of a large position.[4]

Imagine we bought yesterday a butterfly spread that had negative market value and today's market value is even more negative: we have lost money on yesterday's position. Our management may demand that we cut our position – at a loss – despite our plea that we will eventually make money if allowed to hold on to our position, that the arbitrage has actually become more attractive, and that we should in fact increase the size of our position.

2.3 From implied volatilities to local volatilities

The Dupire equation (2.3) expresses the local volatility as a function of derivatives of call option prices. Let us assume that there are no dividends or, less strictly, that dividend amounts are expressed as fixed yields applied to the stock value at the dividend payout date.[5] The dividend yield can then be lumped together with the repo and we can use the Black-Scholes formula to express call option prices as a function of implied volatilities. Let us use the parametrization $f(t, y)$ with:

$$y = \ln\left(\frac{K}{F_t}\right) \tag{2.18a}$$

$$f(t, y) = (t - t_0)\,\widehat{\sigma}_{Kt}^2 \tag{2.18b}$$

where F_t is the forward for maturity t: $F_t = S_0 e^{(r-q)(t-t_0)}$. Replacing C in the Dupire equation (2.3) with the Black-Scholes formula with implied volatility $\widehat{\sigma}_{KT}$, computing analytically all derivatives of C, and using f and y rather than $\widehat{\sigma}$ and K yields the following formula:

$$\sigma(t, S)^2 = \left.\frac{\frac{df}{dt}}{\left(\frac{y}{2f}\frac{df}{dy} - 1\right)^2 + \frac{1}{2}\frac{d^2f}{dy^2} - \frac{1}{4}\left(\frac{1}{4} + \frac{1}{f}\right)\left(\frac{df}{dy}\right)^2}\right|_{y=\ln\left(\frac{S}{F_t}\right)} \tag{2.19}$$

---

[4] Also note that, as we take advantage of a maturity arbitrage, we make a bet on the repo level prevailing at T_1 for maturity T_2 – which could turn sour.

[5] While this is reasonable for dividends far into the future, it is a poor assumption for nearby dividends whose cash amount is usually known, either because it has been announced, or through analysts' forecasts. As a result, equities are probably the only asset class for which even vanilla options cannot be priced in closed form.

As mentioned above, option markets typically do not violate the no-arbitrage conditions of Section 2.2.2. Market prices, however, are only available for discrete strikes and maturities: prior to using equation (2.19) we need to build an interpolation in between discrete strikes and maturities – and an extrapolation outside the range of market-traded strikes – of market implied volatilities that comply with no-arbitrage conditions.

The latter take a particularly simple form in the (y, t) coordinates. Let T_i be the discrete maturities for which implied volatilities are available and set $f_i(y) = f(T_i, y)$. The convex order condition (2.15) translates into $\frac{df}{dt} \geq 0$, thus implies the simple rule: $f_{i+1}(y) \geq f_i(y)$.

Once each $f_i(y)$ function has been created by interpolating $\hat{\sigma}^2(K, T_i)T_i$ as a function of $\ln(K/F_{T_i})$ the simple rule that the f_i profiles should not cross ensures the positivity of the numerator in the right-hand side of (2.19).

$f(t, y)$ for $y \in [T_i, T_{i+1}]$ is then generated by affine interpolation:

$$f(t, y) = \frac{T_{i+1} - t}{T_{i+1} - T_i} f_i(y) + \frac{t - T_i}{T_{i+1} - T_i} f_{i+1}(y) \qquad (2.20)$$

Though rustic, interpolation (2.20) ensures that the convex order condition holds over $[T_i, T_{i+1}]$ and that local volatilities $\sigma(t, S)$ for $t \in [T_i, T_{i+1}]$ only depend on implied volatilities for maturities T_i, T_{i+1}. Otherwise – in case a spline interpolation was used, for example – a European option expiring at $T \in [T_i, T_{i+1}]$ would be sensitive to implied volatilities for maturities longer than T_{i+1}, an incongruous and unintended consequence of the interpolation scheme.

As we turn to extrapolating $f(t, y)$ for values of y corresponding to strikes that lie beyond the lowest/highest market-quoted strikes, care must be taken not to create strike arbitrage. Typically an affine extrapolation is used: $f_i(y) = a_i y + b_i$. It is easy to check that $|a_i| \leq 2$ is a necessary condition for positivity of the denominator in (2.19) for large values of y.

Finally, there may be situations – for illiquid underlyings – when one needs to build from scratch a volatility surface; we refer the reader to [49] for a popular example of a parametric volatility surface that, under certain conditions, is arbitrage-free: the SVI formula.

2.3.1 Dividends

In the presence of cash-amount dividends, while the Dupire formula (2.3) with option prices is still valid, its version (2.19) expressing local volatilities directly as a function of implied volatilities cannot be used as is, as option prices are no longer given by the Black-Scholes formula.

We first present an exact solution then an accurate approximate solution.

2.3.1.1 An exact solution

The exact solution is taken from [58] and [19]. It relies on the mapping of S to an asset X that does not jump on dividend dates.

Let us assume that dividends consist of two portions: a fixed cash amount and a proportional part. The dividend d_i falling at time t_i is given by:

$$d_i = y_i S_{t_i^-} + c_i$$

When looking for a security that does not experience dividend jumps the forward naturally comes to mind. However, we would have to pick an arbitrary maturity T for the forward – the local volatility function would change whenever an option with maturity longer than T was priced.

Let us instead use a driftless process X which starts with the same value as S: $X_{t=0} = S_{t=0}$ and define X_t as:

$$S_t = \alpha(t) X_t - \delta(t) \tag{2.21}$$

with $\alpha(t), \delta(t)$ given by:

$$\alpha(t) = e^{(r-q)t} \prod_{t_i < t} (1 - y_i)$$

$$\delta(t) = \sum_{t_i < t} c_i e^{(r-q)(t-t_i)} \prod_{t_i < t_j < t} (1 - y_j)$$

One can check that X_t is driftless and does not jump across dividend dates. Because the relationship of S to X is affine, the price of a vanilla option on X is a multiple of the price of a vanilla option on S, with a shifted strike. We then have all implied volatilities for X and can use equation (2.19) to get the local volatility function for X: $\sigma_X(t, X)$. The local volatility for S is then given by:

$$\sigma(t, S) = \frac{S + \delta(t)}{S} \sigma_X(t, X(S, t)) \tag{2.22}$$

Across dividend dates σ_X is continuous, but σ is not, as $\delta(t)$ jumps. Those taking local volatility seriously may object to this. Consider, however, that just before a dividend date, the portion of S which is the cash dividend is frozen and has no volatility: the volatility of S only comes from the volatility of $S - c$. Consequently, as one crosses the dividend date, it is natural that the lognormal volatility of S jumps, in a fashion that is exactly expressed by (2.22).

Equation (2.21) seems to imply that S can go negative. This would be the case, for example, if X were lognormal. In reality, it does not happen, as the implied volatilities of X are derived from the smile of S which – if extrapolated properly – ensures that S_T cannot go negative, hence X_T cannot go below $\delta(T)/\alpha(T)$. For a typical negatively skewed smile for S, the smile of X will have a similar shape, except implied volatilities for low strikes, of the order of $\delta(T)/\alpha(T)$, will fall off.

2.3.1.2 An approximate solution

We really would like to use an expression relating local volatilities to implied volatilities directly, similar to (2.19). Because of the presence of cash-amount dividends, the definition of y in (2.18a) has to change.

The ingredient in (2.19) is $f(t, y)$, that is a parametrization of the implied volatility surface. When there are dividends, a suitable parametrization must ensure that appropriate matching conditions hold across dividend dates. Consider a dividend d falling at time τ, part cash amount, part yield:

$$S_{\tau+} = (1 - z)S_{\tau-} - c$$

and a call option of strike K, maturity τ^+. Its payoff can be written as a function of $S_{\tau-}$:

$$(S_{\tau+} - K)^+ = ((1 - z)S_{\tau-} - c - K)^+$$
$$= (1 - z)\left(S_{\tau-} - \frac{c + K}{1 - z}\right)^+$$

This option's payoff is proportional to that of a vanilla option of strike $\frac{c+K}{1-z}$, maturity τ^-. Their implied volatilities are thus identical:

$$\widehat{\sigma}_{K\tau+} = \widehat{\sigma}_{\frac{K+c}{1-z}\,\tau-}$$

Equivalently:

$$\widehat{\sigma}_{K\tau-} = \widehat{\sigma}_{(1-z)K-c\,\tau+} \tag{2.23}$$

How can we alter the definition of y in (2.18a) so that (2.23) holds? Our inspiration comes from an approximation for vanilla option prices in the Black-Scholes model when cash-amount dividends are present.

Proportional dividends are readily converted in an adjustment of the initial spot value. With regard to the cash-amount portion of dividends, one typically uses an approximation that condenses them into a smaller number of effective cash-amount dividends.

The most well-known is that published by Michael Bos and Stephen Vandermark – see [16] – where dividends are replaced with just two effective dividends. Each cash-amount dividend is split into two pieces, one falling at $t = 0$, resulting in a negative adjustment δS of the initial spot value, the other at maturity T, which translates into a positive shift δK of the strike. For vanishing interest rate and repo, the proportions are, respectively, $\frac{T-t}{T}$ and $\frac{t}{T}$ where t is the time the dividend falls.

Let y_i and c_i be the yield and cash-amount of the dividend falling at time t_i: $S_{t_i^+} = (1 - y_i)S_{t_i^-} - c_i$. Define the functions $\alpha(T)$, $\delta S(T)$, $\delta K(T)$ as:

$$
\begin{cases}
\alpha(T) &= \displaystyle\prod_{t_i < T} (1 - y_i) \\[2ex]
\delta S(T) &= \displaystyle\sum_{t_i < T} \frac{T - t_i}{T} c_i^* \, e^{-(r-q)t_i} \\[2ex]
\delta K(T) &= \displaystyle\sum_{t_i < T} \frac{t_i}{T} c_i^* \, e^{(r-q)(T-t_i)}
\end{cases}
\tag{2.24}
$$

with the effective cash amounts c_i^* given by:

$$
c_i^* = c_i \prod_{t_i < t_j < T} (1 - y_j)
$$

The price of a vanilla option of strike K, maturity T is given approximately by the Black-Scholes formula with rate r and repo q, with the initial spot value S and strike K replaced, respectively, by $\alpha(T)S - \delta S(T)$ and $K + \delta K(T)$:

$$
C(K,T) = P_{BS}\left(t_0,\ \alpha(T)S_0 - \delta S(T),\ K + \delta K(T),\ \widehat{\sigma}_{KT}\right)
\tag{2.25}
$$

When there are no dividends, $\alpha = 1$, $\delta S = 0$, $\delta K = 0$.

Directly using the Bos & Vandermark approximation for vanilla option prices in the Dupire formula (2.3) does not work. Indeed, (2.3) expresses the square of the local volatility as the ratio of two quantities, each of which becomes very small when $S \ll S_0$ or $S \gg S_0$. An approximation of $C(K,T)$ has to be such that its derivatives with respect to K, T are very accurate and remain so even when they are very small – this is too much to ask from (2.25).

For example consider a flat implied volatility surface: $\widehat{\sigma}_{KT} \equiv \sigma_0$. In order to recover σ_0 as the (constant) local volatility out of (2.3), $C(K,T)$ needs to obey the forward equation (2.7). It is easy to check that expression (2.25) for $C(K,T)$ with $\widehat{\sigma}_{KT} \equiv \sigma_0$ does not fulfill this condition.

While we will not use (2.25), we make use of the expressions of $\alpha(T)$, $\delta S(T)$, $\delta K(T)$. Consider the following amended definition of y and parametrization $f(t,y)$ of the volatility surface:

$$
\begin{cases}
y &= \ln\left(\dfrac{K + \delta K(t)}{\alpha(t)S_0 - \delta S(t)}\right) - (r - q)(t - t_0) \\[2ex]
f(t,y) &= (t - t_0)\,\widehat{\sigma}_{Kt}^2
\end{cases}
\tag{2.26}
$$

where $f(t,y)$ is continuous across dividend dates.

Consider a dividend falling at time τ, with cash-amount c and yield z and a vanilla option of strike K, maturity τ^-. Let us check that condition (2.23) holds.

Since f is continuous across τ, $\widehat{\sigma}_{K\tau^-} = \widehat{\sigma}_{K'\tau^+}$, where K, K' are such that they correspond to the same value of y:

$$\frac{K' + \delta K(\tau^+)}{\alpha(\tau^+)S_0 - \delta S(\tau^+)} = \frac{K + \delta K(\tau^-)}{\alpha(\tau^-)S_0 - \delta S(\tau^-)} \qquad (2.27)$$

From the definition of $\alpha, \delta S, \delta K$:

$$\alpha(\tau^+) = (1 - z)\alpha(\tau^-)$$
$$\delta S(\tau^+) = (1 - z)\delta S(\tau^-)$$
$$\delta K(\tau^+) = (1 - z)\delta K(\tau^-) + c$$

which, once plugged in (2.27) yields:

$$K' = (1 - z)K - c$$

Thus (2.23) is exactly obeyed: using a smooth function $f(t, y)$ with y given by (2.26) automatically takes care of the matching conditions across dividend dates. Our final recipe is thus:

- Build a smooth interpolation of $f = (t - t_0)\widehat{\sigma}_{Kt}^2$ as a function of (t, y) with y defined in (2.26).

- Use formula (2.19) to generate the local volatility function.

In the author's experience this approximate technique is accurate for indexes (many small dividends) and stocks (few large dividends) alike – see Figure 2.1 for an example with the S&P 500 index. As an additional benefit, whenever we input a flat volatility surface – $\widehat{\sigma}_{KT} = \sigma_0, \forall K, T$ – we exactly recover a flat local volatility function: $\sigma(t, S) = \sigma_0, \forall t, S$.[6]

2.4   From local volatilities to implied volatilities

Expression (2.19) gives local volatilities as a function of implied volatilities. For the sake of analyzing the dynamics of the local volatility model, we need to study how, for a set local volatility function $\sigma(t, S)$, implied volatilities respond to a move of S. Rather than solving the forward equation (2.7) for call option prices, we will use an approximate formula that expresses implied volatilities as a function of the local volatility function directly.

We first derive a more general identity.

---

[6]The ratios $\frac{T-t_i}{T}$, $\frac{t_i}{T}$ in the definition of δS, δK, could be replaced by other functions of $\frac{t_i}{T}$, provided these vanish respectively for $t_i = T$ and $t_i = 0$, for the sake of ensuring condition (2.23). We leave this optimization to the reader.

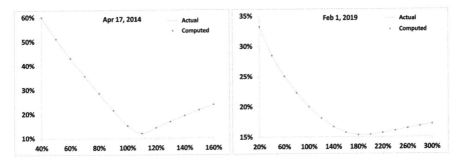

Figure 2.1: Comparison of market implied volatilities (solid line) with implied volatilities computed using the local volatility function generated by the approximate technique of Section 2.3.1.2 (dots), for the S&P 500 index, for two maturities. Market parameters as of February 1, 2014 have been used.

2.4.1 Implied volatilities as weighted averages of instantaneous volatilities

Consider a base model – denoted model I – a local volatility model with volatility function $\sigma_1(t, S)$. Denote by $P_1(t, S)$ the price of a vanilla option of strike K, maturity T in this base model.

Consider another model – denoted model II – such that the dynamics of S_t is given by:

$$dS_t = (r - q) S_t dt + \sigma_{2t} S_t dW_t \qquad (2.28)$$

where the instantaneous volatility σ_{2t} is for now an arbitrary process.

Imagine delta-hedging a short vanilla option position of maturity T using the base model with the actual dynamics of S_t given by the second model. Our final P&L is the sum of all gamma/theta P&Ls between successive delta rehedges, each one equal to: $-S_t^2 \frac{d^2 P_1}{dS^2}\Big|_{S=S_t} \left(\sigma_{2t}^2 - \sigma_1(t, S_t)^2\right) dt.$

The price P_2 we should charge for this option is then the price in model I plus the opposite of an estimate of the sum of such gamma/theta P&Ls. The following derivation makes this trading intuition more precise.

Consider process Q_t defined by:

$$Q_t = e^{-rt} P_1(t, S_t)$$

At $t = 0$, $Q_{t=0}$ is simply the initial price in model I, for the initial spot value S_0. The variation during dt of Q_t reads, in the dynamics of model II:

$$
dQ_t = e^{-rt} \left[\left(-rP_1 + \frac{dP_1}{dt} \right) dt + \frac{dP_1}{dS} dS_t + \frac{1}{2} \frac{d^2 P_1}{dS^2} \langle dS_t^2 \rangle \right]
$$

$$
= e^{-rt} \left[\left(-rP_1 + \frac{dP_1}{dt} \right) dt + \frac{dP_1}{dS} dS_t + \frac{1}{2} \sigma_{2t}^2 S_t^2 \frac{d^2 P_1}{dS^2} dt \right] \qquad (2.29)
$$

where P_1 and its derivatives are evaluated at (t, S_t). Taking the expectation of (2.29) yields:

$$
E_2[dQ_t | t, S_t] = e^{-rt} \left(-rP_1 + \frac{dP_1}{dt} + (r - q) S_t \frac{dP_1}{dS} + \frac{1}{2} \sigma_{2t}^2 S_t^2 \frac{d^2 P_1}{dS^2} \right) dt
$$

$$
= e^{-rt} \frac{S_t^2}{2} \frac{d^2 P_1}{dS^2} \left(\sigma_{2t}^2 - \sigma_1(t, S_t)^2 \right) dt
$$

where we have made use of pricing equation (2.2) for P_1 and where the subscript 2 indicates that the expectation is taken over paths of S_t generated by SDE (2.28). Integrating the above expression on $[0, T]$:

$$
E_2[Q_T] = Q_0 + \int_0^T E_2[dQ_t]
$$

$$
= P_1(0, S_0) + E_2 \left[\int_0^T e^{-rt} \frac{S_t^2}{2} \frac{d^2 P_1}{dS^2} \left(\sigma_{2t}^2 - \sigma_1(t, S_t)^2 \right) dt \right]
$$

Now $Q_T = e^{-rT} P_1(T, S_T) = e^{-rT} f(S_T)$, where f is the European option's payoff, thus $E_2[Q_T] = e^{-rT} E_2[f(S_T)] = P_2(0, S_0, \bullet)$ where P_2 is the pricing function of model 2, which involves its own parameters in addition to t, S. We thus have:

$$
P_2(0, S_0, \bullet) = P_1(0, S_0) + E_2 \left[\int_0^T e^{-rt} \frac{S_t^2}{2} \frac{d^2 P_1}{dS^2} \left(\sigma_{2t}^2 - \sigma_1(t, S_t)^2 \right) dt \right] \qquad (2.30)
$$

which expresses mathematically what trading intuition suggested.

The price of an option for an arbitrary dynamics of S_t is given by its price in a given base model – here a local volatility model – supplemented with the expectation of the discounted gamma/theta P&Ls incurred as one delta-hedges the option using the base model until maturity.

Assume now that model I is the Black-Scholes model with a constant volatility equal to the implied volatility of the vanilla option at hand, backed out of model II price: $\sigma_1(t, S_t) \equiv \hat{\sigma}_{KT}$. By definition of $\hat{\sigma}_{KT}$:

$$
P_1(0, S_0) = P_{\hat{\sigma}_{KT}}(0, S_0) = P_2(0, S_0, \bullet)
$$

where $P_{\hat{\sigma}_{KT}}(t, S)$ denotes the Black-Scholes price with volatility $\hat{\sigma}_{KT}$. (2.30) then yields:

$$\widehat{\sigma}_{KT}^2 = \frac{E_2\left[\int_0^T e^{-rt}S_t^2 \frac{d^2 P_{\widehat{\sigma}_{KT}}}{dS^2}\sigma_{2t}^2 dt\right]}{E_2\left[\int_0^T e^{-rt}S_t^2 \frac{d^2 P_{\widehat{\sigma}_{KT}}}{dS^2}dt\right]} \tag{2.31}$$

This is true for an arbitrary instantaneous volatility σ_{2t}. Specialize now to the case of a local volatility model: $\sigma_{2t} \equiv \sigma(t,S)$ and $\widehat{\sigma}_{KT}$ is the implied volatility corresponding to the local volatility function. We have:

$$\widehat{\sigma}_{KT}^2 = \frac{E_{\sigma(t,S)}\left[\int_0^T e^{-rt}S_t^2 \frac{d^2 P_{\widehat{\sigma}_{KT}}}{dS^2}\sigma(t,S)^2 dt\right]}{E_{\sigma(t,S)}\left[\int_0^T e^{-rt}S_t^2 \frac{d^2 P_{\widehat{\sigma}_{KT}}}{dS^2}dt\right]} \tag{2.32}$$

$\widehat{\sigma}_{KT}^2$ is thus the average value of $\sigma(t,S)^2$, weighted by the dollar gamma computed with the constant volatility $\widehat{\sigma}_{KT}$ itself, over paths generated by the local volatility $\sigma(t,S)$.

Going back to identity (2.30), consider instead that model I is the local volatility model – $\sigma_1(t,S) \equiv \sigma(t,S)$ – and that model II is the Black-Scholes model with volatility $\widehat{\sigma}_{KT}$. Again we have $P_1(0,S_0) = P_2(0,S_0)$, and now get:

$$\widehat{\sigma}_{KT}^2 = \frac{E_{\widehat{\sigma}_{KT}}\left[\int_0^T e^{-rt}S^2 \frac{d^2 P_{\sigma(S,t)}}{dS^2}\sigma(t,S)^2 dt\right]}{E_{\widehat{\sigma}_{KT}}\left[\int_0^T e^{-rt}S^2 \frac{d^2 P_{\sigma(S,t)}}{dS^2}dt\right]} \tag{2.33}$$

where the averaging is now performed using the density generated by a Black-Scholes model with constant volatility $\widehat{\sigma}_{KT}$ and $\sigma(t,S)^2$ is now weighted by the dollar gamma computed in the local volatility model.[7]

In this derivation we have used as base model the local volatility model, with a fixed volatility function – or the Black-Scholes model, a particular version of it – and have considered a delta-hedged vanilla option position.

In Section 8.4, page 316, we derive a different expression for European option prices in diffusive models by considering a delta-hedged, vega-hedged option position, using as base model a Black-Scholes model whose volatility is constantly recalibrated to the terminal VS volatility.

2.4.2 Approximate expression for weakly local volatilities

While the Dupire formula (2.19) explicitly expresses local volatilities as a function of implied volatilities, formulas (2.32) and (2.33) for $\widehat{\sigma}_{KT}$ are implicit, as the right-hand side depends on the unknown implied volatility $\widehat{\sigma}_{KT}$, either through the dollar gamma or through the density used for averaging the numerator and denominator.

---

[7]As far as I can remember, expression (2.32) for implied volatilities in the local volatility model was first presented by Bruno Dupire at a Global Derivatives conference in the late 90s.

In order to get a more usable expression, we will assume that $\sigma(t, S)$ is only weakly local.

We will use formula (2.32) with the Black-Scholes model with time-dependent volatility $\sigma_0(t)$ as base model.

Let us use the local variance $u(t, S) = \sigma(t, S)^2$ and assume that

$$u(t, S) = u_0(t) + \delta u(t, S)$$

where $u_0 = \sigma_0^2(t)$ and δu is a small perturbation. If $\delta u = 0$, $\hat{\sigma}_{KT} = \hat{\sigma}_{0T}$ where $\hat{\sigma}_{t_1 t_2}$ is defined as:

$$\hat{\sigma}_{t_1 t_2}^2 = \frac{1}{t_2 - t_1} \int_{t_1}^{t_2} \sigma_0^2(t) dt$$

Let us use expression (2.32) and expand $\hat{\sigma}_{KT}^2$ at first order in δu: $\hat{\sigma}_{KT}^2 + \delta(\hat{\sigma}_{KT}^2)$. $u(t, S)$ appears explicitly in the numerator as well as implicitly in the density used for computing expectations in both numerator and denominator. For the sake of computing the order-one perturbation in δu, however, the contribution generated by the density vanishes: from equation (2.32) and using a compact notation:

$$\hat{\sigma}_{KT}^2 + \delta(\hat{\sigma}_{KT}^2) = \frac{E_{u_0 + \delta u}[(u_0 + \delta u) \bullet]}{E_{u_0 + \delta u}[\bullet]} = u_0 + \frac{E_{u_0 + \delta u}[\delta u \bullet]}{E_{u_0 + \delta u}[\bullet]}$$

$$= u_0 + \frac{E_{u_0}[\delta u \bullet]}{E_{u_0}[\bullet]} \tag{2.34}$$

where $\bullet$ is computed at order zero in δu. We thus have:

$$\delta(\hat{\sigma}_{KT}^2) = \frac{E_{\sigma_0}\left[\int_0^T e^{-rt} \, \delta u(t, S) \, S^2 \frac{d^2 P_{\sigma_0}}{dS^2} dt\right]}{E_{\sigma_0}\left[\int_0^T e^{-rt} \, S^2 \frac{d^2 P_{\sigma_0}}{dS^2} dt\right]} \tag{2.35}$$

The right-hand side of equation (2.35) now only requires the density and the dollar gamma of a call or put option, evaluated in the Black-Scholes model with deterministic volatility $\sigma_0(t)$ – both analytically known. The denominator in (2.35) involves the discounted dollar gamma of a European option, averaged over its lifetime. In the Black-Scholes model, $e^{-rt} S^2 \frac{d^2 P_{\sigma_0}}{dS^2}$ is a martingale – see a proof in Appendix A of Chapter 5, page 181. Thus:

$$E_{\sigma_0}\left[e^{-rt} S^2 \frac{d^2 P_{\sigma_0}}{dS^2}\right] = S_0^2 \frac{d^2 P_{\sigma_0}}{dS^2}\bigg|_{t=0, S=S_0} \tag{2.36}$$

where S_0 denotes today's spot price. The denominator is then equal to $T S_0^2 \frac{d^2 P_{\sigma_0}}{dS^2}\big|_{t=0, \, S=S_0}$.

Focus now on the numerator in (2.35). It reads:

$$\int_0^T dt \int_0^\infty dS \, \rho_{\sigma_0}(t, S) \, e^{-rt} \, \delta u(t, S) \, S^2 \frac{d^2 P_{\sigma_0}}{dS^2} \tag{2.37}$$

where $\rho_{\sigma_0}(t, S)$ is the lognormal density with deterministic volatility $\sigma_0(t)$. Define $x = \ln(S/F_t)$ where F_t is the forward for maturity t: $F_t = S_0 e^{(r-q)t}$. $\rho_{\sigma_0}(t, S)$ and $S^2 \frac{d^2 P_{\sigma_0}}{dS^2}$ are given by:

$$\rho_{\sigma_0}(t, S) = \frac{1}{\sqrt{2\pi\omega_t}S} e^{-\frac{\left(x+\frac{\omega_t}{2}\right)^2}{2\omega_t}} \tag{2.38}$$

$$S^2 \frac{d^2 P_{\sigma_0}}{dS^2} = S\frac{F_T}{F_t} e^{-r(T-t)} \frac{1}{\sqrt{2\pi(\omega_T-\omega_t)}} e^{-\frac{\left(-x_K + x + \frac{(\omega_T-\omega_t)}{2}\right)^2}{2(\omega_T-\omega_t)}} \tag{2.39}$$

where $x_K = \ln(\frac{K}{F_T})$ and we have introduced the integrated variance ω_t, defined as:

$$\omega_t = \int_0^t \sigma_0^2(\tau) d\tau$$

The numerator then reads:

$$F_T e^{-rT} \int_0^T dt \int_{-\infty}^{+\infty} dx \, \frac{\delta u(t, F_t e^x)}{\sqrt{2\pi(\omega_T-\omega_t)}\sqrt{2\pi\omega_t}} e^x e^{-\frac{\left(-x_K + x + \frac{(\omega_T-\omega_t)}{2}\right)^2}{2(\omega_T-\omega_t)}} e^{-\frac{\left(x+\frac{\omega_t}{2}\right)^2}{2\omega_t}}$$

Combining the exponentials, we can rewrite this expression as:

$$F_T e^{-rT} \int_0^T dt \int_{-\infty}^{+\infty} dx \, \frac{\delta u(t, F_t e^x)}{\sqrt{2\pi(\omega_T-\omega_t)}\sqrt{2\pi\omega_t}} e^{-\left(\frac{1}{2(\omega_T-\omega_t)}+\frac{1}{2\omega_t}\right)\frac{(\omega_T x - \omega_t x_K)^2}{\omega_T^2}} e^{-\frac{\left(x_K-\frac{\omega_T}{2}\right)^2}{2\omega_T}}$$

We now divide this by (2.36), the dollar gamma evaluated at $t = 0$ and multiplied by T, which reads:

$$TF_T e^{-rT} \frac{1}{\sqrt{2\pi\omega_T}} e^{-\frac{\left(-x_K + \frac{\omega_T}{2}\right)^2}{2\omega_T}}$$

and get for the ratio in (2.35):

$$\frac{1}{T}\int_0^T dt \int_{-\infty}^{+\infty} dx \, \frac{\sqrt{\omega_T}}{\sqrt{\omega_t}} \frac{\delta u(t, F_t e^x)}{\sqrt{2\pi(\omega_T-\omega_t)}} e^{-\left(\frac{1}{2(\omega_T-\omega_t)}+\frac{1}{2\omega_t}\right)\frac{(\omega_T x - \omega_t x_K)^2}{\omega_T^2}}$$

Switching now from x to a new coordinate y:

$$x = \frac{\omega_t}{\omega_T}x_K + \frac{\sqrt{(\omega_T-\omega_t)\omega_t}}{\sqrt{\omega_T}}y$$

yields our final formula for $\delta(\hat{\sigma}_{KT}^2)$ at order one in δu:

$$\delta(\hat{\sigma}_{KT}^2) = \frac{1}{T}\int_0^T dt \int_{-\infty}^{+\infty} dy \, \frac{e^{-\frac{y^2}{2}}}{\sqrt{2\pi}} \delta u\left(t, F_t e^{\frac{\omega_t}{\omega_T}x_K + \frac{\sqrt{(\omega_T-\omega_t)\omega_t}}{\sqrt{\omega_T}}y}\right)$$

Noting that, at order zero, $\hat{\sigma}^2_{KT} = \frac{1}{T}\int_0^T \sigma_0^2(t)dt = \frac{1}{T}\int_0^T u_0(t)dt$, this can be rewritten, at order one in $\delta u = \sigma^2(t,S) - \sigma_0^2(t)$, as:

$$\hat{\sigma}^2_{KT} = \frac{1}{T}\int_0^T dt \int_{-\infty}^{+\infty} dy\, \frac{e^{-\frac{y^2}{2}}}{\sqrt{2\pi}}\, u\left(t, F_t e^{\frac{\omega_t}{\omega_T}x_K + \frac{\sqrt{(\omega_T - \omega_t)\omega_t}}{\sqrt{\omega_T}}y}\right) \tag{2.40}$$

This is the expression of $\hat{\sigma}_{KT}$ at order one in the perturbation of σ around a time-dependent volatility $\sigma_0(t)$. The square of the implied volatility is expressed as an integral of the square of the instantaneous volatility – thus (2.40) is exact when u is a function of t only.[8]

2.4.3   Expanding around a constant volatility

Consider as base case the Black-Scholes model with constant volatility σ_0: $\sigma_0(t) = \sigma_0$, thus $\omega_t = \sigma_0^2 t$ and write:

$$\sigma(t,S) = \sigma_0 + \delta\sigma(t,S)$$

thus $\delta u = 2\sigma_0\delta\sigma(t,S)$. Using (2.40) together with $\hat{\sigma}^2_{KT} = \sigma_0^2 + 2\sigma_0\delta(\hat{\sigma}_{KT})$ yields:

$$\delta\hat{\sigma}_{KT} = \frac{1}{T}\int_0^T dt \int_{-\infty}^{+\infty} dy\, \frac{e^{-\frac{y^2}{2}}}{\sqrt{2\pi}}\, \delta\sigma\left(t, F_t e^{\frac{t}{T}x_K + \sigma_0\sqrt{\frac{(T-t)t}{T}}y}\right) \tag{2.41}$$

Adding σ_0 on both sides yields, at order one in $\delta\sigma(t,S)$:

$$\hat{\sigma}_{KT} = \frac{1}{T}\int_0^T dt \int_{-\infty}^{+\infty} dy\, \frac{e^{-\frac{y^2}{2}}}{\sqrt{2\pi}}\, \sigma\left(t, F_t e^{\frac{t}{T}x_K + \sigma_0\sqrt{\frac{(T-t)t}{T}}y}\right) \tag{2.42}$$

Even though (2.40) may be slightly more accurate when the term-structure of volatilities is strong, we will use (2.42) in the sequel, since resulting expressions for the ATMF skew and the SSR are simpler.

2.4.4   Discussion

For $t = 0$, $F_t \exp\left(\frac{t}{T}\ln\frac{K}{F_T} + \frac{\sqrt{\sigma_0^2(T-t)t}}{\sqrt{T}}y\right) = S_0$. Thus values of $\sigma\,(t=0,S)$ for $S \neq S_0$ do not contribute in (2.42) to $\hat{\sigma}_{KT}$.

Likewise, for $t = T$, $F_t \exp\left(\frac{t}{T}\ln\frac{K}{F_T} + \frac{\sqrt{\sigma_0^2(T-t)t}}{\sqrt{T}}y\right) = K$: values of $\sigma\,(t=T,S)$ for $S \neq K$ do not contribute either. This is natural, upon inspection of expression

---

[8]This is comforting but should not cause too much rejoicing – (2.40) is only an order-one approximation, besides we will see further below that for $T \to 0$ there is an exact expression of the *inverse* of $\hat{\sigma}_{KT}$ as an average of the *inverse* of $\sigma\,(0,S)$.

(2.37): for $t = 0$, the density ρ_{σ_0} vanishes unless $S = S_0$, while for $t = T$ the dollar gamma $S^2 \frac{d^2 P_{\sigma_0}}{dS^2}$ vanishes, unless $S = K$.

In (2.42), the largest weight is obtained for $y = 0$: this singles out a path for $\ln(S)$ which is a straight line starting at $\ln(S_0)$ for $t = 0$ and ending at $\ln(K)$ at $t = T$. Replacing the integral over y by the value for $y = 0$ would give an approximation of $\widehat{\sigma}_{KT}$ as a uniform average of the local volatility along this line:

$$\widehat{\sigma}_{KT} \simeq \frac{1}{T} \int_0^T \sigma \left(t, F_t e^{\frac{t}{T} \ln \frac{K}{F_T}} \right) dt \tag{2.43}$$

Summing over values of $y \neq 0$, includes other paths in the average, with their ends pinned down at S_0 at $t = 0$ and at K for $t = T$ by the factor $\sqrt{\sigma_0^2 (T - t) t}$.

However appealing expression (2.42) for $\widehat{\sigma}_{KT}$ may be, market smiles on equity underlyings are strong enough, and bid-offer spreads on vanilla option prices are tight enough, that its numerical accuracy is not sufficient for practical trading purposes: an order one expansion in $\delta\sigma(t, S)$ is simply not adequate.[9]

Can we do better? Expression (2.42) for $\widehat{\sigma}_{KT}$ amounts to merely setting $\widehat{\sigma}_{KT} = \sigma_0$ and using the lognormal density with volatility σ_0 for computing both averages in the numerator and denominator of the right-hand side of equation (2.32).

A number of tricks have been proposed under the loose name of "most likely path" techniques for approximating the right-hand side of (2.32), using a lognormal density but with a different implied volatility for each time slice – still, their accuracies are not adequate.

The reason for this is that formula (2.32) seems to suggest that the main contribution of $\sigma(t, S)$ is embodied in the explicit gamma term in the numerator and that using an approximate density – for example lognormal – for computing both averages in the numerator and the denominator will do. This is not the case. In practice, taking into account the dependence of the density itself on $\sigma(t, S)$ is mandatory for achieving the accuracy needed in trading applications.[10] For realistic equity smiles, there isn't yet a computationally efficient alternative to numerically solving the forward equation (2.7).

Is formula (2.42) then of any use? While a good approximation of absolute volatility levels is hard to come by, the skew – which is the difference of two volatilities – is more easily approximated. We now use equation (2.42) to calculate the skew and curvature of the smile, as a function of parameters of the local volatility function.

---

[9] The cruder version (2.43) is even less usable.
[10] Julien Guyon and Pierre Henry-Labordère provide in [51] a nice summary and comparison of different "most likely path" approximations, along with a technique based on a short-time heat-kernel expansion for ρ.

2.4.5 The smile near the forward

Let us assume that the local volatility is smooth and given by:

$$\sigma(t, S) = \overline{\sigma}(t) + \alpha(t) x + \frac{\beta(t)}{2} x^2 \tag{2.44}$$

where x – which we call moneyness – is given by $x = \ln(S/F_t)$.

We assume that $\alpha(t)$, $\beta(t)$ are small, and that $\overline{\sigma}(t)$ does not vary too much, so that the difference $\overline{\sigma}(t) - \sigma_0$ is small, where σ_0 is the constant volatility level around which the order-one expansion in (2.35) is performed.

We could as well perform the expansion around the time-deterministic volatility $\overline{\sigma}(t)$ – calculations are similar. For the sake of simplicity we carry out the expansion around a constant σ_0 – expressions for the more general case appear in (2.60), page 51.

Equation (2.42) gives :

$$\hat{\sigma}_{KT} = \frac{1}{T} \int_0^T \overline{\sigma}(t)\, dt \tag{2.45}$$

$$+ \frac{1}{T} \int_0^T dt \int_{-\infty}^{+\infty} dy\, \frac{e^{-\frac{y^2}{2}}}{\sqrt{2\pi}} \left[\alpha(t) X(t,y) + \frac{\beta(t)}{2} X(t,y)^2 \right]$$

where

$$X(t,y) = \frac{t}{T} x_K + \frac{\sqrt{\sigma_0^2 (T-t) t}}{\sqrt{T}} y \tag{2.46}$$

Doing the integrals over y we get at order 1 in α, β:

$$\hat{\sigma}_{KT} = \frac{1}{T} \int_0^T \overline{\sigma}(t)\, dt + \frac{\sigma_0^2 T}{2} \frac{1}{T} \int_0^T \frac{(T-t) t}{T^2} \beta(t) dt \tag{2.47}$$

$$+ \left(\frac{1}{T} \int_0^T \frac{t}{T} \alpha(t) dt \right) x_K + \frac{1}{2} \left(\frac{1}{T} \int_0^T \frac{t^2}{T^2} \beta(t) dt \right) x_K^2$$

Thus, for a sufficiently smooth local volatility function, the skew and curvature of the smile near the forward are related to the skew and curvature of the local volatility function through:

$$\mathcal{S}_T = \left. \frac{d\hat{\sigma}_{KT}}{d\ln K} \right|_{F_T} = \frac{1}{T} \int_0^T \frac{t}{T} \alpha(t) dt \tag{2.48}$$

$$\left. \frac{d^2 \hat{\sigma}_{KT}}{d\ln K^2} \right|_{F_T} = \frac{1}{T} \int_0^T \left(\frac{t}{T} \right)^2 \beta(t) dt \tag{2.49}$$

where we have introduced $\mathcal{S}_T$ as a notation for the ATMF (at the money forward) skew.[11]

---

[11]These approximate formulas for the implied ATMF skew and curvature in the local volatility model can be obtained in a number of ways – see [78] for an alternative derivation of the "skew-averaging" expression (2.48).

2.4.5.1 A constant local volatility function

Assume that α and β are constant. We get:

$$\left.\frac{d\hat{\sigma}_{KT}}{d\ln K}\right|_{F_T} = \frac{1}{T}\int_0^T \frac{t}{T}dt\,\alpha = \frac{\alpha}{2} \tag{2.50a}$$

$$\left.\frac{d^2\hat{\sigma}_{KT}}{d\ln K^2}\right|_{F_T} = \frac{1}{T}\int_0^T \left(\frac{t}{T}\right)^2 \beta dt = \frac{\beta}{3} \tag{2.50b}$$

Thus, for a local volatility function of the form (2.44) with α and β constant, at order one in α, β, the ATMF skew of the implied volatility is half the skew of the local volatility function, while its curvature is one third the curvature of the local volatility function.

2.4.5.2 A power-law-decaying ATMF skew

Let us assume a power-law form for $\alpha(t)$. To prevent divergence as $t \to 0$ we take:

$$\begin{cases} \alpha(t) = \alpha_0 \left(\frac{\tau_0}{t}\right)^\gamma & t > \tau_0 \\ \alpha(t) = \alpha_0 & t \le \tau_0 \end{cases} \tag{2.51}$$

where τ_0 is a cutoff – typically $\tau_0 = 3$ months – and γ is the characteristic exponent of the long-term decay of $\alpha(t)$. We get, from (2.48):

$$\begin{cases} \mathcal{S}_T = \frac{\alpha_0}{2} & T \le \tau_0 \\ \mathcal{S}_T = \frac{1}{2-\gamma}\alpha_0 \left(\frac{\tau_0}{T}\right)^\gamma - \frac{\gamma}{2(2-\gamma)}\alpha_0 \left(\frac{\tau_0}{T}\right)^2 & T \ge \tau_0 \end{cases} \tag{2.52}$$

γ is typically smaller than 2. For (very) long maturities the second piece in (2.52) can be ignored and we get:

$$\mathcal{S}_T \simeq \frac{1}{2-\gamma}\alpha_0 \left(\frac{\tau_0}{T}\right)^\gamma \tag{2.53}$$

The long-term ATMF skew thus decays with the same exponent γ as the local volatility function. For typical equity smiles, $\gamma \simeq \frac{1}{2}$ – see examples in Figure 2.3, page 58. With respect to the local volatility skew $\alpha(t)$, $\mathcal{S}_T$ is rescaled by a factor $\frac{1}{2-\gamma}$.

2.4.6 An exact result for short maturities

We consider here the case $T \to 0$, for which a particularly simple relationship linking local and implied volatilities exists, which we now derive.

Let us recall the definition of y and f, which appear in expression (2.19):

$$y = \ln\left(\frac{K}{F_T}\right)$$

$$f(T, y) = (T - t_0)\,\hat{\sigma}^2_{K=F_Te^y,T}$$

To lighten the notation we use $\widehat{\sigma}\,(T, y)$ instead of $\widehat{\sigma}_{K\,=\,F_T e^y, T}$ and take $t_0 = 0$. Expression (2.19) reads:

$$\sigma^2 = \frac{\widehat{\sigma}^2 + 2T\widehat{\sigma}\widehat{\sigma}_T}{\left(\frac{y}{\widehat{\sigma}^2}\widehat{\sigma}\widehat{\sigma}_y - 1\right)^2 + T\left(\widehat{\sigma}_y^2 + \widehat{\sigma}\widehat{\sigma}_{yy}\right) - \left(\frac{1}{4} + \frac{1}{T\widehat{\sigma}^2}\right)\widehat{\sigma}^2\widehat{\sigma}_y^2 T^2}$$

where $\widehat{\sigma}_y, \widehat{\sigma}_T$ denote derivatives of $\widehat{\sigma}$ with respect to y and T. Take the limit $T \to 0$ and keep the leading term in the numerator and denominator in an expansion in powers of T, assuming that $\widehat{\sigma}$ is a smooth function of T and y as $T \to 0$. We get:

$$\sigma^2 = \frac{\widehat{\sigma}^2}{\left(\frac{y}{\widehat{\sigma}}\widehat{\sigma}_y - 1\right)^2} = \frac{1}{\left(\frac{y}{\widehat{\sigma}^2}\widehat{\sigma}_y - \frac{1}{\widehat{\sigma}}\right)^2}$$

which yields

$$\frac{1}{\sigma} = \pm\left(\frac{y}{\widehat{\sigma}^2}\widehat{\sigma}_y - \frac{1}{\widehat{\sigma}}\right) = \mp\left(y\left(\frac{1}{\widehat{\sigma}}\right)_y + \frac{1}{\widehat{\sigma}}\right) = \mp\left(\frac{y}{\widehat{\sigma}}\right)_y$$

Thus

$$\int_0^y \frac{du}{\sigma\,(T = 0,\, Se^u)} = \mp\frac{y}{\widehat{\sigma}\,(T = 0, Se^y)}$$

Following the usual convention of using positive volatilities:

$$\frac{1}{\widehat{\sigma}\,(T = 0, Se^y)} = \frac{1}{y}\int_0^y \frac{du}{\sigma\,(T = 0,\, Se^u)} \qquad (2.54)$$

or, equivalently:

$$\frac{1}{\widehat{\sigma}\,(T = 0, K)} = \frac{1}{\ln\frac{K}{S}}\int_S^K \frac{1}{\sigma\,(T = 0,\, S)}\frac{dS}{S}$$

This result was first published by Henri Beresticki, Jérôme Busca and Igor Florent [6]. The fact that the inverse of $\widehat{\sigma}$ should be given by the average of the inverse of σ may surprise at first.

The squared volatility that one may have expected appears usually in averages only because they are temporal averages, akin to a quadratic variation. In our case $T \to 0$ and there is no temporal averaging.

Then note that the harmonic average complies with the basic requirement that if the local volatility vanishes in a region between the initial spot level and the strike, the implied volatility for that strike should vanish, as for $T \mapsto 0$, the effect of the drift is immaterial and the spot would be unable to cross that region.[12]

---

[12]The motivation for the harmonic average is that it appears naturally in the density of $\ln{(S_T)}$ for short maturities as the change of variable $S \to z = \int_{S_0}^{S}\frac{dS}{S\sigma(S)}$ results in a process for z_t which is Gaussian at short times. A derivation of (2.54) using the zeroth order of the heat-kernel expansion can be found in Section 5.2.2 of [56].

2.5 The dynamics of the local volatility model

Once the volatility function $\sigma\,(t, S)$ is set, expression (2.32) shows that, in the local volatility model, $\hat{\sigma}_{KT}$ changes only if time advances or S moves. Mathematically, this is a consequence of the fact that the local volatility model is a market model for spot and implied volatilities that has a one-dimensional Markovian representation in terms of t, S.

Thus, practically, to characterize the joint dynamics of spot and implied volatilities, we only need to analyze how implied volatilities respond to a move of S.

As will be made clear in Section 2.7 below, the volatilities of volatilities and spot/volatility covariances, or SSRs, that we derive below are exactly the break-even levels of the P&L of a delta and vega-hedged position – inclusive of recalibration of the local volatility function – an aspect that looks counter-intuitive at first.

2.5.1 The dynamics for strikes near the forward

How much do implied volatilities move when S moves? Let us take the derivative of equation (2.42) with respect to $\ln(S_0)$, introducing the notation $S\,(t, y) = F_t \exp(\frac{t}{T} x_K + \frac{\sqrt{\sigma_0^2(T-t)t}}{\sqrt{T}} y)$ and remembering that F_t and x_K depend on S_0:

$$F_t = S_0 e^{(r-q)t}; \; x_K = \ln\left(\frac{K}{S_0 e^{(r-q)T}}\right).$$

$$\frac{d\hat{\sigma}_{KT}}{d\ln S_0} = \frac{1}{T}\int_0^T dt \int_{-\infty}^{+\infty} dy \, \frac{e^{-\frac{y^2}{2}}}{\sqrt{2\pi}} \frac{d\sigma}{d\ln S_0}(t, S(t, y))$$

$$= \frac{1}{T}\int_0^T dt \int_{-\infty}^{+\infty} dy \, \frac{e^{-\frac{y^2}{2}}}{\sqrt{2\pi}} \frac{d\sigma}{dS}(t, S(t, y)) \frac{dS(t, y)}{d\ln S_0}$$

$$= \frac{1}{T}\int_0^T dt \int_{-\infty}^{+\infty} dy \, \frac{e^{-\frac{y^2}{2}}}{\sqrt{2\pi}} \frac{d\sigma}{dS}(t, S(t, y)) \left(1 - \frac{t}{T}\right) S(t, y)$$

$$= \frac{1}{T}\int_0^T \left(1 - \frac{t}{T}\right) dt \int_{-\infty}^{+\infty} dy \, \frac{e^{-\frac{y^2}{2}}}{\sqrt{2\pi}} \frac{d\sigma}{d\ln S}(t, S(t, y))$$

Using expression (2.44) for $\sigma\,(t, S)$ and the definition of $X\,(t, y)$ in (2.46) we get:

$$\frac{d\hat{\sigma}_{KT}}{d\ln S_0} = \frac{1}{T}\int_0^T \left(1 - \frac{t}{T}\right) dt \int_{-\infty}^{+\infty} dy \, \frac{e^{-\frac{y^2}{2}}}{\sqrt{2\pi}} \Big(\alpha(t) + \beta(t)X\,(t, y)\Big)$$

$$= \frac{1}{T}\int_0^T \left(1 - \frac{t}{T}\right) \alpha(t) dt + \left[\frac{1}{T}\int_0^T \left(1 - \frac{t}{T}\right) \frac{t}{T} \beta(t) dt\right] x_K$$

Let us consider the special case of the ATMF volatility, that is the implied volatility for a strike equal to the forward: $x_K = 0$. We get:

$$\left.\frac{d\hat{\sigma}_{KT}}{d\ln S_0}\right|_{K=F_T} = \frac{1}{T}\int_0^T \left(1 - \frac{t}{T}\right)\alpha(t)dt$$

This formula quantifies how the implied volatility for a fixed strike equal to the forward F_T moves when the spot moves. It resembles equation (2.48) except the weight is $1 - \frac{t}{T}$ rather than $\frac{t}{T}$. This is natural, as when S_0 moves while K stays fixed, for $t = T$ only the *value* $\sigma(T, S = K)$ contributes to formula (2.32), thus $\alpha(t = T)$ is immaterial.

Symmetrically, for calculating how the implied volatility changes with strike K for a fixed spot S_0, knowledge of $\alpha(t)$ for $t = 0$ is not needed as only the *value* $\sigma(t = 0, S_0)$ is contributing – hence the vanishing weight for $\alpha(t = 0)$ in equation (2.48).

Consider now the motion of the ATMF implied volatility $\hat{\sigma}_{K=F_T T}$ keeping in mind that, as S_0 moves, strike K moves as well, so as to track the change in the forward F_T. We get the sum of two contributions:

$$\begin{aligned}
\frac{d\hat{\sigma}_{F_T T}}{d\ln S_0} &= \left.\frac{d\hat{\sigma}_{KT}}{d\ln K}\right|_{K=F_T} + \left.\frac{d\hat{\sigma}_{KT}}{d\ln S_0}\right|_{K=F_T} \\
&= \frac{1}{T}\int_0^T \frac{t}{T}\alpha(t)dt + \frac{1}{T}\int_0^T \left(1 - \frac{t}{T}\right)\alpha(t)dt \\
&= \frac{1}{T}\int_0^T \alpha(t)dt
\end{aligned}$$

(2.55)

Thus the rate at which the ATMF volatility varies as the spot moves is simply given by the uniform time average of the skew of the local volatility function at the forward. In practice, given a market smile, $\alpha(t)$ is not accessible, but $\mathcal{S}_T$ is. Inverting equation (2.48) gives:

$$\alpha(t) = \frac{d}{dt}(t\mathcal{S}_t) + \mathcal{S}_t$$

(2.56)

Inserting this expression in equation (2.55) yields:

$$\frac{d\hat{\sigma}_{F_T T}}{d\ln S_0} = \mathcal{S}_T + \frac{1}{T}\int_0^T \mathcal{S}_t dt$$

(2.57)

The rate at which the ATMF volatility moves when the spot moves is purely dictated by the term structure of the ATMF skew for maturities from 0 to T.

Let us assume that $\mathcal{S}$ is constant. Formula (2.57) then gives:

$$\frac{d\hat{\sigma}_{F_T T}}{d\ln S_0} = 2\mathcal{S}_T$$

(2.58)

We get the property that, for weak skews that do not depend on maturity, the rate at which the ATMF implied volatility moves as the spot moves is exactly twice the rate at which the implied volatility varies as a function of the strike, near the forward.

This is a fundamental feature of the dynamics of implied volatilities in the local volatility model: their dynamics is entirely determined by the implied smile to which the model has been calibrated.

Recalling (2.48), let us summarize the three key properties for implied volatilities near the forward that we have derived at order one in $\alpha(t)$, in an expansion around a constant volatility σ_0:

$$\left.\frac{d\widehat{\sigma}_{KT}}{d\ln K}\right|_{K=F_T} = \frac{1}{T}\int_0^T \frac{t}{T}\alpha(t)dt \tag{2.59a}$$

$$\left.\frac{d\widehat{\sigma}_{KT}}{d\ln S_0}\right|_{K=F_T} = \frac{1}{T}\int_0^T \left(1-\frac{t}{T}\right)\alpha(t)dt \tag{2.59b}$$

$$\frac{d\widehat{\sigma}_{F_TT}}{d\ln S_0} = \frac{1}{T}\int_0^T \alpha(t)dt = \mathcal{S}_T + \frac{1}{T}\int_0^T \mathcal{S}_t dt \tag{2.59c}$$

Expanding around a time-dependent volatility

What about expanding around a time-dependent volatility $\overline{\sigma}(t)$ rather than a constant volatility σ_0? Starting from expression (2.40), page 44, and local volatility function (2.44): $\sigma(t,S) = \overline{\sigma}(t) + \alpha(t)x + \frac{\beta(t)}{2}x^2$, and expanding around $\sigma_0(t) = \overline{\sigma}(t)$ yields the following expressions:

$$\left.\frac{d\widehat{\sigma}_{KT}}{d\ln K}\right|_{K=F_T} = \frac{1}{T}\int_0^T \frac{\widehat{\sigma}_t^2 t}{\widehat{\sigma}_T^2 T}\frac{\overline{\sigma}(t)}{\widehat{\sigma}_T}\alpha(t)dt \tag{2.60a}$$

$$\left.\frac{d\widehat{\sigma}_{KT}}{d\ln S_0}\right|_{K=F_T} = \frac{1}{T}\int_0^T \left(1-\frac{\widehat{\sigma}_t^2 t}{\widehat{\sigma}_T^2 T}\right)\frac{\overline{\sigma}(t)}{\widehat{\sigma}_T}\alpha(t)dt \tag{2.60b}$$

$$\frac{d\widehat{\sigma}_{F_TT}}{d\ln S_0} = \frac{1}{T}\int_0^T \frac{\overline{\sigma}(t)}{\widehat{\sigma}_T}\alpha(t)dt = \mathcal{S}_T + \frac{1}{T}\int_0^T \frac{\overline{\sigma}^2(t)}{\widehat{\sigma}_t\widehat{\sigma}_T}\mathcal{S}_t dt \tag{2.60c}$$

where $\widehat{\sigma}_\tau = \sqrt{\frac{1}{\tau}\int_0^\tau \overline{\sigma}^2(u)du}$. $\overline{\sigma}(t)$ is arbitrary – a natural choice is to calibrate $\overline{\sigma}(t)$ to the term structure of ATMF volatilities.

2.5.2 The Skew Stickiness Ratio (SSR)

It is useful to normalize $\frac{d\widehat{\sigma}_{F_TT}}{d\ln S_0}$ by the ATMF skew of maturity T, $\mathcal{S}_T$, thus defining a dimensionless number $\mathcal{R}_T$ which we call the Skew Stickiness Ratio (SSR) for maturity T:

$$\mathcal{R}_T = \frac{1}{\mathcal{S}_T}\frac{d\widehat{\sigma}_{F_TT}}{d\ln S_0} \tag{2.61}$$

$\mathcal{R}_T$ quantifies how much the ATMF volatility for maturity T responds to a move of the spot, *in units of the ATMF skew*.

$\mathcal{R}_T$ will be given in Chapter 9 a more general definition as the regression coefficient of the ATMF volatility on $\ln S$, normalized by the ATMF skew:

$$\mathcal{R}_T = \frac{1}{\mathcal{S}_T} \frac{\langle d\widehat{\sigma}_{F_T T} d\ln S_0\rangle}{\langle (d\ln S_0)^2\rangle} \tag{2.62}$$

$\mathcal{R}_T$ essentially quantifies the spot/volatility covariance in the model at hand.

In the local volatility model, $\widehat{\sigma}_{F_T T}$ is a *function* of (t, S), hence expression (2.62) for $\mathcal{R}_T$ simplifies to (2.61).[13]

Practitioners routinely refer to two archetypical regimes:

- The "sticky-strike" regime corresponds to $\mathcal{R}_T = 1$. As the spot moves, implied volatilities *for fixed strikes* near the money stay frozen – the ATMF volatility slides along the smile.

- The "sticky-delta" regime corresponds to $\mathcal{R}_T = 0$. The whole smile experiences a translation alongside the spot: volatilities *for fixed log-moneyness* are frozen.

While the sticky-delta regime is observed for all T in models with iid increments for $\ln S$ – such as jump-diffusion models – sticky-strike behavior is only observed as a limiting regime for long maturities for certain types of stochastic volatility models – see Section 9.5.

Using expression (2.57) for $\frac{d\widehat{\sigma}_{F_T T}}{d\ln S_0}$ we get:

$$\langle d\widehat{\sigma}_{F_T T} d\ln S_0\rangle = \left(\mathcal{S}_T + \frac{1}{T}\int_0^T \mathcal{S}_t dt\right)\langle (d\ln S_0)^2\rangle\, dt \tag{2.63}$$

which yields the following approximate expression for the SSR in the local volatility model, at order one in $\delta\sigma(t, S)$:

$$\mathcal{R}_T = 1 + \frac{1}{T}\int_0^T \frac{\mathcal{S}_t}{\mathcal{S}_T} dt \tag{2.64}$$

For strong skews (2.64) typically overestimates the SSR – see the examples in Figure 2.4, page 59. This is due to the omission of higher-order contributions from $\alpha(t)$ in the expansion.

Expanding around a time-dependent volatility

Expanding around a time-dependent volatility $\overline{\sigma}(t)$, rather than a constant, yields the following expression for $\mathcal{R}_T$:

$$\mathcal{R}_T = 1 + \frac{1}{T}\int_0^T \frac{\overline{\sigma}^2(t)}{\widehat{\sigma}_t\widehat{\sigma}_T} \frac{\mathcal{S}_t}{\mathcal{S}_T} dt \tag{2.65}$$

This follows directly from (2.60c) – we use the same notations.[14]

---

[13]See Section 9.2 for a study of the SSR in stochastic volatility models.

[14]Expression (2.65), with $\overline{\sigma}(t)$ calibrated to the term structure of ATMF volatilities, is more accurate than its counterpart (2.64). (2.64) is, however, already a good approximation. It owes its robustness to

2.5.2.1 The $\mathcal{R} = 2$ rule

In case $\mathcal{S}$ does not depend on maturity, or equivalently when $\alpha(t)$ is constant, (2.57) – or (2.64) – yields:

$$\mathcal{R}_T = 2 \tag{2.66}$$

for all T. This is also true in the limit $T \to 0$ if $\alpha(t)$ is smooth:

$$\lim_{T \to 0} \frac{d\hat{\sigma}_{F_T T}}{d \ln S_0} = 2 \lim_{T \to 0} \frac{d\hat{\sigma}_{KT}}{d \ln S_0}\bigg|_{K=F_T} = 2 \lim_{T \to 0} \frac{d\hat{\sigma}_{KT}}{d \ln K}\bigg|_{K=F_T} \tag{2.67}$$

Thus for short maturities, in the local volatility model:

$$\lim_{T \to 0} \mathcal{R}_T = 2 \tag{2.68}$$

We will see in Chapter 9 that this property is shared by stochastic volatility models.

2.5.3 The $\mathcal{R} = 2$ rule is exact

While we have just derived them using approximation (2.42), we now show that the properties

- $\mathcal{R}_T = 2 \ \forall T$, if $\alpha(t)$ is a constant

- $\lim_{T \to 0} \mathcal{R}_T = 2$

are in fact exact.

2.5.3.1 Time-independent local volatility functions

Imagine that the local volatility function is a function of $\frac{S}{F_t}$ only:

$$\sigma(t, S) \equiv \sigma\left(\frac{S}{F_t}\right) \tag{2.69}$$

with $F_t = S^\star e^{(r-q)t}$, where $S^\star$ is some fixed reference spot level. The time dependence is embedded in the moneyness and σ has no explicit time dependence: we call this a time-independent local volatility function.

Let $C(tS; KT)$ be the price of a call option of maturity T and strike K, computed at time t and spot value S. C solves the following usual backward equation:

$$\frac{dC}{dt} + (r - q) S \frac{dC}{dS} + \frac{1}{2}\sigma^2\left(\frac{S}{F_t}\right) S^2 \frac{d^2 C}{dS^2} = rC$$

with terminal condition: $C(t = T, S; KT) = (S - K)^+$. Consider now the change of variables: $\tau = T - t$, $s = S/F_t$, $k = K/F_T$ and let $f(\tau s; k)$ be the solution of the following *forward* equation:

---

the fact that it does not involve σ_0, the constant volatility around which the order-one expansion is performed – see Figure 2.4, page 59.

$$\frac{df}{d\tau} = \frac{1}{2}\sigma^2(s)s^2\frac{d^2f}{ds^2} \tag{2.70}$$

with initial condition: $f(\tau = 0, s; k) = (s - k)^+$. C can be expressed as:

$$C(tS; KT) = e^{-r\tau}F_T f(\tau s; k) \tag{2.71}$$

Let now $P(tS; KT)$ be the price of a *put* option of maturity T and strike K, computed at time t and spot value S and let us now express the fact that P solves the forward equation:

$$\frac{dP}{dT} + (r - q)K\frac{dP}{dK} - \frac{1}{2}\sigma^2\left(\frac{K}{F_T}\right)K^2\frac{d^2P}{dK^2} = -qP$$

with initial condition $P(tS; K, T = t) = (K - S)^+$. P can be written as:

$$P(tS; KT) = e^{-r\tau}F_T f(\tau k; s) \tag{2.72}$$

Notice how the right-hand sides of (2.71) and (2.72) are identical, except s and k are exchanged. For a constant σ, the Black-Scholes solution of (2.70) is denoted $f_{BS}(\tau s; k; \sigma)$. Given a general solution of (2.70) with initial condition $(s - k)^+$ let us denote $\Sigma_{k\tau}(s)$ its Black-Scholes implied volatility; $\Sigma_{k\tau}(s)$ is such that:

$$f(\tau s; k) = f_{BS}(\tau s; k; \Sigma_{k\tau}(s))$$

(2.71) and (2.72) can be rewritten as:

$$C(tS; KT) = e^{-r\tau}F_T f_{BS}(\tau s; k; \Sigma_{k\tau}(s)) \tag{2.73}$$

$$P(tS; KT) = e^{-r\tau}F_T f_{BS}(\tau k; s; \Sigma_{s\tau}(k)) \tag{2.74}$$

The following identity holds for f_{BS}:

$$f_{BS}(\tau s; k; \Sigma) = (s - k) + f_{BS}(\tau k; s; \Sigma) \tag{2.75}$$

Using (2.75) and the call/put parity, we derive from (2.74) the following expression for the value of the *call* option:

$$C(tS; KT) = e^{-r\tau}F_T f_{BS}(\tau s; k; \Sigma_{s\tau}(k)) \tag{2.76}$$

The right-hand sides of equations (2.73) and (2.76) are Black-Scholes formulas for the price of a call option with the same strike and maturity, computed for the same initial spot value. Their implied volatilities are then identical:

$$\Sigma_{k\tau}(s) = \Sigma_{s\tau}(k) \tag{2.77}$$

Standard implied volatilities $\hat{\sigma}_{KT}(S)$ are given by: $\hat{\sigma}_{KT}(S) = \Sigma_{\frac{K}{F_T},T}\left(\frac{S}{F_t}\right)$. Using (2.77) we get our final result:[15]

$$\hat{\sigma}_{S\frac{F_T}{F_t},T}\left(K\frac{F_t}{F_T}\right) = \hat{\sigma}_{KT}(S) \tag{2.78}$$

---

[15] I am grateful to Julien Guyon for pointing out this symmetry property to me – see also exercise 9.1 in [56].

For zero interest rate and repo this simplifies to:

$$\hat{\sigma}_{ST}(K) = \hat{\sigma}_{KT}(S)$$

Thus knowledge of implied volatilities of all strikes for a given initial spot level S (the right-hand side) supplies information on the implied volatility of a particular strike equal to the initial spot S, for all values of the spot level (the left-hand side).

The $R = 2$ rule

Taking the derivative of both sides of equation (2.78) with respect to $\ln(K)$ and setting $t = 0$, $K = F_T$ and $S = F_t = S_0$ then yields:

$$\left.\frac{d\hat{\sigma}_{KT}}{d\ln K}\right|_{K=F_T} = \left.\frac{d\hat{\sigma}_{KT}}{d\ln S_0}\right|_{K=F_T} \tag{2.79}$$

From this we derive the relationship linking the rate at which the ATMF implied volatility moves as the spot moves to the ATMF skew:

$$\frac{d\hat{\sigma}_{F_T T}}{d\ln S_0} = \left.\frac{d\hat{\sigma}_{KT}}{d\ln K}\right|_{K=F_T} + \left.\frac{d\hat{\sigma}_{KT}}{d\ln S_0}\right|_{K=F_T} = 2\left.\frac{d\hat{\sigma}_{KT}}{d\ln K}\right|_{K=F_T}$$

Hence:

$$\mathcal{R}_T = 2$$

This is an important result. The rule that the rate at which the ATMF implied volatility moves when the spot moves is twice the ATMF skew – or that the SSR equals 2 – is in fact exact for local volatilities that are a function of S/F_t only.

The reason why the order-one expansion of $\hat{\sigma}_{KT}$ in equation (2.42) yields this result is that it complies with the symmetry condition (2.78). Remember that in (2.42), F_t is the forward associated to S, not the reference spot $S^\star$. For a local volatility of the form (2.69), equation (2.42) reads:

$$\hat{\sigma}_{KT}(S) = \frac{1}{T-t}\int_t^T du \int_{-\infty}^{+\infty} dy \frac{e^{-\frac{y^2}{2}}}{\sqrt{2\pi}} \sigma\left(\frac{S}{S^\star}e^{\frac{u-t}{T-t}\ln\left(\frac{KF_t}{SF_T}\right) + \frac{\sqrt{\sigma_0^2(T-u)(u-t)}}{\sqrt{T-t}}y}\right)$$

where we have set the initial time equal to t. One can check that, replacing in this expression S with $K\frac{F_t}{F_T}$ and K with $S\frac{F_T}{F_t}$ and making the change of variables $u \to T + t - u$ leaves the integrand unchanged and yields:

$$\hat{\sigma}_{S\frac{F_T}{F_t},T}\left(K\frac{F_t}{F_T}\right) = \hat{\sigma}_{KT}(S)$$

The diligent reader will have noticed that the backward-forward symmetry condition that yields equations (2.73) and (2.74) still holds if the local volatility function σ is allowed to depend on t such that $\sigma(t, s)$ is symmetric on $[0, T]$ with respect to $\frac{T}{2}$. Again one can check that if this holds, expression (2.42) yields identity (2.79).

2.5.3.2   Short maturities

Consider the case $T \to 0$. Implied volatilities are then given by the exact formula 2.54. One can check using (2.54) that the identity (2.79) holds, hence:

$$\mathcal{R}_T = 2$$

holds, for any local volatility function: for $t \to 0$ the local volatility function becomes in effect "time-independent".

2.5.4   SSR for a power-law-decaying ATMF skew

Let us use the power-law benchmark (2.53) for $\mathcal{S}_T$, with characteristic exponent γ and vanishing cutoff: $\tau_0 = 0$. (2.64) yields the following maturity-independent value of $\mathcal{R}_T$:

$$\frac{2 - \gamma}{1 - \gamma} \tag{2.80}$$

That $\mathcal{R}_T$ does not depend on T is due to our assumption of a vanishing cutoff. In practice $\mathcal{S}_T$ does not diverge as $T \to 0$.

Assume that $\mathcal{S}_T$ is given by (2.52), which is derived from expression (2.51) for $\alpha(t)$ with characteristic exponent γ and cutoff τ_0. Evaluation of the integral in (2.64) is straightforward. The resulting profile of $\mathcal{R}_T$ appears in Figure 2.2 for $\gamma = \frac{1}{2}$ and $\tau_0 = 0.05$ and 0.25.

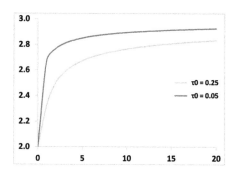

Figure 2.2: $\mathcal{R}_T$ as a function of T (years) as given by formula (2.64), using expression (2.52), page 47, for $\mathcal{S}_T$, with $\gamma = \frac{1}{2}$ and two values of τ_0. $\mathcal{R}_\infty = 3$.

$\mathcal{R}_T$ is very sensitive to τ_0. The limiting value (2.80) for long maturities

$$\mathcal{R}_\infty = \frac{2 - \gamma}{1 - \gamma} \tag{2.81}$$

may thus be reached for outrageously large maturities only – the limiting value in Figure 2.2 is $\mathcal{R}_\infty = 3$ ($\gamma = \frac{1}{2}$). Stated differently, $\mathcal{R}_T$ is very dependent on the ATMF skew for short maturities.

What happens when $\gamma = 1$? From (2.81) $\mathcal{R}_\infty = \infty$. How fast does $\mathcal{R}_T$ diverge? Going back to expression (2.64) we can see that, for large T, $\mathcal{R}_T \propto \ln T$. Again, the precise value of $\mathcal{R}_T$ depends on the short end of the smile – see the example in Section 2.5.6 below and Figure 2.6, page 61, for a smile whose ATMF skew decays like $\frac{1}{T}$ for long maturities.

2.5.5 Volatilities of volatilities

In the local volatility model implied volatilities are a function of S_t.

$$d\widehat{\sigma}_{F_T T} = \frac{d\widehat{\sigma}_{F_T T}}{d\ln S} \, d\ln S_t + \bullet \, dt$$

From (2.61):

$$d\widehat{\sigma}_{F_T T} = \mathcal{R}_T \mathcal{S}_T \, d\ln S_t + \bullet \, dt \tag{2.82}$$

Let us now set $t = 0$ and note that the instantaneous volatility $\sigma(0, S_0)$ is equal to the short ATMF volatility $\widehat{\sigma}_{F_0 0}$: $\langle d\ln S^2 \rangle = \widehat{\sigma}_{F_0 0}^2 dt$. The instantaneous (lognormal) volatility of $\widehat{\sigma}_{F_T T}$ is given by:

$$\mathrm{vol}(\widehat{\sigma}_{F_T T}) = \mathcal{R}_T \mathcal{S}_T \frac{\widehat{\sigma}_{F_0 0}}{\widehat{\sigma}_{F_T T}}$$

- Inserting expression (2.64) for $\mathcal{R}_T$:

$$\mathrm{vol}(\widehat{\sigma}_{F_T T}) = \left(\mathcal{S}_T + \frac{1}{T} \int_0^T \mathcal{S}_t dt \right) \frac{\widehat{\sigma}_{F_0 0}}{\widehat{\sigma}_{F_T T}} \tag{2.83}$$

- If instead we use expression (2.65) for $\mathcal{R}_T$:

$$\mathrm{vol}(\widehat{\sigma}_{F_T T}) = \left(\mathcal{S}_T + \frac{1}{T} \int_0^T \frac{\overline{\sigma}^2(t)}{\widehat{\sigma}_t \widehat{\sigma}_T} \mathcal{S}_t dt \right) \frac{\widehat{\sigma}_{F_0 0}}{\widehat{\sigma}_{F_T T}} \tag{2.84}$$

For short maturities $\mathcal{R}_T = 2$ and we get:

$$\mathrm{vol}(\widehat{\sigma}_{F_T T}) = 2\mathcal{S}_T \tag{2.85}$$

The (lognormal) volatility of a short volatility is just twice the ATMF skew.

For typical equity index smiles, whose ATMF skews decrease with T, the longer the maturity, the lower the instantaneous volatility of the ATMF volatility. For a power-law decay of the ATMF skew with a characteristic exponent γ, we get, for long maturities and ignoring the factor $\frac{\widehat{\sigma}_{F_0 0}}{\widehat{\sigma}_{F_T T}}$, which is only dependent on the term structure of ATMF volatilities:

$$\mathrm{vol}(\widehat{\sigma}_{F_T T}) = \frac{2 - \gamma}{1 - \gamma} \mathcal{S}_T \tag{2.86}$$

For long maturities the volatility of $\widehat{\sigma}_{F_T T}$ thus approximately decays as a function of T with the same exponent as the ATMF skew.[16]

2.5.6 Examples and discussion

We now illustrate what we have just discussed with the example of two Euro Stoxx 50 smiles, then end with a remark on local volatility considered as a stochastic volatility model.

We use the Euro Stoxx 50 smiles of October 4, 2010 (a strong smile) and May 16, 2013 (a mild smile). Let us first consider implied volatilities for, respectively, September 16, 2011 and June 20, 2014 – roughly a 1-year maturity in both cases – and only use implied volatility data for this single maturity for calibrating the local volatility function.

As we have a single maturity, we take the local volatility function to be a function of $\frac{S}{F_t}$ only, so that it falls in the class of time-independent local volatilities. This is easily achieved by using the parametrization $f(t, y)$ in equations (2.18a) and (2.18b) where $f(T, y)$ is given for T, our single maturity. f is defined for $t < T$ by $f(t, y) = \frac{t}{T} f(T, y)$. We are then in the setting of Section 2.5.3.1. Once we have calibrated a time-independent local volatility function, we move the initial spot value S_0 and reprice vanilla options.

The resulting smiles, along with the initial smile, are shown in Figure 2.3. In each smile the marker highlights the ATMF volatility.

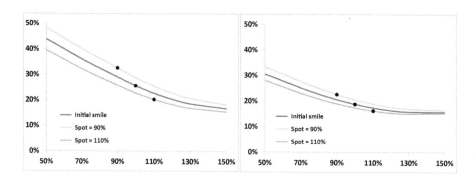

Figure 2.3: Smiles of the Euro Stoxx 50 index for a maturity $\simeq 1$ year (see text), observed on October 4, 2010 (left) and May 16, 2013 (right), along with smiles produced by the local volatility model – calibrated on these initial smiles – for two other values of S_0.

---

[16]Typically the ATMF skews of index smiles decay like $\frac{1}{\sqrt{T}}$. (2.86) implies that in the local volatility model, vol($\widehat{\sigma}_{F_T T}$) decays approximately like $\frac{1}{\sqrt{T}}$ as well.

That the rate at which the ATMF volatility varies when S_0 varies is twice the ATMF skew – or equivalently that $\mathcal{R}_T = 2$ – is apparent to the eye.[17]

We now use implied volatilities for all of the available maturities. The local volatility function cannot be assumed to be time-independent anymore and the SSR will be different than 2. We use formula (2.64):

$$\mathcal{R}_T = 1 + \frac{1}{T}\int_0^T \frac{\mathcal{S}_t}{\mathcal{S}_T} dt \qquad (2.87)$$

$\mathcal{R}_T$ as a function of T is shown in Figure 2.4, together with the actual value of $\mathcal{R}_T$ obtained by shifting the spot value and repricing vanilla options.

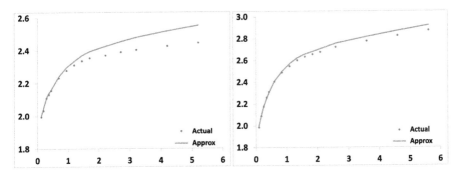

Figure 2.4: $\mathcal{R}_T$ for the Euro Stoxx 50 index as a function of T computed: (a) directly in the local volatility model (actual), (b) using expression (2.87) (approx), for the smiles of October 4, 2010 (left) and May 16, 2013 (right).

Agreement is good except for the long end of the smile of October 4, 2010; (2.87) overestimates the SSR as the order-one expansion that leads to (2.87) ignores contributions from higher orders, which become material for strong skews.

Still the relative error in the estimation of the SSR – or equivalently in the level of volatility of volatility, or spot/volatility covariance – is about 5%.

For the smile of May 16, 2013, which displays an appreciable (increasing) term-structure of ATMF volatilities, using (2.65) rather than (2.87) results in a slightly higher value for $\mathcal{R}_T$ – about 0.05. For smiles with a strong term-structure of ATMF volatilities, that are not too steep, (2.65) is in practice more accurate than (2.87).

That ATMF skews of equity smiles are well captured by the power-law benchmark (2.52) is illustrated in Figure 2.5.

In our two examples the SSR values given by (2.87) using either the actual market ATMF skew or expression (2.52) are similar.

---

[17]We have used zero repo and interest rate for simplicity.

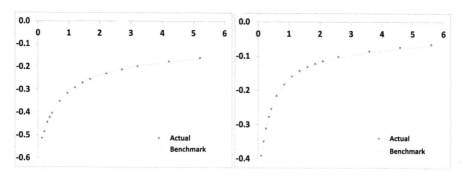

Figure 2.5: $\mathcal{S}_T$ for the Euro Stoxx 50 index as a function of T (in years) as read off the market smile (actual) and as given by the power-law benchmark (2.52), for the smiles of October 4, 2010 (left: $\tau_0 = 0.15$, $\gamma = 0.37$) and May 16, 2013 (right: $\tau_0 = 0.12$, $\gamma = 0.52$).

The long-maturity value of the Âš is given by (2.81):

$$\mathcal{R}_\infty = \frac{2 - \gamma}{1 - \gamma}$$

$\mathcal{R}_\infty = 2.6$ for the October 4, 2010 smile and $\mathcal{R}_\infty = 3.1$ for the May 16, 2013 smile – we know that $\mathcal{R}_\infty$ is only reached for very long maturities.

In the local volatility model, because implied volatilities are a *function* of (t, S) the SSR provides substantial information: it determines both the break-even levels of the spot/volatility cross-gamma *and* of the volatility gamma. Expressions (2.64) for $\mathcal{R}_T$ and (2.83) for $\mathrm{vol}(\hat{\sigma}_{F_T T})$ are useful for sizing up these break-even levels and comparing them to realized levels.

Remember that the SSR involves the ratio of the spot/volatility covariance to the ATMF skew. Large values of the SSR may only be due to weak or vanishing ATMF skews and may not be a signal of particularly large volatilities of volatilities or spot/volatility covariances.

This applies to the following example of a fast-decaying ATMF skew.

A $\frac{1}{T}$ decay for the ATMF skew

We now consider the case of a smile whose long-term ATMF skew decays like $\frac{1}{T}$. This is the case of stochastic volatility models of Type I – see Section 9.5 in Chapter 9.

For $\gamma = 1$ formula (2.81) yields $\mathcal{R}_\infty = \infty$. From (2.87), it can be checked that if $\mathcal{S}_T \propto \frac{1}{T}$ for large T, then $\mathcal{R}_T \propto \ln T$.

Figure 2.6 shows that this is indeed the case. The local volatility function is calibrated on a smile generated by a one-factor stochastic volatility model of the type discussed in Chapter 7, with $k = 6.0$; for $T \gg \frac{1}{k}$ the resulting ATMF skew decays like $\frac{1}{T}$.[18]

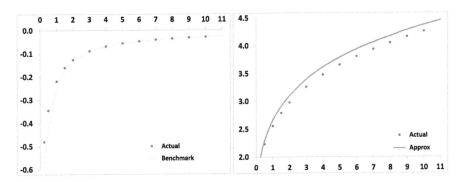

Figure 2.6: Left: $\mathcal{S}_T$ as a function of T, as read off the smile used as input (actual) and as given by the power-law benchmark (2.52), with $\gamma = 0.999$. Right: $\mathcal{R}_T$ as a function of T computed: (a) directly in the local volatility model (actual), (b) using expression (2.87) (approx). The smile used as input has been generated by a stochastic volatility model of Type I – its ATMF skew decays like $\frac{1}{T}$ for large T.

2.5.7 SSR in local and stochastic volatility models

Consider the instantaneous covariance of $\ln S_t$ and $\widehat{\sigma}_{F_T T}$ – denoted more compactly by $\widehat{\sigma}_T(t)$ – observed at t. Because $\widehat{\sigma}_T(t)$ is a function of $\ln S_t$ we have simply:

$$\frac{\langle d\ln S_t d\widehat{\sigma}_T(t)\rangle}{dt} = \frac{d\widehat{\sigma}_T(t)}{d\ln S_t}\sigma^2(t, S_t)$$

$$= \widehat{\sigma}_t^2\left(\frac{1}{T-t}\int_t^T \alpha(\tau)d\tau\right)$$

where we have used expression (2.59c) for $\frac{d\widehat{\sigma}_T}{d\ln S}$, applied to time t rather than 0, and the fact that $\sigma(t, S_t) = \widehat{\sigma}_t$.

We are using formulas at order one in $\alpha(t)$, perturbing around a constant volatility σ_0. At order one in $\alpha(t)$, we can use the order-zero value for $\widehat{\sigma}_{F_t t}$, that is σ_0. We thus have:

$$\frac{\langle d\ln S_t d\widehat{\sigma}_T(t)\rangle}{dt} = \sigma_0^2\left(\frac{1}{T-t}\int_t^T \alpha(\tau)d\tau\right) \qquad (2.88)$$

---

[18]We are discussing the SSR of the local volatility model as calibrated on the smile generated by a stochastic volatility model. The SSR of the stochastic volatility model is different as the instantaneous spot/volatility covariance is different. Type I models, whose ATMF skew decays like $\frac{1}{T}$ or faster are such that $\mathcal{R}_T \to 1$ for $T \to \infty$ – see Section 9.5, page 361.

Multiplying both sides by $(T - t)$ and integrating with respect to t on $[0, T]$ yields:

$$
\int_0^T (T-t)\frac{\langle d\ln S_t d\widehat{\sigma}_T(t)\rangle}{dt}dt = \sigma_0^2 \int_0^T dt \int_t^T \alpha(\tau)d\tau
$$
$$
= \sigma_0^2 \int_0^T \tau\alpha(\tau)d\tau
$$
$$
= \sigma_0^2 T^2 \mathcal{S}_T
$$

where the last line follows from the order-one expression of the ATMF skew in (2.59a). We thus get:

$$
\mathcal{S}_T = \frac{1}{\widehat{\sigma}_T^2 T} \int_0^T \frac{T-t}{T}\frac{\langle d\ln S_t\, d\widehat{\sigma}_T(t)\rangle}{dt}dt \tag{2.89}
$$

where we have replaced in the denominator σ_0^2 with $\widehat{\sigma}_T^2$, still preserving the order-one accuracy in $\alpha(t)$.

As a formula for $\mathcal{S}_T$ (2.89) is useless – we may just as well use (2.59a). What it expresses though – that the ATMF skew for maturity T is given by the integrated instantaneous covariance of $\ln S_t$ and the ATMF volatility for the residual maturity, $\widehat{\sigma}_T(t)$, weighted by $\frac{T-t}{T}$ – has wider relevance.

We have derived it here in the context of local volatility at order one in $\alpha(t)$ but this result is more general.[19]

As will be proven in Section 8.4 of Chapter 8, page 316, formula (2.89) holds for any diffusive model, at order one in volatility of volatility, whenever the instantaneous spot/variance covariation $\langle d\ln S_t d\widehat{\sigma}_T(t)\rangle$ does not depend on S_t. This is the case for a local volatility function linear in $\ln S$, hence (2.89).

Expression (2.89) accounts for why local volatility and stochastic volatility models calibrated to the same smile may have different SSRs, or, equivalently, generate different break-even levels for the spot/volatility cross-gamma.

From (2.89), the ATMF skew sets the integrated value of the covariance of $\ln S$ and $\widehat{\sigma}_T(t)$, the implied ATMF volatility for the residual maturity. Changing the distribution of this covariance on $[0, T]$ without changing its integrated value leaves the ATMF skew unchanged, but changes $\langle d\ln S_t d\widehat{\sigma}_T(t)\rangle_{t=0}$ – which sets the value of the SSR.

With respect to time-homogeneous stochastic volatility models, the local volatility model tends to generate larger covariances at short times – and consequently smaller covariances at future times. This translates into:

- larger SSRs than in time-homogeneous stochastic volatility models

- weaker future skews

See also related discussions in Section 9.11.1 of Chapter 9, page 379, and Section 12.6 of Chapter 12, page 482.

---

[19]Had we started from the order-one expansion (2.40) around a deterministic volatility $\sigma_0(t)$ rather than around a constant volatility σ_0, we would have obtained the exact same formula.

2.6 Future skews and volatilities of volatilities

In Section 1.3.1 we showed how the price of a barrier digital option is mostly determined by the magnitude of the local skew at the barrier for the residual maturity, as generated by the model used for pricing. To assess how a given model prices barrier options, one needs to investigate the ATMF skews generated by a model for a given residual maturity, at future dates, and for different future spot levels. We do this now, for the local volatility model.

Imagine we are using a local volatility model calibrated to the market smile, with a local volatility function given by (2.44).

Let us assume that we are sitting at the forward date $\tau > 0$ with spot S_τ. What is the ATMF skew generated by the local volatility model? Expression (2.48) for the skew was derived for $t = 0$ and a local volatility function given by (2.44) where F_t is the forward at time t for the initial spot level: $F_t = S_0 e^{(r-q)t}$.

We can reuse the results above, but first need to express the local volatility function as a function of $y = \ln\left(\frac{S}{F_t(S_\tau)}\right)$, where $F_t(S_\tau) = S_\tau e^{(r-q)(t-\tau)}$. We have:

$$x = \ln\left(\frac{S}{F_t}\right) = \ln\left(\frac{S}{F_t(S_\tau)}\right) + \ln\left(\frac{F_t(S_\tau)}{F_t}\right)$$

$$= y + x_\tau$$

where $x_\tau = \ln\left(\frac{S_\tau}{F_\tau}\right)$. Sitting at time τ and using S_τ as reference spot level, the local volatility function for $t > \tau$ is given by:

$$\sigma(t, S) = \overline{\sigma}(t) + \alpha(t)(y + x_\tau) + \frac{\beta(t)}{2}(y + x_\tau)^2$$

$$= \overline{\sigma}_\tau(t) + \alpha_\tau(t) y + \frac{\beta_\tau(t)}{2} y^2$$

with

$$\begin{cases} \overline{\sigma}_\tau(t) &= \overline{\sigma}(t) + \alpha(t) x_\tau + \frac{\beta(t)}{2} x_\tau^2 \\ \alpha_\tau(t) &= \alpha(t) + \beta(t) x_\tau \\ \beta_\tau(t) &= \beta(t) \end{cases}$$

If $\beta(t) \neq 0$, since $\alpha_\tau(t)$ depends on x_τ, the ATMF skew at time τ will depend on the spot level S_τ. Let us, however, set $x_\tau = 0$ – that is $S_\tau = F_\tau$ – and focus instead on how the ATMF skew at τ for a given residual maturity θ depends on τ; or equivalently consider that $\beta(t) = 0$.

Using (2.48), the ATMF skew at time τ for a residual maturity θ – that is for maturity $\tau + \theta$ – is given by:

$$S_\theta(\tau) = \left.\frac{d\widehat{\sigma}_{K_{\tau+\theta}}(S_\tau, \tau)}{d\ln K}\right|_{K=F_{\tau+\theta}(S_\tau)} = \frac{1}{\theta}\int_\tau^{\tau+\theta} \frac{t-\tau}{\theta}\alpha(t)\,dt \qquad (2.90)$$

Using now expression (2.56), page 50, for $\alpha(t)$ we get the following expression of the foward-starting ATMF skew as a function of the term structure of the ATMF skew read off the vanilla smile used for calibration:

$$\mathcal{S}_\theta(\tau) = \mathcal{S}_{\tau+\theta} - \frac{\tau}{\theta}\left(\frac{1}{\theta}\int_\tau^{\tau+\theta} \mathcal{S}_t dt - \mathcal{S}_{\tau+\theta}\right) \qquad (2.91)$$

where $\mathcal{S}_t$ is the spot-starting ATMF skew for maturity t.

- Formula (2.91) for $\mathcal{S}_\theta(\tau)$ involves the ATMF skew of the initial vanilla smile, but only for maturities in $[\tau, \tau + \theta]$. Thus there is no reason that $\mathcal{S}_\theta(\tau)$ should bear any resemblance with $\mathcal{S}_\theta$.

- The second piece in (2.91) involves $\mathcal{S}_{\tau+\theta}$ minus the average of $\mathcal{S}_t$ on the interval $[\tau, \tau + \theta]$. For a decreasing term structure of the ATMF skew – which is typical – the latter is larger, in absolute value. We then have the property that:

$$|\mathcal{S}_\theta(\tau)| \leq |\mathcal{S}_{\tau+\theta}| \ll |\mathcal{S}_\theta|$$

The future skew is weaker than the spot-starting skew for maturity $\tau + \theta$, thus (much) weaker than the spot-starting skew for the same residual maturity.

- Consider the case of a power-law decaying skew with $\alpha(t)$ given by (2.51), without any cutoff for simplicity: $\alpha(t) \propto \left(\frac{\tau_0}{t}\right)^\gamma$. Focus on the case of a very short residual maturity θ. From (2.90) for θ small, we have:

$$\mathcal{S}_\theta = \frac{1}{2-\gamma}\alpha_0\left(\frac{\tau_0}{\theta}\right)^\gamma$$

$$\mathcal{S}_\theta(\tau) = \frac{\alpha(\tau)}{2} = \frac{\alpha_0}{2}\left(\frac{\tau_0}{\tau}\right)^\gamma$$

thus

$$\mathcal{S}_\theta(\tau) \propto \left(\frac{\theta}{\tau}\right)^\gamma \mathcal{S}_\theta \qquad (2.92)$$

For typical equity smiles $\gamma \simeq \frac{1}{2}$. Thus, in the local volatility model, future ATMF skews will be much weaker than spot-starting skews for the same residual maturity: $\mathcal{S}_\theta(\tau) \ll \mathcal{S}_\theta$. This is apparent in Figure 2.7 which shows ATMF skews for the two smiles considered in the examples of Section 2.5.6, for different forward dates, calculated using (2.91).

- In the local volatility model, volatilities of ATMF volatilities are determined by the ATMF skew – see formula (2.83), page 57. Thus low levels of future ATMF skews translate also into low future levels of volatility of volatility.

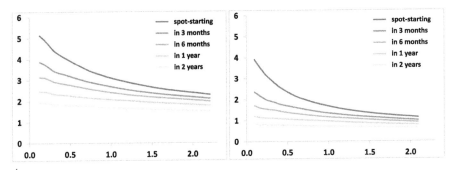

Figure 2.7: $S_\theta(\tau)$ as a function of residual maturity θ for different future dates: $\tau = 0$ (forward-starting), 3 months, 6 months, 1 year and 2 years. $S_\theta(\tau)$ is evaluated using (2.91) and multiplied by $\ln(95/105)$ to convert it into the difference of implied volatilities for the 95% and 105% moneynesses. Smiles of the Euro Stoxx 50 index on October 4, 2010 (left) and May 16, 2013 (right) have been used.

2.6.1 Comparison with stochastic volatility models

Our analysis has concentrated on the price as generated by the local volatility model at $t = 0$. We derive in the following section the expression of the daily P&L of a delta and vega-hedged position. As we risk-manage our forward-start or barrier option together with its vanilla hedge, the daily P&L of the hedged position reads as in (2.105). In case gammas and cross-gammas of the hedged position are sizeable, these P&Ls will be large and unpredictable, randomly polluting our final P&L – this is the price we pay for using the local volatility model.

In contrast, the stochastic volatility models of Chapter 7 are time-homogeneous. Future skews and future volatilities of volatilities are determined by the model's parameters and are commensurate with their values at $t = 0$ – they are not altered by recalibration to the term structure of VS volatilities.

When using the local volatility model, it is then essential to realize that the model price at $t = 0$ incorporates assumptions on future skews and break-even levels of spot/volatility gammas and cross-gammas that cannot be locked and will change as the model is recalibrated to future market smiles.

We refer the reader to Section 3.2 of the following chapter, page 119, where we continue our investigation of forward-start options in the local volatility model; the case of a forward-start call is covered in detail.

See also Section 12.6.1 of Chapter 12 where future smiles generated by local, stochastic and local volatility models are compared.

2.7   Delta and carry P&L

We now consider two practically important issues that need to be answered before we can even consider using a model for trading purposes:

- Which delta should we trade?

- What is the carry P&L of a hedged position?

We will show that the local volatility model is indeed a legitimate market model.

2.7.1   The "local volatility delta"

The price P^{LV} of a derivative in the local volatility model is given by:

$$P^{\mathrm{LV}}(t, S, \sigma)$$

where σ denotes the local volatility function.

Imagine using the "local volatility delta", that is the delta computed with a fixed local volatility function:

$$\Delta^{\mathrm{LV}} = \left.\frac{dP^{\mathrm{LV}}}{dS}\right|_{\sigma} \tag{2.93}$$

The pricing equation (2.2) implies that the P&L during δt of a delta-hedged position reads – see expression (1.5):

$$P\&L = -\frac{1}{2}S^2 \frac{d^2 P^{\mathrm{LV}}}{dS^2} \left(\frac{\delta S^2}{S^2} - \sigma^2(t, S)\,\delta t\right) \tag{2.94}$$

In the local volatility model, implied volatilities of vanilla options are functions of t and S. For strike K and maturity T:

$$\hat{\sigma}_{KT}(t, S) \equiv \Sigma^{\mathrm{LV}}_{KT}(t, S, \sigma) \tag{2.95}$$

where σ denotes the local volatility function used.

As S moves by δS during δt, only if market implied volatilities $\hat{\sigma}_{KT}$ move as prescribed by (2.95) does the P&L of the delta-hedged position read as in (2.94). Expression (2.94) has thus little usefulness.

In reality, implied volatilities will move however they wish. Changes in market implied volatilities $\delta\hat{\sigma}_{KT}$ will be arbitrary, thus the local volatility function calibrated to the market smile at $t + \delta t$ differs from that calibrated at time t: our P&L will include additional terms reflecting this change.

Practically, we will be using the local volatility model in a way that it wasn't meant to be used – that is, recalibrating daily the local volatility function on market smiles. Is this nonsensical? Is it nevertheless possible to express simply our carry P&L? Do payoff-independent break-even levels for volatilities of implied volatilities and correlations of spot and implied volatilities exist?

2.7.2 Using implied volatilities – the sticky-strike delta Δ^{SS}

Denote by $P\left(t, S, \widehat{\sigma}_{KT}\right)$ the option price given by the local volatility model, at time t, for a spot value S and implied volatilities $\widehat{\sigma}_{KT}$:

$$P\left(t, S, \widehat{\sigma}_{KT}\right) \equiv P^{LV}\left(t, S, \sigma\left[t, S, \widehat{\sigma}_{KT}\right]\right) \tag{2.96}$$

where the notation $\sigma\left[t, s, \widehat{\sigma}_{KT}\right]$ signals that the local volatility function is calibrated at time t, spot S, to the volatility surface $\widehat{\sigma}_{KT}$. Our state variables are thus S and the $\widehat{\sigma}_{KT}$. Equivalently, we could use option prices O_{KT} rather than implied volatilities – see the following section in this respect.

$P\left(t, S, \widehat{\sigma}_{KT}\right)$ is our pricing function. Assume we are short the option; the P&L during the interval $[t, t + \delta t]$ of our option position – without its delta hedge – is simply:

$$P\&L = -\left(P(t + \delta t, S + \delta S, \widehat{\sigma}_{KT} + \delta\widehat{\sigma}_{KT}) - (1 + r\delta t)P(t, S, \widehat{\sigma}_{KT})\right)$$

Expanding at order one in δt and two in δS and $\delta\widehat{\sigma}_{KT}$:

$$
\begin{aligned}
P\&L = {}& rP\delta t \\
& - \frac{dP}{dt}\delta t - \frac{dP}{dS}\delta S - \frac{dP}{d\widehat{\sigma}_{KT}} \bullet \delta\widehat{\sigma}_{KT} \\
& - \left(\frac{1}{2}\frac{d^2P}{dS^2}\delta S^2 + \frac{d^2P}{dSd\widehat{\sigma}_{KT}} \bullet \delta\widehat{\sigma}_{KT}\delta S + \frac{1}{2}\frac{d^2P}{d\widehat{\sigma}_{KT}d\widehat{\sigma}_{K'T'}} \bullet \delta\widehat{\sigma}_{KT}\delta\widehat{\sigma}_{K'T'}\right)
\end{aligned}
\tag{2.97}
$$

The notation $\bullet$ stands for:[20]

$$\frac{df}{d\widehat{\sigma}_{KT}} \bullet \delta\widehat{\sigma}_{KT} \equiv \iint dKdT \frac{\delta f}{\delta\widehat{\sigma}_{KT}}\delta\widehat{\sigma}_{KT}$$

We have made no model assumption so far – (2.97) is a basic accounting statement.

The derivatives $\frac{dP}{dS}$, $\frac{dP}{dt}$ are computed keeping the $\widehat{\sigma}_{KT}$ fixed – the underlying local volatility function is *not* fixed. Let us call $\frac{dP}{dS}$ the sticky-strike delta and denote it by Δ^{SS}

$$\Delta^{SS} = \left.\frac{dP}{dS}\right|_{\widehat{\sigma}_{KT}} \tag{2.98}$$

---

[20]The finicky reader will observe – with reason – that the $\widehat{\sigma}_{KT}$ cannot be considered as independent variables as shifting by a finite amount one single point of the volatility surface creates arbitrage. Similarly, for $T \to 0$, in-the-money options are redundant with respect to S.

In practice, implied volatilities $\widehat{\sigma}_{K_iT_j}$ for discrete sets of strikes and maturities are used, out of which the full volatility surface $\widehat{\sigma}_{KT}$ is interpolated/extrapolated. Our pricing function indeed takes as inputs a finite number of parameters.

$\frac{df}{d\widehat{\sigma}_{KT}} \bullet \delta\widehat{\sigma}_{KT}$ should thus really be understood as $\Sigma_{ij}\frac{df}{\widehat{\sigma}_{K_iT_j}}\delta\widehat{\sigma}_{K_iT_j}$.

Let us now utilize the fact that our "black box" valuation function P is in fact the local volatility price. We will express the derivatives of P^{LV} with respect to t, S – that is with a fixed local volatility function – in terms of derivatives of P and use the pricing equation of the local volatility model to derive an identity involving derivatives of P.

By definition of $\Sigma_{KT}^{\mathrm{LV}}(t, S, \sigma)$, $P\left(t, S, \widehat{\sigma}_{KT} = \Sigma_{KT}^{\mathrm{LV}}(t, S, \sigma)\right)$ is the local volatility price:

$$P^{\mathrm{LV}}(t, S, \sigma) = P\left(t, S, \widehat{\sigma}_{KT} = \Sigma_{KT}^{\mathrm{LV}}(t, S, \sigma)\right)$$

We will not carry the three arguments of $\Sigma_{KT}^{\mathrm{LV}}$ anymore, unless necessary. Taking the derivative of this expression with respect to t and S yields:

$$\frac{dP^{\mathrm{LV}}}{dt} = \frac{dP}{dt} + \frac{dP}{d\widehat{\sigma}_{KT}} \bullet \frac{d\Sigma_{KT}^{\mathrm{LV}}}{dt} \tag{2.99}$$

$$\frac{dP^{\mathrm{LV}}}{dS} = \frac{dP}{dS} + \frac{dP}{d\widehat{\sigma}_{KT}} \bullet \frac{d\Sigma_{KT}^{\mathrm{LV}}}{dS} \tag{2.100}$$

Taking once more the derivative of $\frac{dP^{\mathrm{LV}}}{dS}$ with respect to S we get:

$$\frac{d^2 P^{\mathrm{LV}}}{dS^2} = \left(\frac{d^2 P}{dS^2} + 2\frac{d^2 P}{dS d\widehat{\sigma}_{KT}} \bullet \frac{d\Sigma_{KT}^{\mathrm{LV}}}{dS} + \frac{d^2 P}{d\widehat{\sigma}_{KT} d\widehat{\sigma}_{K'T'}} \bullet \frac{d\Sigma_{KT}^{\mathrm{LV}}}{dS}\frac{d\Sigma_{K'T'}^{\mathrm{LV}}}{dS}\right)$$
$$+ \frac{dP}{d\widehat{\sigma}_{KT}} \bullet \frac{d^2 \Sigma_{KT}^{\mathrm{LV}}}{dS^2}$$

$$\tag{2.101}$$

We now use (2.99) to express $\frac{dP}{dt}$ in terms of $\frac{dP^{\mathrm{LV}}}{dt}$ and then use the pricing equation of the local volatility model:

$$\frac{dP^{\mathrm{LV}}}{dt} + (r-q)S\frac{dP^{\mathrm{LV}}}{dS} + \frac{1}{2}\sigma^2(t, S)S^2\frac{d^2 P^{\mathrm{LV}}}{dS^2} = rP^{\mathrm{LV}} \tag{2.102}$$

to express $\frac{dP^{\mathrm{LV}}}{dt}$ as a function of $\frac{dP^{\mathrm{LV}}}{dS}$, $\frac{P^{\mathrm{LV}}}{dS^2}$. We then use (2.101) and (2.100) to write everything in terms of derivatives of P. This yields the following expression of $\frac{dP}{dt}$:

$$\frac{dP}{dt} = rP - (r-q)S\frac{dP}{dS} - \frac{dP}{d\widehat{\sigma}_{KT}} \bullet \mu_{KT}$$
$$-\frac{1}{2}\sigma^2(t, S)S^2\left(\frac{d^2 P}{dS^2} + 2\frac{d^2 P}{dS d\widehat{\sigma}_{KT}} \bullet \frac{d\Sigma_{KT}^{\mathrm{LV}}}{dS} + \frac{d^2 P}{d\widehat{\sigma}_{KT} d\widehat{\sigma}_{K'T'}} \bullet \frac{d\Sigma_{KT}^{\mathrm{LV}}}{dS}\frac{d\Sigma_{K'T'}^{\mathrm{LV}}}{dS}\right)$$

where μ_{KT} is given by:

$$\mu_{KT} = \frac{d\Sigma_{KT}^{\mathrm{LV}}}{dt} + \frac{1}{2}\sigma^2(t, S)S^2\frac{d^2 \Sigma_{KT}^{\mathrm{LV}}}{dS^2} + (r-q)S\frac{d\Sigma_{KT}^{\mathrm{LV}}}{dS} \tag{2.103}$$

Inserting now this expression of $\frac{dP}{dt}$ in (2.97) yields the following expression for our P&L:

$$
\begin{aligned}
P\&L &= -\frac{dP}{dS}(\delta S - (r-q)S\delta t) - \frac{dP}{d\widehat{\sigma}_{KT}} \bullet (\delta\widehat{\sigma}_{KT} - \mu_{KT}\delta t) \\
&+ \frac{1}{2}\sigma^2(t,S)S^2 \left(\frac{d^2P}{dS^2} + 2\frac{d^2P}{dSd\widehat{\sigma}_{KT}} \bullet \frac{d\Sigma_{KT}^{LV}}{dS} + \frac{d^2P}{d\widehat{\sigma}_{KT}d\widehat{\sigma}_{K'T'}} \bullet \frac{d\Sigma_{KT}^{LV}}{dS}\frac{d\Sigma_{K'T'}^{LV}}{dS} \right)\delta t \\
&- \left(\frac{1}{2}\frac{d^2P}{dS^2}\delta S^2 + \frac{d^2P}{dSd\widehat{\sigma}_{KT}} \bullet \delta\widehat{\sigma}_{KT}\delta S + \frac{1}{2}\frac{d^2P}{d\widehat{\sigma}_{KT}d\widehat{\sigma}_{K'T'}} \bullet \delta\widehat{\sigma}_{KT}\delta\widehat{\sigma}_{K'T'} \right)
\end{aligned}
$$

$$(2.104)$$

In the first line of (2.104), the linear contributions of δS and $\delta\widehat{\sigma}_{KT}$ to the P&L are accompanied by their respective financing costs – or risk-neutral drifts. While the financing cost of S is model-independent, μ_{KT} is not as it depends on the dynamics of implied volatilities assumed by the model – in our case the local volatility model – hence the LV superscript.

This is not an issue – it only happens because $\widehat{\sigma}_{KT}$ is not a tradeable asset; when we use option prices rather than implied volatilities, the model-dependence of the drift disappears – see Section 2.7.3 below.

The P&L in (2.104) can be rewritten so as to make the break-even volatilities and correlations of δS and $\delta\widehat{\sigma}_{KT}$ apparent. Denote by ν_{KT} the instantaneous (lognormal) volatility of $\widehat{\sigma}_{KT}$ in the local volatility model – that is of Σ_{KT}^{LV}:

$$
\nu_{KT} = \frac{1}{\Sigma_{KT}^{LV}}\frac{d\Sigma_{KT}^{LV}}{dS}S\sigma(t,S)
$$

Our P&L during δt can be rewritten as:

$$P\&L =$$

$$
-\frac{dP}{dS}(\delta S - (r-q)S\delta t) - \frac{dP}{d\widehat{\sigma}_{KT}} \bullet (\delta\widehat{\sigma}_{KT} - \mu_{KT}\delta t) \tag{2.105a}
$$

$$
-\frac{1}{2}S^2\frac{d^2P}{dS^2}\left[\frac{\delta S^2}{S^2} - \sigma^2(t,S)\,\delta t\right] \tag{2.105b}
$$

$$
-\frac{d^2P}{dSd\widehat{\sigma}_{KT}} \bullet S\widehat{\sigma}_{KT}\left[\frac{\delta S}{S}\frac{\delta\widehat{\sigma}_{KT}}{\widehat{\sigma}_{KT}} - \sigma(t,S)\nu_{KT}\delta t\right] \tag{2.105c}
$$

$$
-\frac{1}{2}\frac{d^2P}{d\widehat{\sigma}_{KT}d\widehat{\sigma}_{K'T'}} \bullet \widehat{\sigma}_{KT}\widehat{\sigma}_{K'T'}\left[\frac{\delta\widehat{\sigma}_{KT}}{\widehat{\sigma}_{KT}}\frac{\delta\widehat{\sigma}_{K'T'}}{\widehat{\sigma}_{K'T'}} - \nu_{KT}\nu_{K'T'}\delta t\right] \tag{2.105d}
$$

$\frac{dP}{dS}$ in (2.105a) is the sticky-strike delta Δ^{SS}. This is our total P&L – (2.105) incorporates the P&L generated by the change in local volatility function over $[t, t+\delta t]$ – up to second order – except it is expressed in terms of changes in implied volatilities.

2.7.3 Using option prices – the market-model delta Δ^{MM}

While we have used implied volatilities $\widehat{\sigma}_{KT}$ as state variables, there is nothing special about them. The reader can check that the fact that $\widehat{\sigma}_{KT}$ is an implied volatility has played no part in the derivation leading to (2.105).

We could have used a different representation of the vanilla smile – for example straight option prices. Let us then replace $\widehat{\sigma}_{KT}$ with O_{KT}, the price of the vanilla option (say, a call) of strike K, maturity T and replace $\Sigma_{KT}^{\text{LV}}(t, S, \sigma)$ with $\Omega_{KT}^{\text{LV}}(t, S, \sigma)$, the price of the same vanilla option in the local volatility model, as a function of time t, spot S, and the local volatility function σ.

Denote by $\mathcal{P}(t, S, O_{KT})$ the price of our exotic option, now a function of vanilla option prices, rather than implied volatilities. $\mathcal{P}$ is related to P through:

$$P(t, S, \widehat{\sigma}_{KT}) = \mathcal{P}\left(t, S, O_{KT} = P_{KT}^{\text{BS}}(t, S, \widehat{\sigma}_{KT})\right) \qquad (2.106)$$

The expression of our P&L is similar to (2.105), with $\widehat{\sigma}_{KT}$ replaced with O_{KT}. Drift μ_{KT}, according to expression (2.103), is now given by:

$$\mu_{KT} = \frac{d\Omega_{KT}^{\text{LV}}}{dt} + \frac{1}{2}\sigma^2(t, S) S^2 \frac{d^2\Omega_{KT}^{\text{LV}}}{dS^2} + (r - q)S \frac{d\Omega_{KT}^{\text{LV}}}{dS}$$

Because Ω_{KT}^{LV} is the price in the local volatility model, with a fixed local volatility function, it obeys (2.102), thus we have:

$$\mu_{KT} = r\Omega_{KT}^{\text{LV}} = rO_{KT}$$

– which we knew in the first place: O_{KT} is the price of an asset, thus its drift is model-independent and is simply its financing cost.

The expression of our P&L during δt using option prices thus reads:

$$P\&L =$$

$$- \frac{d\mathcal{P}}{dS}\left(\delta S - (r - q)S\delta t\right) - \frac{d\mathcal{P}}{dO_{KT}} \bullet \left(\delta O_{KT} - rO_{KT}\delta t\right) \qquad (2.107\text{a})$$

$$- \frac{1}{2}\frac{d^2\mathcal{P}}{dS^2}\left[\delta S^2 - \sigma^2(t, S) S^2 \delta t\right] \qquad (2.107\text{b})$$

$$- \frac{d^2\mathcal{P}}{dSdO_{KT}} \bullet \left[\delta S\delta O_{KT} - \sigma^2(t, S) S^2 \frac{d\Omega_{KT}^{\text{LV}}}{dS}\delta t\right] \qquad (2.107\text{c})$$

$$- \frac{1}{2}\frac{d^2\mathcal{P}}{dO_{KT}dO_{K'T'}} \bullet \left[\delta O_{KT}\delta O_{K'T'} - \sigma^2(t, S) S^2 \frac{d\Omega_{KT}^{\text{LV}}}{dS}\frac{d\Omega_{K'T'}^{\text{LV}}}{dS}\delta t\right]$$

$$\qquad (2.107\text{d})$$

Three observations are in order:

- Expression (2.107) for our carry P&L is typical of a market model – remember our discussion in Section 1.1. Second-order gamma contributions involving prices of all hedge instruments – the spot and vanilla options – appear together with their offsetting thetas. The corresponding break-even levels are payoff-independent.

- (2.107) makes plain that the hedge ratios of our exotic option are simply given by $\frac{d\mathcal{P}}{dS}$ and $\frac{d\mathcal{P}}{dO_{KT}}$. In particular, the delta – that is the sensitivity of $\mathcal{P}$ to a move of S that is not offset by the vanilla option hedge – is given by $\frac{d\mathcal{P}}{dS}$. This is obtained by moving S while keeping vanilla option prices fixed. This is the natural delta as generated in any market model: move one asset value keeping all others unchanged. We call it the market-model delta, denoted by Δ^{MM}:

$$\Delta^{\mathrm{MM}} = \left. \frac{d\mathcal{P}}{dS} \right|_{O_{KT}} \tag{2.108}$$

- In a market model – which local volatility is – the delta of a vanilla option is an irrelevant notion. S and O_{KT} are prices of two different assets, which are both hedge instruments. The incongruity of asking a model to output a hedge ratio of one hedge instrument on another is made manifest in the two-asset example of Section 2.7.6 below.

The local volatility delta of vanilla options, $\frac{d\Omega_{KT}^{\mathrm{LV}}}{dS}$, only appears in the expressions of the break-even levels as option prices in the local volatility model are functions of (t, S). The instantaneous covariance of two securities O_1, O_2 in the model – be they the underlying or vanilla options – is then given by $\frac{d\Omega_1^{\mathrm{LV}}}{dS} \frac{d\Omega_2^{\mathrm{LV}}}{dS} \sigma^2(t, S) S^2 \delta t$.

2.7.4 Consistency of Δ^{SS} and Δ^{MM}

Δ^{MM} is the "real" delta of the local volatility model: it is the number of units of S that need to be held in the hedge portfolio, alongside a position $\frac{d\mathcal{P}}{dO_{KT}}$ in vanilla options of strike K, maturity T.

Δ^{SS} on the other hand is tied to a specific representation of vanilla option prices – in terms of Black-Scholes lognormal implied volatilities. Should we use it? What is its connection to Δ^{MM}?

Our hedge portfolio Π comprises $\Delta^{\mathrm{MM}} = \frac{d\mathcal{P}}{dS}$ units of the underlying and $\frac{d\mathcal{P}}{dO_{KT}}$ vanilla options of strike K, maturity T:

$$\Pi = \frac{d\mathcal{P}}{dS} S + \frac{d\mathcal{P}}{dO_{KT}} \bullet O_{KT}$$

Typically, the convention on an exotic desk is to use delta-hedged – rather than naked – vanilla options, where the delta hedge is the vanilla option's Black-Scholes delta computed using the option's implied volatility $\hat{\sigma}_{KT}$. The composition of our hedge portfolio can thus be rewritten differently:

$$\Pi = \left[\frac{dP}{dS} + \frac{dP}{dO_{KT}} \bullet \frac{dP_{KT}^{BS}}{dS} \right] S + \frac{dP}{dO_{KT}} \bullet \left[O_{KT} - \frac{dP_{KT}^{BS}}{dS} S \right]$$

What does $\frac{dP}{dS} + \frac{dP}{dO_{KT}} \bullet \frac{dP_{KT}^{BS}}{dS}$ correspond to? It is the sensitivity of P to a simultaneous move of S and a variation of vanilla option prices generated by their Black-Scholes deltas. Saying that a vanilla option's price varies by its Black-Scholes delta times the variation of S is equivalent to saying that its Black-Scholes implied volatility stays fixed. This is the sticky-strike delta: $\frac{dP}{dS} + \frac{dP}{dO_{KT}} \bullet \frac{dP_{KT}^{BS}}{dS} = \Delta^{SS}$, which we can rewrite as:

$$\Delta^{MM} + \frac{dP}{dO_{KT}} \bullet \frac{dP_{KT}^{BS}}{dS} = \Delta^{SS}$$

Thus Δ^{SS} is equal to the market-model delta Δ^{MM} augmented by the Black-Scholes deltas of the hedging vanilla options. Once Black-Scholes deltas of the hedging options are accounted for, we recover Δ^{MM} as the aggregate delta in our hedge portfolio.

Had we used a different representation of vanilla option prices,[21] we would have obtained a different "sticky-strike" delta. Yet, once the hedge porfolio is broken down into underlying + *naked* vanilla options the delta is always equal to Δ^{MM}.

Δ^{SS} thus has no special status. It is simply the delta an exotic desk should trade if – as is customary – the delta hedges of delta-hedged vanilla options used as vega hedges are their Black-Scholes deltas.

2.7.5 Local volatility as the simplest market model

Expression (2.107) for the carry P&L is so natural from a trading point of view that the derivation in Section 2.7.2 seems unnecessarily cumbersome on one hand and on the other hand does not shed much light on why things pan out so neatly – indeed we have obtained (2.107) by using the local volatility model in an unorthodox manner, as we recalibrate the local volatility function at $t + \delta t$.

Consider a stochastic volatility market model[22] defined by the following joint SDEs for the spot and vanilla option prices:

$$dS_t = (r - q) S_t dt + \sigma_t S_t dW_t^S \qquad (2.109a)$$

$$dO_{KT,t} = r O_{KT,t} dt + \lambda_{KT,t} dW_t^{KT} \qquad (2.109b)$$

and the following condition:

$$O_{KT,t=T} = (S_T - K)^+ \qquad (2.110)$$

---

[21] For example the implied intensity of a Poisson process.
[22] By stochastic volatility market model, we mean here a market model driven by diffusive processes.

together with the initial values $S_{t=0}$, $O_{KT,t=0}$. Processes σ_t (resp. λ_{KT}^t) are the instantaneous lognormal (resp. normal) volatilities of S_t (resp. O_{KT}). Imagine we are able to build such a model – that is a solution to (2.109) that satisfies (2.110). Then, the carry P&L during δt in such a model is exactly of the form in (2.107), with the respective break-even levels given by:

- $S_t^2 \sigma_t^2$ for the spot/spot gamma

- $\sigma_t S_t \lambda_{KT,t} \rho_{S,KT,t}$ for spot/vanilla option cross-gammas

- $\lambda_{KT,t} \lambda_{K'T',t} \rho_{KT,K'T',t}$ for the option/option cross-gammas.

$\rho_{S,KT,t}$ and $\rho_{KT,K'T',t}$ are, respectively, the instantaneous correlations of S_t and $O_{KT,t}$ and of $O_{KT,t}$ and $O_{K'T',t}$.

Building market models from scratch is difficult. Furthermore only market models possessing a Markov representation in terms of a (small) finite number of state variables can realistically be considered. Failing that, pricing requires simultaneous simulation of S_t as well as all of the O_{KT}, a task which is unfeasible numerically.

Denote by $\sigma[O_{KT}, t, S]$ the local volatility function calibrated at time t using vanilla option prices O_{KT} and spot value S, and denote by $\sigma[O_{KT}, t, S](\tau, \mathcal{S})$ its value for time τ and spot value $\mathcal{S}$. Denote by $\Omega_{KT}^{LV}(t, S, \sigma)$ the local volatility price of a vanilla option of strike K, maturity T, as a function of time t, spot S and local volatility function σ.

Set:

$$W_t^{KT} \equiv W_t \tag{2.111a}$$

$$\sigma_t = \sigma[O_{KT,t}, t, S_t](t, S_t) \tag{2.111b}$$

$$\lambda_{KT,t} = \sigma_t S_t \left. \frac{d\Omega_{KT}^{LV}}{dS} \right|_{t,\ S=S_t,\ \sigma=\sigma[O_{KT,t},t,S_t]} \tag{2.111c}$$

SDEs (2.109) together with (2.111) define a market model that starts from the initial condition S_0, $O_{KT,0}$.

σ_t and $\lambda_{KT,t}$ are functions of S_t, $O_{KT,t}$ only – information available at time t. Our model is Markovian in these state variables.

It is in fact the local volatility model and we know that it has a Markov representation in terms of t, S_t. We know that in the model defined by (2.109) and (2.111) – which is the local volatility model – the local volatility function $\sigma[O_{KT,t}, t, S_t]$ is in fact constant and equal to $\sigma[O_{KT,0}, 0, S_0]$. $O_{KT,t}$ can thus be written as:

$$O_{KT,t} = \Omega_{KT}^{LV}(t, S_t, \sigma[O_{KT,0}, 0, S_0])$$

An alternative definition of local volatility is thus that it is a[23] diffusive market model that has a one-dimensional Markov representation in terms of (t, S). Because

---

[23]Presumably the only one.

the local volatility model is a diffusive market model – that is its SDEs are of the form in (2.109) – the carry P&L is automatically of the form in (2.107) – no need to go through the rigmarole of Section 2.7.2.

In Chapter 12 we examine local-stochastic volatility models. They have a Markov representation in terms of t, S, plus a few other state variables – X_t and Y_t if one uses the two-factor model of Chapter 7. This additional flexibility can be exploited to produce different break-even levels for gamma/theta P&Ls.

As we will see, only a few of them are market models, that is can actually be used to risk-manage a derivatives book.

2.7.6   A metaphor of the local volatility model

We now illustrate the irrelevance of the local volatility delta Δ^{LV} and the inconsequentiality of the "recalibration" of the local volatility function, using the simple example of a basket option on two underlyings S_1, S_2 – say the Euro Stoxx 50 and S&P 500 indexes. In our analogy with the local volatility model, S_1 is the actual underlying while S_2 is a vanilla option.

Consider the example of an ATM call option on an equally weighted basket of S_1, S_2 in a Black-Scholes model with identical volatilities and correlation ρ. We assume that the initial values $S_{1,\tau_0}, S_{2,\tau_0}$ of both underlyings are equal. The option price is $P\left(t, S_1, S_2\right)$.

The delta of an ATM option is about 50% thus:

$$\Delta_1 \simeq 25\%, \quad \Delta_2 \simeq 25\%$$

Imagine now raising ρ until it reaches 100%.[24] For $\rho = 100\%$, we still have $\Delta_1 \simeq 25\%, \Delta_2 \simeq 25\%$.

For $\rho = 100\%$, however, rather than solving the two-dimensional PDE for the option price, we can express S_2 as a function of (t, S_1). If the volatilities of S_1, S_2 are equal, $S_2 = \left(\frac{S_{2,\tau_0}}{S_{1,\tau_0}}\right) S_1$. Our model has a Markov representation in terms of (t, S_1) – the counterpart of the fact that the local volatility model has a one-dimensional Markov representation in terms of (t, S).

We can then solve a one-dimensional equation for the option price which we denote by $P^{\mathrm{LV}}\left(t, S_1\right)$.

- In the local volatility model P^{LV} does not depend explicitly on $\widehat{\sigma}_{KT}$ anymore, because $\widehat{\sigma}_{KT}$ is a function of t, S: $\widehat{\sigma}_{KT} = \Sigma_{KT}\left(t, S, \sigma\right)$.

- Likewise, in our example P^{LV} does not depend on S_2 because S_2 is a function of S_1: $S_2 = \left(\frac{S_{2,\tau_0}}{S_{1,\tau_0}}\right) S_1$ – hence the notation P^{LV}. The ratio $\frac{S_{2,\tau_0}}{S_{1,\tau_0}}$ plays the role of the local volatility function.

---

[24]We may use such high correlation for setting conservatively the break-even level of the cross-gamma $\frac{d^2 P}{dS_1 dS_2}$.

We now compute deltas as derivatives of $P^{\text{LV}}(t, S_1)$. We have:

$$\Delta_1^{\text{LV}} = \frac{dP^{\text{LV}}}{dS_1} \simeq 50\%, \quad \Delta_2^{\text{LV}} = \frac{dP^{\text{LV}}}{dS_2} = 0$$

- In the local volatility model, hedge ratios computed by taking the derivatives of P^{LV} imply that the only hedge instrument is S and delta is $\Delta^{\text{LV}} = \frac{dP^{\text{LV}}}{dS}$. Vanilla options are not needed as hedges.

- Likewise, in our example, $\Delta_2 = 0$ means there's no need for S_2 in our hedge portfolio.

Δ_1^{LV}, Δ_2^{LV} are obviously ludicrous – in particular any move of S_2 generates a change in option price that our (nonexistent) delta Δ_2^{LV} is unable to offset.

Likewise, in the local volatility model, as discussed on page 66, Δ^{LV} is useless since any move of $\widehat{\sigma}_{KT}$ that is not equal to that specified by $\Sigma_{KT}(t, S, \sigma)$ is not hedged.

What about the significance of recalibrating the local volatility function at $\tau_0 + \delta\tau$ to take into account the fact that, at $\tau_0 + \delta\tau$, $\widehat{\sigma}_{KT} \neq \Sigma_{KT}(\tau_0 + \delta\tau, S + \delta S, \sigma)$?

In our example, at $\tau + \delta\tau$, $S_{2,\tau_0+\delta\tau}$ will likely not be equal to $\left(\frac{S_{2,\tau_0}}{S_{1,\tau_0}}\right) S_{1,\tau_0+\delta\tau}$.

Thus by using the actual value $S_{2,\tau_0+\delta\tau}$ of S_2, we "recalibrate" the ratio $\left(\frac{S_{2,\tau_0}}{S_{1,\tau_0}}\right)$ which, in our one-dimensional Markov representation, relates S_2 to S_1.

Yet, this "recalibration" of S_2 at $\tau_0 + \delta\tau$ is not an issue: we just enter at $\tau_0 + \delta\tau$ the new values of S_1, S_2 in our pricing function $P(t, S_1, S_2)$. The fact that we are using $\rho = 100\%$ in our model is of no consequence, other than that of setting the break-even level for the P&L generated by the cross-gamma $\frac{d^2 P}{dS_1 dS_2}$.

It is important to stress that (a) the delta, (b) the covariance structure of the model, are unrelated issues – a point we make again in Section 7.3.3.

The purpose of delta-hedging is to immunize a position at first order against arbitrary moves of the underlying assets, not just those allowed by the covariance structure of the model.

2.7.7 Conclusion

- The local volatility model is a diffusive market model – a somewhat special one as (a) it is a one-factor model, (b) it possesses a Markov representation in terms of t, S.

  It can be used for risk-managing options, recalibrating on a daily basis the local volatility function to market smiles. The gamma/theta P&L is well-defined: break-even levels for volatilities and correlations of S and $\widehat{\sigma}_{KT}$ exist and are *payoff-independent*.

Obviously, one may wish that break-even correlations were different than 100% and that volatilities of implied volatilities could be controlled exogenously and not be dictated by the smile used for calibration, but this is how much we can get with a one-factor model.

- The fact that volatilities of implied volatilities, as generated by the model, are neither chosen by the user, nor even deterministic but depend at each point in time on the then-prevailing market smile is however an issue. There is no guarantee that these levels will be adequate with respect to realized levels. Moreover, they will vary unpredictably: in case future market smiles happen to be flat, so that $\frac{d\Sigma_{KT}^{LV}}{dS} \simeq 0$, hence $\nu_{KT} = 0$, we will find ourselves risk-managing our exotic option with a model that is locally pricing vanishing volatilities of implied volatilities.

  In a stochastic volatility model of the type studied in Chapter 7, in contrast, these levels depend on parameters that are set at inception.

- In practice, unlike delta hedging, vega hedging is typically not performed on a daily basis, because of larger bid/offer costs. When rehedging frequencies for delta and vega hedges differ, what gamma/theta P&L do we materialize? This question is answered in Section 9.11.3, page 383.

- The "local volatility delta" Δ^{LV}, computed with a fixed local volatility function, has no special significance or usefulness.

  The delta of the local volatility model is the market-model delta Δ^{MM} in (2.108), that is the sensitivity of the option's price to a move of the spot, with fixed vanilla option prices: $\frac{d\mathcal{P}}{dS}$. The hedge then consists of a position in $\frac{d\mathcal{P}}{dS}$ shares and $\frac{d\mathcal{P}}{dO_{KT}}$ *naked* vanilla options of strike K, maturity T.

  If, as is customary, we use implied volatilities $\hat{\sigma}_{KT}$ as a representation of market prices of vanilla options, rather than prices O_{KT}, we need to trade the sticky-strike delta Δ^{SS} defined in (2.98) computed by moving S and keeping the $\hat{\sigma}_{KT}$ fixed. This is not the total delta in our hedge, as we also need to delta-hedge the vanilla options used as vega hedges. Their deltas are computed in the Black-Scholes model using their respective implied volatilities. The aggregate delta is thus:

$$\Delta^{SS} - \frac{d\mathcal{P}}{dO_{KT}} \bullet \frac{dP_{KT}^{BS}}{dS} = \left[\frac{d\mathcal{P}}{dS} + \frac{d\mathcal{P}}{dO_{KT}} \bullet \frac{dP_{KT}^{BS}}{dS} \right] - \frac{d\mathcal{P}}{dO_{KT}} \bullet \frac{dP_{KT}^{BS}}{dS}$$
$$= \frac{d\mathcal{P}}{dS} = \Delta^{MM}$$

that is, we recover as the aggregate delta of our hedge portfolio the market-model delta Δ^{MM}.

- The very notion of a vanilla option's delta *does not make sense* in the context of a market model, such as the local volatility model. A market model takes as inputs vanilla option prices, in addition to the spot. The former are treated as hedge instruments, on the same footing as the underlying itself. Just as the notion of the delta of one asset with respect to a different asset in a multi-asset model makes no sense, the delta of a vanilla option in a market model is an irrelevant notion.

We refer the reader to Section 2.9 for the practical numerical calculation of vega hedge ratios $\frac{\delta P}{\delta \hat{\sigma}_{KT}}$.

2.7.8 Appendix – delta-hedging only

Consider a position that is only delta-hedged. The P&L of the delta-hedged position is the P&L in (2.105) supplemented with the contribution from the delta position, equal to $\Delta\left(\delta S - (r-q)S\delta t\right)$ where Δ is the delta we trade.

Trading a delta given by $\Delta = \frac{dP}{dS}$ leaves us with a residual vega position. At order one in $\delta\hat{\sigma}_{KT}, \delta S, \delta t$ the P&L or our delta-hedged position is:

$$P\&L = -\frac{dP}{d\hat{\sigma}_{KT}} \bullet \left(\delta\hat{\sigma}_{KT} - \mu_{KT}^{LV}\delta t\right) \qquad (2.112)$$

Consider instead trading the local volatility delta Δ^{LV} – computed with a fixed local volatility function. This is the delta generated by the local volatility model *for a fixed volatility function*. From (2.100):

$$\Delta^{LV} = \frac{dP}{dS} + \frac{dP}{d\hat{\sigma}_{KT}} \bullet \frac{d\Sigma_{KT}^{LV}}{dS}$$

The P&L at order one of our delta-hedged position now reads:

$$P\&L = -\frac{dP}{d\hat{\sigma}_{KT}} \bullet \left(\delta\hat{\sigma}_{KT} - \frac{d\Sigma_{KT}^{LV}}{dS}(\delta S - (r-q)S\delta t) - \mu_{KT}^{LV}\delta t\right) \qquad (2.113)$$

In the dynamics of the local volatility model, S and O_{KT} – or equivalently S and $\hat{\sigma}_{KT}$ – are perfectly correlated. One can be written as a function of the other: $\hat{\sigma}_{KT} = \Sigma_{KT}^{LV}(t, S)$. The SDE for $\hat{\sigma}_{KT}$ reads:

$$d\hat{\sigma}_{KT} = \frac{d\Sigma_{KT}^{LV}}{dS}dS + \left(\frac{1}{2}\frac{d^2\Sigma_{KT}^{LV}}{dS^2}\sigma^2(t, S)S^2 + \frac{d\Sigma_{KT}^{LV}}{dt}\right)dt$$

$$= \frac{d\Sigma_{KT}^{LV}}{dS}(dS - (r-q)S\delta t) + \mu_{KT}^{LV}dt$$

thus P&L (2.113) vanishes with probability one.

This property, however, has no practical relevance. In real life S and $\hat{\sigma}_{KT}$ are separate instruments that will move about freely.

- Just as in the case of two equity underlyings – see the example in Section 2.7.6 – it makes no sense to try to offset P&L (2.112) by spreading a vega position against a delta position. Doing so using the ratio $\frac{d\Sigma_{KT}^{LV}}{dS}$ prescribed by the local volatility model is even less defensible: the realized regression coefficient of $\delta\hat{\sigma}_{KT}$ on δS will likely not match the regression coefficient implied by the local volatility model $\beta_{KT}^{LV} = \frac{d\Sigma_{KT}^{LV}}{dS}$.

- This is seen in the fact that realized values of the SSR (see for example Figure 9.3, page 367) are lower than their value in the local volatility model (see Figure 2.4, page 59 – typically $\mathcal{R} > 2$ for equity smiles) – and also higher than what the sticky-strike delta implies ($\mathcal{R} = 1$). Neither sticky-strike nor local-volatility deltas are good proxies for hedging the residual vega position.

- Local and stochastic volatility models, and more generally diffusive models, share the property that the SSR for short maturities – thus the value of β_{KT} – is model-independent: $\mathcal{R}_{T \to 0} = 2$. This implies that the *implied* spot/ATMF volatility instantaneous covariance can be read off the market smile in model-independent fashion, just as a short ATM option's implied volatility supplies the implied instantaneous break-even variance of $\delta \ln S$. However, this is irrelevant to delta-hedging: the way we compute deltas has nothing to do with the covariance structure – the break-even levels of correlations and volatilities – of the particular model at hand. We also refer the reader to the discussions in Sections 7.3.3, page 225, and 9.11.1, page 379.

- In case we are adamant about delta-hedging the residual vega exposure (2.112) we should use the delta that minimizes the standard deviation of the order-one contribution to the P&L:

$$\Delta = \frac{dP}{dS} + \frac{dP}{d\hat{\sigma}_{KT}} \bullet \beta_{KT}$$

where β_{KT} is the *historical* – rather than *implied* – regression coefficient of $\delta\hat{\sigma}_{KT}$ on δS.

2.7.9 Appendix – the drift of V_t in the local volatility model

Local volatility is but a special kind of stochastic volatility. Let V_t be the instantaneous variance. Two features single out the local volatility model:

- V_t is 100% correlated with S_t.

- V_t is not just a functional of the path of S_t, it is actually a function of S_t.

In the local volatility model $V_t = \sigma^2(t, S_t)$. The dynamics of S_t, V_t then reads:

$$dS_t = (r - q) S_t dt + \sqrt{V_t} S_t dW_t$$
$$dV_t = \left((r - q) S_t \frac{d\sigma^2}{dS} + \frac{d\sigma^2}{dt} + \frac{S_t^2}{2} \frac{d^2\sigma^2}{dS^2} \sigma^2 \right) dt + S_t \frac{d\sigma^2}{dS} \sqrt{V_t} dW_t$$

where σ^2, $\frac{d\sigma^2}{dS}$, $\frac{d^2\sigma^2}{dS^2}$ are evaluated with arguments t, S_t. Notice again that the instantaneous volatility of V_t is determined by the skew of the local volatility function, $S\frac{d\sigma}{dS}$.

Also note how complicated the drift of V_t is. Yet, while the drift of V_t – historically called the "market price of risk" – has been the subject of much ado in stochastic volatility papers, no one seems to get quite as concerned about the drift of V_t in the local volatility model.

We come back to the issue of the drift of V_t and its significance in Section 6.3.

2.8 Digression – using payoff-dependent break-even levels

We have stressed in the previous section that the local volatility model is a market model: break-even levels of volatilities and correlations of hedging instruments – spot and vanilla options – are payoff-independent. This has an important consequence: if the gamma of a position locally vanishes, so does its theta.

What if we do not use a real model?

Imagine for example being short a one-year 90/110 call spread: short a call option struck at $K_1 = 90$ and long a call option struck at $K_2 = 110$ and assume that the initial value of S is 100. Let us also make the assumption that we are free to risk-manage these options until maturity without remarking them to market at intermediate times.

Figure 2.8 shows the price of this call spread in the Black-Scholes model as a function of volatility for zero interest rate and repo.

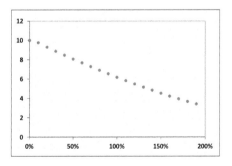

Figure 2.8: Black-Scholes price of a one-year 90/110 call spread as a function of volatility, for $S = 100$, zero interest rate and repo.

Notice that the price is maximum for zero volatility. For other configurations of K_1, K_2 it would have peaked for a different volatility: because of the varying sign of the payoff's convexity and unlike vanilla options, the Black-Scholes price of a call spread is not necessarily a monotonic function of volatility.

The market price of the call spread is simply $P_{BS}(tS, K_1T, \widehat{\sigma}_{K_1T}) - P_{BS}(tS, K_2T, \widehat{\sigma}_{K_2T})$. The smile of equity underlyings is usually strong enough ($\widehat{\sigma}_{K_1T}$ larger than $\widehat{\sigma}_{K_2T}$) that the market price of the call spread lies above the highest price attainable in the Black-Scholes model: the notion of an implied volatility vanishes. For example, taking $\widehat{\sigma}_{K_1T} = 22.5\%, \widehat{\sigma}_{K_2T} = 17.5\%$ gives a price equal to 11.04.

Thus we cannot risk-manage our short call spread position in the Black-Scholes model using a single implied volatility – what about delta-hedging each vanilla option using its own implied volatility? This gives rise during δt to the following P&L:

$$P\&L = -\frac{1}{2}S^2\frac{d^2P_{BS}^1}{dS^2}\left(\frac{\delta S^2}{S^2} - \widehat{\sigma}_{K_1T}^2\delta t\right) + \frac{1}{2}S^2\frac{d^2P_{BS}^2}{dS^2}\left(\frac{\delta S^2}{S^2} - \widehat{\sigma}_{K_2T}^2\delta t\right)$$
(2.114)

Now, consider risk-managing this position with the local volatility model, keeping the local volatility function fixed, that is using the delta Δ^{LV} in (2.93). There is no contradiction with our discussion in the previous section: since we are not marking our call spread to market, we are free to keep the local volatility function fixed. The P&L over $[t, t+\delta t]$ is then given by (2.94) rather than (2.105). Our P&L reads:

$$P\&L = -\frac{1}{2}S^2\frac{d^2P_\sigma^1}{dS^2}\left(\frac{\delta S^2}{S^2} - \sigma(t,S)^2\delta t\right) + \frac{1}{2}S^2\frac{d^2P_\sigma^2}{dS^2}\left(\frac{\delta S^2}{S^2} - \sigma(t,S)^2\delta t\right)$$
$$= -\frac{1}{2}S^2\left(\frac{d^2P_\sigma^1}{dS^2} - \frac{d^2P_\sigma^2}{dS^2}\right)\left(\frac{\delta S^2}{S^2} - \sigma(t,S)^2\delta t\right)$$
(2.115)

In the Black-Scholes model, for a vanilla option, $\frac{d^2P_{BS}}{dS^2}$ is positive and peaks in the vicinity of the option's strike. There is then a particular value of S such that $\frac{d^2P_{BS}^1}{dS^2} - \frac{d^2P_{BS}^2}{dS^2} = 0$. For this value of S, the P&L in equation (2.114) becomes:

$$P\&L = \frac{1}{2}S^2\frac{d^2P_{BS}^1}{dS^2}\left(\widehat{\sigma}_{K_1T}^2 - \widehat{\sigma}_{K_2T}^2\right)\delta t$$

Our delta strategy pays us "free" money ($\widehat{\sigma}_{K_1T} > \widehat{\sigma}_{K_2T}$) in a region of spot prices where there is no gamma risk. Note that this is not the case if we risk-manage both options in the same model – see (2.115) – as cancellation of gamma implies cancellation of the associated theta as well, which is much more reasonable. Note that prices of the call spread in both delta-hedging strategies are identical. Depending on which one is used, however, the theta is distributed differently.

It seems more judicious to use a hedging strategy that pays more theta in regions where gamma is large and pays no theta wherever gamma vanishes – which is what the local volatility model, or any model for that matter, does – rather than squander theta in regions of S where there is little or no risk. We examine this issue further below in Appendix A in the context of the Uncertain Volatility Model.

2.9 The vega hedge

In the Black-Scholes model there is only one vega, as there is only one volatility parameter: any option can be used to vega-hedge any other option. The situation improves somewhat in the Black-Scholes model with deterministic time-dependent volatility $\sigma(t)$. The relationship linking $\sigma(t)$ to implied volatilities $\hat{\sigma}_T$ is:

$$\sigma^2(t) = \left. \frac{d}{dT}\left(T\hat{\sigma}_T^2\right)\right|_{T=t}$$

Thus an option's sensitivity to $\delta\sigma(t)$ can be hedged – within the model – by trading narrow calendar spreads of vanilla options. Equivalently, given the sensitivity of an option to $\sigma(t)$ for all t up to its maturity, we can derive the maturity distribution of vanilla options to be used as hedges.

In the local volatility model, the price of an exotic option is a functional of the whole volatility surface $\hat{\sigma}_{KT}$. Alongside a position in the underlying, the hedge portfolio consists of $\frac{dP}{dO_{KT}}$ vanilla options of strike K, maturity T. Equivalently, the vanilla option hedge immunizes the global position at order one against fluctuations of the local volatility function $\sigma(t, S)$.

How do we calculate $\frac{dP}{dO_{KT}}$? This question was first tackled by Bruno Dupire [41]. What follows is based on his and Pierre Henry-Labordère's work – see [59].

2.9.1 The vanilla hedge portfolio

Let $P(t, S, \bullet)$ be the price of an exotic option, where $\bullet$ stands for all path-dependent variables, whose number usually varies during the option's life.[25] Path-dependent variables only change discontinuously at times when S is observed, as mandated by the term sheet of our exotic payoff. While the option price is continuous across each of these dates, the *function* P changes discontinuously so as to absorb the discontinuity in the path-dependent variables, but in between obeys the usual pricing equation:

$$\frac{dP}{dt} + (r-q)S\frac{dP}{dS} + \frac{\sigma^2(t,S)S^2}{2}\frac{d^2P}{dS^2} - rP = 0 \qquad (2.116)$$

Let us perturb $\sigma^2(t, S)$ by $\delta\sigma^2(t, S)$ and let us call δP the resulting perturbation for P – working with variances is equivalent to working with volatilities and lightens the notation. Replacing σ^2 with $\sigma^2 + \delta\sigma^2$ and P with $P + \delta P$ in (2.116) and expanding at order one in $\delta\sigma^2$ yields the following equation for δP:

$$\frac{d\delta P}{dt} + (r-q)S\frac{d\delta P}{dS} + \frac{\sigma^2(t,S)S^2}{2}\frac{d^2\delta P}{dS^2} - r\delta P = -\frac{1}{2}S^2\frac{d^2P}{dS^2}\delta\sigma^2(t,S)$$

[25]In the case of payoff $\left(S_{T_2}/S_{T_1} - 1\right)^+$ for example, P is a function of t, S for $t < T_1$ and of t, S, S_{T_1} for $t \in [T_1, T_2]$.

which is similar to (2.116) except it has a source term. At maturity $\delta P = 0$. Application of the Feynman-Kac theorem yields:

$$\delta P = \frac{1}{2} E_\sigma \left[\int_0^T dt \, e^{-rt} S^2 \frac{d^2 P}{dS^2} (t, S, \bullet) \, \delta\sigma^2(t, S) \right]$$

where E_σ denotes the expectation taken over paths of S_t generated by the local volatility $\sigma(t, S)$. Conditioning now with respect to the value of S at time t:

$$\delta P = \frac{1}{2} \int_0^T dt \, e^{-rt} \int_0^\infty dS \, \rho(t, S) \, E_\sigma \left[S^2 \frac{d^2 P}{dS^2} (t, S, \bullet) \, |S, t \right] \delta\sigma^2(t, S)$$

$$= \frac{1}{2} \int_0^T dt \, e^{-rt} \int_0^\infty dS \, \rho(t, S) \, \phi(t, S) \, \delta\sigma^2(t, S) \tag{2.117}$$

where $\rho(t, S)$ is the density of S at time t and $\phi(t, S)$ is the expectation of the dollar gamma conditional on the underlying having the value S at time t:

$$\phi(t, S) = E_\sigma \left[S^2 \frac{d^2 P}{dS^2} (t, S, \bullet) \, |S, t \right] \tag{2.118}$$

Equation (2.117) expresses δP as an average of $\delta\sigma^2$, weighted by the product of the density and ϕ. We now look for a portfolio Π of call options of all strikes and maturities:

$$\Pi = \int_0^T d\tau \int_0^\infty dK \, \mu(\tau, K) \, C_{K\tau} \tag{2.119}$$

that hedges our exotic option at order one against any perturbation $\delta\sigma^2(t, S)$. $\mu(\tau, K)$ is the density of vanilla options of strike K, maturity τ. Equation (2.117) implies:

$$\phi_\Pi(t, S) = \phi(t, S)$$

where ϕ_Π is the dollar gamma of portfolio Π. How can we choose μ so that the resulting dollar gamma is ϕ? Imagine that the exotic option was in fact a straight vanilla option – could we tell by just looking at ϕ?

First note that, because a vanilla option is European, it is not path-dependent and ϕ is simply the dollar gamma:

$$\phi(t, S) = S^2 \frac{d^2 P}{dS^2} (t, S)$$

Starting from the pricing equation:

$$\frac{dP}{dt} + (r - q) S \frac{dP}{dS} + \frac{\sigma^2(t, S) S^2}{2} \frac{d^2 P}{dS^2} = rP$$

and applying the operator $S^2 \frac{d^2}{dS^2} \equiv \left(S \frac{d}{dS} \right)^2 - S \frac{d}{dS}$:

$$\frac{d\left(S^2 \frac{d^2 P}{dS^2} \right)}{dt} + (r - q) S \frac{d\left(S^2 \frac{d^2 P}{dS^2} \right)}{dS} + \frac{1}{2} S^2 \frac{d^2}{dS^2} \left(\sigma^2(t, S) S^2 \frac{d^2 P}{dS^2} \right) = r \left(S^2 \frac{d^2 P}{dS^2} \right)$$

yields:

$$\frac{d\phi}{dt} + (r - q)\, S \frac{d\phi}{dS} + \frac{1}{2}S^2 \frac{d^2}{dS^2}\left(\sigma^2(t, S)\,\phi\right) = r\phi$$

Let us define operator $\mathcal{L}$ as:

$$\mathcal{L}f = \frac{df}{dt} + (r - q)\, S \frac{df}{dS} + \frac{1}{2}S^2 \frac{d^2}{dS^2}\left(\sigma^2(t, S)\,f\right) - rf \qquad (2.120)$$

We get the property that $\mathcal{L}\phi = 0$ for a European option. Consider a discrete portfolio Π of vanilla options of strikes (K_i, τ_i), in quantities μ_i, and let the corresponding dollar gamma be ϕ_Π. For $t \in]\tau_{i-1},\, \tau_i[$, $\mathcal{L}\phi_\Pi = 0$. However, as we cross – forward – time τ_i, ϕ_Π is discontinuous since for $t > \tau_i$ it does not include the dollar gamma of option i anymore. This discontinuity contributes to $\mathcal{L}\phi_\Pi$, through the $\frac{d}{dt}$ operator in $\mathcal{L}$. The dollar gamma of option i vanishes at $t = \tau_i^-$ except for $S = K_i$. It is then equal to $\mu_i K_i^2 \delta\,(S - K_i)$ where δ denotes the Dirac distribution: this is the distinguishing feature of call and put options. For our discrete portfolio we then have:

$$(\mathcal{L}\phi_\Pi)\,(t, S) = -\sum_i \mu_i K_i^2 \delta(t - \tau_i)\,\delta(S - K_i)$$

Consider now a portfolio consisting of a continuous density of call options as in (2.119):

$$\mathcal{L}\phi_\Pi\,(\tau, K) = -K^2 \mu\,(\tau, K)$$

We then have our final result. μ is simply given by:

$$\mu\,(\tau, K) = -\frac{1}{K^2}\mathcal{L}\phi\,(\tau, K) \qquad (2.121)$$

To be able to use (2.121) we need an estimate for ϕ that is sufficiently smooth so that we can apply operator $\mathcal{L}$. Practically ϕ will be evaluated on a grid, in a Monte Carlo simulation, using Malliavin techniques as it is a conditional expectation: this is numerically delicate.

Our task is made a lot easier if we simply carry out the perturbation analysis around a flat local volatility function, as the forward transition densities are all analytically known and S is simulated in the Black-Scholes model. In practice, for exotic options that do not depend explicitly on realized variance, the nature of the vega hedge will not depend much on the precise shape of the local volatility function around which perturbation is performed.

Consider a path-dependent payoff $f\,(\mathbf{S})$ where $\mathbf{S}$ is the vector of spot observations $S_i \equiv S_{t=t_i}$. Consider a time t and let t_{k-1}, t_k be spot observation dates such that $t \in]t_k, t_{k+1}[$. $\phi\,(t, S)$ is given by:

$$\phi\,(t, S) = \frac{\int \prod_{i<k}(p_{i-1,i}dS_i)\,p\,(t_{k-1}S_{k-1}, tS)\,S^2 \frac{d^2 P}{dS^2}\,(t, S, \bullet)}{p\,(t_0 S_0, tS)}$$

$$= \frac{\int \prod_{i<k} (p_{i-1,i}dS_i)\, p\left(t_{k-1}S_{k-1},tS\right) S^2 \frac{d^2}{dS^2} \left[p\left(tS,t_kS_k\right)dS_k \prod_{j>k} (p_{j-1,j}dS_j)\, f(\mathbf{S}) \right]}{p\left(t_0S_0,tS\right)}$$

$$= \frac{\int \prod_{i\neq k} (p_{i-1,i}dS_i)\, p\left(t_{k-1}S_{k-1},tS\right) S^2 \frac{d^2p}{dS^2} \left(tS,t_kS_k\right) dS_k\, f(\mathbf{S})}{p\left(t_0S_0,tS\right)} \tag{2.122}$$

where $p_{i,i+1}$ is a shorthand notation for the transition density $p\left(t_iS_i,t_{i+1}S_{i+1}\right)$ in the Black-Scholes model:

$$p\left(t_iS_i,t_{i+1}S_{i+1}\right) \;=\; \frac{1}{S_{i+1}\sqrt{2\pi\sigma_0^2\left(t_{i+1}-t_i\right)}} e^{-\frac{\left(\ln(S_{i+1}/S_i)-\left(r-q-\frac{\sigma_0^2}{2}\right)\left(t_{i+1}-t_i\right)\right)^2}{2\sigma_0^2\left(t_{i+1}-t_i\right)}}$$

In equation (2.122), $S^2 \frac{d^2}{dS^2}$ only acts on $p\left(tS,t_kS_k\right)$: the calculation can be done analytically and we get our final expression for ϕ:

$$\phi\left(t,S\right) \;=\; \frac{\int \prod_{i\neq k} (p_{i-1,i}dS_i)\, p\left(t_{k-1}S_{k-1},tS\right) w\left(tS,t_kS_k\right) p\left(tS,t_kS_k\right)dS_k f(\mathbf{S})}{p\left(t_0S_0,tS\right)}$$

$$= \frac{1}{p\left(t_0S_0,tS\right)} E\left[\frac{p\left(t_{k-1}S_{k-1},tS\right) w\left(tS,t_kS_k\right) p\left(tS,t_kS_k\right)}{p\left(t_{k-1}S_{k-1},t_kS_k\right)} f(\mathbf{S}) \right] \tag{2.123}$$

where w is given by:

$$w\left(tS,t_kS_k\right) \;=\; \frac{1}{\sigma_0^2\left(t_k-t\right)} \left(\frac{z^2}{\sigma_0^2\left(t_k-t\right)} - 1 - z \right)$$

$$z \;=\; \ln\left(S_k/S\right) - \left(r-q-\frac{\sigma_0^2}{2}\right)\left(t_k-t\right)$$

Equation (2.123) expresses ϕ as an expectation of the option's payoff multiplied by a weight that involves S_{k-1}, S, S_k. We recognize in w the classical expression of the weight for computing gamma in a Monte Carlo simulation, in the Black-Scholes model.

As is well known, in practice it provides noisy estimates of gamma, especially when $t_k - t$ is small, as the variance of w blows up. In our context this will be the case whenever the spot observation dates of our exotic option are closely spaced. This issue is compounded by the fact that in (2.123) w is sandwiched in between two transition densities that contribute their fair share of the variance of our estimator for ϕ. Getting an accurate estimate for ϕ is then computationally expensive but presents no special difficulty.

2.9.2 Calibration and its meaningfulness

What do we do once we have the hedge portfolio Π? Can we use it in practice?

Consider a constant volatility $\widehat{\sigma}_0$ and call P^0 the corresponding Black-Scholes price. Let us assume that the market smile is not too strong so that $\widehat{\sigma}_{K\tau} - \widehat{\sigma}_0$ is small. Using the above results and expanding at order one in $\widehat{\sigma}_{K\tau} - \widehat{\sigma}_0$:

$$P = P^0 + \int_0^T d\tau \int_0^\infty dK \, \mu\left(\tau, K\right)\left(C_{K\tau} - C_{K\tau}^0\right) \tag{2.124}$$

where $C_{K\tau}^0$ is the Black-Scholes price with volatility $\widehat{\sigma}_0$ and $C_{K\tau}$ the market price of the vanilla option of strike K, maturity τ.

(2.124) can be interpreted as expressing the following:[26]

- Choose an implied volatility $\widehat{\sigma}_0$ for risk-managing the exotic option. This generates price P^0.

- Determine the quantities $\mu\left(\tau, K\right)$ of vanilla options to be used as hedges.

- Seting up the hedging portfolio entails paying market prices $C_{K\tau}$ rather than model prices $C_{K\tau}^0$ for the hedging vanilla options. The corresponding mismatch is passed on to the client as a hedging cost – the price we quote for the exotic option is given by (2.124).

The conclusion is that the price produced by a calibrated model is as credible as the hedge it implies. Is the latter really a statement on the exotic option or does it reflect model-specific features? This issue needs to be assessed on a case-by-case basis.

The impatient reader can jump to Section 3.2.4 where the case of a forward-start call is analyzed in detail.

2.10 Markov-functional models

In the local volatility model S_t is generally a function of the path of W_t up to time t, where W_t is the driving Brownian motion. Are there special forms of the local volatility function such that S_t can be written as a function of t and W_t, hence can be simulated without any time-stepping? The Black-Scholes model is one example:

$$S_t = S_0 e^{\left(r - q - \frac{\sigma^2}{2}\right)t + \sigma W_t}$$

---

[26]This is how trading desks use to price exotic options in the second half of the '90s, before models were available and/or (mis)understood.

Imagine there exists $f(t,x)$ such that

$$S_t = f(t, W_t) \tag{2.125}$$

The condition that the drift of S be $r - q$ translates into the following PDE for f:

$$\frac{df}{dt} + \frac{1}{2}\frac{d^2 f}{dx^2} = (r - q)f \tag{2.126}$$

and the instantaneous (lognormal) volatility of S is given by:

$$\sigma(t, S) = \frac{d\ln f}{dx}\bigg|_{x=f^{-1}(t,S),t} \tag{2.127}$$

which makes it clear that it is a local volatility model.

For $\sigma(t, S)$ to be well-defined, f has to be a monotonic function of x. Markov-functional Models (MFM) for equities were first introduced by Peter Carr and Dilip Madan – see [23] – who provide some analytic non-trivial solutions to (2.126).

Given now a market smile, is it possible to find a function f such that market prices of vanilla options are recovered?

If f is known for a given time T, equation (2.126) generates f for times $t \leq T$, thus determining smiles for maturities less than T. This implies that we can at most calibrate the smile for *one* maturity T. Smiles for maturities shorter than T are dictated by the smile at T.

We have shown at the beginning of this chapter that, given a full vanilla smile that is free of arbitrage, there exists *one* local volatility function $\sigma(t, S)$ that is able to generate it. If instead we only have a volatility smile for a single maturity, there exist generally many different local volatility functions that are able to recover it. What we have just shown is that one of them corresponds to a Markov-functional model: the process for S_t in this particular local volatility model can be simulated without any time-stepping by simply drawing W_t and setting $S_t = f(t, W_t)$.

Assume we are given the market smile for maturity T: we can price digital options for all strikes, which gives access to the cumulative distribution function of S_T, $\mathcal{F}(S)$. Denoting $\mathcal{N}$ the cumulative distribution of the centered normal distribution, $f(t = T, x)$ is given by:

$$f(T, x) = \mathcal{F}^{-1}\left(\mathcal{N}\left(\frac{x}{\sqrt{T}}\right)\right) \tag{2.128}$$

$f(t = T, x)$ is monotonic by construction and so is $f(t, x)$ for $t < T$.[27]

---

[27] Assume $f(T, x)$ is monotonic in x – say increasing: $\frac{df}{dx}(T, x) \geq 0$, $\forall x$. Take the derivative of both sides of (2.126) with respect to x: $\frac{df}{dx}$ obeys the same PDE as f, thus $\frac{df}{dx}(t, x) = e^{-(r-q)(T-t)}E\left[\frac{df}{dx}(T, W_T)|W_t = x\right]$ where W_t is a Brownian motion. $\frac{df}{dx}(T, x) \geq 0$, $\forall x$ then implies $\frac{df}{dx}(t, x) \geq 0$, $\forall x$.

Note that by using other processes than a straight Brownian motion in (2.125) one can generate different smiles for intermediate maturities, for example by taking $S_t = f(t, Z_t)$ where $dZ_t = \sigma(t)\,dW_t$.

In the context of equities MFMs are very rarely used: one usually needs to calibrate a set of maturities simultaneously. In fixed income markets, on the other hand, MFMs are natural, as cap/floors on LIBOR rates, swaptions, have maturities that match the fixing date of the underlying rate – see [64] for Markov-functional interest rate models. This is also the case of futures in commodity markets and VIX futures – see Section 7.8.2 for an example of an MFM in this context.

2.10.1 Relationship of Gaussian copula to multi-asset local volatility prices

MFMs can be used for European options on a basket of equities S^i. Let us call T the option's maturity: one draws the (correlated) Gaussian random variables W_T^i, applies the mapping in (2.128) and evaluates the payoff. This is exactly equivalent to using the marginal densities supplied by the market smile for maturity T for each asset, and then using a Gaussian copula function to generate the multivariate density for the S_T^i.

This is worth noting as, usually, given a multivariate density ρ generated by an arbitrary copula function, one is unable to characterize the dynamics that underlies ρ: it is not even clear that there exists a diffusive process that is able to generate ρ – one then has no idea of what the gamma/theta break-even levels of his/her position are: the model is unusable.

Because MFMs are a particular instance of local volatility, in the case of a multi-asset European option, pricing with a Gaussian copula thus exactly boils down to using a particular[28] multi-asset local volatility model calibrated on implied volatilities of the option's maturity, with constant correlations, equal to the correlations of the Gaussian copula.

Appendix A – the Uncertain Volatility Model

Treating the Uncertain Volatility Model (UVM) as a local volatility model is not quite natural, as it is typically used in situations when there are no market implied volatilities. We still cover it as it is a (very) special and useful instance of local volatility: in its basic version the local volatility function is not determined by the market smile, but set by a trading criterion.

---

[28]Because we only calibrate implied volatilities of maturity T, there exist many other local volatility functions that achieve exact calibration. Prices generated by these other volatility functions will differ from the Gaussian copula price.

Imagine selling an option on an underlying for which no option market exists – typically a fund share. For example, consider a short position in a call option, whose dollar gamma is always positive. We should typically sell it for a Black-Scholes implied volatility $\hat{\sigma}$ that is sufficiently higher than the expected realized volatility σ_r to ensure that our gamma/theta P&L is mostly positive. Conversely, we would buy this option for a value of $\hat{\sigma}$ lower than σ_r. What about an option whose gamma can be positive or negative, depending on S, t?

For example, imagine being short a call spread: we sell the call struck at K_1 and buy the call struck at K_2 ($K_2 > K_1$), and assume that we price the K_1 call with $\hat{\sigma}_1$ and the K_2 call with $\hat{\sigma}_2$, with $\hat{\sigma}_1 > \hat{\sigma}_2$. Our P&L during δt is:

$$P\&L = -\frac{\Gamma_1}{2}\left(\sigma_r^2 - \hat{\sigma}_1^2\right) + \frac{\Gamma_2}{2}\left(\sigma_r^2 - \hat{\sigma}_2^2\right) \tag{2.129}$$

where Γ_1, Γ_2 are the (positive) dollar gammas of both calls.

Whenever σ_r is such that $\hat{\sigma}_1 < \sigma_r < \hat{\sigma}_2$, both contributions in (2.129) are positive. For $S \ll K_1$, $\Gamma_2 \ll \Gamma_1$ and $P\&L \simeq -\Gamma_1\left(\sigma_r^2 - \hat{\sigma}_1^2\right)/2$. Similarly, for $S \gg K_2$, $P\&L \simeq \Gamma_2\left(\sigma_r^2 - \hat{\sigma}_2^2\right)/2$: our gamma/theta break-even levels are $\hat{\sigma}_1$ (resp. $\hat{\sigma}_2$) for very low (res. high) values of S. Now, as discussed in Section 2.8, for S such that the residual gamma $\Gamma_1 - \Gamma_2$ vanishes, our P&L is uselessly positive. Can we use a pricing and hedging scheme such that this P&L is redistributed to regions where Γ is sizeable, thus improving our gamma/theta break-even levels?

This is exactly what the Uncertain Volatility Model (UVM), introduced by Marco Avellaneda, Arnon Levy, Antonio Paras in [3] and Terry Lyons in [71], does.

The UVM is a local volatility model where the instantaneous volatility σ is a function of the dollar gamma of the option being priced. In the original version of the UVM, σ can take two values: $\sigma_{\min}$, $\sigma_{\max}$, depending on the sign of the dollar gamma: $\sigma(t, S) = \sigma_{\max}$ if $\frac{d^2 P}{dS^2} > 0$, $\sigma(t, S) = \sigma_{\min}$ if $\frac{d^2 P}{dS^2} < 0$. Our P&L thus reads:

$$P\&L = -\frac{\Gamma}{2}\left(\sigma_r^2 - \sigma_{\max}^2\right) \text{ if } \Gamma > 0$$
$$= -\frac{\Gamma}{2}\left(\sigma_r^2 - \sigma_{\min}^2\right) \text{ if } \Gamma < 0$$

The pricing equation for P is non-linear as $\sigma(t, S)$ is a function of P :

$$\frac{dP}{dt} + (r - q) S \frac{dP}{dS} + \frac{1}{2}\sigma^2\left(\frac{d^2 P}{dS^2}\right) S^2 \frac{d^2 P}{dS^2} = rP$$

which can be written as:

$$\frac{dP}{dt} + (r - q) S \frac{dP}{dS} + \frac{1}{2} \max_{\sigma = \sigma_{\min}, \sigma_{\max}}\left(\sigma^2 S^2 \frac{d^2 P}{dS^2}\right) = rP \tag{2.130}$$

Equation (2.130) is the Hamilton-Jacobi-Bellman equation for the following stochastic control problem:

$$P = \max_{\sigma_t \in [\sigma_{\min}, \sigma_{\max}]} E[f(S_T)] \tag{2.131}$$

where f is the option's payoff, the maximum is taken over all processes for the instantaneous volatility σ_t such that $\sigma_t \in [\sigma_{\min}, \sigma_{\max}]$ and the expectation is taken over paths of S_t.

This implies that for an option whose payoff is a sum of two payoffs, the UVM price P satisfies the inequality $P \leq P_1 + P_2$, where P_1, P_2 are the respective UVM prices of both payoffs. Going back to our example, the price of the call spread in the UVM with $\sigma_{\max} = \hat{\sigma}_1$, $\sigma_{\min} = \hat{\sigma}_2$ is lower than the difference of the Black-Scholes prices of each call, computed with volatilities $\hat{\sigma}_1, \hat{\sigma}_2$.

Equivalently, we can find better levels $\sigma_{\max} > \hat{\sigma}_1$ and $\sigma_{\min} < \hat{\sigma}_2$ that still allow us to match the market price. Risk-managing our call spread with the UVM pays zero theta wherever gamma vanishes but provides better break-even volatility levels elsewhere.

For a payoff whose dollar gamma has varying sign, there exist generally many different couples $(\sigma_{\min}, \sigma_{\max})$ such that the UVM price matches a given price level.

A.1 An example

Take the example of a 3-year maturity call spread with $K_1 = 90\%$, $K_2 = 160\%$ with $\hat{\sigma}_1 = 38\%$, $\hat{\sigma}_2 = 31\%$. Examples of $(\sigma_{\min}, \sigma_{\max})$ couples that match the Black-Scholes price of the call spread are given in Table 2.1.

$\sigma_{\min}$	20%	22%	24%	26%	28%
$\sigma_{\max}$	28%	32%	36%	41%	46%

Table 2.1: Examples of $(\sigma_{\min}, \sigma_{\max})$ couples yielding the same price for a 3-year 90%/160% call spread, in the UVM.

From a trading point of view, for the same price charged, it is more reasonable to not have any theta P&L whenever gamma vanishes and have break-even levels 26%/41%, for example, than getting positive theta P&L in a region of vanishing gamma and having break-even levels 31%/38% in regions where gamma is sizeable.

In practice, rather than taking $\sigma = \sigma_{\max}\theta(\Gamma) + \sigma_{\min}(1 - \theta(\Gamma))$, we can use smoother functions of Γ, for example requiring more comfortable break-even levels as the dollar gamma increases.

It is however necessary to ensure that $\sigma(\Gamma)^2 \Gamma$ is an increasing function of Γ to preclude arbitrage, that is the possibility that given two payoffs $u(S)$, $v(S)$ such that $u(S) \geq v(S)$, u might be cheaper than v.

A.2   Marking to market

The UVM was originally designed for underlyings for which no volatility market exists. Is it suited to underlyings for which implied volatilities exist?

A.2.1   An unhedged position

Consider the case of a large trade in a call spread. While the liquidity of vanilla options is not sufficient for us to hedge ourselves in the market, it may be sufficient enough that we decide[29] to mark our position to market.

With respect to the previous situation, the benefits of the wider break-even levels are wiped out because of the mark-to-market constraint: $\sigma_{min}, \sigma_{max}$ have to be moved throughout time so that the UVM price of the call spread always matches its market value.

Let us assume that, during the option's lifetime, implied volatilities $\widehat{\sigma}_1, \widehat{\sigma}_2$ do not move: we start initially with $\sigma_{max} > \widehat{\sigma}_1$ and $\sigma_{min} < \widehat{\sigma}_2$. As we reach maturity, the dollar gammas Γ_1, Γ_2 become localized near their respective strikes and do not overlap anymore. Thus, at maturity $\sigma_{max} = \widehat{\sigma}_1$ and $\sigma_{min} = \widehat{\sigma}_2$. As time advances, σ_{max} (resp. σ_{min}) will converge to $\widehat{\sigma}_1$ (resp. $\widehat{\sigma}_2$). This daily remarking of $\sigma_{min}, \sigma_{max}$ will generate additional theta.

By doing this, we extract more theta from the UVM than we need with the consequence that, near the option's maturity, when dollar gammas are largest, our break-even levels become identical to the Black-Scholes ones ($\sigma_{max} = \widehat{\sigma}_1$, $\sigma_{min} = \widehat{\sigma}_2$). This defeats the purpose of the UVM.

A.2.2   A hedged position – the λ-UVM

Let us assume here that we have traded an exotic option F that can be reasonably – but not perfectly – hedged with a portfolio of vanilla options. Pricing the exotic at hand in the UVM is a very conservative approach: rather than pricing the full gamma of F with $\sigma_{min}, \sigma_{max}$, depending on its sign, it is preferable to first assemble a portfolio of vanilla options that best offsets the gamma profile of our exotic, and then price the package consisting of the exotic option minus its hedge in the UVM. This idea was first proposed in [4] under the name of λ-UVM model – or Lagrangian UVM; see also [47] for more recent work.

The price $\mathcal{P}(F)$ we quote for the exotic is then given by:

$$\mathcal{P}(F) = \mathcal{P}_{UVM}(F - \Sigma\lambda_i O_i) + \Sigma\lambda_i \mathcal{P}_{Mkt}(O_i) \qquad (2.132)$$

where F is the exotic option's payoff, $\mathcal{P}_{UVM}$ and $\mathcal{P}_{Mkt}$ denote, respectively, the UVM price and the market price, and λ_i the quantities of vanilla options O_i traded. The second piece represents the cost of buying the vanilla portfolio at market price.

How should we choose vector λ? The λ_i should be chosen so that for given values of $\sigma_{min}, \sigma_{max}$, we quote the most competitive – i.e. lowest possible – price

---

[29] or that the Risk Control department decides.

$\overline{\mathcal{P}}(F)$ for our exotic option:

$$\lambda = \arg\min_{\lambda}\left[\mathcal{P}_{\text{UVM}}(F - \Sigma\lambda_i O_i) + \Sigma\lambda_i \mathcal{P}_{\text{Mkt}}(O_i)\right]$$

$$\overline{\mathcal{P}}(F) = \min_{\lambda}\left[\mathcal{P}_{\text{UVM}}(F - \Sigma\lambda_i O_i) + \Sigma\lambda_i \mathcal{P}_{\text{Mkt}}(O_i)\right] \tag{2.133}$$

Optimality conditions

Let us express that the derivative of $\overline{\mathcal{P}}(F)$ with respect to λ_i vanishes. Consider equation (2.130) for the UVM price P of the package $F - \Sigma\lambda_i O_i$ and assume a small perturbation of the λ_i which results in a variation δP of P and a variation $\delta(\sigma^2)$ of σ^2:

$$\frac{d(P + \delta P)}{dt} + (r - q)S\frac{d(P + \delta P)}{dS} + \frac{\sigma^2 + \delta(\sigma^2)}{2}S^2\frac{d^2(P + \delta P)}{dS^2} = r(P + \delta P) \tag{2.134}$$

where σ is the local volatility function that maximizes the UVM price of $F - \Sigma\lambda_i O_i$. Expanding (2.134) at order one in δP and $\delta(\sigma^2)$ gives for δP the following PDE:

$$\frac{d\delta P}{dt} + (r - q)S\frac{d\delta P}{dS} + \frac{\sigma^2}{2}S^2\frac{d^2\delta P}{dS^2} = r\delta P - \frac{\delta(\sigma^2)}{2}S^2\frac{d^2 P}{dS^2} \tag{2.135}$$

with the terminal condition $\delta P(T, S) = -\Sigma\delta\lambda_i O_i(S)$. The solution of (2.135) is given by:

$$\delta P = e^{-rT}E_\sigma[\delta P(T, S_T)] + E_\sigma\left[\int_0^T e^{-rt}S^2\frac{d^2 P}{dS^2}\frac{\delta(\sigma^2)}{2}dt\right] \tag{2.136}$$

Consider the second piece of (2.136). It would be the only contribution to δP if $\delta P(T, S) = 0$, that is if the $\delta\lambda_i$ were vanishing. It represents the effect of a small perturbation of $\delta\sigma^2$ on P – for an unchanged payoff. By definition, σ is the local volatility function that maximizes the price $\mathcal{P}_{\text{UVM}}(F - \Sigma\lambda_i O_i)$: at order one any perturbation $\delta(\sigma^2)$ leaves P unchanged: the second piece in (2.136) vanishes.

We are then left with:

$$\delta P = e^{-rT}E_\sigma[\delta P(T, S_T)]$$

which expresses that δP is given by a standard local volatility PDE where the local volatility is fixed, given by the solution of the UVM price for $F - \Sigma\lambda_i O_i$. In other words:

$$\delta\mathcal{P}_{\text{UVM}}(F - \Sigma\lambda_i O_i) = -\Sigma\delta\lambda_i\mathcal{P}_{\text{UVM}}^{F-\Sigma\lambda_i O_i}(O_i)$$

where $\mathcal{P}_{\text{UVM}}^{F-\Sigma\lambda_i O_i}(G)$ is defined as the price of payoff G calculated with a local volatility function σ that maximizes the UVM price *of the package* $F - \Sigma\lambda_i O_i$.

Taking now the derivative of $\mathcal{P}(F)$ with respect to λ_i yields:

$$\frac{d\mathcal{P}(F)}{d\lambda_i} = -\mathcal{P}_{\text{UVM}}^{F-\Sigma\lambda_i O_i}(O_i) + \mathcal{P}_{\text{Mkt}}(O_i)$$

Condition $\frac{d\mathcal{P}(F)}{d\lambda_i} = 0$ implies:

$$\mathcal{P}_{\text{UVM}}^{F-\Sigma\lambda_i O_i}(O_i) = \mathcal{P}_{\text{Mkt}}(O_i) \tag{2.137}$$

Thus, for λ such that $\mathcal{P}(F)$ is extremal, prices of vanilla options used as hedges calculated using the local volatility that maximize the UVM price of $F - \Sigma\lambda_i O_i$ match their market prices. The λ-UVM thus ensures that if F should collapse to a vanilla option O_i, $\overline{\mathcal{P}}(O_i)$ would match the market price $\mathcal{P}_{\text{Mkt}}(O_i)$. $\overline{\mathcal{P}}(F)$ can then be called a mark-to-market price.

Going back to the definition of $\overline{\mathcal{P}}(F)$ in (2.133) and using identity (2.137) yields:

$$\overline{\mathcal{P}}(F) = \mathcal{P}_{\text{UVM}}^{F-\Sigma\lambda_i O_i}(F)$$

Alternatively $\overline{\mathcal{P}}(F)$ can also be characterized as the solution of the following stochastic control problem:

$$\overline{\mathcal{P}}(F) = \max_{\substack{\sigma_t \in [\sigma_{\min},\sigma_{\max}] \\ E_\sigma[O_i] = \mathcal{P}_{\text{Mkt}}(O_i)}} E_\sigma[F] \tag{2.138}$$

which generalizes criterion (2.131): we have added the additional constraint that market prices of options used as hedges have to be matched.[30]

A.2.3   Discussion

Note that, for given market prices $\mathcal{P}_{\text{Mkt}}(O_i)$, $\sigma_{\min}$, $\sigma_{\max}$ have to be chosen so that $\mathcal{P}_{\text{Mkt}}(O_i)$ is attainable in the UVM. For example, imagine that the market implied volatilities are all equal to 25% and that we have chosen $\sigma_{\min} = 10\%$, $\sigma_{\max} = 20\%$. Obviously UVM prices of vanilla options are such that their implied volatilities cannot exceed 20%. Problem (2.138) has no solution. Practically this would manifest itself in the fact that $\mathcal{P}(F)$ in (2.132) can be made as negative as we wish by making λ sufficiently negative.

The above derivation was carried out for the situation when we are selling payoff F: $\overline{\mathcal{P}}(F)$ is our offer price. If instead we are buying the exotic option we can follow a similar derivation, defining our bid price as $\underline{\mathcal{P}}(F)$ given by:

$$\underline{\mathcal{P}}(F) = \max_\lambda \left[\mathcal{P}_{\text{UVM}}(F - \Sigma\lambda_i O_i) + \Sigma\lambda_i \mathcal{P}_{\text{Mkt}}(O_i) \right]$$

$\underline{\mathcal{P}}(F)$ is also characterized as:

$$\underline{\mathcal{P}}(F) = \min_{\substack{\sigma_t \in [\sigma_{\min},\sigma_{\max}] \\ E_\sigma[O_i] = \mathcal{P}_{\text{Mkt}}(O_i)}} E_\sigma[F]$$

In practice, whenever F can be suitably hedged with vanilla options – this is assessed by checking that for a comfortably wide interval $[\sigma_{\min},\sigma_{\max}]$, $\overline{\mathcal{P}}(F)$ is

---

[30]The λ_i introduced in (2.132) are Lagrange multipliers for these constraints.

not inconsiderately expensive, or better that our bid/offer spread $\overline{\mathcal{P}}(F) - \underline{\mathcal{P}}(F)$ is not too large – the λ-UVM model is an effective tool for pricing and risk-managing exotic options. Unfortunately this isn't often the case, which is testament to the fact that most exotic risks are of a different nature than vanilla risks. We refer the reader to our experiment with forward-start options in Section 3.1.7.

When using vanilla options as hedges, $\overline{\mathcal{P}}(F) - \underline{\mathcal{P}}(F)$ provides an indication of how vanilla-like the risk of our exotic option is. We can, however, also use exotic options as hedges – for example cliquets. $\overline{\mathcal{P}}(F) - \underline{\mathcal{P}}(F)$ then supplies a measure of the kinship of exotic risks of the hedged and hedging options. Practically though, solving the stochastic control problem (2.138) in the situation of an exotic option hedged with other exotic options is typically not possible, as the high degree of path-dependence of exotic options results in the high dimensionality of equation (2.130).

Note that, as time elapses and we update λ so as to keep $\overline{\mathcal{P}}(F)$ in (2.133) minimal, we never lose any money, as, by construction, shifting from a previously optimized to a currently optimal vector λ lowers $\overline{\mathcal{P}}(F)$.

What if we set $\sigma_{\min} = 0$, $\sigma_{\max} = +\infty$? $\underline{\mathcal{P}}(F)$ (resp. $\overline{\mathcal{P}}(F)$) are then model-independent lower (resp. upper) bounds for the price of payoff F, given market prices of payoffs O_i – see [47] for more on this and the connection with the dual problem of model-independent sub- (resp. super-) replication.

A.3 Using the UVM to price transaction costs

Consider selling a call option on a security S, risk-managed in the Black-Scholes model with an implied volatility $\hat{\sigma}$, but assume that bid/offer costs are incurred as we adjust our delta – say on a daily basis. Assume the relative bid-offer spread is k: we pay $\left(1 + \frac{k}{2}\right) S$ to buy one unit of the security, and receive $\left(1 - \frac{k}{2}\right) S$ when we sell it.

The P&L over $[t, t + \delta t]$ of a delta-hedged short option position reads as in (1.5), page 4, with the additional contribution of bid/offer costs:

$$P\&L = -\frac{S^2}{2} \frac{d^2 P}{dS^2} \left(\frac{\delta S^2}{S^2} - \hat{\sigma}^2 \delta t\right) - \frac{k}{2} S |\delta \Delta|$$

where $\delta \Delta$ is the variation of our delta during δt; we use an absolute value as the impact of bid/offer is always a cost. $\delta \Delta$ is generated by (a) time advancing by δt, (b) S moving by δS.

Since δS if or order $\sqrt{\delta t}$ we keep the latter contribution only: $\delta \Delta = \frac{d^2 P}{dS^2} \delta S$.

$$\begin{aligned} P\&L &= -\frac{S^2}{2} \frac{d^2 P}{dS^2} \left(\frac{\delta S^2}{S^2} - \hat{\sigma}^2 \delta t\right) - \frac{k}{2} S \left|\frac{d^2 P}{dS^2} \delta S\right| \\ &= -\frac{S^2}{2} \frac{d^2 P}{dS^2} \left(\frac{\delta S^2}{S^2} - \hat{\sigma}^2 \delta t + \varepsilon_\Gamma k \left|\frac{\delta S}{S}\right|\right) \end{aligned}$$

where ε_Γ is the sign of $\frac{d^2 P}{dS^2}$.

If the realized volatility of S is indeed $\widehat{\sigma}$, the gamma and theta contributions cancel out and the third piece will make our P&L persistently negative. Can we find an implied volatility $\widehat{\sigma}^*$ such that risk-managing the option position at $\widehat{\sigma}^*$ generates, on average, zero P&L? $\widehat{\sigma}^*$ must be such that:

$$\left\langle \frac{\delta S^2}{S^2} \right\rangle - \widehat{\sigma}^{*2}\delta t + \varepsilon_\Gamma k \left\langle \left| \frac{\delta S}{S} \right| \right\rangle = 0$$

Assuming a lognormal dynamics for S with realized volatility $\widehat{\sigma}$ and δt small, $\frac{\delta S}{S} = \widehat{\sigma}\sqrt{\delta t}Z$ where Z is a standard normal variable, thus $\left\langle \frac{\delta S^2}{S^2} \right\rangle = \widehat{\sigma}^2 \delta t$ and $\left\langle \left| \frac{\delta S}{S} \right| \right\rangle = \gamma \widehat{\sigma}\sqrt{\delta t}$ with $\gamma = \sqrt{\frac{2}{\pi}}$. $\widehat{\sigma}^*$ is given by:

$$\widehat{\sigma}^* = \sqrt{\widehat{\sigma}^2 + \varepsilon_\Gamma k \gamma \frac{\widehat{\sigma}}{\sqrt{\delta t}}}$$

Depending on the sign of the option's gamma, we need to use either $\widehat{\sigma}^*_{\Gamma+}$ or $\widehat{\sigma}^*_{\Gamma-}$, given by:

$$\widehat{\sigma}^*_{\Gamma+} = \sqrt{\widehat{\sigma}^2 + k\gamma \frac{\widehat{\sigma}}{\sqrt{\delta t}}} \qquad \widehat{\sigma}^*_{\Gamma-} = \sqrt{\widehat{\sigma}^2 - k\gamma \frac{\widehat{\sigma}}{\sqrt{\delta t}}} \qquad (2.139)$$

with $\widehat{\sigma}^*_{\Gamma+} \geq \widehat{\sigma}^*_{\Gamma-}$.

Expression (2.139) for $\widehat{\sigma}^*_{\Gamma+}/\widehat{\sigma}^*_{\Gamma-}$ was first published by Hayne E. Leland in [68]. When trading a call option, whose gamma is always positive, we use $\widehat{\sigma}^*_{\Gamma+}$ when selling it and $\widehat{\sigma}^*_{\Gamma-}$ when buying it.

What if we trade a call spread, or generally an option payoff whose gamma has varying sign? This is where the UVM is called for. We use the UVM with:

$$\sigma_{\min} = \widehat{\sigma}^*_{\Gamma-}, \qquad \sigma_{\max} = \widehat{\sigma}^*_{\Gamma+}$$

Two final observations are in order:

- γ depends on the distribution we assume for daily returns. It equals $\sqrt{\frac{2}{\pi}}$ for Gaussian returns. In practice, daily returns of equities are better modeled with a Student distribution – see Chapter 10 for examples of actual distributions of index returns. In this respect, $\gamma = \sqrt{\frac{2}{\pi}}$ is an over-estimation.

- The period δt of our delta rehedging schedule appears explicitly in (2.139). As $\delta t \to 0$, rehedging costs become prohibitive, to the point where $\widehat{\sigma}^*_{\Gamma-}$ does not exist anymore. When rehedging is frequent, or bid/offer spreads are large, rather than use the Black-Scholes delta and charge for the costs this incurs, it is preferable to go back to square one and cast the delta-hedging strategy as a stochastic control problem that maximizes a utility function which balances the costs generated by hedging with the benefit of a reduced uncertainty of our final P&L. This was done by Mark Davis, Vassilios Panas and Thaleia Zhariphopoulou in [35]. In practice, an exponential utility function $e^{-\lambda P\&L}$ is well suited as the ensuing delta strategy is independent on the initial wealth.[31]

---

[31]Solving this stochastic control problem numerically is tricky, as its solution is of the bang-bang type. The optimal hedging strategy is a function of t, S and the current delta: Δ. The (S, Δ) plane splits into three zones separated by two lines $\Delta_{\pm}(t, S)$. For $\Delta_-(t, S) < \Delta < \Delta_+(t, S)$, no adjustment is needed, if $\Delta < \Delta_-(t, S)$ (resp. $\Delta > \Delta_+(t, S)$) we need to (instantaneously) increase (resp. decrease) our delta so that it equals $\Delta_-(t, S)$ (resp. $\Delta_+(t, S)$). In the limit $k \to 0$ both Δ_- and Δ_+ tend to the Black-Scholes delta Δ_{BS}. For small values of k, Elizabeth Walley and Paul Wilmott show in [84] that $\Delta_{\pm} = \Delta_{BS} \pm \frac{1}{S} \left(\frac{3}{4} e^{-r(T-t)} \frac{k}{\lambda} \Gamma_\$^2 \right)^{\frac{1}{3}}$. $\Gamma_\$ = S^2 \frac{d^2 P_{BS}}{dS^2}$ is the dollar gamma – remember k is the total bid/offer spread.

Chapter's digest

2.2 From prices to local volatilities

▶ Given a full volatility surface $\hat{\sigma}_{KT}$ that complies with the following no-arbitrage conditions:

$$e^{qT_1}C\left(\alpha F_{T_1}, T_1\right) \leq e^{qT_2}C\left(\alpha F_{T_2}, T_2\right)$$

$$\frac{d^2C\left(K,T\right)}{dK^2} \geq 0$$

there exists one volatility function $\sigma(t, S)$ given by the Dupire formula (2.3):

$$\sigma\left(t,S\right)^2 = 2 \left.\frac{\frac{dC}{dT} + qC + (r-q)K\frac{dC}{dK}}{K^2\frac{d^2C}{dK^2}}\right|_{\substack{K=S \\ T=t}}$$

More generally, for any stochastic volatility model that recovers the market smile, whose instantaneous volatility is σ_t, the following condition holds:

$$E[\sigma_t^2 | S_t = S] = \sigma\left(t, S\right)^2$$

▶ In the absence of arbitrage, implied volatilities of vanilla options obey the convex order condition:

$$T_2 \hat{\sigma}_{\alpha F_{T_2}, T_2}^2 \geq T_1 \hat{\sigma}_{\alpha F_{T_1}, T_1}^2$$

This condition is shared by any family of convex payoffs $f(S_T) = h(\frac{S_T}{F_T})$.

☙ ☙ ☙ ☙ ☙

2.3 From implied volatilities to local volatilities

▶ When there are no cash-amount dividends, local volatilities are obtained directly as a function of implied volatilities through (2.19):

$$\sigma\left(t,S\right)^2 = \left.\frac{\frac{df}{dt}}{\left(\frac{y}{2f}\frac{df}{dy} - 1\right)^2 + \frac{1}{2}\frac{d^2f}{dy^2} - \frac{1}{4}\left(\frac{1}{4} + \frac{1}{f}\right)\left(\frac{df}{dy}\right)^2}\right|_{y=\ln\left(\frac{S}{F_t}\right)}$$

▶ When cash-amount dividends are present, there exists (a) an exact solution based on a mapping from S_t to an asset that does not jump across dividend dates, (b) an approximate solution that allows one to use the same formula as in the no-dividend case, except the definition of y changes so that, in particular, it complies with the matching conditions for implied volatilities across dividend dates.

🐦 🐦 🐦 🐦 🐦

2.4 From local volatilities to implied volatilities

▶ Given a local volatility function $\sigma(t, S)$, implied volatilities satisfy the following condition: (2.32):

$$\hat{\sigma}_{KT}^2 = \frac{E_{\sigma(S,t)}\left[\int_0^T e^{-rt} \, \sigma(t,S)^2 \, S^2 \frac{d^2 P_{\hat{\sigma}_{KT}}}{dS^2} dt\right]}{E_{\sigma(S,t)}\left[\int_0^T e^{-rt} \, S^2 \frac{d^2 P_{\hat{\sigma}_{KT}}}{dS^2} dt\right]}$$

▶ For weakly local volatilities $\hat{\sigma}_{KT}$ is given, at order one in the perturbation with respect to a time-dependent volatility $\sigma_0(t)$ by (2.40):

$$\hat{\sigma}_{KT}^2 = \frac{1}{T} \int_0^T dt \int_{-\infty}^{+\infty} dy \frac{e^{-\frac{y^2}{2}}}{\sqrt{2\pi}} u\left(t, F_t e^{\frac{\omega_t}{\omega T} x_K + \frac{\sqrt{(\omega_T - \omega_t)\omega_t}}{\sqrt{\omega T}} y}\right)$$

where $\omega_t = \int_0^t \sigma_0^2(u) du$. When expanding around a constant volatility σ_0, this simplifies to (2.42):

$$\hat{\sigma}_{KT} = \frac{1}{T} \int_0^T dt \int_{-\infty}^{+\infty} dy \frac{e^{-\frac{y^2}{2}}}{\sqrt{2\pi}} \sigma\left(t, F_t e^{\frac{t}{T} x_K + \frac{\sqrt{(T-t)t}}{\sqrt{T}} y}\right)$$

▶ For a local volatility function of the form

$$\sigma(t, S) = \overline{\sigma}(t) + \alpha(t) x + \frac{\beta(t)}{2} x^2, \quad x = \ln \frac{S}{F_t}$$

the ATMF skew and curvature are given, at order one in α and β by (2.48) and (2.49):

$$\mathcal{S}_T = \left.\frac{d\hat{\sigma}_{KT}}{d\ln K}\right|_{F_T} = \frac{1}{T}\int_0^T \frac{t}{T}\alpha(t)dt$$

$$\left.\frac{d^2\hat{\sigma}_{KT}}{d\ln K^2}\right|_{F_T} = \frac{1}{T}\int_0^T \left(\frac{t}{T}\right)^2 \beta(t)dt$$

If $\alpha(t)$ decays as a power law, so does $\mathcal{S}_T$ for large T, with the same exponent: the exponent of the decay of $\alpha(t)$ can be read off the market smile.

▶ For $T \to 0$ we have the exact result:

$$\frac{1}{\hat{\sigma}(T=0,K)} = \frac{1}{\ln \frac{K}{S}} \int_S^K \frac{1}{\sigma(T=0,S)} \frac{dS}{S}$$

🐦 🐦 🐦 🐦 🐦

2.5 The dynamics of the local volatility model

▶ Given a fixed local volatility function of the above form, as the spot moves, implied volatilities move. At first order in $\alpha(t)$ the sensitivity of the ATMF volatility to a spot move is given by

$$\frac{d\widehat{\sigma}_{F_T T}}{d\ln S_0} = \frac{1}{T}\int_0^T \alpha(t)dt$$

which can be expressed using the term-structure of the ATMF skew:

$$\frac{d\widehat{\sigma}_{F_T T}}{d\ln S_0} = \mathcal{S}_T + \frac{1}{T}\int_0^T \mathcal{S}_t dt$$

▶ How the ATMF volatility moves when the spot moves is quantified by the Skew Stickiness Ratio (SSR) – a dimensionless number defined as:

$$\mathcal{R}_T = \frac{1}{\mathcal{S}_T}\frac{d\widehat{\sigma}_{F_T T}}{d\ln S_0}$$

For a weakly local volatility function:

$$\mathcal{R}_T = 1 + \frac{1}{T}\int_0^T \frac{\mathcal{S}_t}{\mathcal{S}_T}dt$$

which implies $\lim_{T\to 0}\mathcal{R}_T = 2$, an exact result shared by all diffusive models.

For an ATMF skew decaying as a power law with exponent γ, in the local volatility model, for large T, for a weakly local volatility function:

$$\mathcal{R}_T \to \frac{2-\gamma}{1-\gamma}$$

For typical equity smiles $\gamma = \frac{1}{2}$, thus $\mathcal{R}_\infty = 3$.

▶ We have the exact result that $\lim_{T\to 0}\mathcal{R}_T = 2$. In addition, for a local volatility function that is a function of $\frac{S}{F_t}$ only, $\mathcal{R}_T = 2, \forall T$.

▶ Expressions above for $\left.\frac{d\widehat{\sigma}_{KT}}{d\ln K}\right|_{F_T}$, $\left.\frac{d^2\widehat{\sigma}_{KT}}{d\ln K^2}\right|_{F_T}$, $\mathcal{R}_T$ are obtained in an order-one expansion around a constant volatility σ_0. See (2.60) and (2.65) for formulas in an expansion around a deterministic volatility $\bar{\sigma}(t)$.

▶ In the local volatility model, implied volatilities are a function of S. The instantaneous volatility of the ATMF volatility is thus proportional to the instantaneous volatility of S, which is equal to $\widehat{\sigma}_{F_t t}$. $\mathrm{vol}(\widehat{\sigma}_{F_T T})$ is given by:

$$\mathrm{vol}(\widehat{\sigma}_{F_T T}) = \mathcal{R}_T \mathcal{S}_T \frac{\widehat{\sigma}_{F_t 0}}{\widehat{\sigma}_{F_T T}} = \left(1 + \frac{1}{T}\int_0^T \frac{\mathcal{S}_t}{\mathcal{S}_T}dt\right)\mathcal{S}_T \frac{\widehat{\sigma}_{F_t t}}{\widehat{\sigma}_{F_T T}}$$

Thus, for short maturities, $\text{vol}(\hat{\sigma}_{F_T T}) = 2\mathcal{S}_T$ while for long maturities: $\text{vol}(\hat{\sigma}_{F_T T}) = \frac{2-\gamma}{1-\gamma}\mathcal{S}_T$, for a flat term structure of ATMF volatilities where γ is the characteristic exponent of the decay of the ATMF skew.

▶ At order one in $\alpha(t)$ the ATMF skew is related to the weighted average of the instantaneous covariance of $\ln S$ and the ATMF volatility for the residual maturity:

$$\mathcal{S}_T = \frac{1}{\hat{\sigma}_T^2 T}\int_0^T \frac{T-t}{T}\langle d\ln S_t \, d\hat{\sigma}_{F_T T}(t)\rangle$$

This relationship is derived more generally in Chapter 8, Section 8.4. One consequence of this formula is that, with respect to time-homogeneous stochastic volatility models, a local volatility model calibrated to the same smile generates larger SSRs and weaker future skews.

❧ ❧ ❧ ❧ ❧

2.6 Future skews and volatilities of volatilities

▶ In the local volatility model, skews observed at future dates for a given residual maturity are typically weaker than spot-starting skews. For a spot-starting skew $\mathcal{S}_\theta(\tau = 0)$ that decays as a function of residual maturity θ with characteristic exponent γ, the skew at a future date τ for the same residual maturity θ scales like, for small θ:

$$\mathcal{S}_\theta(\tau) \propto \left(\frac{\theta}{\tau}\right)^\gamma \mathcal{S}_\theta(\tau=0)$$

▶ Investigating model-generated future skews is useful for assessing local-volatility prices of options that are subject to forward-smile risk. One should bear in mind that these future skews – and future levels of volatility of volatility – cannot be locked and will vary as the model is recalibrated to market smiles. Thus, unpredictable gamma/theta carry P&Ls will impact substantially the P&L of a hedged position, in case residual gammas and cross-gammas are sizeable.

❧ ❧ ❧ ❧ ❧

2.7 Delta and carry P&L

▶ The carry P&L of a delta and vega-hedged option position in the local volatility model can be expressed in the usual gamma-theta form, with payoff-independent break-even levels for spot variance, spot/volatility covariances and volatility/volatility covariances. The local volatility model is a genuine diffusive market model.

▶ The carry P&L of a delta-hedged, vega-hedged option position can be equivalently expressed either in terms of spot and implied volatilities (2.105), or as a function of spot and option prices (2.107). The "real" delta of the local volatility model is given by the derivative of the price with respect to S, keeping *vanilla option prices* fixed. The delta computed by keeping *fixed implied volatilities* is called the sticky-strike delta. Regardless of the particular parametrization used for vanilla option prices, once the deltas of the hedging vanilla options are included, the "real" delta is recovered.

▶ The delta of the local volatility model – computed with a fixed local volatility function – has no particular significance or usefulness. Furthermore the delta of a vanilla option in the local volatility model is an irrelevant notion. More generally, the issue of outputting a delta of one asset (a vanilla option) on another (the spot) is irrelevant in a market model.

🦢 🦢 🦢 🦢 🦢

2.9 The vega hedge

▶ Which portfolio of vanilla options immunizes our derivative position at order one against all perturbations of the local volatility function? Denote by $\mu\left(\tau, K\right)$ the density of vanilla options of maturity τ, strike K, in the hedging portfolio. $\mu\left(\tau, K\right)$ is given by formula (2.121):

$$\mu\left(\tau, K\right) \;=\; -\frac{1}{K^2}\mathcal{L}\phi\left(\tau, K\right)$$

where $\phi\left(t, S\right)$ is the conditional dollar gamma, given by formula (2.118):

$$\phi\left(t, S\right) \;=\; E_\sigma\left[S^2\frac{d^2P}{dS^2}\left(t, S, \bullet\right)|S, t\right]$$

and operator $\mathcal{L}$ is defined by:

$$\mathcal{L}f = \frac{df}{dt} + (r-q)\,S\frac{df}{dS} + \frac{1}{2}S^2\frac{d^2}{dS^2}\left(\sigma^2\left(t, S\right)f\right)\;-rf$$

In the case of a flat local volatility function, $\phi\left(t, S\right)$ is easily computed in a Monte Carlo simulation – see expression (2.123).

🦢 🦢 🦢 🦢 🦢

2.10 Markov-functional models

▶ In Markov-functional models, S_t is a function of a process W_t: $S_t = f(t, W_t)$, where W_t is typically a Brownian motion or an Ornstein-Ühlenbeck process. Markov-functional models are special instances of local volatility and can be calibrated at most to the smile of a single maturity. Smiles for intermediate maturities depend on the choice for the underlying process W_t.

▶ Pricing a multi-asset European derivative using a Gaussian copula together with marginals supplied by each asset's respective vanilla smile is equivalent to pricing with a multi-asset local volatility model, with the correlation matrix equal to the Gaussian copula's correlation matrix.

❧ ❧ ❧ ❧ ❧

Appendix A – the Uncertain Volatility Model

▶ In the UVM, minimum and maximum volatility levels σ_{min}, σ_{max} are specified. The UVM ensures that no money is lost as long as the realized volatility lies in the interval $[\sigma_{min}, \sigma_{max}]$. The seller's price of a derivative with payoff $f(S_T)$ in the UVM solves PDE (2.130) – it is also characterized by:

$$P = \max_{\sigma_t \in [\sigma_{min}, \sigma_{max}]} E[f(S_T)]$$

▶ Rather than pricing a derivative fully in the UVM, one can avail oneself of market-traded options to lower as much as possible the sensitivity of the hedged position to realized volatility, and instead use the UVM for the hedged position: this is the idea in the Lagrangian UVM. The seller's price for the derivative can be defined as:

$$\overline{\mathcal{P}}(F) = \max_{\substack{\sigma_t \in [\sigma_{min}, \sigma_{max}] \\ E_\sigma[O_i] = \mathcal{P}_{Market}(O_i)}} E_\sigma[F]$$

where O_i are the market-traded options. Likewise the buyer's price is defined by:

$$\underline{\mathcal{P}}(F) = \min_{\substack{\sigma_t \in [\sigma_{min}, \sigma_{max}] \\ E_\sigma[O_i] = \mathcal{P}_{Market}(O_i)}} E_\sigma[F]$$

Options that can be exactly replicated by means of a static position in market-traded options are such that $\overline{\mathcal{P}}(F) = \underline{\mathcal{P}}(F)$.

$\overline{\mathcal{P}}(F) - \underline{\mathcal{P}}(F)$ otherwise quantifies the non-vanilla character of the payoff at hand.

▶ The UVM can be used to price-in transaction costs on the delta-hedge: σ_{min}, σ_{max} are given by Leland's formula.

Chapter 3

Forward-start options

Before embarking on stochastic volatility, we pause to study the case of forward-start options – also called cliquets, which we briefly touched upon in Section 1.3. Characterizing the risks of cliquets and how their pricing should be approached provides precious clues as to which aspects of the dynamics of implied volatilities are relevant for pricing these (popular) options, and which features of the vanilla smile a model should be calibrated to.

That cliquet prices are in fact loosely constrained by vanilla smiles is made plain in Section 3.1.7 where we compute lower and upper bounds on the price of a forward-start call option, given vanilla smiles.

We then assess how forward-smile risk is handled in the local volatility model. We work out the example of a forward-start call option in detail; this sheds light on the suitability of local volatility with regard to forward-start options.

3.1 Pricing and hedging forward-start options

Many exotic options are sensitive to forward-smile risk, that is risk associated with the uncertainty about market implied volatilities observed in the future, or future smiles – the case of barrier options was briefly examined in the introduction. Among exotics, forward-start options, or cliquets, form a very popular class of derivatives whose prices are purely determined by the distribution of *forward* returns in the pricing model.

The payoffs of cliquets involve the ratio of a security's price observed at two different dates T_1, T_2: the payoff at time T_2 is $g\left(\frac{S_{T_2}}{S_{T_1}}\right)$. It is a function of the *forward* return $\frac{S_{T_2}}{S_{T_1}} - 1$. As shown below in Section 3.1.3, any European payoff can be expressed as a linear combination of call and put option payoffs. Market smiles for maturities T_1 (resp. T_2) determine prices of payoffs of the form $f(S_{T_1})$, (resp. $f(S_{T_2})$), but payoffs of the form $g\left(\frac{S_{T_2}}{S_{T_1}}\right)$ require modeling assumptions. It is not clear at this stage what, if any, implied volatility data of maturities T_1 and T_2 are relevant for pricing payoff g.

The following analysis of the risks of cliquets is typical of how one approaches the problem of pricing and hedging an exotic option, by:

- first finding a pricing model *and* a hedge portfolio that, within the chosen model, hedges the gamma/theta and vega risks,

- then estimating the costs of rebalancing the vanilla hedge.

The forward smile

Throughout this book, we use the expression *forward-smile risk* to designate the risk associated with the realization of *future smiles*. We do not use the notion of *forward smile*. The forward smile $\hat{\sigma}_k^{T_1 T_2}$ for two dates $T_1, T_2 > T_1$ is obtained, in a given model, by:

- pricing a forward-start call for different values of moneyness k, whose payoff is: $\left(\frac{S_{T_2}/F_{T_2}}{S_{T_1}/F_{T_1}} - k\right)^+$, whose undiscounted price is $P(k)$, where F_T is the forward for maturity T.

- implying a Black-Scholes volatility $\hat{\sigma}_k^{T_1 T_2}$ through:

$$P(k) = P_{BS}\left(S = 1, K = k, T = T_2 - T_1, r = 0, q = 0; \hat{\sigma}_k^{T_1 T_2}\right)$$

$\hat{\sigma}_k^{T_1 T_2}$ is a well-defined function of k, but is an impractical object that has no historical counterpart. Indeed, $\hat{\sigma}_k^{T_1 T_2}$ is the future implied volatility for moneyness k, averaged over all realizations of future smiles in the model: it is an aggregate of future smile risk and volatility-of-volatility risk.

Therefore, it does not make sense to assess the suitability of a model by comparing $\hat{\sigma}_k^{T_1 T_2}$ with typical market smiles of maturity $T_2 - T_1$, if anything because the forward smile is invariably more convex, due to its being an average and the connection between price and implied volatility being nonlinear.[1]

The forward smile is thus a notion of limited usefulness – we do not use in the sequel.

3.1.1 A Black-Scholes setting

In the standard Black-Scholes model, implied volatilities have no term structure. Since a cliquet involves S_{T_1} and S_{T_2}, it is natural to use a Black-Scholes model with time-dependent instantaneous volatility $\sigma(t)$: in such a model, implied volatilities are maturity-dependent but the smile for any given maturity is flat and the implied volatility for maturity T, $\hat{\sigma}_T$, is given by:

$$\hat{\sigma}_T^2 = \frac{1}{T - t} \int_t^T \sigma(u)^2 \, du$$

Because of homogeneity, the price of a cliquet in this model does not depend on S and, besides interest and repo rates, only depends on the integrated variance over the interval $[T_1, T_2]$: it is a function of the forward volatility $\hat{\sigma}_{T_1 T_2}$:

$$P = e^{-r(T_1 - t)} G\left(\hat{\sigma}_{T_1 T_2}\right) \tag{3.1}$$

---

[1]See the discussion of the forward smile of the Heston model in [8].

defined by:

$$\widehat{\sigma}^2_{T_1 T_2} = \frac{1}{T_2 - T_1} \int_{T_1}^{T_2} \sigma\left(t\right)^2 dt = \frac{(T_2 - t)\widehat{\sigma}^2_{T_2} - (T_1 - t)\widehat{\sigma}^2_{T_1}}{T_2 - T_1} \quad (3.2)$$

We need to answer two questions:

- How can this forward-start option be hedged?

- Which payoffs of maturities T_1 and T_2 should $\widehat{\sigma}_{T_1}$ and $\widehat{\sigma}_{T_2}$ be calibrated to? Note that implied volatilities can be defined for any payoff that is convex. These payoffs need to be such that their implied volatilities satisfy the convex order condition:

$$(T_2 - t)\widehat{\sigma}^2_{T_2} \geq (T_1 - t)\widehat{\sigma}^2_{T_1} \quad (3.3)$$

so that $\widehat{\sigma}_{T_1 T_2}$ in (3.2) is well-defined.

Prior to T_1, the cliquet's delta and gamma vanish and the cliquet is only sensitive to $\widehat{\sigma}_{T_1 T_2}$. At time $t = T_1$, S_{T_1} is known: the cliquet becomes a standard European option of maturity T_2.

The above formula shows that $\widehat{\sigma}_{T_1 T_2}$ is a function of implied volatilities $\widehat{\sigma}_{T_1}$, $\widehat{\sigma}_{T_2}$. Trading single vanilla options of maturities T_1, T_2 is inappropriate as, unlike the forward-start option, their sensitivities to $\widehat{\sigma}_{T_1}$, $\widehat{\sigma}_{T_2}$ will vary as S moves and their combined gamma will likely not vanish. Does there exist a portfolio of vanilla options of maturities T_1, T_2 whose vega is spot-independent?

3.1.2 A vanilla portfolio whose vega is independent of S

Let $\rho\left(K\right)$ be the density of vanilla options of strike K in the portfolio. We make no distinction between call or put options struck at the same strike as they have the same vega. Because of the homogeneity of degree 1 in S and K of the Black-Scholes formula, the price of a vanilla option of strike K can be written as: $P = Kf\left(\frac{S}{K}\right)$ and likewise for the vega:

$$\text{Vega}_K\left(S\right) = K\varphi\left(\frac{S}{K}\right)$$

The vega of our portfolio thus reads:

$$\text{Vega}_\Pi\left(S\right) = \int dK \rho(K) K\varphi\left(\frac{S}{K}\right)$$

After switching to variable $u = \frac{S}{K}$ this becomes:

$$\text{Vega}_\Pi\left(S\right) = \int \frac{du}{u^3} S^2 \rho\left(\frac{S}{u}\right)\varphi(u) \quad (3.4)$$

Vega$_\Pi$ (S) is independent of S only if we choose:

$$\rho(K) \propto \frac{1}{K^2} \tag{3.5}$$

A portfolio of European options with a density proportional to $\frac{1}{K^2}$ is such that its vega in the Black-Scholes model does not depend on S. Up to an affine function of S the corresponding payoff is in fact $\ln S$ – it is called the *log contract* and was first proposed by Anthony Neuberger in [75].

We now verify this by recalling a representation of an arbitrary European payoff in terms of cash, forwards and a portfolio of vanilla options.

3.1.3 Digression: replication of European payoffs

The derivation of the well-known formula expressing this decomposition[2] starts from the identity:

$$f(S) = f(K_0) + \int_{K_0}^{S} \frac{df}{dK} dK$$

$$= f(K_0) + \int_{K_0}^{\infty} \theta(S-K) \frac{df}{dK} dK - \int_{0}^{K_0} \theta(K-S) \frac{df}{dK} dK$$

where $\theta(x)$ is the Heaviside function ($\theta(x) = 1$ if $x \geq 0$, $\theta(x) = 0$ otherwise) and where the second line can be checked by taking either the case $S > K_0$ or the case $S < K_0$. Integrate now by parts using $(K-S)^+$ as primitive of $\theta(K-S)$ and $-(S-K)^+$ as primitive of $\theta(S-K)$. We get:

$$f(S) = f(K_0) + \int_{K_0}^{\infty} \frac{d^2 f}{dK^2} (S-K)^+ dK + \int_{0}^{K_0} \frac{d^2 f}{dK^2} (K-S)^+ dK$$

$$+ [-\frac{df}{dK} (S-K)^+]_{K_0}^{\infty} - [\frac{df}{dK} (K-S)^+]_{0}^{K_0}$$

which after simplification gives:

$$f(S) = f(K_0) + \frac{df}{dK}\Big|_{K_0} (S-K_0)$$

$$+ \int_{0}^{K_0} \frac{d^2 f}{dK^2} (K-S)^+ dK + \int_{K_0}^{\infty} \frac{d^2 f}{dK^2} (S-K)^+ dK \tag{3.6}$$

This expresses f as a linear combination of an affine function and a continuous density of calls struck above K_0 and puts struck below K_0. Let P_f be the price of the

---

[2]See the article by Peter Carr and Dilip Madan ([25]) who trace this result back to the work of Breeden & Litzenberger ([18]), Green & Jarrow ([50]) and Nachman ([74]).

options that pays $f(S_T)$. P_f is the sum of the prices of the different contributions to f :

$$P_f = f(K_0) e^{-r(T-t)} + \left.\frac{df}{dK}\right|_{K_0} \left(Se^{-q(T-t)} - K_0 e^{-r(T-t)}\right)$$

$$+ \int_0^{K_0} \frac{d^2 f}{dK^2} P_K dK + \int_{K_0}^{\infty} \frac{d^2 f}{dK^2} C_K dK$$

where C_K, P_K denote, respectively, the prices of a call and a put struck at K. Choosing for K_0 the forward for maturity T, F_T, yields:

$$P_f = f(F_T) e^{-r(T-t)} + \int_0^{F_T} \frac{d^2 f}{dK^2} P_K dK + \int_{F_T}^{\infty} \frac{d^2 f}{dK^2} C_K dK \qquad (3.7)$$

This is an equality of prices – one should not forget the forwards in the replication of f.

Conversely, consider a portfolio consisting of a density $\rho(K)$ of vanilla options of strike K. Identity (3.6) shows that the resulting payoff $-\int_0^{\infty} \rho(K)(S-K)^+ dK$, or $\int_0^{\infty} \rho(K)(K-S)^+ dK$ if we use put options – is obtained, up to an affine function of S, by simply integrating $\rho(K)$ twice.

3.1.4 A vanilla hedge

Starting with the density $\frac{1}{K^2}$ derived above and integrating twice, we recover the payoff of the log contract: $-\ln S$.

For reasons that will become clear when we discuss variance swaps, we prefer to work with payoff $-2\ln S$ rather than $\ln S$. We will henceforth denote by "log contract" the payoff $-2\ln S$. It is replicated with a density of vanilla options equal to $\frac{2}{K^2}$.

The price $Q^T(t, S)$ of a log contract of maturity T in the Black-Scholes model is given by:

$$Q^T(t, S) = -2e^{-r(T-t)} \left(\ln S + (r-q)(T-t) - \frac{\hat{\sigma}_T^2}{2}(T-t)\right) \qquad (3.8)$$

and its sensitivity to $\hat{\sigma}_T$ is:

$$\frac{dQ^T}{d\hat{\sigma}_T} = 2e^{-r(T-t)}(T-t)\,\hat{\sigma}_T$$

We now resume our discussion of the vega hedge of our forward-start option P. Using equations (3.1) and (3.2) we get the sensitivities of P to $\hat{\sigma}_{T_1}$ and $\hat{\sigma}_{T2}$:

$$\frac{dP}{d\hat{\sigma}_{T_2}} = e^{-r(T_1-t)}(T_2-t)\,\hat{\sigma}_{T_2}\mathcal{N}$$

$$\frac{dP}{d\hat{\sigma}_{T_1}} = -e^{-r(T_1-t)}(T_1-t)\,\hat{\sigma}_{T_1}\mathcal{N}$$

where prefactor $\mathcal{N}$ is:

$$\mathcal{N} = \frac{1}{(T_2 - T_1)\,\widehat{\sigma}_{T_1 T_2}} \frac{dG}{d\widehat{\sigma}_{T_1 T_2}}$$

Thus the portfolio

$$\Pi = -P + \frac{\mathcal{N}}{2}\left(e^{r(T_2 - T_1)}Q^{T_2} - Q^{T_1}\right) \tag{3.9}$$

has zero sensitivity to both $\widehat{\sigma}_{T_1}$ and $\widehat{\sigma}_{T_2}$. Notice that the hedge ratios for Q^{T_1} and Q^{T_2} depend neither on t, nor on S: the hedge is stable as time elapses and S moves, and will only need to be readjusted whenever the forward volatility $\widehat{\sigma}_{T_1 T_2}$ varies.

   What about the gamma of the vanilla hedge? Taking twice the derivative of equation (3.8) we get:

$$S^2 \frac{d^2 Q^T}{dS^2} = 2e^{-r(T-t)} \tag{3.10}$$

Thus the dollar gamma of the log contract – up to the usual discounting factor – is equal to 2. The combination $e^{r(T_2 - T_1)}Q^{T_2} - Q^{T_1}$ has thus vanishing gamma and consequently, in our deterministic volatility model, vanishing theta as well.

   We have been able to choose a hedging model and have assembled a static[3] vanilla portfolio that perfectly hedges at order one our cliquet against variations of $\widehat{\sigma}_{T_1 T_2}$.

Checking that $\widehat{\sigma}_{T_1 T_2}$ is well-defined

Implied volatilities of log contracts of maturities T_1 and T_2 satisfy the convex order condition (3.3) because of the convexity of the log contract. The log-contract falls in the class of payoffs considered in Section 2.2.2.2, page 32, for which (a) an implied volatility can be defined, (b) the convex order condition (3.3) holds.

   When there are no cash-amount dividends, log contract implied volatilities $\widehat{\sigma}_T$ are expressed directly as an average of implied volatilities of vanilla options – see formula (4.21), page 142.

3.1.5   Using the hedge in practice – additional P&Ls

3.1.5.1   Before T_1– volatility-of-volatility risk

   Market implied volatilities are not only maturity- but also strike-dependent: how should we define $\sigma(t)$ or $\widehat{\sigma}_T$?

   We *decide* to define $\widehat{\sigma}_T$ as the implied volatility of the log contract of maturity T. $\widehat{\sigma}_T$ is well-defined as the function $\ln S$ is concave: we only need to invert equation (3.8). Using this definition for $\widehat{\sigma}_T$ ensures that our hedge instruments have their

---

[3]In the sense that it does not need to be readjusted as t advances and S moves.

right market prices. The P&L over $[t,\ t+\delta t]$ of portfolio Π reads:

$$
\begin{aligned}
P\&L_\Pi = & -\left(e^{-r(T_1-t-\delta t)}G\left(\widehat{\sigma}_{T_1T_2}+\delta\widehat{\sigma}_{T_1T_2}\right)-(1+r\delta t)\,e^{-r(T_1-t)}G\left(\widehat{\sigma}_{T_1T_2}\right)\right)\\
& +\frac{\mathcal{N}}{2}e^{r(T_2-T_1)}\left(Q^{T_2}_{t+\delta t}-(1+r\delta t)\,Q^{T_2}_t+\frac{dQ^{T_2}}{dS}\left(\delta S-rS\delta t\right)\right)\\
& -\frac{\mathcal{N}}{2}\left(Q^{T_1}_{t+\delta t}-(1+r\delta t)\,Q^{T_1}_t+\frac{dQ^{T_1}}{dS}\left(\delta S-rS\delta t\right)\right)
\end{aligned}
$$

where Q^T_t (resp. $Q^T_{t+\delta t}$) is a shorthand notation for $Q^T(t,S,\widehat{\sigma}_T)$ (resp. $Q^T(t+\delta t,S+\delta S,\widehat{\sigma}_T+\delta\widehat{\sigma}_T)$). Let us expand this P&L at order 1 in δt and order 2 in δS.

If $\delta\widehat{\sigma}_{T_2}=\delta\widehat{\sigma}_{T_1}=0$, since the cliquet has vanishing gamma/theta and so does our log contract hedge, we get zero P&L.

The only contribution to the P&L is then generated by $\delta\widehat{\sigma}_{T_1},\delta\widehat{\sigma}_{T_2}$. Using expression (3.8) for Q^T, we get:

$$
\begin{aligned}
P\&L_\Pi = & -e^{-r(T_1-t)}\left(G\left(\widehat{\sigma}_{T_1T_2}+\delta\widehat{\sigma}_{T_1T_2}\right)-G\left(\widehat{\sigma}_{T_1T_2}\right)\right)\\
& +\frac{\mathcal{N}}{2}e^{-r(T_1-t)}\left(T_2-t\right)\left(\left(\widehat{\sigma}_{T_2}+\delta\widehat{\sigma}_{T_2}\right)^2-\widehat{\sigma}^2_{T_2}\right)\\
& -\frac{\mathcal{N}}{2}e^{-r(T_1-t)}\left(T_1-t\right)\left(\left(\widehat{\sigma}_{T_1}+\delta\widehat{\sigma}_{T_1}\right)^2-\widehat{\sigma}^2_{T_1}\right)
\end{aligned}
$$

which yields:

$$
\begin{aligned}
P\&L_\Pi = & -e^{-r(T_1-t)}\left(G\left(\widehat{\sigma}_{T_1T_2}+\delta\widehat{\sigma}_{T_1T_2}\right)-G\left(\widehat{\sigma}_{T_1T_2}\right)\right)\\
& +\frac{1}{2\widehat{\sigma}_{T_1T_2}}\frac{dG}{d\widehat{\sigma}_{T_1T_2}}e^{-r(T_1-t)}\left(\left(\widehat{\sigma}_{T_1T_2}+\delta\widehat{\sigma}_{T_1T_2}\right)^2-\widehat{\sigma}^2_{T_1T_2}\right)\\
= & \ e^{-r(T_1-t)}\left[-\left(G\left(\widehat{\sigma}_{T_1T_2}+\delta\widehat{\sigma}_{T_1T_2}\right)-G\left(\widehat{\sigma}_{T_1T_2}\right)\right)+\frac{dG}{d(\widehat{\sigma}^2_{T_1T_2})}\delta(\widehat{\sigma}^2_{T_1T_2})\right]\quad(3.11)
\end{aligned}
$$

In the derivation of (3.11) we have used variances $\widehat{\sigma}^2_{T_1},\widehat{\sigma}^2_{T_2},\widehat{\sigma}^2_{T_1T_2}$ rather than volatilities $\widehat{\sigma}_{T_1},\widehat{\sigma}_{T_2},\widehat{\sigma}_{T_1T_2}$. The natural reason for analyzing our P&L using variances rather than volatilities is that the price of the hedge instruments – log-contracts – is affine in $\widehat{\sigma}^2_{T_1},\widehat{\sigma}^2_{T_2}$, rather than $\widehat{\sigma}_{T_1},\widehat{\sigma}_{T_2}$; just as we use S rather than $\sqrt{S}$ for the sake of writing out the P&L of a delta-hedged option.

Expression (3.11) shows that $P\&L_\Pi$ is a function of $\delta(\widehat{\sigma}^2_{T_1T_2})$, not $\delta(\widehat{\sigma}^2_{T_1})$ and $\delta(\widehat{\sigma}^2_{T_2})$ separately.

(3.11) is *not* an expansion of $P\&L_\Pi$ at order two in $\delta\widehat{\sigma}_{T_1T_2}$: it is the actual P&L generated by a change of $\widehat{\sigma}_{T_1T_2}$. One can check on equation (3.11) that the order-one contribution from $\delta(\widehat{\sigma}^2_{T_1T_2})$ vanishes – as it should. This expression highlights the fact that the value of the log contract hedge – the second piece inside the brackets – is exactly quadratic in $\widehat{\sigma}_{T_1T_2}$: this is also apparent in (3.8).

If G is an affine function of $\widehat{\sigma}^2_{T_1T_2}$, the hedge is perfect and our P&L is exactly zero until we reach T_1. This is the case for two particular payoffs: one is given by

$g\left(\frac{S_{T_2}}{S_{T_1}}\right) = \ln\left(\frac{S_{T_2}}{S_{T_1}}\right)$, which is a linear combination of our two log contracts and has no commercial interest whatsoever. The other one is the same payoff, but delta-hedged on a daily basis: it is called a forward variance swap and pays the realized quadratic variation over the interval $[T_1, T_2]$.[4]

Usually G will *not* be an affine function of $\widehat{\sigma}^2_{T_1 T_2}$. P&L (3.11) thus starts with a term of order two in $\delta(\widehat{\sigma}^2_{T_1 T_2})$. Keeping this term only yields:

$$P\&L_\Pi = -\frac{e^{-r(T_1 - t)}}{2}\frac{d^2 G}{d(\widehat{\sigma}^2_{T_1 T_2})^2}\left(\delta(\widehat{\sigma}^2_{T_1 T_2})\right)^2 \qquad (3.12)$$

For an at-the-money forward call, g is given by:

$$g\left(\frac{S_{T_2}}{S_{T_1}}\right) = \left(\frac{S_{T_2}}{S_{T_1}} - 1\right)^+ \qquad (3.13)$$

If $T_2 - T_1$ is small, for vanishing interest rate and repo, G is approximately linear in $\widehat{\sigma}$:

$$G\left(\widehat{\sigma}_{T_1 T_2}\right) \simeq \frac{\widehat{\sigma}_{T_1 T_2}}{\sqrt{2\pi}}\sqrt{T_2 - T_1}$$

Using (3.12), at order 2 in $\delta(\widehat{\sigma}^2_{T_1 T_2})$, $P\&L_\Pi$ reads:

$$P\&L_\Pi \simeq e^{-r(T_1 - t)}\frac{\sqrt{T_2 - T_1}}{\sqrt{2\pi}}\frac{1}{2\widehat{\sigma}_{T_1 T_2}}\left(\delta\widehat{\sigma}_{T_1 T_2}\right)^2 \qquad (3.14)$$

We thus make money every time $\widehat{\sigma}_{T_1 T_2}$ moves – this will occur generally for all payoffs whose value is a *concave* function of $\widehat{\sigma}^2_{T_1 T_2}$.

In the general case, $P\&L_\Pi$ will then not vanish – it is generated by the dynamics of $\widehat{\sigma}_{T_1 T_2}$: we call this volatility-of-volatility risk. The estimation of $P\&L_\Pi$ over $[0, T_1]$ entails making an assumption for the volatility of $\widehat{\sigma}_{T_1 T_2}$. The resulting extra charge – or gain – has to be added to the price $P = e^{-rT_1}G\left(\widehat{\sigma}_{T_1 T_2}\left(t = 0\right)\right)$ quoted at time $t = 0$. In the case of the at-the-money forward option, this charge will be negative, as we reduce the price charged to the client by an estimation of the positive P&Ls (3.14) pocketed every time $\widehat{\sigma}_{T_1 T_2}$ moves.[5]

Up to $t = T_1$, our pricing and hedging scheme works, provided we have included this extra charge – or gain – in our price.

3.1.5.2 At T_1 – forward-smile risk

At T_1, the cliquet turns into a European option of maturity T_2. As seen in Section 3.1.3, it can be replicated with vanilla options of maturity T_2, hence its value is a function of implied volatilities for maturity T_2 observed at T_1: $\widehat{\sigma}_{K T_2}\left(T_1\right)$.

---

[4]Variance swaps are abundantly discussed in Chapter 5.
[5]Making random positive P&L may seem less serious than randomly losing money. However, not adjusting the price for this random gain will result in the loss of the trade – this is similar to trying to buy a vanilla option for its intrinsic value.

Imagine that the cliquet is an at-the-money forward call, whose payoff is given in equation (3.13). At T_1 the *market* price of the cliquet is then:

$$P_{BS}\left(S_{T_1}, K = S_{T_1}, T_2; \widehat{\sigma}_{K=S_{T_1}T_2}(T_1)\right) \tag{3.15}$$

In our hedging scheme both the cliquet and its hedge are risk-managed in a Black-Scholes model with volatilities $\widehat{\sigma}(T)$ defined as implied volatilities of log contracts. Equation (3.2) shows that, at $t = T_1$, $\widehat{\sigma}_{T_1 T_2} = \widehat{\sigma}_{T_2}$, thus the *model* price of the cliquet at T_1 is:

$$P_{BS}\left(S_{T_1}, K = S_{T_1}, T_2; \widehat{\sigma}_{T_2}(T_1)\right) \tag{3.16}$$

Compare this formula with expression (3.15). They are identical, except in our pricing and hedging scheme the at-the-money call is valued using the implied volatility of the log contract of the same maturity, T_2, rather than the market implied volatility of the at-the-money call. The price we quote at $t = 0$ for our forward-start at-the-money option has then to include a provision to cover for the difference between (3.15) and (3.16).

The risk created by the uncertainty as to the smile prevailing at T_1 for maturity T_2 – given a known level of the log-contract implied volatility $\widehat{\sigma}_{T_2}(T_1)$ – is called forward-smile risk. For a general cliquet payoff $g\left(S_{T_2}/S_{T_1}\right)$, we need to supplement the initial price $P = e^{-rT_1} G\left(\widehat{\sigma}_{T_1 T_2}(t = 0)\right)$ with an extra charge to cover for the difference between the market price of the payoff at T_1 and the Black-Scholes price computed with volatility $\widehat{\sigma}_{T_2}(T_1)$.

What if our forward-start option has no – or hardly any – sensitivity to $\widehat{\sigma}_{T_2}(T_1)$? This is the case for a digital payoff:

$$g\left(\frac{S_{T_2}}{S_{T_1}}\right) = 1_{\frac{S_{T_2}}{S_{T_1}} > 1}$$

or a narrow call spread or put spread struck around S_{T_1}. In this case G has negligible or vanishing sensitivity to $\widehat{\sigma}_{T_1 T_2}$. Using the term structure of log-contracts as hedge instruments does not make sense anymore, and the adjustment for forward-smile risk represents in fact the bulk of the forward-start option price: our cliquet is a pure forward-smile instrument.

3.1.6 Cliquet risks and their pricing: conclusion

- We *choose* to price and risk-manage our cliquet in a Black-Scholes model with time-dependent volatility, recalibrated every day on market implied volatilities $\widehat{\sigma}(T)$ of log contracts. We are then able to exactly gamma-hedge and vega-hedge our cliquet, and have to make two adjustments to our price: δP_1 to cover for volatility-of-volatility risk over $[0, T_1]$, that is P&Ls (3.12), and δP_2 to cover for forward-smile risk at T_1, that is the difference between (3.15) and (3.16). The price we quote for this cliquet at $t = 0$ is then:

$$P = e^{-rT_1} G\left(\widehat{\sigma}_{T_1 T_2}(t = 0)\right) + (\delta P_1 + \delta P_2) \tag{3.17}$$

How should we estimate δP_1 and δP_2? In case there is no market for volatility of volatility and forward smile we can do the following:

- δP_1: use historical data of log contract implied volatilities to estimate conservatively the expected realized volatility of $\hat{\sigma}_{T_1 T_2}$ over $[0, T_1]$.
- δP_2: use historical data of market smiles of residual maturity $T_2 - T_1$ to estimate conservatively the difference between implied volatilities of our payoff and that of the log contract.

If instead we are able to offset these risks on the market – for example by trading other cliquets – δP_1 and δP_2 are computed using *implied* values in place of *historical* values. Practically, whether we decide to use calibrated or chosen values for volatility of volatility or the future smile, we will use a stochastic volatility model to estimate $\delta P_1 + \delta P_2$. The model generates a global price that aggregates all risks – as priced by the model. δP_1 and δP_2 are evaluated initially at $t = 0$.

- As we risk-manage the cliquet from $t = 0$ to $t = T_1$, we need to ensure that (a) δP_1 converges to zero at T_1, (b) δP_2 converges to the adjustment corresponding to the exact difference between the implied volatilities of our forward-start payoff and that of the log contract of maturity T_2, as observed at T_1 – whether δP_1, δP_2 have been calculated by hand or evaluated in a model.

- In a model, δP_1 will automatically converge to zero as $t \to T_1$. For δP_2 to converge to the right value, though, the model has to be such that it recovers at T_1 the smile of maturity T_2.

δP_2 makes up the bulk of volatility risk for options that have mostly forward-smile risk – forward ATM call spreads or digital payoffs: G has hardly any sensitivity to $\hat{\sigma}_{T_1 T_2}$ and $\delta P_1 \simeq 0$.

Pricing forward-smile risk in a model

- In the *continuous* forward variance models of Chapter 7, there are no separate handles on (a) the spot-starting vanilla smile and (b) future smiles. Thus, at inception, the model should be parametrized so that the desired future smile at T_1 is obtained, while as $t \to T_1$, the model is calibrated to the vanilla smile.

- The local-stochastic volatility models of Chapter 12, on the other hand, are calibrated to the vanilla smile, thus $\delta P_2 \to 0$ as $t \to T_1$. The price we pay for this is less control on future smiles generated by the model – that is the value of δP_2 at $t = 0$.

We refer the reader to Section 12.6.1 of Chapter 12 for a comparison of prices of forward-start options in different models calibrated to the same vanilla smile.

- The *discrete* forward variance models discussed in Section 7.8 of Chapter 7 offer maximum flexibility: we have a handle on the term-structure of forward skew, while still retaining the capability of matching the short market spot-starting smile. In addition they also afford separate control of δP_1 (volatility-of-volatility risk) and δP_2 (forward-smile risk).

Our analysis has shown that the risk that can be hedged using vanilla options is the forward volatility $\hat{\sigma}_{T_1 T_2}$. Other risks – forward smile and volatility-of-volatility risk – cannot be hedged with vanilla options and have to be priced-in using exogenous parameters.

Contrary to what is sometimes heard on trading desks, an at-the-money forward call option *does* have forward smile sensitivity. The fact that it is at-the-money has no special significance, as forward implied volatilities of vanilla options – be they at-the-money or not – cannot be locked, unlike forward log-contract implied volatilities.[6] Consequently, for cliquets with payoff $g\left(\frac{S_{T_2}}{S_{T_1}}\right)$, the only information available in vanilla option prices to be used in the calibration of our model is the term structure of implied volatilities $\hat{\sigma}_T$ of log contracts, as these are our hedge instruments.

We have no reason to use other vanilla option data: market prices of other instruments should be included in the calibration only if one is able to exactly pinpoint which instruments to use and which risk they offset.

We study in Section 3.1.9 further below the interesting example of payoffs $S_{T_1} g\left(\frac{S_{T_2}}{S_{T_1}}\right)$ which call for different hedging instruments, thus requiring calibration on a different set of European payoffs, leading to the definition of yet another class of forward volatilities.

The notion that a pricing model should generally be calibrated to the vanilla smile on the grounds that, hopefully, vanilla option prices provide information that should, somehow, be used, is fallacious at best and dangerous at worst. Going back to our forward-start option, this amounts to letting the model arbitrarily link the values of $\delta P_1, \delta P_2$ to the vanilla smile, when in practice no trading strategy is able to lock this dependency.

We now provide an illustration of the fact that cliquet prices are in fact rather loosely confined by vanilla smiles.

3.1.7   Lower/upper bounds on cliquet prices from the vanilla smile

As a simple example of a cliquet consider a forward ATM call that pays $\left(\frac{S_{T_2}}{S_{T_1}} - 1\right)^+$ at T_2. Let the hedging instruments be the underlying itself and call options of maturities T_1 and T_2. Assume zero interest rate and repo for simplicity.

---

[6]As will be discussed further on, log contracts are not traded, but variance swaps – which for practical purposes can be considered delta-hedged log contracts – are. Their gamma and vega have the same properties as those of the log contract.

Can the information in market smiles of maturities T_1 and T_2 be used to bound the price of our cliquet? We would like to derive model-independent lower and upper bounds such that if we were quoted a price that lay outside these bounds – say a lower price than the lower bound – a strategy consisting of (a) a long position in the cliquet entered at that price, (b) a static position in vanilla options of maturities T_1 and T_2 entered at market price, (c) a delta strategy over $[T_1, T_2]$, would generate positive P&L at T_2, for all (S_{T_1}, S_{T_2}) configurations.

The spread between lower and upper bound quantifies how much the cliquet differs from a statically replicable payoff and provides a measure of how vanilla-like – or unlike – the cliquet is. In case our cliquet payoff could be in fact synthesized by a combination of: (a) a fixed cash amount, (b) a static position in vanilla options of maturities T_1 and T_2, (c) a delta strategy set up at T_1 and unwound at T_2, then lower and upper bounds would coincide.

Consider thus a trading strategy that consists of (a) a cash amount c, (b) a static position in λ_i call options of strikes K_i / maturity T_1 and μ_j call options of strikes K_j / maturity T_2, (c) a delta position $\Delta(S_{T_1})$ starting at T_1 and unwound at T_2, such that it super-replicates the cliquet's payoff.

Since entering the delta position $\Delta(S_{T_1})$ at T_1 does not require any cash outlay, the initial amount of cash needed for setting up this strategy is:

$$c + \Sigma_i \lambda_i C_{K_i T_1} + \Sigma_j \mu_j C_{K_j T_2}$$

Among all super-replicating strategies, that with the lowest initial cost provides the upper model-independent bound UB for the cliquet's price that is compatible with market prices of hedging instruments. Mathematically:

$$UB = \min_{\lambda, \mu, \Delta(S_1), c} \left(c + \sum_i \lambda_i C_{K_i T_1} + \sum_j \mu_j C_{K_j T_2} \right)$$

such that:

$$c + \sum_i \lambda_i (S_1 - K_i)^+ + \sum_j \mu_j (S_2 - K_j)^+ + \Delta(S_1)(S_2 - S_1) \geq \left(\frac{S_2}{S_1} - 1 \right)^+$$

$$\forall (S_1, S_2)$$

The model-independent lower bound can be defined analogously.[7]

We choose a discrete set of strikes K_i for options of maturity T_1, a discrete set of strikes K_j for options of maturity T_2 and a finite-dimensional basis for $\Delta(S_1)$. We

---

[7] The dual formulation of this problem corresponds to finding a joint distribution $\rho(S_{T_1}, S_{T_2})$ such that (a) the marginal distributions of S_{T_1} and S_{T_2} match the respective market smiles, (b) the martingality condition $E[S_{T_2}|S_{T_1}] = S_{T_1}$ holds and (c) the cliquet price is maximal.

See also the related discussion in the context of the λ-UVM in Appendix A of Chapter 2, page 90. In case prices of market instruments used for constraining the joint distribution are arbitrageable, this will manifest itself in the result that bounds are infinite.

then need to minimize an affine function of the vectors λ, μ and the components of $\Delta(S_1)$ on the basis we have chosen, subject to a set of constraints that are linear as well in λ, μ, $\Delta(S_1)$: one constraint for each couple (S_1, S_2). Practically we choose a (large) discrete set of such couples.

Numerically, this problem is solved with the simplex algorithm – we refer the reader to [57] for a more general account of sub and super-replicating strategies.

- Let us take $T_1 = 1$ year, $T_2 = 2$ years, vanishing interest rate and repo, and assume that the vanilla smiles for maturities T_1 and T_2 are flat with implied volatilities all equal to 20%. The model-independent lower and upper bounds for the implied volatility of the forward ATM call are: $\widehat{\sigma}_{\min} \simeq 9\%$, $\widehat{\sigma}_{\max} \simeq 25\%$.[8]

- Now, keeping the same smiles at T_1 and T_2 let us compute lower ($CS_{\min}$) and upper bounds ($CS_{\max}$) for the price of a 95%/105% forward call spread: $\left(\frac{S_{T_2}}{S_{T_1}} - 95\%\right)^+ - \left(\frac{S_{T_2}}{S_{T_1}} - 105\%\right)^+$.

  We get $CS_{\min} \simeq 1.6\%$, $CS_{\max} \simeq 7.7\%$. These prices correspond approximately to skews $(\widehat{\sigma}_{95\%} - \widehat{\sigma}_{105\%})_{\min} = -8\%$ and $(\widehat{\sigma}_{95\%} - \widehat{\sigma}_{105\%})_{\max} \simeq 8\%$. Notice how wide this range is – the typical order of magnitude of an index skew for a one-year maturity is $\widehat{\sigma}_{95\%} - \widehat{\sigma}_{105\%} = 3\%$.

- Let us compute again lower and upper bounds for our call spread, this time adding as an extra constraint the price of the forward ATM call, computed with an implied volatility of 20%: 7.96%.

  We now get: $CS_{\min} \simeq 1.8\%$, $CS_{\max} \simeq 7.2\%$. The additional information on the joint distribution of S_{T_1}, S_{T_2} supplied by the forward ATM call has narrowed the price range only slightly.

- This time let us use as extra constraint the price of the forward call spread $\left(\frac{S_{T_2}}{S_{T_1}} - 90\%\right)^+ - \left(\frac{S_{T_2}}{S_{T_1}} - 110\%\right)^+$ computed with implied volatilities $\widehat{\sigma}_{90\%} = 22\%$, $\widehat{\sigma}_{110\%} = 18\%$. The price of this call spread is 10.7%.

  We now get: $CS_{\min} \simeq 4.25\%$, $CS_{\max} \simeq 6.7\%$, which corresponds approximately to $(\widehat{\sigma}_{95\%} - \widehat{\sigma}_{105\%})_{\min} = -1\%$ and $(\widehat{\sigma}_{95\%} - \widehat{\sigma}_{105\%})_{\max} \simeq 5.5\%$. Observe how inclusion of the price of a payoff whose risk is congruent with that of our 95%/105% call spread has tightened significantly the price range of the latter.

---

[8]I thank Pierre Henry-Labordère for sharing these results, which are obtained numerically – hence the symbol $\simeq$. While results depend somewhat on the discretization chosen for the (S_1, S_2) couples, the same parameters have been used throughout. Note that the simplex algorithm is able to deal with a very large number of constraints.

3.1.8   Calibration on the vanilla smile – conclusion

The conclusion from the above results is that vanilla smiles hardly constrain cliquet prices. Only market prices of payoffs whose risks are congruent with those of the exotic at hand are able to narrow down the price range of the latter.

Our example illustrates the fact that an exotics business is not a brokerage business – otherwise exotics could be statically hedged by vanillas and hence simply priced with models whose parameters are fully determined by the vanilla smile. Rather, running a book of exotics entails taking controlled risks on exotic parameters such as: forward smile, volatility of volatility, smile of volatility of volatility, spot/volatility covariance. A suitable model should offer the capability of specifying these parameter levels exogenously.

Ideally the model should also be able to match market prices of vanilla options (really) used as hedges.

When this proves impossible, it is more reasonable to adjust the exotic's price to cover for the difference between model and market prices of hedging instruments, than to corrupt the dynamics in the model in order to calibrate vanilla option prices, with the prospect of mispricing future carry P&Ls.

3.1.9   Forward volatility agreements

So far we have defined cliquets as payoffs involving ratios of a spot price observed at dates T_1 and T_2. As a consequence their price in the Black-Scholes model does not depend on S and, besides interest and repo rates, is a function of forward volatility only.

One may also be interested in payoffs involving *absolute*, rather than *relative* performances of an asset – typically for assets that move in narrow ranges, such as FX exchange rates.

Consider for example the payoff $(S_{T_2} - kS_{T_1})^+$: this is known in FX as a *forward volatility agreement* (FVA).[9] More generally, consider payoffs of type $S_{T_1} g\left(\frac{S_{T_2}}{S_{T_1}}\right)$. In a Black-Scholes framework with deterministic time-dependent volatility the value at T_1 of this option is $S_{T_1} G(\widehat{\sigma}_{T_1 T_2})$ where $\widehat{\sigma}_{T_1 T_2}$ is defined in (3.2). G also involves interest and repo rates over $[T_1, T_2]$. The value of our forward-start option at time $t < T_1$ is:

$$P(t, S) = Se^{-q(T_1 - t)}G(\widehat{\sigma}_{T_1 T_2}) \tag{3.18}$$

Just as in Section (3.1.1) we need to answer the following questions:

- Which vanilla payoffs should be used as hedges and whose implied volatilities should be used for defining $\widehat{\sigma}_{T_1 T_2}$?

- Which additional P&Ls do we need to estimate?

---

[9]This is one type of FVA – other types of FVAs include the regular forward-start vanilla option on the relative performance of the spot.

3.1.9.1 A vanilla portfolio whose vega is linear in S

In comparison with (3.1), (3.18) involves S as a prefactor. While in Section 3.1.2 we looked for European hedges of maturities T_1, T_2 whose vegas do not depend on S, we now look for payoffs whose vegas are proportional to S. From expression (3.4) for the vega of a portfolio of vanilla options in the Black-Scholes model, $\text{Vega}_\Pi (S)$ is proportional to S only if $\rho(K)$ has the form:

$$\rho(K) \propto \frac{1}{K}$$

Integrating twice with respect to K yields the payoff:

$$f(S) = S \ln S$$

which we call the $S \ln S$ contract. In the Black-Scholes model the price $R^T(t, S)$ of the $S \ln S$ contract of maturity T is:

$$R^T(t, S) = Se^{-q(T-t)} \left(\ln S + (r - q)(T - t) + \frac{(T - t)\hat{\sigma}_T^2}{2} \right) \qquad (3.19)$$

and its sensitivity to $\hat{\sigma}_T$ is given by:

$$\frac{dR^T}{d\hat{\sigma}_T} = Se^{-q(T-t)} (T - t) \hat{\sigma}_T \qquad (3.20)$$

The sensitivities of P in (3.18) to $\hat{\sigma}_{T_1}, \hat{\sigma}_{T_2}$ are given by:

$$\frac{dP}{d\hat{\sigma}_{T_2}} = Se^{-q(T_1-t)} (T_2 - t) \hat{\sigma}_{T_2} \mathcal{N} \qquad (3.21)$$

$$\frac{dP}{d\hat{\sigma}_{T_1}} = -Se^{-q(T_1-t)} (T_1 - t) \hat{\sigma}_{T_1} \mathcal{N} \qquad (3.22)$$

where $\mathcal{N}$ is given by:

$$\mathcal{N} = \frac{1}{(T_2 - T_1) \hat{\sigma}_{T_1 T_2}} \frac{dG}{d\hat{\sigma}_{T_1 T_2}}$$

Observe how the vega of the $S \ln S$ contract in (3.20) exactly matches the $\hat{\sigma}_{T_1}$ and $\hat{\sigma}_{T_2}$ vegas of P, both in its dependence on S and the repo rate. The portfolio

$$\Pi = -P + \mathcal{N} \left(e^{q(T_2-T_1)} R^{T_2} - R^{T_1} \right) \qquad (3.23)$$

has vanishing sensitivity to $\hat{\sigma}_{T_1}$ and $\hat{\sigma}_{T_2}$. As is apparent in (3.23) the hedge ratios for $S \ln S$ payoffs of maturities T_1 and T_2 do not depend on S or t: our hedge remains stable if S moves or time advances. Only when $\hat{\sigma}_{T_1 T_2}$ moves will it need to be readjusted.

Taking the second derivatives of (3.18) with respect to S shows that the gamma of our forward-start option vanishes. What about the gamma of our hedge? From expression (3.19) of R^T we can see that the coefficient of $S \ln S$ in the portfolio $e^{q(T_2-T_1)} R^{T_2} - R^{T_1}$ vanishes: our hedge has vanishing gamma and vanishing theta as well.

Is $\widehat{\sigma}_{T_1 T_2}$ well-defined?

The answer is yes. Implied volatilities of $S \ln S$ payoffs of maturities T_1 and T_2 satisfy the convex order condition (3.3) because of the convexity of the $S \ln S$ payoff. The $S \ln S$ payoff falls in the class of payoffs considered in Section 2.2.2.2, page 32, for which (a) an implied volatility can be defined, (b) the convex order condition (3.3) holds.

When there are no cash-amount dividends, implied volatilities $\widehat{\sigma}_T$ of $S \ln S$ payoffs are expressed directly as an average of implied volatilities of vanilla options – see formula (4.22), page 143.

3.1.9.2   Additional P&Ls and conclusion

Consider portfolio Π and let us write the P&L during δt, at order two in $\delta S, \widehat{\sigma}_{T_1}, \widehat{\sigma}_{T_2}$. In contrast with the forward-start option in Section 3.1.5, S appears explicitly in P and R^{T_1}, R^{T_2}. In addition to the forward-start option and the $S \ln S$ contracts of maturities T_1, T_2, our hedge portfolio also includes a delta position.

Our P&L comprises:

- no order-one contributions from $\delta S, \delta \widehat{\sigma}_{T_1}, \delta \widehat{\sigma}_{T_2}$ as, by construction, the sensitivities of Π to $\widehat{\sigma}_{T_1}, \widehat{\sigma}_{T_2}$ vanish and Π is delta-hedged.

- no δt and δS^2 terms as the portfolio's theta and gamma vanish.

- no order-two $\delta S \delta \widehat{\sigma}_{T_1}$ and $\delta S \delta \widehat{\sigma}_{T_2}$ contributions. Indeed, the sensitivities of the forward-start option's price P in (3.21), (3.22) and of the $S \ln S$ contract's price R^T in (3.20) to $\widehat{\sigma}_{T_1}, \widehat{\sigma}_{T_2}$ are proportional to S. If $\frac{d\Pi}{d\widehat{\sigma}_{T_1}}, \frac{d\Pi}{d\widehat{\sigma}_{T_2}}$ vanish for a particular value of S, they do so for all values of S: $\frac{d^2\Pi}{dS d\widehat{\sigma}_{T_1}} = \frac{d^2\Pi}{dS d\widehat{\sigma}_{T_2}} = 0$.

At order two our P&L thus only comprises terms in $\delta \widehat{\sigma}_{T_1}^2, \delta \widehat{\sigma}_{T_2}^2, \delta \widehat{\sigma}_{T_1} \delta \widehat{\sigma}_{T_2}$. Moreover, from (3.19), it is apparent that $e^{q(T_2-T_1)} R^{T_2} - R^{T_1}$ is a function of $\widehat{\sigma}_{T_1 T_2}$, rather than a separate function of $\widehat{\sigma}_{T_1}, \widehat{\sigma}_{T_2}$. Note that it is simply an affine function of $\widehat{\sigma}_{T_1 T_2}^2$. Using this variable, rather than $\widehat{\sigma}_{T_1 T_2}$, we can write down our P&L at order two in $\delta(\widehat{\sigma}_{T_1 T_2}^2)$ directly – note the similarity with (3.12).

$$\text{P\&L}_\Pi = -\frac{Se^{-q(T_1-t)}}{2} \frac{d^2G}{d(\widehat{\sigma}_{T_1 T_2}^2)^2} \left(\delta(\widehat{\sigma}_{T_1 T_2}^2) \right)^2 \tag{3.24}$$

Conclusion

Again we have been able to find a portfolio of European payoffs that provides a suitable vega, gamma and theta-hedge for our forward-start option. The implied volatilities we use are those of $S \ln S$ contracts.

The price we quote consists of the three pieces in (3.17). δP_1 is an estimation of P&Ls (3.24) over $[0, T_1]$. Because S appears as a prefactor, the estimation of δP_1 may also involve an assumption about the correlation of S and the realized volatility of the forward variance $\widehat{\sigma}^2_{T_1 T_2}$. δP_2 represents an estimation of the difference between the market price at T_1 of a vanilla option of strike $k S_{T_1}$, maturity T_2, and its price computed with the implied volatility at T_1 of the $S \ln S$ contract of maturity T_2.

A hasty assessment of our forward-start option starting with expression (3.18) in the Black-Scholes model would have singled out the cross spot/forward volatility covariance as one of the main risks in our product. Our analysis demonstrates, however, that spot/volatility covariance risk is not relevant; it is offset by utilizing the right hedge instruments – $S \ln S$ payoffs.

3.2 Forward-start options in the local volatility model

We now assess how the local volatility model prices forward-smile risk.

Once the local volatility model is calibrated on a given smile, log contract volatilities $\widehat{\sigma}_T$ will be exactly calibrated, as the log contract is a European payoff. However, calibration to the market smile also determines the dynamics of $\widehat{\sigma}_T$ – and especially that of $\widehat{\sigma}_{T_1 T_2}$. The local volatility model will price volatility-of-volatility and forward-smile risks – that is adjustments $\delta P_1, \delta P_2$ – according to its own dynamics, which we now characterize assuming that the local volatility function is weakly local.

3.2.1 Approximation for $\widehat{\sigma}_T$

Let us approximate the log contract implied volatility $\widehat{\sigma}_T$ by starting from expression (2.35). The reader can check that the derivation leading to equation (2.35) applies generally to any European option for which an implied volatility can be defined, i.e. applies to any payoff that is convex or concave, in particular the log contract.

Calculation is simpler now than it was in Section 2.4.2 for vanilla options as the dollar gamma of the log contract does not depend on S. The numerator in (2.35) reads:

$$\int_0^T dt \int_0^\infty dS \, \rho_{\sigma_0}(t, S) \, e^{-rt} \, \delta u\,(t, S) \, S^2 \frac{d^2 P_{\sigma_0}}{dS^2}$$

$$= e^{-rT} \int_0^T dt \int_0^\infty dS \, \rho_{\sigma_0}(t, S) \, \delta u\,(t, S)$$

where we have used (3.10). Using now expression (2.38) for $\rho_{\sigma_0}(t, S)$ we get for the numerator:

$$e^{-rT} \int_0^T dt \int_{-\infty}^{\infty} dy \, \frac{1}{\sqrt{2\pi}} \, e^{-\frac{y^2}{2}} \delta u \left(t, \, F_t e^{-\frac{\sigma_0^2 t}{2} + \sigma_0 \sqrt{t} y} \right)$$

where F_t is the forward for maturity t: $F_t = S_0 e^{(r-q)t}$. Dividing now by the denominator in (2.35), which is simply equal to Te^{-rT}, yields:

$$\hat{\sigma}_T^2 = \frac{1}{T} \int_0^T dt \int_{-\infty}^{\infty} dy \, \frac{1}{\sqrt{2\pi}} \, e^{-\frac{y^2}{2}} u \left(t, \, F_t e^{-\frac{\sigma_0^2 t}{2} + \sigma_0 \sqrt{t} y} \right) \qquad (3.25)$$

(3.25) should be compared to formula (2.40), page 44, previously derived for vanilla options. While (3.25) could be used directly, for consistency with Section 2.5 we derive an equation for volatilities rather than variances. Let us assume that $\sigma(t, S) = \sigma_0 + \delta\sigma(t, S)$. At order one in $\delta\sigma$, $\delta u = 2\sigma_0 \delta\sigma$. (3.25) together with $\hat{\sigma}_T^2 = \sigma_0^2 + 2\sigma_0 \delta\hat{\sigma}_T$ yields:

$$\delta\hat{\sigma}_T = \frac{1}{T} \int_0^T dt \int_{-\infty}^{\infty} dy \, \frac{1}{\sqrt{2\pi}} \, e^{-\frac{y^2}{2}} \delta\sigma \left(t, \, F_t e^{-\frac{\sigma_0^2 t}{2} + \sigma_0 \sqrt{t} y} \right)$$

Adding σ_0 to both sides, we get:

$$\hat{\sigma}_T = \frac{1}{T} \int_0^T dt \int_{-\infty}^{\infty} dy \, \frac{1}{\sqrt{2\pi}} \, e^{-\frac{y^2}{2}} \sigma \left(t, \, F_t e^{-\frac{\sigma_0^2 t}{2} + \sigma_0 \sqrt{t} y} \right) \qquad (3.26)$$

Assume that σ is given by the smooth form of the same type as in (2.44):

$$\sigma(t, S) = \bar{\sigma}(t) + \alpha(t) x + \frac{\beta(t)}{2} x^2 \qquad (3.27)$$

where $x = \ln(S/F_t)$. We get:

$$\hat{\sigma}_T = \frac{1}{T} \int_0^T \bar{\sigma}(t) \, dt - \frac{1}{T} \int_0^T \alpha(t) \frac{\sigma_0^2 t}{2} dt + \frac{1}{T} \int_0^T \frac{\beta(t)}{2} \left(\sigma_0^2 t + \frac{\sigma_0^4 t^2}{4} \right) dt$$

$$= \frac{1}{T} \int_0^T \bar{\sigma}(t) \, dt - \frac{\sigma_0^2 T}{2} \mathcal{S}_T + \frac{1}{T} \int_0^T \frac{\beta(t)}{2} \left(\sigma_0^2 t + \frac{\sigma_0^4 t^2}{4} \right) dt$$

where we have used formula (2.48) for the ATMF skew $\mathcal{S}_T$.

We also need an expression for how $\hat{\sigma}_T$ moves when S_0 moves; for this we go back to the general expression of $\hat{\sigma}_T$ in (3.26). Let us use the notation $S(t, y) = F_t \exp\left(-\frac{\sigma_0^2 t}{2} + \sigma_0 \sqrt{t} y \right)$ and take the derivative of (3.26) with respect to $\ln S_0$.

$$\frac{d\widehat{\sigma}_T}{d\ln S_0} = \frac{1}{T}\int_0^T dt \int_{-\infty}^{\infty} dy \frac{1}{\sqrt{2\pi}} e^{-\frac{y^2}{2}} \frac{d\sigma\left(t,S\left(t,y\right)\right)}{d\ln S_0}$$

$$= \frac{1}{T}\int_0^T dt \int_{-\infty}^{\infty} dy \frac{1}{\sqrt{2\pi}} e^{-\frac{y^2}{2}} \frac{d\sigma}{d\ln S}\left(t,S\left(t,y\right)\right) \frac{d\ln S\left(t,y\right)}{d\ln S_0}$$

$$= \frac{1}{T}\int_0^T dt \int_{-\infty}^{\infty} dy \frac{1}{\sqrt{2\pi}} e^{-\frac{y^2}{2}} \frac{d\sigma}{d\ln S}\left(t,S\left(t,y\right)\right)$$

Substituting now in this equation the smooth form (3.27) for $\sigma\left(t,S\right)$ yields:

$$\frac{d\widehat{\sigma}_T}{d\ln S_0} = \frac{1}{T}\int_0^T dt \int_{-\infty}^{\infty} dy \frac{1}{\sqrt{2\pi}} e^{-\frac{y^2}{2}} \left(\alpha\left(t\right) + \beta(t)\left(-\frac{\sigma_0^2 t}{2} + \sigma_0\sqrt{t}y\right)\right)$$

$$= \frac{1}{T}\int_0^T \alpha\left(t\right) dt - \frac{\sigma_0^2 T}{2}\frac{1}{T}\int_0^T \frac{t}{T}\beta(t)dt$$

We now drop the contribution from the curvature term β as we will not use it. Setting $\beta\left(t\right) = 0$ in the expressions for $\widehat{\sigma}_T$ and $\frac{d\widehat{\sigma}_T}{d\ln S_0}$ above yields the following simpler expressions:

$$\widehat{\sigma}_T = \frac{1}{T}\int_0^T \overline{\sigma}\left(t\right) dt - \frac{\sigma_0^2 T}{2}S_T \tag{3.28}$$

$$\frac{d\widehat{\sigma}_T}{d\ln S_0} = \frac{1}{T}\int_0^T \alpha\left(t\right) dt \tag{3.29}$$

Taking $\beta\left(t\right) = 0$ in expression (2.47) for implied volatilities of vanilla options yields the following simple result for the ATMF implied volatility: $\widehat{\sigma}_{F_T T} = \frac{1}{T}\int_0^T \overline{\sigma}\left(t\right) dt$. Let us choose σ_0, the constant volatility level around which the order-one expansion in (2.35) is performed, as $\sigma_0 = \widehat{\sigma}_{F_T T}$. Equation (3.28) now reads:

$$\widehat{\sigma}_T = \widehat{\sigma}_{F_T T} - \frac{\widehat{\sigma}_{F_T T}^2 T}{2}S_T \tag{3.30}$$

The integral in the right-hand side in (3.29) also appears in (2.59c). (3.29) can be rewritten, with no explicit reference to local volatility σ anymore, as:

$$\frac{d\widehat{\sigma}_T}{d\ln S_0} = \frac{d\widehat{\sigma}_{F_T T}}{d\ln S_0} \tag{3.31}$$

While we have derived equation (3.30) using local volatilities, it is a relationship involving implied volatilities only: it relates the implied volatility of the log contract, a European payoff, to the smile of vanilla options for the same maturity at order one in the perturbation around a flat local volatility function.[10] We could have derived it in a model-independent fashion, by assuming that the smile for maturity T is given by:

$$\hat{\sigma}_{KT} = \hat{\sigma}_{F_T T} + \mathcal{S}_T \ln\left(\frac{K}{F_T}\right)$$

At order one in $\mathcal{S}_T$ we would have recovered (3.30).

Equation (3.30) implies that for downward sloping smiles ($\mathcal{S}_T < 0$), which are typical for equities, $\hat{\sigma}_T > \hat{\sigma}_{F_T T}$. Equation (3.31) states that the rate at which the log contract implied volatility moves as S moves matches that of the ATMF volatility for the same maturity. Equations (3.30) and (3.31) have been obtained for a weakly local volatility of the form

$$\sigma(t, S) = \overline{\sigma}(t) + \alpha(t) \ln\left(\frac{S}{F_t}\right) \tag{3.32}$$

How accurate are they for real smiles?

Figure 3.1 shows $\hat{\sigma}_T$, $\hat{\sigma}_{F_T T}$, as well as $\hat{\sigma}_T$ computed with formula (3.30) for two Euro Stoxx 50 smiles – with zero rates and repos. Figure 3.2 shows the ratio $\frac{d\hat{\sigma}_T}{d\ln S_0}/\frac{d\hat{\sigma}_{F_T T}}{d\ln S_0}$ for the same smiles.

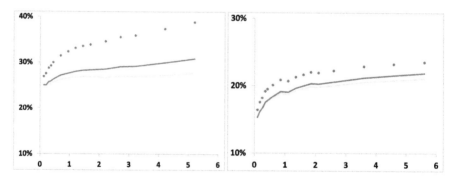

Figure 3.1: Implied volatilities as a function of maturity – in years – for the Euro Stoxx 50 smiles of October 4, 2010 (left) and May 16, 2013 (right); dots: $\hat{\sigma}_T$, light line: $\hat{\sigma}_{F_T T}$, dark line: approximation (3.30) for $\hat{\sigma}_T$.

The 1-year smiles of the Euro Stoxx 50 index appear in Figure 2.3, page 58.

It is apparent that formula (3.30) is inaccurate. This is not surprising: the local volatility of the Euro Stoxx 50 market smile is not of type (3.32). In particular the difference between $\hat{\sigma}_T$ and $\hat{\sigma}_{F_T T}$ is not entirely determined by the ATMF skew $\mathcal{S}_T$, a very local feature of the smile.

Then note that an expansion of implied volatilities at order one in the local volatility is expected to be less accurate for log contracts than for calls and puts. We

---

[10]Equivalently, at order one in the perturbation around a flat *implied* volatility surface.

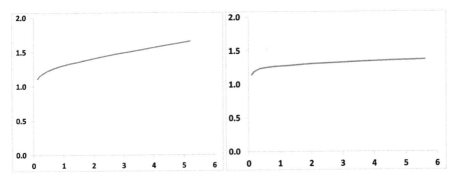

Figure 3.2: $\frac{d\hat{\sigma}_T}{d\ln S_0}\Big/\frac{d\hat{\sigma}_{F_T T}}{d\ln S_0}$ ratio in the local volatility model, as a function of maturity – in years – for the Euro Stoxx 50 index smiles of October 4, 2010 (left) and May 16, 2013 (right).

already mentioned that, because of the steepness of market smiles, unless one takes into account – at least at order one – the correction to the density in equation (2.32), the approximation is inaccurate.

This issue is magnified in the case of the log contract: for vanilla options, the dollar gamma restricts integration in (2.32) to a region of spot prices around K, thus reducing the dependence of implied volatilities of vanilla options to the density for spot values far away from K. In contrast, the dollar gamma of the log contract is constant: its implied volatility depends on the density for values of S far away from the initial spot price. Typically densities implied from market smiles are skewed: for low values of S, for which $\sigma\left(t,S\right)$ is large, the density is larger than the lognormal density used in the approximation.

Formula (3.30) for $\hat{\sigma}_T$ will then only be acceptable for smooth smiles that are not too steep.

Figure 3.2 shows that the ratio $\frac{d\hat{\sigma}_T}{d\ln S_0}\Big/\frac{d\hat{\sigma}_{F_T T}}{d\ln S_0}$ is, in our example, larger than its value of 1 predicted by approximation (3.32), especially for the October 4, 2010 smile, which is particularly steep, presumably because approximation (3.30) underestimates $\hat{\sigma}_T$ in the first place. While numerically inaccurate, approximation (3.31) does give however the right order of magnitude for $\frac{d\hat{\sigma}_T}{d\ln S_0}$.

We now use it to estimate the volatility of $\hat{\sigma}_{T_1 T_2}$, assuming for simplicity that the term structure of $\hat{\sigma}_T$ is flat: $\hat{\sigma}_{T_1} = \hat{\sigma}_{T_2}$. From the definition (3.2) and using approximation (3.29) we have:

$$
\begin{aligned}
d\hat{\sigma}_{T_1 T_2} &= \frac{T_2 - t}{T_2 - T_1}\frac{\hat{\sigma}_{T_2}}{\hat{\sigma}_{T_1 T_2}}d\hat{\sigma}_{T_2} - \frac{T_1 - t}{T_2 - T_1}\frac{\hat{\sigma}_{T_1}}{\hat{\sigma}_{T_1 T_2}}d\hat{\sigma}_{T_1} \\
&\simeq \left(\frac{1}{T_2 - T_1}\int_t^{T_2}\alpha\left(u\right)du - \frac{1}{T_2 - T_1}\int_t^{T_1}\alpha\left(u\right)du\right)d\ln S_t \\
&\simeq \left(\frac{1}{T_2 - T_1}\int_{T_1}^{T_2}\alpha\left(u\right)du\right)d\ln S_t \qquad (3.33)
\end{aligned}
$$

The volatility of $\widehat{\sigma}_{T_1 T_2}$ over $[0, T_1]$ is then entirely set by the skew of the local volatility function for $t \in [T_1, T_2]$, and presumably will bear no resemblance to historical or implied levels of volatility of volatility. In particular, (3.33) shows that if the local volatility is flat over the interval $[T_1, T_2]$, $\widehat{\sigma}_{T_1 T_2}$ is frozen.

Moreover, for $T_2 - T_1$ fixed, the volatility of $\widehat{\sigma}_{T_1 T_2}$ is not a function of $T_1 - t$: the local volatility model is not time-homogeneous, in contrast to the stochastic volatility models of Chapter 7.

3.2.2   Future skews in the local volatility model

We now turn to the forward skew in the local volatility model. The smile prevailing at T_1 for maturity T_2 is fully determined by $\sigma(t, S)$ for $t \in [T_1, T_2]$. The local volatility function in (3.32) falls in the class studied in Section 2.5, with $\beta(t) \equiv 0$.

Let us use approximation (3.30) for the difference between the log contract and ATMF implied volatilities. We have:

$$\widehat{\sigma}_{T_2}(S_{T_1}, T_1) - \widehat{\sigma}_{F_{T_2}(S_{T_1})T_2}(S_{T_1}, T_1) = \frac{(T_2 - T_1)\,\widehat{\sigma}^2_{F_{T_2}(S_{T_1})T_2}}{2}\, \mathcal{S}_{T_2 - T_1}(S_{T_1}, T_1)$$

where $\mathcal{S}_{T_2 - T_1}(S_{T_1}, T_1)$ is the ATMF skew for the residual maturity $T_2 - T_1$, at time T_1, as a function of S_{T_1}.

Let us denote the residual maturity by: $\theta = T_2 - T_1$. $\mathcal{S}_\theta(S_{T_1}, T_1)$ is given by expression (2.90), page 63:

$$\mathcal{S}_\theta(S_{T_1}, T_1) = \int_0^1 \alpha(T_1 + u\theta)u\,du \tag{3.34}$$

which can be contrasted with the spot-starting ATMF skew for the same residual maturity:

$$\mathcal{S}_\theta(S_0, 0) = \int_0^1 \alpha(u\theta)u\,du$$

For typical equity smiles, $\alpha(t)$ is of the form (2.51) and decays with an exponent $\gamma \simeq \frac{1}{2}$. For short residual maturities ($\theta \ll T_1$), result (2.92), page 64, implies that:

$$\mathcal{S}_\theta(S_{T_1}, T_1) \propto \frac{1}{T_1^\gamma}$$

This formula shows that the ATMF skew generated by the local volatility model at time T_1 for residual maturity θ decays like $1/T_1^\gamma$, thus will be much lower than the ATMF skew observed at $t = 0$ for the same residual maturity:

$$\mathcal{S}_\theta(S_{T_1}, T_1) \ll \mathcal{S}_\theta(S_0, 0)$$

Imagine that our cliquet is an at-the-money forward call whose payoff is (3.13). At T_1 we are exposed to the difference between the implied volatilities of the log contract and the at-the-money call for maturity T_2. If T_1 is far into the future, the local volatility model generates a smile at T_1 for maturity T_2 which is very weak compared to the smile we are likely to witness on the market.

The local volatility model will underestimate the difference between the log contract and at-the-money implied volatilities, and more generally the differences between implied volatilities of vanilla options with different strikes: it misprices forward-smile risk.

Remember that in the local volatility model, instantaneous volatilities of volatilities are also determined by the ATM skew. From equation (2.83), page 57,

$$\text{vol}(\widehat{\sigma}_{F_T T}) = \left(S_T + \frac{1}{T} \int_0^T S_\tau d\tau \right) \frac{\widehat{\sigma}_{F_0 0}}{\widehat{\sigma}_{F_T T}}$$

Thus future volatilities of volatilities are smaller than their spot-starting values: the local volatility model also misprices volatility-of-volatility risk.

3.2.3 Conclusion

The conclusion is that the local volatility model is not suitable for pricing forward-start options, or more generally options that involve volatility-of-volatility and forward-smile risks.

On one hand, volatilities of forward volatilities generated by the model will depend exclusively on the steepness of the skews prevailing at the time of calibration and may lie arbitrarily above or below historical or implied volatilities of volatilities – thus δP_1 in (3.17), page 111, is mispriced.

On the other hand, forward skews generated in the model will invariably be too low with respect to both market implied forward skews and historical skews – thus δP_2 in (3.17) is mispriced as well.

To further understand how the local volatility model uses the information in vanilla smiles to price a cliquet, we now consider a forward-start call and apply the technique of Section 2.9 to explicitly derive the vega hedge – as generated by the local volatility model.

3.2.4 Vega hedge of a forward-start call in the local volatility model

Consider the following payoff:

$$\left(\frac{S_{T_2}}{S_{T_1}} - k \right)^+ = \frac{1}{S_{T_1}} (S_{T_2} - k S_{T_1})^+ \tag{3.35}$$

To compute the vega hedge in the local volatility model, all we need is the conditional gamma notional, defined in (2.118), page 82:

$$\phi(t, S) = E_\sigma \left[S^2 \frac{d^2 P}{dS^2} (t, S, \bullet) | S, t \right]$$

We consider for simplicity the case of flat local volatility function. Calculations can then be carried out analytically, besides, we do not expect the structure of the hedge portfolio to depend much on this assumption.

We also use vanishing interest and repo rates.

Let us call σ_0 the constant level of our flat local volatility function. Because of homogeneity, for $t < T_1$ the Black-Scholes price does not depend on S, hence $\phi = 0$. At $t = T_1^+$, our option becomes $\frac{1}{S_{T_1}}$ times a standard call of maturity T_2 whose strike is kS_{T_1}.

Hedge portfolio for maturity T_1

At $t = T_1^+$, $S_{T_1} = S$ and the dollar gamma is given by:

$$\phi\left(T_1^+, S\right) = \frac{1}{S}S^2 \left.\frac{d^2 P_{BS}\left(T_1 S; K T_2; \sigma_0\right)}{dS^2}\right|_{K=kS}$$

where the $\frac{1}{S}$ prefactor is the $\frac{1}{S_{T_1}}$ in (3.35). We now use the relationship connecting the vega and dollar gamma of a European option in the Black-Scholes model: $\frac{dP}{d\sigma_0} = S^2 \frac{d^2 P}{dS^2}\sigma_0 T$ to rewrite this as:

$$\phi\left(T_1^+, S\right) = \frac{1}{\sigma_0\left(T_2 - T_1\right)}\frac{1}{S}\left.\frac{dP_{BS}\left(T_1 S; K T_2; \sigma_0\right)}{d\sigma_0}\right|_{K=kS}$$

Because $P_{BS}\left(T_1 S; kST_2; \sigma_0\right)$ is homogeneous in S, $\phi\left(T_1^+, S\right)$ does not depend on S. ϕ thus has a discontinuity at $t = T_1$ that does not depend on S, which generates a discrete portfolio of vanilla options of maturity T_1. Applying operator $\mathcal{L}$ defined in (2.120) on ϕ – only $\frac{d}{dt}\phi$ contributes – we get the following expression for the (discrete) density of vanilla options of maturity T_1, struck at K:

$$\Psi_1\left(K\right) = -\frac{1}{K^2}\phi\left(T_1^+, S\right)$$

$$= -\frac{1}{K^2}\frac{1}{\sqrt{2\pi\sigma_0^2\left(T_2 - T_1\right)}}e^{-\frac{\left(-\ln k + \frac{\sigma_0^2(T_2-T_1)}{2}\right)^2}{2\sigma_0^2(T_2-T_1)}} \tag{3.36}$$

This is an interesting result: μ is proportional to $\frac{1}{K^2}$. Compare (3.36) with expression (3.9) that specifies the number of log contracts of maturity T_1 and T_2 needed to hedge a forward-start option, in the framework of Section 3.1.4. Once we recall that one log contract can be synthesized with a continuous density $\frac{2}{K^2}$ of vanilla options, we realize that result (3.36) is exactly identical: perturbation around a flat volatility yields a vega hedge for maturity T_1 which is exactly what we used in Section 3.1.4.

Notice that the vanilla portfolio struck at T_1 is static, in that it depends only on $T_2 - T_1$: it depends neither on T_1, nor on the initial spot level S_0.

Hedge portfolio for maturity T_2

Let us compute ϕ for $t \in]T_1, T_2[$. Using formula (2.39) for the dollar gamma of a call option in the Black-Scholes model, we get the dollar gamma for payoff (3.35):

$$S^2 \frac{d^2 P}{dS^2} = \frac{k}{\sqrt{2\pi\sigma_0^2 (T_2 - t)}} e^{-\frac{\left(\ln\left(\frac{S}{kS_{T_1}}\right) - \frac{\sigma_0^2(T_2-t)}{2}\right)^2}{2\sigma_0^2(T_2-t)}} \qquad (3.37)$$

which we need to average, conditional on the underlying's value being S at time t. The Brownian motion W_{T_1} can be written as a function of W_t and an independent Gaussian random variable Z:

$$W_{T_1} = \frac{T_1}{t} W_t + \sqrt{T_1\left(1 - \frac{T_1}{t}\right)} Z$$

which gives:

$$\ln \frac{S_{T_1}}{S_0} = \frac{T_1}{t} \ln \frac{S_t}{S_0} + \sigma_0 \sqrt{T_1\left(1 - \frac{T_1}{t}\right)} Z$$

Inserting this expression for S_{T_1} in (3.37) yields the following expression for ϕ:

$$\phi = \int_{-\infty}^{+\infty} \frac{e^{-\frac{z^2}{2}}}{\sqrt{2\pi}} \frac{k}{\sqrt{2\pi\sigma_0^2 (T_2 - t)}} e^{-\frac{\left(\left(1-\frac{T_1}{t}\right)\ln\frac{S}{S_0} - \ln k - \frac{\sigma_0^2(T_2-t)}{2} - \sigma_0\sqrt{T_1\left(1-\frac{T_1}{t}\right)}z\right)^2}{2\sigma_0^2(T_2-t)}} dZ$$

Computing the integral over Z yields:

$$\phi = \frac{k}{\sqrt{2\pi\sigma_0^2 \left(T_2 - t + T_1\left(1 - \frac{T_1}{t}\right)\right)}} e^{-\frac{\left(\left(1-\frac{T_1}{t}\right)\ln\frac{S}{S_0} - \ln k - \frac{\sigma_0^2(T_2-t)}{2}\right)^2}{2\sigma_0^2\left(T_2 - t + T_1\left(1-\frac{T_1}{t}\right)\right)}}$$

Let us compute ϕ for $t = T_2^-$:

$$\phi\left(t = T_2^-, S\right) = \frac{k}{\sqrt{2\pi\sigma_0^2 T_1 \left(1 - \frac{T_1}{T_2}\right)}} e^{-\frac{\left(\left(1-\frac{T_1}{T_2}\right)\ln\frac{S}{S_0} - \ln k\right)^2}{2\sigma_0^2 T_1\left(1-\frac{T_1}{T_2}\right)}}$$

For $t > T_2$ $\phi = 0$: the discontinuity of ϕ in T_2 generates a discrete quantity of vanilla options struck at T_2, whose density $\Psi_2(K)$ is $\frac{1}{K^2}\phi\left(T_2^-, K\right)$:

$$\Psi_2(K) = \frac{1}{K^2} \frac{k}{\sqrt{2\pi\sigma_0^2 T_1 \left(1 - \frac{T_1}{T_2}\right)}} e^{-\frac{\left(\left(1-\frac{T_1}{T_2}\right)\ln\frac{K}{S_0} - \ln k\right)^2}{2\sigma_0^2 T_1\left(1-\frac{T_1}{T_2}\right)}} \qquad (3.38)$$

First note that, unlike the T_1 portfolio, the T_2 hedge portfolio is not static: T_1, T_2 and S appear explicitly. As time advances, T_1 and T_2 shrink, and S moves. The number of vanilla options of maturity T_2 struck at K, as given by (3.38) will need to be readjusted. This does not correspond to the T_2 hedge we assembled in Section 3.1.4 which, with zero interest rates, consists of a number of log contracts for maturity T_2 that is exactly the opposite of that of maturity T_1 – see equation (3.9).

Intermediate maturities

Beside the discrete portfolios of vanilla options for maturities T_1 and T_2 that are generated by the discontinuity of ϕ at T_1 and T_2, application of the operator $\mathcal{L}$ defined in (2.120) on ϕ generates a continuous density of options for intermediate maturities that can be easily computed numerically.

Our hedge portfolio in Section 3.1.4, on the contrary, only consists of options of maturity T_1 and T_2.

3.2.5   Discussion and conclusion

The density of vanilla options generated by the log contract hedge in Section 3.1.4 was $|\Psi_1(K)|$ for T_2 and $-|\Psi_1(K)|$ for T_1.

Rather than work directly with the (discrete) density of options struck at T_2 and the (continuous) density for $T \in [T_1, T_2]$, let us normalize the former by $|\Psi_1(K)|$ and the latter by $|\Psi_1(K)|/(T_2-T_1)$, to highlight the deviation with respect to the log contract hedge. As (3.36) shows, $|\Psi_1(K)|$ is proportional to $1/K^2$.

Consider the following at-the-money forward call: $T_1 = 1$ year, $T_2 = 2$ years, $\sigma_0 = 20\%$, $k = 100\%$, and pick $T = 1.5$ years. Figure 3.3 shows the following quantities:

$$\frac{\Psi_2(K)}{|\Psi_1(K)|} \quad , \quad \frac{(T_2-T_1)\mu(T,K)}{|\Psi_1(K)|} \tag{3.39}$$

where we multiply μ by (T_2-T_1) since μ, unlike Ψ_1 and Ψ_2, is a continuous density.

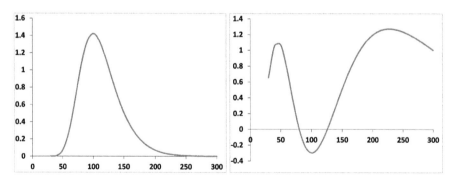

Figure 3.3: Densities of vanilla options struck at T_2 (left) and at an intermediate maturity T, normalized as in (3.39) (right).

Notice how the T_2-portfolio deviates from the log contract hedge we used in Section 3.1.4. The local volatility model suggests that we should trade more options struck near the current value of the spot price (100 in our example) and fewer out-of-the-money options. It makes up for this by asking us to trade options for intermediate maturities: a short position for strikes near the current value of the spot price and a long position for far out-of-the-money strikes. Inspection of Figure 3.3 shows that the vanilla hedge for intermediate maturities approximately makes up for the difference between the T_2-portfolio and the log contract hedge, which – because of the normalization we use – would correspond in the left-hand graph to a constant value equal to 1.

While the T_1-portfolio is static, portfolios for T_2 and intermediate maturities in $]T_1, T_2[$ depend explicitly on T_1, T_2, S_0: as the spot moves and time advances, they will need to be readjusted. Using this vega hedge in practice would expose us to variations of S and market implied volatilities, thus generating extra P&Ls – of order two in δS and $\delta \widehat{\sigma}_{KT}$ – which we cannot expect the local volatility model to have priced in properly.

Carrying out the same analysis for a forward call option struck at $k \neq 100\%$, would have given similar curves, now centered on $K^* = S_0 k^{\frac{T_2}{T_2-T_1}}$ – as is clear from (3.38).

The conclusion is that it is much more reasonable to use the log contract hedge developed in Section 3.1.4: only when the forward volatility $\widehat{\sigma}_{12}$ moves does the hedge need to be readjusted and the volatility-of-volatility and forward-smile risks can be cleanly isolated and priced.

What about the *price* in the local volatility model? Consider again Figure 3.3 which expresses the dependence of the price of the forward call option to implied volatilities of vanilla options – as seen by the local volatility model. It is difficult to imagine a plausible justification for such peculiar sensitivities, especially as they change when S and t move. They may be more a statement on the local volatility model itself than on the forward call option: besides the *hedge*, the *price* generated by the local volatility model is suspicious as well.

Quite generally, whenever the vega hedge suggested by the local volatility model is *not* static, both the usefulness of the hedge *and* the reliability of the model's price are questionable. We are exposed to the cost of future vega rehedging at then-prevailing market conditions without the ability to gauge the size and sign of these future P&Ls, and cannot trust the local volatility model to have priced them correctly.

Chapter's digest

3.1 Pricing and hedging forward-start options

▶ Cliquets whose payoffs are of the form $g\left(\frac{S_{T_2}}{S_{T_1}}\right)$ are hedged by trading dynamically log-contracts. Implied volatilities of log contracts obey the convex order condition, hence forward volatilities are well-defined, and the sensitivities of log-contracts to forward volatilities are spot-independent. The carry P&L of a vega-hedged cliquet consists of (a) gamma P&L on forward volatility and (b) an adjustment for forward-smile risk, that is uncertainty about the value at T_1 of the difference between the implied volatility of the log-contract and that of the payoff at hand.

▶ Calibration on the vanilla smile has little relevance for the pricing of cliquets. Calibrating a model on the vanilla smile in the hope that information contained therein should somehow be reflected in the cliquet price is unreasonable. Indeed, this may result in a price that is overly dependent on the arbitrariness of the connection that a given model establishes between today's smile and its future dynamics. The local volatility model is a case in point.

▶ That vanilla option smiles do not constrain much cliquet prices is confirmed in Section 3.1.7. Model-independent lower and upper bounds are computed for the price of a forward-start call. The farther apart the lower and upper bounds are, the least vanilla-like the cliquet is.

▶ Forward-start options whose payoffs are of the form $S_{T_1} g\left(\frac{S_{T_2}}{S_{T_1}}\right)$ – which include the case of FVAs in the FX world – have a Vega in the Black-Scholes model that is linear in S. This calls for a new family of hedging instruments, European payoffs $S_T \ln S_T$ whose vegas are linear in S. The implied volatilities of these payoffs obey the convex order condition, thus allowing the definition of forward volatilities. As with payoffs $g\left(\frac{S_{T_2}}{S_{T_1}}\right)$, the carry P&L includes a gamma P&L on these particular forward volatilities.

🙚 🙚 🙚 🙚 🙚

3.2 Forward-start options in the local volatility model

▶ For a local volatility function of the form $\sigma(t, S) = \overline{\sigma}(t) + \alpha(t) \ln \frac{S}{F_t}$, the dynamics of a forward VS volatility is given at order one in α by (3.33):

$$d\widehat{\sigma}_{T_1 T_2} = \left(\frac{1}{T_2 - T_1} \int_{T_1}^{T_2} \alpha(u)\, du\right) d\ln S_t$$

The instantaneous volatility of $d\widehat{\sigma}_{T_1 T_2}$ is set by the slope of the local volatility function for times $t \in [T_1, T_2]$. The volatility of a very short volatility $(T_2 - T_1 \ll T_1)$

is proportional to $\alpha(T_1)$: the model is not time-homogeneous. For typical equity smiles $\alpha(u)$ is decreasing, thus instantaneous volatilities of VS volatilities at future dates will be systematically smaller than those of spot-starting VS volatilities with the same maturity and will also depend on future spot levels. The local volatility model misprices volatility-of-volatility risk.

▶ The skew at a future date T_1 for residual maturity θ is given by expression (3.34):

$$\mathcal{S}_\theta\left(S_{T_1}, T_1\right) = \int_0^1 \alpha(T_1 + u\theta)u\,du$$

For typical equity smiles $\alpha(t)$ is decreasing, thus $\mathcal{S}_\theta\left(S_{T_1}, T_1\right) \ll \mathcal{S}_\theta\left(S_0, t=0\right)$: future skews in the local volatility model are weaker than spot-starting skews. The local volatility model misprices forward smile risk.

▶ Deriving the vega-hedge in the local volatility model of a forward-start call option with payoff $(\frac{S_{T_2}}{S_{T_1}} - k)^+$ yields a hedge portfolio consisting of a discrete quantity of options of maturity T_1 – which exactly matches the log-contract hedge studied in Section 3.1, a discrete portfolio of vanilla options of maturity T_2 and a continuous density of options with intermediate maturities.

In the discussion of Section 3.2.5 we argue why this is not a reasonable hedge. Hoping to immunize ourselves from the local volatility model's idiosyncrasies by hedging against perturbations of the local volatility function is not a viable route.

Chapter 4

Stochastic volatility – introduction

Chapter 2 was devoted to local volatility, a very constrained form of stochastic volatility: implied volatilities have 100% correlation among themselves and with S, and their volatilities are determined by the market smile used for calibration of the model. Volatilities of implied volatilities – as generated by the model – will likely bear no resemblance to conservative levels based on historically observed volatilities of implied volatilities, or to implied volatilities of volatilities, whenever they are observable.

Can we design models that let us freely specify the dynamics of implied volatilities? We tackle this issue, first starting from a general point of view, then specializing to a modeling framework that can be used practically, based on forward variances of specific European payoffs.

4.1 Modeling vanilla option prices

It is natural to try and model the dynamics of the prices of vanilla options directly. Let C_{KT} be the price of a call option with strike K, maturity T, and let us assume without loss of generality zero interest rates and repos. Because an option is an asset that can be bought or sold, its pricing drift is r – which we have taken to be zero. The dynamics of C_{KT}, along with S can be written as:

$$\begin{cases} dS = \bar{\sigma} S dW^S \\ dC_{KT} = \Lambda_{KT}\, dW^{KT} \end{cases} \tag{4.1}$$

with the initial conditions:

$$\begin{cases} S_{t=0} = S_{t=0}^{\text{market}} \\ C_{KT}(t=0) = C_{KT}^{\text{market}}(t=0) \end{cases}$$

and the terminal condition:

$$C_{KT}(t=T) = (S_T - K)^+ \tag{4.2}$$

where W^S, W^{KT} are Brownian motions, Λ_{KT} is the volatility of C_{KT} and $\bar{\sigma}$ is the instantaneous volatility of S. We will omit the t subscripts in the processes whenever possible.

4.1.1   Modeling implied volatilities

Let us try to work with implied volatilities $\widehat{\sigma}_{KT}$. In the pricing equation, $\widehat{\sigma}_{KT}$ acquires a drift that represents the cost of financing a delta position on $\widehat{\sigma}_{KT}$. Taking a directional position on $\widehat{\sigma}_{KT}$ entails trading the corresponding delta-hedged vanilla option. Expressing that the pricing drift of C_{KT} is the interest rate – here zero – will then determine the pricing – or "risk-neutral" – drift of $\widehat{\sigma}_{KT}$. The joint SDEs of S and $\widehat{\sigma}_{KT}$ read:

$$\begin{cases} dS = \overline{\sigma} S dW^S \\ d\widehat{\sigma}_{KT} = \mu_{KT} dt + \lambda_{KT}\, dW^{KT} \end{cases} \tag{4.3}$$

where W^{KT} is a Brownian motion. $\widehat{\sigma}_{KT}$ is related to C_{KT} through: $C_{KT} = P_{BS}(t, S, \widehat{\sigma}_{KT}, T)$. The terminal condition (4.2) then becomes:

$$\lim_{t \to T} (T - t)\, \widehat{\sigma}^2_{KT,\, t} = 0 \tag{4.4}$$

and the initial condition is simply:

$$\widehat{\sigma}_{KT,\, t=0} = \widehat{\sigma}^{\text{market}}_{KT,\, t=0}$$

The SDE for C_{KT} reads:

$$dC_{KT} = \left(\frac{dP_{BS}}{dt} + \frac{\overline{\sigma}^2}{2} S^2 \frac{d^2 P_{BS}}{dS^2} + \frac{\lambda^2_{KT}}{2} \frac{d^2 P_{BS}}{d\widehat{\sigma}^2_{KT}} + \rho \lambda_{KT} \overline{\sigma} S \frac{d^2 P_{BS}}{dS d\widehat{\sigma}_{KT}} \right. $$
$$\left. + \frac{dP_{BS}}{d\widehat{\sigma}_{KT}} \mu_{KT} \right) dt + S \frac{dP_{BS}}{dS} \overline{\sigma} dW^S + \frac{dP_{BS}}{d\widehat{\sigma}_{KT}} \lambda_{KT}\, dW^{KT}$$

where ρ is the correlation between W^S and W^{KT}. Expressing that the drift of C^{KT} vanishes determines μ_{KT}. The derivatives of P_{BS} are all available analytically – the resulting expression for μ^{KT} is:

$$\mu_{KT} = \frac{1}{\widehat{\sigma}_{KT}} \left(\frac{\widehat{\sigma}_{KT} - \overline{\sigma}^2}{2(T-t)} - \frac{1}{2} d_1 d_2 \lambda^2_{KT} + \frac{d_2}{\sqrt{T-t}} \rho \overline{\sigma} \lambda_{KT} \right)$$

which we directly quote from [80], where d_1, d_2 are standard expressions appearing in Black-Scholes formulas:

$$d_1 = \frac{1}{\widehat{\sigma}_{KT}\sqrt{T-t}} \ln \frac{F_T(S_t)}{K} + \frac{\widehat{\sigma}_{KT}\sqrt{T-t}}{2}, \quad d_2 = d_1 - \widehat{\sigma}_{KT}\sqrt{T-t}$$

The joint dynamics of $(S, \widehat{\sigma}_{KT})$ in (4.3) would now be completely specified if the process $\overline{\sigma}$ was known. The processes for $\overline{\sigma}$ and $\widehat{\sigma}_{KT}$ cannot be chosen arbitrarily, for a solution to (4.3) to exist. For example it is possible to prove that, for short maturities, the at-the-money implied volatility should tend to the instantaneous volatility – see [43] for a proof:[1]

$$\lim_{T \to t} \widehat{\sigma}_{ST,t} = \overline{\sigma}_t \tag{4.5}$$

---

[1]Surprisingly, the proof of this simple and natural result is rather technical. Also note that the non-explosion condition imposed on μ_{KT} in [80] is not needed: as long as (4.4) holds, how fast or how slowly the left-hand side tends to zero does not matter.

One may think that once some technical conditions on processes $\overline{\sigma}$ and $\widehat{\sigma}_{KT}$ are satisfied, we should be able to explicitly construct – at least numerically – a solution to (4.3) that complies with (4.4). The trouble is that once we have a solution to (4.3) over $[0, T]$, the ensuing dynamics for S determines prices of European options for maturities up to T, hence the implied volatilities $\widehat{\sigma}_{k\tau}$ of all vanilla options with $\tau < T$.

This is problematic, as our intention in modeling the $\widehat{\sigma}_{KT}$ was to be able to choose their initial values freely. What this means is that information about the initial smile has to be embedded in the process for $\overline{\sigma}$.[2] Moreover, such embedding is presumably convoluted: an arbitrary configuration of the $\widehat{\sigma}_{KT}$ is consistent with a dynamics of S – hence a process $\overline{\sigma}$ may exist – only if the no-arbitrage conditions highlighted in Section 2.2.2 are satisfied. Whenever these conditions are violated, $\overline{\sigma}$ does not exist and this has to manifest itself in the structural impossibility of constructing a process for $\overline{\sigma}$.

The upshot is that direct modeling of implied volatilities of vanilla options is impractical.

4.2 Modeling the dynamics of the local volatility function

We try here a different line of approach: given a non-arbitrageable configuration of implied volatilities $\widehat{\sigma}_{KT}$ a local volatility function $\sigma(t, S)$ exists, given by the Dupire formula (2.3). There is a one-to-one mapping between the volatility surface and the local volatility function. We are not using a local volatility *model*; we are using the local volatility *function* to represent at any time t the full set of implied volatilities.

We can then generate a dynamics for implied volatilities by generating a dynamics for this local volatility function which we denote by σ_t: $\sigma_t(\tau, S)$ is the local volatility function associated with the volatility surface at time t. For a fixed couple (τ, S), $\sigma_t(\tau, S)$ is a *process* that exists for $t \leq \tau$.

No-arbitrage restrictions on $\widehat{\sigma}_{KT}$ translate into the simple condition that $\sigma_t^2(\tau, S)$ be positive for all τ, S. In what follows we work with variances σ^2. In the local volatility model, the local volatility function is fixed and the instantaneous volatility $\overline{\sigma}_t$ of S_t is given by:

$$\overline{\sigma}_t = \sigma_{t_0}(t, S_t)$$

where the t_0 subscript indicates that $\sigma_{t_0}(t, S_t)$ was obtained from implied volatilities observed at time t_0. In a model where the local volatility function itself is dynamic, $\overline{\sigma}$ is given by:

$$\overline{\sigma}_t = \sigma_t(t, S_t) \tag{4.6}$$

---

[2]This is precisely what occurs in the local volatility model, the simplest of all market models.

This follows from the definition of σ_t. From equation (2.6) the local volatility is given by:

$$\sigma_t^2(\tau, S) = \frac{E_t[\sigma_\tau^2 \delta(S_\tau - S)]}{E_t[\delta(S_\tau - S)]} = E_t[\sigma_\tau^2 \mid S_\tau = S]$$

Setting $S = S_t$ and taking the limit $\tau \to t$ removes the conditionality in the expectation and yields (4.6). In contrast to the local volatility model, the dynamics of S_t is generated by the short end of the volatility function σ_t only. Local volatilities $\sigma_t(\tau, S)$ for $\tau > t$ do not appear explicitly in the SDE for S_t. Their role is to encode information on implied volatilities $\hat{\sigma}_{KT}$: prices of vanilla options for maturities $\tau > t$ are derived from $\sigma_t(\tau, S)$ by solving the forward equation (2.7). In our notation:

$$\frac{dC_t^{S\tau}}{d\tau} + (r - q)\, S \frac{dC_t^{S\tau}}{dS} - \frac{\sigma_t^2(\tau, S)}{2} S^2 \frac{d^2 C_t^{S\tau}}{dS^2} = -q C_t^{S\tau}$$

with the initial condition $C_t^{St} = (S_t - S)^+$, where $C_t^{S\tau}$ is the value at time t of a call option of maturity τ, strike S.

We now use notation $\sigma_{\tau S}$ for $\sigma_t(\tau, S)$ and omit the t-subscript in processes whenever possible. The pricing SDEs for S, $\sigma_{\tau S}^2$ are given by:

$$\begin{cases} dS = (r - q)\, S dt + \sigma_{tS} S dW^S \\ d\sigma_{\tau S}^2 = \mu_{\tau S} dt + \lambda_{\tau S} dW^{\tau S} \end{cases} \tag{4.7}$$

where $W^{\tau S}$ is a Brownian motion. Just as in the previous section, we use zero interest rates and repos without loss of generality and determine $\mu_{\tau S}$ by imposing that option prices have vanishing drift. $\sigma_{\tau S}^2$ is given by

$$\sigma_{\tau S}^2 = 2 \frac{\frac{dC_{KT}}{dT}}{K^2 \frac{d^2 C_{KT}}{dK^2}} \Bigg|_{\substack{K=S \\ T=\tau}} \tag{4.8}$$

Both numerator and denominator are prices of linear combinations of vanilla options – a calendar spread for the numerator, a butterfly spread for the denominator – hence their drifts vanish. Remembering from equation (2.5), page 27, that

$$\frac{d^2 C_{S\tau}}{dS^2} = e^{-r(\tau - t)} E_t[\delta(S_\tau - S)] = e^{-r(\tau - t)} \rho_{\tau S}$$

where $\rho_{\tau S}$ is the probability density that $S_\tau = S$, yields:

$$S^2 \sigma_{\tau S}^2 \rho_{\tau S} = 2 e^{r(\tau - t)} \frac{dC_{KT}}{dT} \Bigg|_{\substack{K=S \\ T=\tau}} \tag{4.9}$$

Since the right-hand side of (4.9) has zero drift, so must the left-hand side: $\sigma_{\tau S}^2 \rho_{\tau S}$ has vanishing drift.[3]

---

[3]In the left-hand side of (4.9) S is not a process – only $\sigma_{\tau S}$ and $\rho_{\tau S}$ are processes.

$\rho_{\tau S}$ is a process: it is the (undiscounted) price of a sharp butterfly spread: this implies that its drift vanishes as well. $\rho_{\tau S}$ is a *function* of S and t and a *functional* of the local volatility function:

$$\rho_{\tau S} \equiv \rho_{\tau S}\left(t, S, \sigma^2\right)$$

Its SDE then simply reads:

$$d\rho_{\tau S} = \frac{d\rho_{\tau S}}{dS}dS + \int\int \frac{\delta\rho_{\tau S}}{\delta\sigma_{ux}^2}d\sigma_{ux}^2$$

where the double integral stands for $\int_t^\tau du \int_0^\infty dx$.

Using now expression (4.7) for $d\sigma_{\tau S}^2$ and $d\sigma_{ux}^2$, we get the drift of $\sigma_{\tau S}^2 \rho_{\tau S}$:

$$\rho_{\tau S}\mu_{\tau S}dt + \lambda_{\tau S}\left(S\frac{d\rho_{\tau S}}{dS}\sigma_{tS}\left\langle dW^S dW^{\tau S}\right\rangle + \int\int \frac{\delta\rho_{\tau S}}{\delta\sigma_{ux}^2}\lambda_{ux}\left\langle dW^{ux} dW^{\tau S}\right\rangle\right)$$

Expressing that it vanishes yields:

$$\mu_{\tau S} = -\frac{\lambda_{\tau S}}{\rho_{\tau S}}\left(S\frac{d\rho_{\tau S}}{dS}\sigma_{tS}\frac{\left\langle dW^S dW^{\tau S}\right\rangle}{dt} + \int\int \frac{\delta\rho_{\tau S}}{\delta\sigma_{ux}^2}\lambda_{ux}\frac{\left\langle dW^{ux} dW^{\tau S}\right\rangle}{dt}\right)$$

$$(4.10)$$

We now derive a more explicit expression for $\frac{\delta\rho_{\tau S}}{\delta\sigma_{ux}^2}$, which appears in the right-hand side of (4.10). $\rho_{\tau S}\left(t, S, \sigma^2\right)$ is an expectation: $\rho_{\tau S} = E_t[\delta\left(S_\tau - S\right)]$. It solves the usual backward equation:

$$\frac{d\rho_{\tau S}}{dt} + \frac{\sigma_{tS}^2}{2}S^2\frac{d^2\rho_{\tau S}}{dS^2} = 0 \qquad (4.11)$$

with the terminal condition $\rho\left(t = \tau\, S, \tau S\right) = \delta\left(S - S\right)$. By definition the functional derivative $\frac{\delta\rho_{\tau S}}{\delta\sigma_{ux}^2}$ is such that, at order one in a perturbation $\delta\sigma^2$ of the local volatility function,

$$\delta\rho_{\tau S} = \int\int \frac{\delta\rho_{\tau S}}{\delta\sigma_{ux}^2}\delta\sigma_{ux}^2$$

Consider a perturbation $\delta\sigma^2$ of σ^2. At first order in $\delta\sigma^2$, the perturbation $\delta\rho_{\tau S}$ solves the following equation:

$$\frac{d\delta\rho_{\tau S}}{dt} + \frac{\sigma_{tS}^2}{2}S^2\frac{d^2\delta\rho_{\tau S}}{dS^2} = -\frac{\delta\sigma_{tS}^2}{2}S^2\frac{d^2\rho_{\tau S}}{dS^2} \qquad (4.12)$$

with the terminal condition at $t = \tau$: $\delta\rho_{\tau S} = 0$. The solution to (4.12) is given by the Feynman-Kac theorem:

$$\delta\rho_{\tau S} = E\left[\int_t^\tau \frac{\delta\sigma_{uS_u}^2}{2}S_u^2\frac{d^2\rho_{\tau S}}{dS^2}\bigg|_{S=S_u} du\right]$$

where the expectation is taken with respect to the dynamics generated by the local volatility function σ_{ux}^2. Let us rewrite this expectation using the probability density for S at time u – which is ρ_{ux}:

$$\delta\rho_{\tau S} = \frac{1}{2} \int_t^\tau du \int_0^\infty dx\, \rho_{ux} x^2 \frac{d^2 \rho_{\tau S}\left(ux, \sigma^2\right)}{dx^2} \delta\sigma_{ux}^2$$

which yields:

$$\frac{\delta\rho_{\tau S}}{\delta\sigma_{ux}^2} = \frac{1}{2}\rho_{ux} x^2 \frac{d^2 \rho_{\tau S}\left(ux, \sigma^2\right)}{dx^2} \tag{4.13}$$

The sagacious reader will have noticed that we have already derived this result in Section 2.9, in the context of the vega hedge in the local volatility model. $\rho_{\tau S}$ is the price of a European option of maturity τ: its sensitivity to σ_{ux}^2 is then given by equation (2.117) which involves the product of the probability density and the dollar gamma: (4.13) is identical to (2.117), page 82. We have our final expression for $\mu_{\tau S}$:

$$\mu_{\tau S} = -\frac{\lambda_{\tau S}}{\rho_{\tau S}} \left(S \frac{d\rho_{\tau S}\left(tS, \sigma^2\right)}{dS} \sigma_{tS} \frac{\left\langle dW^S dW^{\tau S}\right\rangle}{dt} \right.$$

$$\left. + \frac{1}{2} \int_t^\tau du \int_0^\infty dx\, \rho_{ux} x^2 \frac{d^2 \rho_{\tau S}\left(ux, \sigma^2\right)}{dx^2} \lambda_{ux} \frac{\left\langle dW^{ux} dW^{\tau S}\right\rangle}{dt} \right) \tag{4.14}$$

This expression for $\mu_{\tau S}$ was first published by Iraj Kani and Emanuel Derman in [39] – albeit without the first piece in the right-hand side – and more recently rederived by René Carmona and Sergey Nadtochiy in [21], who also prove that the instantaneous volatility $\overline{\sigma}_t$ of S_t is indeed $\sigma_t\left(t, S_t\right)$: this establishes the connection between the SDEs for $\sigma_{\tau S}^2$ and S in (4.7).

Drift $\mu_{\tau S}$ in (4.14) is computationally expensive: not only is it non-local in the sense that it depends on the whole local volatility function, but it involves *forward* transition densities $\rho_{\tau S}\left(ux, \sigma^2\right)$ for all u in $[t, \tau]$. It seems difficult to come up with explicit non-trivial solutions to (4.7) based on the direct modeling of local volatilities.

Inspection of expression (4.14) suggests however two simple solutions for which $\mu_{\tau S}$ vanishes:

- $\lambda_{\tau S} \equiv 0$: this implies $\mu_{\tau S} \equiv 0$. The local volatility function is frozen: this recovers the local volatility model.

- $\left\langle dS dW^{\tau S}\right\rangle \equiv 0$ and $\left\langle dW^{ux} dW^{\tau S}\right\rangle \equiv 0$: local volatilities have zero correlation among themselves and with S. All points of the local volatility function have their own uncorrelated dynamics: such a model amounts to randomly drawing the instantaneous volatility of S as time advances in such a way that the expectation of it square matches the square of the local volatility function calibrated on the initial smile: $E[\overline{\sigma}_t^2] = \sigma_{t_0}^2 (t, S_t)$. While this appears to generate a non-trivial dynamics for implied volatilities it recovers in fact the local volatility model.[4]

Indeed, imagine sitting at time t with a spot value S and let us compute prices of a vanilla option of maturity T using expression (2.30), derived in Section 2.4.1, where we use as base model the local volatility model with local volatility function σ_{t_0}:

$$P_{\overline{\sigma}}(t) = P_{\sigma_0}(t, S) + E_{\overline{\sigma}}\left[\int_t^T \frac{1}{2} e^{-ru} S_u^2 \frac{d^2 P_{\sigma_0}}{dS^2} \left(\overline{\sigma}_u^2 - \sigma_0(u, S_u)^2\right) du \right]$$

As, by construction, $E_{\overline{\sigma}}[\overline{\sigma}_u^2] = \sigma_0(u, S_u)^2$ we get $P_{\overline{\sigma}}(t) = P_{\sigma_0}(t, S)$. This can also be established by noting that, in our model, because each point of the local volatility function is driven by an independent random variable, $E[\overline{\sigma}_u^2 \mid S_u = S] = \sigma_0^2(u, S)$: the general Dupire equation (2.6) then shows that prices in both models are identical. Over any finite time interval the randomness of σ_t averages out and the model behaves as though the local volatility function was fixed, equal to $\sigma_0(t, S)$.

Johannes Wissel proposes in [86] an approach that consists in modeling a discrete set of vanilla options prices: he introduces "local implied volatilities", whose relationship to vanilla option prices is more direct than that of local volatilities. The drifts of these local implied volatilities are non-local as well, except we now have a large but finite number of local implied volatilities. Still, an explicit non-trivial example of a model has not been available yet.

4.2.1 Conclusion

Again, we hit a snag in our attempt to model the dynamics of implied volatilities. Implied volatilities of vanilla options are unwieldy objects, as are their associated local volatilities: because they are related indirectly to option prices, their drifts are complex. In addition, we are handling the dynamics of a two-dimensional set of processes – reducing the dimensionality may be the price to pay to gain some tractability.

Observe that implied volatilities can be defined for any European payoff, as long as it is convex or concave. Are there particular payoffs whose implied volatilities are easier to handle?

---

[4]I thank Bruno Dupire for pointing this out to me.

4.3   Modeling implied volatilities of power payoffs

Consider payoff S_T^p, which we call a *power payoff*, following the terminology of Schweizer and Wissel in [81], and denote its price by Q^{pT}.[5]

In the absence of cash-amount dividends, which is the assumption we make throughout this section, in the Black-Scholes model with implied volatility $\hat{\sigma}$, Q^{pT} is given by:

$$Q^{pT} \; = \; e^{-r(T-t)} F_T^p e^{\frac{p(p-1)}{2}(T-t)\hat{\sigma}^2} \tag{4.15}$$

where F_T is the forward for maturity T.

Given the market price Q^{pT} of a power payoff, its implied volatility $\hat{\sigma}_{pT}$ is obtained by inverting (4.15). Power payoffs are concave for $p \in]0, 1[$ and convex otherwise: $\hat{\sigma}_{pT}$ is well-defined as long as $p \neq 0$ and $p \neq 1$.

4.3.1   Implied volatilities of power payoffs

A power payoff is a European option: its market price can be calculated through replication on vanilla options, using expression (3.7), page 107. However, given a market smile, existence of arbitrary moments of S_T is not guaranteed.

For $p \in [0, 1]$ power payoffs have finite prices, as S^p is bounded above by an affine function of S. For values of $p > 1$ or values of $p < 0$, market prices of power payoffs may not be finite.

In [67] Roger Lee relates the existence of moments of S_T to the asymptotic behavior of implied volatilities for large and small strikes. Specifically, he shows that:

- When $K \to 0$ or $K \to \infty$, $T\hat{\sigma}_{KT}^2$ grows at most linearly in $\ln K$, with a slope that cannot be higher than 2 or lower than -2.[6] Moreover, the asymptotic slope of $T\hat{\sigma}_{KT}^2$ as a function of $\ln K$ is related to the largest/lowest index for which moments of S_T are finite:

- Let $p_+ = \sup\{p\colon E[S_T^{1+p}] < \infty\}$. Then

$$\limsup_{K \to \infty} \frac{T\hat{\sigma}_{KT}^2}{\ln K} \; = \; 2 - 4 \left(\sqrt{p_+^2 + p_+} - p_+ \right)$$

- Let $p_- = \sup\{p\colon E[S_T^{-p}] < \infty\}$. Then

$$\limsup_{K \to 0} \frac{T\hat{\sigma}_{KT}^2}{\ln K} \; = \; -2 + 4 \left(\sqrt{p_-^2 + p_-} - p_- \right)$$

---

[5]This section is based on joint work with Pierre Henry-Labordère.

[6]These bounds on the slope of the integrated variance f as a function of log-moneyness can be derived by imposing that the denominator in the Dupire formula (2.19), page 33, is positive. Assuming that, asymptotically, $f = ay + b$, positivity of the denominator (a butterfly spread) requires that $|a| < 2$.

In practice, even for indexes, implied volatilities exist only in a limited range of strikes: the asymptotic behavior of the smile is then set by the extrapolation chosen by the trader.

Which power payoffs have finite prices then depends on non-observable implied volatility data. Typically $T\hat{\sigma}_{KT}^2$ is parametrized as a function of $\ln(K/F_T)$ as the Dupire equation acquires a simple form (see expression (2.19), page 33), and no-arbitrage conditions are more easily handled. Choosing an affine extrapolation in these units amounts – through its slope – to deciding which power payoffs have finite prices.

Calculating implied volatilities of power payoffs

Implied volatilities $\hat{\sigma}_{pT}$ are obtained from market prices Q^{pT} by inverting (4.15). For $p \in]0, 1[$ prices of power payoffs are always finite; there exists a direct formula of $\hat{\sigma}_{pT}$ as a weighted average of vanilla implied volatilities, which we now derive.

Undiscounted prices of power payoffs are related to the characteristic function $L(p)$ of $x = \ln \frac{S_T}{F_T}$:

$$e^{r(T-t)}\frac{Q^{pT}}{F_T^p} = E\left[\left(\frac{S_T}{F_T}\right)^p\right] = E\left[e^{px}\right] = L(p) \qquad (4.16)$$

$L(p)$ is obtained from the market smile through:

$$L(p) = \int_0^\infty e^{r(T-t)}\frac{d^2C_{KT}}{dK^2}e^{p\ln\frac{K}{F_T}}dK$$

where we have used the fact that the probability density of S_T is given by: $\rho(S_T) = e^{r(T-t)}\frac{d^2C_{KT}}{dK^2}\Big|_{K=S_T}$

Andrew Matytsin – see [73] – introduces a measure of moneyness z defined by:

$$z(K) = \frac{\ln\left(\frac{F_T}{K}\right)}{\hat{\sigma}_{KT}\sqrt{T}} - \frac{\hat{\sigma}_{KT}\sqrt{T}}{2} \qquad (4.17)$$

Replacing C_{KT} with its expression as a function of $\hat{\sigma}_{KT}$, he gives the following formula for $L(p)$:

$$L(p) = \int_{-\infty}^{+\infty}\frac{1}{\sqrt{2\pi}}e^{-\frac{z^2}{2}}e^{-p\left(\frac{\omega^2}{2}+z\omega\right)}\left(1+p\frac{d\omega}{dz}\right)dz \qquad (4.18)$$

where $\omega(z) = \hat{\sigma}_{K(z)T}\sqrt{T}$.

Let us introduce a p-dependent measure of moneyness, y, defined by $y = z + p\omega(z)$:

$$y(K) = \frac{\ln\left(\frac{F_T}{K}\right)}{\hat{\sigma}_{KT}\sqrt{T}} + \left(p - \frac{1}{2}\right)\hat{\sigma}_{KT}\sqrt{T} \qquad (4.19)$$

For $p = 0$, $y(K)$ is the Black-Scholes d_2, while for $p = 1$, it is d_1. Mapping $K \to y$ maps $[0, +\infty]$ into $[-\infty, +\infty]$ and, most importantly, is monotonic.[7] Thus, $K(y, p)$ is well-defined.

From (4.18), performing a change of variable from z to y yields:

$$L(p) = \int_{-\infty}^{+\infty} dy \frac{e^{-\frac{y^2}{2}}}{\sqrt{2\pi}} e^{\frac{p(p-1)}{2} \widehat{\sigma}_{K(y,p)}^2 T}$$

where $\widehat{\sigma}_{K(y,p)T}$ is the implied volatility for "moneyness" y, that is for strike K such that y and K are related through (4.19).

Using now (4.16) and (4.15) we get the following direct relationship between vanilla and power-payoff implied volatilities:

$$e^{\frac{p(p-1)}{2} \widehat{\sigma}_p^2 T} = \int_{-\infty}^{+\infty} dy \frac{e^{-\frac{y^2}{2}}}{\sqrt{2\pi}} e^{\frac{p(p-1)}{2} \widehat{\sigma}_{K(y,p)}^2 T} \tag{4.20}$$

which holds for all $p \in [0, 1]$. $\widehat{\sigma}_{K(y,p)T}$ is easily obtained numerically.[8]

Taking the limits $p \to 0$ and $p \to 1$ yields the implied volatilities of two practically important payoffs.

Implied volatility of the log contract

Let us take the limit $p \to 0$. For small p, $S^p = 1 + p \ln S$, thus $\lim_{p \to 0} \widehat{\sigma}_p$ is the implied volatility of the log contract, which is replicated with a density of vanilla options proportional to $\frac{1}{K^2}$ – see Section 3.1.2. It has the property that its Black-Scholes dollar gamma and vega are independent of S. It is closely related to the variance swap, extensively studied in Chapter 5.

Expanding each side of (4.20) at order one in p yields:

$$1 - \frac{p}{2} \widehat{\sigma}_0^2 T = \int_{-\infty}^{+\infty} dy \frac{e^{-\frac{y^2}{2}}}{\sqrt{2\pi}} \left(1 - \frac{p}{2} \widehat{\sigma}_{K(y,0)T}^2 T \right)$$

which supplies the following formula for the log-contract implied volatility:

$$\widehat{\sigma}_{\ln S}^2 = \int_{-\infty}^{+\infty} dy \frac{e^{-\frac{y^2}{2}}}{\sqrt{2\pi}} \widehat{\sigma}_{K(y)T}^2 \tag{4.21a}$$

$$y(K) = \frac{\ln\left(\frac{F_T}{K}\right)}{\widehat{\sigma}_{KT}\sqrt{T}} - \frac{\widehat{\sigma}_{KT}\sqrt{T}}{2} \tag{4.21b}$$

---

[7] From Roger Lee's work we know that for $K \to 0$ and $K \to \infty$, $\widehat{\sigma}_{KT}^2 T$ grows at most linearly in $\ln K$: $\widehat{\sigma}_{KT}^2 T \propto a \ln K$ with $|a| < 2$. Using these two properties one easily shows that $\lim_{K \to 0} z(K) = +\infty$ and $\lim_{K \to \infty} z(K) = -\infty$.
In [46], Masaaki Fukasawa shows that the $K \to y$ mapping is monotonically decreasing for both $p = 0$ and $p = 1$. Because $y(K)$ is affine in p this implies that $y(K)$ is monotonic for all $p \in [0, 1]$ – I am indebted to Ling Ling Cao for this observation.

[8] Choose a set of strikes K_i. For each K_i, calculate the corresponding value y_i of y using (4.19) and record the couple $(y_i, \widehat{\sigma}_{K_i T})$. Then build an interpolation, for example a spline, of these couples to generate the function $\widehat{\sigma}_{K(y,p)T}$.

This expression was first published by Neil Chriss and William Morokoff in [32]. In comparison with formula (3.7), which expresses the price of the power payoff as a weighted integral of vanilla option prices, (4.21a) is less sensitive to numerical discretization of the integral.

For example, if $\widehat{\sigma}_{KT}$ is constant, equal to $\widehat{\sigma}_0$, a Gauss-Hermite quadrature yields $\widehat{\sigma}_{\ln S} = \widehat{\sigma}_0$ no matter how few points we use. In practice using about 10 points provides good accuracy.

Implied volatility of the $S \ln S$ contract

Now set $p = 1 - \varepsilon$ and take the limit $\varepsilon \to 0$. For small ε, $S^p = S - \varepsilon S \ln S$, thus $\lim_{p \to 1} \widehat{\sigma}_p$ is the implied volatility of the $S \ln S$ contract, which is replicated with a density of vanilla options proportional to $\frac{1}{K}$ – see Section 3.1.9.1.

This payoff has the property that its dollar gamma – hence its vega – is proportional to S. Proceeding as above, we get:

$$\widehat{\sigma}^2_{S \ln S} = \int_{-\infty}^{+\infty} dy \, \frac{e^{-\frac{y^2}{2}}}{\sqrt{2\pi}} \widehat{\sigma}^2_{K(y)T} \tag{4.22a}$$

$$y(K) = \frac{\ln\left(\frac{F_T}{K}\right)}{\widehat{\sigma}_{KT}\sqrt{T}} + \frac{\widehat{\sigma}_{KT}\sqrt{T}}{2} \tag{4.22b}$$

4.3.2 Forward variances of power payoffs

Power payoffs are of the type $h\left(\frac{S_T}{F_T}\right)$ with h a convex (resp. concave) function for $p < 0$ or $p > 1$ (resp. $p \in]0, 1[$). They fall in the class considered in Section 2.2.2.2 for which (a) an implied volatility can be defined, (b) the convex order condition for implied volatilities (2.17) holds.

We then have:

$$(T_2 - t)\widehat{\sigma}^2_{pT_2} \geq (T_1 - t)\widehat{\sigma}^2_{pT_1} \tag{4.23}$$

This allows us to define *positive* forward variances ξ, either discrete, or continuous.

• Discrete forward variances $\xi^{pT_1 T_2}$ are defined by:

$$\xi^{pT_1 T_2} = \frac{(T_2 - t)\widehat{\sigma}^2_{pT_2} - (T_1 - t)\widehat{\sigma}^2_{pT_1}}{T_2 - T_1} \tag{4.24}$$

• Continuous forward variances ξ^{pT} are defined by:

$$\xi^{pT} = \frac{d}{dT}\left((T - t)\widehat{\sigma}^2_{pT}\right) \tag{4.25}$$

The set of ξ^{pT} for all T is called the variance curve for index p; these are the state variables whose dynamics we will model.

We could have defined as well forward variances for implied volatilities of vanilla options, for a given moneyness, in Section 4.1.1, but their unwieldiness makes it a poitnless exercise.[9]

4.3.3   The dynamics of forward variances

In what follows, we work with continuous forward variances – discrete forward variances will reappear in Chapter 7.8. Again, we work with zero interest rates without loss of generality, and omit the t subscripts for processes.

$$d\xi^{pT} = \lambda^{pT}dW^{pT} + \bullet\, dt$$

We determine the drift of ξ^{pT} so that the drift of Q^{pT} vanishes. The SDEs for ξ^{pT} and S are:

$$\begin{cases} dS = \overline{\sigma}SdW^S \\ d\xi^{pT} = \mu^{pT}dt + \lambda^{pT}dW^{pT} \end{cases} \tag{4.26}$$

with initial conditions $S_{t=0} = S_{t=0}^{\text{market}}$ and $\xi_{t=0}^{pT} = \xi_{t=0}^{pT\ \text{market}}$. With zero interest rates, repos, Q^{pT} is given by:

$$Q^{pT} = S^p e^{\frac{p(p-1)}{2}\int_t^T \xi^{p\tau}d\tau} \tag{4.27}$$

In what follows we deal with a single variance curve, for a given index p – we omit it in the notation. The SDE for Q^T is given by:

$$\frac{dQ^T}{Q^T} = \left[\frac{p(p-1)}{2}\left(\overline{\sigma}^2 - \xi^t\right) + \frac{p(p-1)}{2}\int_t^T \mu^\tau d\tau \right. \tag{4.28}$$
$$\left. + \frac{p^2(p-1)}{2}\overline{\sigma}\int_t^T \lambda^\tau \rho^{S\tau}d\tau + \frac{p^2(p-1)^2}{8}\int_t^T\int_t^T dudv\rho^{uv}\lambda^u\lambda^v\right]dt$$
$$+ \left[p\overline{\sigma}dW^S + \frac{p(p-1)}{2}\int_t^T \lambda^\tau dW^\tau d\tau\right]$$

where

$$\rho^{S\tau} = \frac{\langle dW^S dW^\tau\rangle}{dt}, \rho^{uv} = \frac{\langle dW^u dW^v\rangle}{dt}$$

---

[9]Strike K cannot be kept constant as the maturity is varied. Provided K scales linearly with the forward F_T: $K_T = \alpha F_T$, the convex order condition for option prices (2.9) holds and translates into the condition (2.15) on implied volatilities:

$$(T_2 - t)\widehat{\sigma}^2_{\alpha F_{T_2},T_2} \geq (T_1 - t)\widehat{\sigma}^2_{\alpha F_{T_1},T_1}$$

The counterpart of equation (4.25) for vanilla option implied volatilities would thus have read:

$$\xi_t^{\alpha,T} = \frac{d}{dT}\left((T-t)\widehat{\sigma}^2_{\alpha F_T,T}\right)$$

The first two lines in equation (4.28) are the drift of Q^T, which has to vanish for all T. Taking the derivative of this drift with respect to T yields:

$$\frac{p(p-1)}{2}\mu^T + \frac{p^2(p-1)}{2}\bar{\sigma}\lambda^T\rho^{ST} + \frac{p^2(p-1)^2}{4}\int_t^T \rho^{Tu}\lambda^T\lambda^u du = 0$$

from which we get the expression of μ^T, already given in [81]:

$$\mu^T = -\left(p\bar{\sigma}\lambda^T\rho^{ST} + \frac{p(p-1)}{2}\int_t^T \rho^{Tu}\lambda^T\lambda^u du\right) \qquad (4.29)$$

Taking $T \to t$ in (4.28) leaves only the first term in the drift of Q^T: $\frac{p(p-1)}{2}\left(\bar{\sigma}^2 - \xi^t\right)$. We then get the extra condition:

$$\xi_t^t = \bar{\sigma}_t^2 \qquad (4.30)$$

Thus, the short end of the variance curve is equal to the instantaneous variance of S, *for all values of* p.

Using now the expression of μ^T in (4.29) yields the final SDEs for S and ξ^T – reinstating the p and t indices:

$$dS_t = \sqrt{\xi_t^t}S_t dW_t^S \qquad (4.31a)$$

$$d\xi_t^{pT} = -\left(p\sqrt{\xi_t^t}\lambda_t^T\rho^{ST} + \frac{p(p-1)}{2}\int_t^T \rho^{Tu}\lambda_t^T\lambda_t^u du\right)dt + \lambda_t^T dW_t^T$$

$$\qquad (4.31b)$$

4.3.4 Markov representation of the variance curve

Generally, the solution of (4.31b) requires evolving each forward variance ξ^T individually in a Monte-Carlo simulation – this is impractical. It may be that, for a well chosen covariance structure of forward variances, each ξ^T can be written as a function of a finite number of state variables, i.e. possesses a Markov representation. Then one only needs to evolve a finite set of state variables to generate the full variance curve at time t.

The question of building Markov representations has been especially addressed in the context of yield curve modeling. Formula (4.27) for Q^{pT} is similar to the expression of a zero-coupon bond, up to the factor S^p, with ξ^T playing the role of the forward rate for date T. It is then tempting, following the work of Oren Cheyette in [30] in the context of the HJM framework for the yield curve, to try and derive a class of solutions to (4.31b) that have a Markov representation.

We carry out the typical derivation one would go through in a yield curve context – for the sake of it, since we must warn the reader that the outcome is fruitless in the case of power payoffs.

While we do not carry the p index, we will now carry t subscripts. Let us assume, following [30], that the covariance structure of the ξ_t^T is such that SDE (4.31b) for

ξ^T has the following particular form:

$$dS_t = \sqrt{\xi_t^t} S_t dW_t^S \tag{4.32a}$$

$$d\xi_t^T = \mu_t^T dt + \sum_{i=0}^{n} \alpha_i(T) \beta_{it} dW_t^i \tag{4.32b}$$

where W_i, $i = 0 \ldots n$ are n correlated Brownian motions, β_{it} are processes, and $\alpha_i(T)$ are functions of T. The correlations of W_t^i and W_t^j is ρ_{ij} and the correlation of W_t^i and W_t^S is ρ_{iS}.

The volatility structure in the above equations is not inapt. For example, taking $\beta_{it} = e^{k_i t}$, a function, and $\alpha_i(T) = \alpha_i e^{-k_i T}$ yields:

$$d\xi_t^T = \mu^T dt + \sum_{i=0}^{n} \alpha_i e^{-k_i(T-t)} dW_t^i$$

The dynamics of the ξ^T is time-homogeneous – volatilities and correlations of $\xi^T, \xi^{T'}$ are only a function of $T - t$ and $T - t, T' - t$, respectively – the "volatilities" of ξ^T being expressed as a linear combination of exponentials.

Mirroring the derivation in [30], we integrate (4.32b) and try to express ξ_t^T as a function of as few processes as possible. We have:

$$\xi_t^T = \xi_0^T - p\Sigma_i \int_0^t \alpha_i(T)\beta_{i\tau}\rho_{iS}\sqrt{\xi_\tau^T} d\tau$$

$$- \frac{p(p-1)}{2} \int_0^t \left(\int_\tau^T \Sigma_{ij}\rho_{ij}\alpha_i(T)\beta_{i\tau}\alpha_j(u)\beta_{j\tau} du \right) d\tau + \Sigma_i \int_0^t \alpha_i(T)\beta_{i\tau} dW_\tau^i$$

$$= \xi_0^T + \Sigma_i \alpha_i(T) \left[-p\rho_{iS} \int_0^t \beta_{i\tau}\sqrt{\xi_\tau^T} d\tau \right.$$

$$\left. - \frac{p(p-1)}{2}\Sigma_j \int_0^t \rho_{ij}\beta_{i\tau}\beta_{j\tau}(A_j(T) - A_j(\tau)) d\tau + \int_0^t \beta_{i\tau} dW_\tau^i \right]$$

where we introduce $A_j(\tau) = \int_0^\tau \alpha_j(u) du$.

We now define process $B_{ij,t} = \rho_{ij} \int_0^t \beta_{it}\beta_{jt}$. A little manipulation yields:

$$\xi_t^T = \xi_0^T + \Sigma_i \alpha_i(T) \left[-\frac{p(p-1)}{2}\Sigma_j B_{ij,t}(A_j^T - A_j^t) \right.$$

$$\left. + \int_0^t \left(-p\rho_{iS}\beta_{i\tau}\sqrt{\xi_\tau^T} d\tau - \frac{p(p-1)}{2}\Sigma_j B_{ij,t}\alpha_j(\tau) d\tau + \beta_{i\tau} dW_\tau^i \right) \right]$$

On top of processes $B_{ij,t}$ we thus need to define n processes x_{it}:

$$x_{i0} = 0$$

$$dx_{it} = -p\rho_{iS}\beta_{it}\sqrt{\xi_t^t} dt - \frac{p(p-1)}{2}\Sigma_j B_{ij,t}\alpha_j(t) dt + \beta_{it} dW_t^i$$

The variance curve is given at time t by:

$$\xi_t^T = \xi_0^T + \Sigma_i \alpha_i(T)\left[-\frac{p(p-1)}{2}\Sigma_j B_{ij,t}\left(A_j^T - A_j^t\right) + x_{it}\right] \qquad (4.33)$$

Setting $T = t$ gives the expression of the short end of the curve – the instantaneous variance of S_t:

$$\xi_t^t = \xi_0^t + \Sigma_i \alpha_i(t)x_{it} \qquad (4.34)$$

Thus, in a Monte-Carlo simulation of our model, we only need to evolve processes x_{it}, $V_{ij,t}$ and S_t, according to the following SDEs:

$$dx_{it} = -p\rho_{iS}\beta_{it}\sqrt{\xi_0^t + \Sigma_i \alpha_i(t)x_{it}}dt - \frac{p(p-1)}{2}\Sigma_j B_{ij,t}\alpha_j(t)dt \qquad (4.35a)$$
$$+ \beta_{it}dW_t^i$$
$$dB_{ij,t} = \rho_{ij}\beta_{it}\beta_{jt}\, dt \qquad (4.35b)$$
$$dS_t = \sqrt{\xi_0^t + \Sigma_i \alpha_i(t)x_{it}}\, S_t\, dW_t^S \qquad (4.35c)$$

The variance curve at time t is given by (4.33). Inspection of equations (4.35) shows that processes β_{it} can depend arbitrarily on the x_{it} and $V_{ij,t}$, thus β_{it} can, for example, be an arbitrary function of the variance curve ξ_t at time t. This would generate the equivalent of a "local volatility" model for the ξ_t^T.

Unlike forward rates though, the instantaneous variance cannot be negative. Looking at the SDE for processes x_i above it is not clear that it is possible to define processes β_i and functions α_i that ensure that $\xi_t^t = \xi_0^t + \Sigma_i \alpha_i(t)x_{it} \geq 0$.[10]

The conclusion is that, unfortunately the ansatz (4.32b) used in [30] cannot be transposed to the framework of forward variances. This does not mean there are no low-dimensional Markov representations of the variance curve.

Indeed, any stochastic volatility model written on the instantaneous variance $V_t = \xi_t^t$ does provide a Markov representation of the variance curve for any value of p, though it may not be explicit. Think for example of the Heston model.

We refer the reader to Chapter 7 for examples of models with low-dimensional Markov representations for forward variances associated to $p \to 0$.

4.3.5 Dynamics for multiple variance curves

Given an initial variance curve $\xi_{t=0}$ for a particular value p^*, SDEs (4.31) generate the joint dynamics of $(S_t, \xi_t^{p^*})$. For a given market smile, we will generally have a

---

[10]Why not assume a lognormal dynamics for ξ_t^T: $d\xi_t^T = \xi_t^T\left(\mu_t^T dt + \Sigma_i \alpha_i(T)\beta_{it}dW_t^i\right)$ and look for a Markov representation of $\ln \xi_t^T$? We encourage the reader to try for herself; there does not seem to be a solution unless $p = 0$ or $p = 1$.

set of values of p for which prices of power payoffs are finite, hence the $\xi_{t=0}^{pT}$ are well-defined. Is it possible to generate a joint dynamics for S_t and a set of ξ_t^p?

A solution to SDEs (4.31) for a given p^* provides the full dynamics of S_t, hence prices for power payoffs and a dynamics for variance curves ξ_t^p with $p \neq p^*$. If our objective is to be able to independently set the initial values of variances curves for multiple values of p, this implies that information about initial curves $\xi_{t=0}^p$ has to be embedded in the SDE (4.31b) for $\xi_t^{p^*}$. It is not clear how this can be done practically.

The joint dynamics of multiple variance curves has to comply with condition (4.30) which expresses that the short ends of variance curves collapse one onto another at all times, as ξ_t^{pt} is the instantaneous variance of S_t. The author does not know of an example of direct modeling of the joint dynamics for multiple curves that is able to calibrate to market prices of power payoffs. Obviously the local volatility model provides a solution – albeit not explicit and not very exciting as it is driven by a single Brownian motion.

Even though we are not able to handle the dynamics of multiple curves, we are able to explicitly construct the joint dynamics of S_t and the variance curve, for a particular value p^*. We generate a dynamics for the full volatility surface by modeling the dynamics of one particular variance curve $\xi_t^{p^*}$. This produces a dynamics for other variance curves ξ_t^p with $p \neq p^*$ that obeys (4.31b), except we cannot set their initial condition. We will generally have $\xi_{t=0}^p \neq \xi_{t=0}^{pmarket}$: our models will not provide exact calibration to the vanilla smile.[11]

4.3.6 The log contract, again

Consider the payoff $\frac{S^p-1}{p}$: it has the same implied volatility as the power payoff of index p. Taking the limit $p \to 0$:

$$\lim_{p\to 0} \frac{S^p - 1}{p} = \ln S$$

We have already made this observation in Section 4.3.1 and have derived an expression for the implied volatility of the log contract, in case there are no cash-amount dividends.

We first encountered the log contract in the discussion of cliquet hedges in Chapter 3. Using the same notation as in Section 3.1.4, we simply denote by $\hat{\sigma}_T$ the implied volatility of the log contract of maturity T and by ζ_t the associated variance curve:

$$\zeta_t^T \equiv \xi^{p=0,T} = \frac{d}{dT}\left((T-t)\,\hat{\sigma}_T^2\,(t)\right)$$

Log contracts have finite prices: they do not require anything beyond the non-abitrageability of the smile, thus $\hat{\sigma}_T$ is always well-defined. Power payoffs satisfy the

---

[11]At least we are able to calibrate exactly the term structure of implied volatilities $\hat{\sigma}_{p^*T}$. In the case of implied volatilities of vanilla options, we were not even able to handle the dynamics of *one* $\hat{\sigma}_{KT}$, let alone a term structure $\hat{\sigma}_{KT T}$.

convex order condition, thus ζ_t^T is positive. Taking the limit $p \to 0$ in (4.31) yields the following joint dynamics for (S_t, ξ_t):

$$\begin{cases} dS_t = \sqrt{\zeta_t^t} S_t dW_t^S \\ d\zeta_t^T = \lambda_t^T dW_t^T \end{cases} \qquad (4.36)$$

Thus, forward variances associated to log contracts have no drift.

In practice, log contracts themselves are not traded, as vanilla options are not traded over a sufficiently wide range of strikes to allow for exact replication. Moreover, our analysis does not carry over to the case of dividends with fixed cash amounts – which cannot be represented by a proportional yield q – as the property that $E[S_{T_2}|S_{T_1}] = \frac{F_{T_2}}{F_{T_1}}$ needed to prove the convex order condition (4.23) no longer holds.

Luckily, closely related instruments known as variance swaps are traded; (4.36) will still hold, except the ζ_t^T will be replaced by variance swap forward variances – they are the basic building blocks in the models of Chapter 7.

Before we do this, we pause to study variance swaps in detail.

Chapter's digest

▶ Hoping to construct a market model for vanilla options by modeling implied volatilities of vanilla options directly is a dead end.

▶ There is a one-to-one mapping between a non-arbitrageable vanilla smile and its corresponding local volatility function. Morover, no-arbitrage conditions simply translate in the requirement that local volatilities be real. It is then tempting to specify a dynamics for the local volatility function. It turns out, however, that the drift of local volatilities is non-local and involves forward transition densities, thus is computationally too expensive. Again, we reach an impasse – rather than trying to model the dynamics of implied volatilities of vanilla payoffs – or their associated local volatilities – are there other types of convex payoffs whose implied volatilities are less unwieldy objects?

▶ One good candidate is the family of power payoffs $\left(\frac{S_T}{F_T}\right)^p$. For each value of p such that the market price of the corresponding power payoff is finite for all T, a term structure of forward variances ξ^{pT} can be defined. Setting the volatilities of the ξ^{pT} determines their drifts. For general values of p, it is not clear that there exist particular forms of the volatilities of the ξ^{pT} that give rise to a Markov representation of the variance curve. For $p \to 0$, however, the drift of ξ^{pT} vanishes. $\xi^{p=0T}$ are forward variances associated to the term structure of log contracts, which are closely related to variance swaps.

Chapter 5

Variance swaps

This chapter is devoted to variance swaps (VS) and their connection to delta-hedged log contracts – it is a prerequisite for the chapters that follow.

We show how a VS can be synthesized using European payoffs, characterize the impact of large returns on its replication and assess the relevance of pricing a VS in a jump-diffusion model. Finally we analyze the impact of cash-amount dividends on the VS replication, as well as the effect of interest-rate volatility.

We then study the replication of weighted variance swaps.

This is followed by two appendices – Appendix A on timer options and Appendix B on the perturbation of the lognormal density.

5.1 Variance swap forward variances

A variance swap (VS) contract pays at maturity the realized variance of a financial underlying, computed as the sum of the squares of daily log-returns. The market convention for the VS payoff is:

$$\frac{252}{N} \sum_{i=0}^{N-1} \ln^2\left(\frac{S_{i+1}}{S_i}\right) - \widehat{\sigma}_{VS,T}^2(t) \tag{5.1}$$

where N is the number of trading days for the maturity of the variance swap, S_i are the daily closing quotes of the underlying. $\widehat{\sigma}_{VS,T}(t)$ is set so that the initial value at time t of the VS is zero and is called the VS volatility for maturity T.[1] 252 is the typical number of trading days in a year. Because the distribution of trading days is not uniform throughout the year, the ratio $\frac{N}{252}$ is generally not equal to the year fraction for the maturity of the VS and $\widehat{\sigma}_{VS}$ is a biased estimator of realized volatility, especially for short maturities. We prefer to work with the following convention:

$$\frac{1}{T-t} \sum_{i=0}^{N-1} \ln^2\left(\frac{S_{i+1}}{S_i}\right) - \widehat{\sigma}_{VS,T}^2(t) \tag{5.2}$$

---

[1]VS term sheets include a prefactor $1/(2\widehat{\sigma}_{VS})$ so that, for a small difference between realized volatility σ_r and VS volatility $\widehat{\sigma}_{VS}$, the payout of a VS contract is simply $\sigma_r - \widehat{\sigma}_{VS}$.

where the S_i are observed at dates t_i, such that $t_N = T$, which we rewrite for notational economy as:

$$\frac{1}{T-t}\sum_t^T \ln^2\left(\frac{S_{i+1}}{S_i}\right) - \widehat{\sigma}_{\text{VS},T}^2(t) \tag{5.3}$$

Imagine taking at time t a long position in $(T_2 - t)$ VSs of a maturity T_2 and a short position in $(T_1 - t)\,e^{-r(T_2-T_1)}$ VSs of maturity T_1 with $T_2 > T_1$. The market implied volatilities of these two VSs are, at time t, $\widehat{\sigma}_{\text{VS},T_2}(t)$ and $\widehat{\sigma}_{\text{VS},T_1}(t)$. From (5.3), the payoff of this position, capitalized at T_2 is:

$$\sum_{T_1}^{T_2} \ln^2\left(\frac{S_{i+1}}{S_i}\right) - \left((T_2 - t)\,\widehat{\sigma}_{\text{VS},T_2}^2(t) - (T_1 - t)\,\widehat{\sigma}_{\text{VS},T_1}^2(t)\right)$$

$$= \sum_{T_1}^{T_2} \ln^2\left(\frac{S_{i+1}}{S_i}\right) - (T_2 - T_1)\,\widehat{\sigma}_{\text{VS},T_1 T_2}^2(t) \tag{5.4}$$

where we have introduced the discrete forward variance $\widehat{\sigma}_{\text{VS},T_1 T_2}$ defined as:

$$\widehat{\sigma}_{\text{VS},T_1 T_2}^2(t) = \frac{(T_2 - t)\,\widehat{\sigma}_{\text{VS},T_2}^2(t) - (T_1 - t)\,\widehat{\sigma}_{\text{VS},T_1}^2(t)}{T_2 - T_1}$$

$\widehat{\sigma}_{\text{VS},T_1 T_2}^2(t)$ is positive by construction, as the value at time t of the second-hand side of (5.4) vanishes and its first piece is positive by construction. Imagine we unwind our position by entering at a later time $t' < T_1$ the reverse position: selling $(T_2 - t')$ VSs of maturity T_2 and buying $(T_1 - t')\,e^{-r(T_2-T_1)}$ VSs of maturity T_1 at market implied volatilities prevailing at t'. This cancels the contribution from the realized variance over $[T_1, T_2]$ and the P&L of our strategy capitalized at time T_2 is:

$$(T_2 - T_1)\left(\widehat{\sigma}_{\text{VS},T_1 T_2}^2(t') - \widehat{\sigma}_{\text{VS},T_1 T_2}^2(t)\right) \tag{5.5}$$

It no longer involves the realized variance of S and only depends on the variation of implied VS volatilities over $[t, t']$. (5.5) shows that we are able to generate a P&L that is linear in the variation of $\widehat{\sigma}_{\text{VS},T_1 T_2}^2$ over $[t, t']$ at no cost.

To produce a P&L that is linear in the variation of an equity underlying S, we borrow money to buy the underlying share and need to pay interest while we hold the share, hence the non-vanishing pricing drift of S. In the case of forward VS variances, no money is needed to materialize P&L (5.5): $\widehat{\sigma}_{\text{VS},T_1 T_2}^2$ has vanishing pricing drift.

$\widehat{\sigma}_{\text{VS},T_1 T_2}^2$ is a discrete forward variance. We can similarly define continuous VS forward variances, which we simply denote by ξ_t^T, given by:

$$\xi_t^T = \frac{d}{dT}\left((T-t)\,\widehat{\sigma}_{\text{VS},T}^2(t)\right)$$

The ξ_t^T are driftless as well; in a diffusive setting:

$$d\xi_t^T = \bullet\, dW_t^T \tag{5.6}$$

5.2 Relationship of variance swaps to log contracts

Consider the VS payoff (5.2). As $\widehat{\sigma}_{VS,T}^2$ is a constant, we will focus on the first piece in (5.2) and simply take for the VS payoff:

$$\sum_{i=0}^{N-1} \ln^2\left(\frac{S_{i+1}}{S_i}\right) \tag{5.7}$$

Let us assume for simplicity that dates t_i are equally spaced by Δt. If $\ln\left(\frac{S_{i+1}}{S_i}\right)$ is small, at order two in $\frac{\delta S_i}{S_i}$ the payoff (5.7), discounted at $t = 0$ can be rewritten as:

$$e^{-rT} \sum_{i=0}^{N-1} \left(\frac{\delta S_i}{S_i}\right)^2 = \sum_{i=0}^{N-1} e^{-rt_i} e^{-r(T-t_i)} \left(\frac{\delta S_i}{S_i}\right)^2 \tag{5.8}$$

where $\delta S_i = S_{i+1} - S_i$. We recognize the typical expression of the discounted sum of the gamma portion of the usual daily gamma/theta P&Ls (1.9) of a delta-hedged option risk-managed at zero volatility. The gamma/theta P&L reduces to its gamma part only if we choose a vanishing implied volatility for risk-managing the option.

Is it possible to find a European payoff whose dollar gamma – when risk-managed at vanishing volatility – matches that of the VS? This condition reads:

$$\frac{1}{2}S^2 \frac{d^2 P_{\widehat{\sigma}=0}}{dS^2} = e^{-r(T-t)} \tag{5.9}$$

The answer is yes – it is, up to a factor, the log contract. The value $Q^T(t,S)$ of payoff $-2 \ln S_T$ in the Black-Scholes model with implied volatility $\widehat{\sigma}$, assuming there are no dividends with fixed cash amounts, is given by (3.8):

$$Q^T(t,S) = -2e^{-r(T-t)}\left(\ln S + (r-q)(T-t) - \frac{\widehat{\sigma}^2}{2}(T-t)\right) \tag{5.10}$$

Take $\widehat{\sigma} = 0$ – the delta and gamma of Q^T do not depend on $\widehat{\sigma}$ and we get:

$$\frac{dQ_{\widehat{\sigma}=0}^T}{dS} = -e^{-r(T-t)}\frac{2}{S}, \quad \frac{1}{2}S^2 \frac{d^2 Q_{\widehat{\sigma}=0}^T}{dS^2} = e^{-r(T-t)} \tag{5.11}$$

$Q_{\widehat{\sigma}=0}^T$ indeed fulfills condition (5.9) – we could have obtained (5.10) by straight integration of (5.9). Risk-managing the European payoff $-2\ln S_T$ with zero implied volatility exactly produces – at second order in δS_i – payoff (5.7). How much should we charge for it – or, equivalently, what is $\widehat{\sigma}_{VS,T}^2$ so that the value at $t = 0$ of the VS contract vanishes?

If the market price at $t = 0$ of the log contract were $Q_{\widehat{\sigma}=0}^T$ we would not need to charge anything – we would set $\widehat{\sigma}_{VS,T} = 0$. In reality, the log contract has a market price Q_{market}^T: purchasing the log contract generates a mark-to-market P&L $-(Q_{\text{market}}^T - Q_{\hat{\sigma}=0}^T)$ for us, which we charge to the client as the premium of the variance swap. This premium is $T\hat{\sigma}_{\text{VS},T}^2$ and is paid at maturity. $\hat{\sigma}_{\text{VS},T}$ is then given by:

$$\hat{\sigma}_{\text{VS},T}^2 = \frac{e^{rT}}{T}\left(Q_{\text{market}}^T - Q_{\hat{\sigma}=0}^T\right) \tag{5.12}$$

Given the market price Q_{market}^T we can invert (5.10) to back out the log contract implied volatility $\hat{\sigma}_T$. Substituting expression (5.10) for Q_{market}^T in (5.12) then yields:

$$\hat{\sigma}_{\text{VS},T} = \hat{\sigma}_T \tag{5.13}$$

$$\xi_t^T = \zeta_t^T \tag{5.14}$$

The log contract is replicated with a vanilla portfolio using a density proportional equal to $\frac{2}{K^2}$ – see Section 3.1.3:

$$-2\ln S = -2\ln S_0 - \frac{2}{S_0}(S - S_0) \tag{5.15}$$

$$+ \int_0^{S_0} \frac{2}{K^2}(K-S)^+ \, dK + \int_{S_0}^{\infty} \frac{2}{K^2}(S-K)^+ \, dK$$

At order two in $\frac{\delta S}{S}$ the payoff of a VS is then synthesized by delta-hedging this portfolio until maturity with zero implied volatility: $\hat{\sigma}_T$ is simply computed as the implied volatility of the replicating vanilla portfolio and is model-independent.[2] As a consequence, forward variances of log contracts and VSs are identical objects. (5.12) can be rewritten as:

$$\hat{\sigma}_{\text{VS},T}^2 = \frac{e^{rT}}{T} \int_0^{\infty} \frac{2}{K^2}\left(P_{\text{market}}^{KT} - P_{\hat{\sigma}=0}^{KT}\right) dK \tag{5.16}$$

where P^{KT} is the price of a vanilla option of strike K, maturity T. $P_{\hat{\sigma}=0}^{KT}$, the price for a vanishing volatility is simply the intrinsic value computed for the forward and discounted to $t = 0$; for a call option:

$$P_{\hat{\sigma}=0}^{KT} = e^{-rT}\left(Se^{(r-q)T} - K\right)^+$$

$(P_{\text{market}}^{KT} - P_{\hat{\sigma}=0}^{KT})$ is identical for a call or a put struck at K because of call-put parity – no need to distinguish between both types of vanilla options.[3]

While $\hat{\sigma}_T$ is well-defined whenever the market smile is non-arbitrageable, it is very sensitive to the extrapolation chosen for implied volatilities outside the range

---

[2]It is this property that prompted banks to start offering variance swaps in the nineties: at the time, variance swaps were exotic instruments that trading desks hedged with vanilla options. Since then, on indexes, they have become emancipated from their vanilla replication and exist as independent instruments.

[3](5.12) and (5.16) hold as long as there are no dividends with fixed cash amounts; in the general case, expression (5.47) applies.

of strikes traded on the market. In practice, for very liquid securities such as indexes, market-makers do the reverse and infer implied volatilities for low strikes from market quotes of VSs.[4]

The delta and gamma of the log contract in (5.11) do not depend on the implied volatility $\widehat{\sigma}$: we could as well have chosen to risk-manage the log contract with a non-zero implied volatility. The most natural choice is $\widehat{\sigma} = \widehat{\sigma}_T$: delta-hedging the log contract then generates a gamma P&L *and* a theta P&L that exactly match both pieces in (5.2).

The idea of using zero implied volatility proves useful when analyzing more complex payoffs involving realized variance weighted by a function of spot value, such as conditional variance swaps, for which squared daily returns are accumulated only when S_i lies within an interval, typically $[0, L]$, $[L, H]$ or $[H, \infty]$ – and also in the case of fixed amount dividends.

Weighted variance swaps – and in particular conditional VSs – are dealt with in Section 5.9, page 176.

5.2.1   A simple formula for $\widehat{\sigma}_{\text{VS},T}$

In (5.16) $\widehat{\sigma}_{\text{VS},T}$ is expressed in terms of market prices of vanilla options. (5.16) holds whenever cash-amount dividends are not present. Otherwise it is replaced with (5.47) – see Section 5.6.2 further below.

In the absence of cash-amount dividends, $\widehat{\sigma}_{\text{VS},T}$ can equivalently be computed as the implied volatility of the log contract – or of a set of European payoffs otherwise; see the derivation in Section 5.3.1 below. This is a more efficient method than using (5.16) as it is less sensitive to the discretization of the replicating portfolio in (5.15) – in particular, for a flat volatility surface, we trivially recover the exact value of $\widehat{\sigma}_{\text{VS},T}$.

Still, with no cash-amount dividends present, there is a direct expression of $\widehat{\sigma}_{\text{VS},T}$ as a weighted average of implied volatilities of vanilla options. This was obtained in the context of power payoffs, in Section 4.3 of Chapter 4.

We simply quote result (4.21) and refer the reader to page 142 for its derivation:

$$\widehat{\sigma}_{\text{VS},T}^2 = \int_{-\infty}^{+\infty} dy\, \frac{e^{-\frac{y^2}{2}}}{\sqrt{2\pi}} \widehat{\sigma}_{K(y)T}^2 \tag{5.17a}$$

$$y(K) = \frac{\ln\left(\frac{F_T}{K}\right)}{\widehat{\sigma}_{KT}\sqrt{T}} - \frac{\widehat{\sigma}_{KT}\sqrt{T}}{2} \tag{5.17b}$$

---

[4]For long-dated variance swaps, the identity (5.13) has to be adjusted to take interest-rate volatility into account. Black-Scholes implied volatilities of European payoffs are in fact implied volatilities of the *forward* for the option's maturity. In contrast, a variance swap pays the realized variance of the *spot*. Interest-rate volatility introduces a difference between the realized variances of the spot and the forward, which is material for long-dated variance swaps. See Section 5.8 below for an estimation of this effect.

5.3 Impact of large returns

The key property that synthesizing the VS payoff boils down to delta-hedging a log contract – hence that $\hat{\sigma}_{VS,T} = \hat{\sigma}_T$ – has been derived in an expansion at order 2 in $\frac{\delta S}{S}$ of the VS payoff and the P&L of a delta-hedged log contract. How robust is it? What if returns are large?

We now consider two canonical examples of dynamics for S_t: a diffusive model and a jump-diffusion model. Unlike the former, the latter is able to generate large returns, even at short time scales, with a probability proportional to Δt. We consider the limit of very frequent observations of S which enables us to explicitly compute all quantities of interest.

The case of real underliers is investigated next.

5.3.1 In diffusive models

The price of the log contract of maturity T, risk-managed at zero implied volatility, $P^T_{\hat{\sigma}=0}$, satisfies the following condition:

$$\frac{1}{2}S^2\frac{d^2P^T_{\hat{\sigma}=0}}{dS^2} = e^{-r(T-t)} \tag{5.18}$$

Assume that S follows a diffusive dynamics:

$$dS_t = (r-q)S_tdt + \overline{\sigma}_tS_tdW_t \tag{5.19}$$

where instantaneous volatility $\overline{\sigma}$ is an arbitrary process.

Remember expression (2.30) in Section 2.4.1 relating the price of a payoff in an arbitrary *diffusive* model $P_{\overline{\sigma}}$ with instantaneous volatility $\overline{\sigma}$ to the price P_σ in a local volatility model whose local volatility function is $\sigma(t, S)$:

$$P_{\overline{\sigma}}(t=0) = P_\sigma(0, S_0) + E_{\overline{\sigma}}\left[\int_0^T \frac{1}{2}e^{-rt}S_t^2\frac{d^2P_\sigma}{dS^2}\left(\overline{\sigma}_t^2 - \sigma(t, S_t)^2\right)dt\right] \tag{5.20}$$

We now set $\sigma(t, S) \equiv 0$ and use the notation: $P^T_{\overline{\sigma}}$ for $P_{\overline{\sigma}}$ and $P^T_{\hat{\sigma}=0}$ for P_σ. Using identity (5.18) in (5.20) yields:

$$P^T_{\overline{\sigma}} = P^T_{\hat{\sigma}=0} + e^{-rT}E_{\overline{\sigma}}\left[\int_0^T \overline{\sigma}_t^2dt\right] \tag{5.21}$$

where all prices are evaluated at $t = 0$.

We now turn to the variance swap. We have:

$$d\ln S_t = (r-q-\frac{1}{2}\overline{\sigma}_t^2)dt + \overline{\sigma}_tdW_t$$

where the first term in the right-hand side is of order dt and the second term is of order $\sqrt{dt}$. Square this expression and take the limit $dt \to 0$. The only contribution at order dt comes from the square of the second term and we get:

$$\lim_{dt \to 0} \frac{1}{dt} (d \ln S_t)^2 = \overline{\sigma}_t^2$$

which implies that:

$$\lim_{\Delta t \to 0} \sum_{i=0}^{N-1} \ln^2 \left(\frac{S_{i+1}}{S_i} \right) = \int_0^T \overline{\sigma}_t^2 dt$$

Thus, in the limit of frequent observations, the VS implied volatility $\widehat{\sigma}_{\mathrm{VS},T}$, defined by (5.2), is given by:

$$\widehat{\sigma}_{\mathrm{VS},T}^2 = \frac{1}{T} E \left[\lim_{\Delta t \to 0} \sum_{i=0}^{N-1} \ln^2 \left(\frac{S_{i+1}}{S_i} \right) \right] = E \left[\frac{1}{T} \int_0^T \overline{\sigma}_t^2 dt \right] \qquad (5.22)$$

How frequent is frequent? The log-return over $[t_i, \, t_{i+1}]$ is given by:

$$\ln \left(\frac{S_{i+1}}{S_i} \right) = \int_{t_i}^{t_{i+1}} \overline{\sigma}_t dW_t + \int_{t_i}^{t_{i+1}} \left(r - q - \frac{\overline{\sigma}_t^2}{2} \right) dt$$

The orders of magnitude of the two pieces in the right-hand side are, respectively, $\overline{\sigma}\sqrt{\Delta t}$ and $\overline{\sigma}^2 \Delta t$, with $\Delta t = t_{i+1} - t_i$. The second piece can be safely ignored whenever $\overline{\sigma}^2 \Delta t \ll \overline{\sigma}\sqrt{\Delta t}$, that is $\overline{\sigma}\sqrt{\Delta t} \ll 1$, which is typically the case for volatility levels of equity underlyings.

Assume now that our diffusive model is calibrated to the market smile: $P_{\widehat{\sigma}}^T = P_{\mathrm{Market}}^T$. From (5.21) and (5.22):

$$\widehat{\sigma}_{\mathrm{VS},T}^2 = E_{\widehat{\sigma}} \left[\frac{1}{T} \int_0^T \overline{\sigma}_t^2 dt \right] = \frac{e^{rT}}{T} \left(P_{\mathrm{Market}}^T - P_{\widehat{\sigma}=0}^T \right) \qquad (5.23)$$

This identity implies that any diffusive model calibrated to the vanilla smile prices VSs identically. Denote by $\widehat{\sigma}_T$ the implied volatility of the log contract; $\widehat{\sigma}_T$ is such that $P_{\mathrm{Market}}^T = P_{\widehat{\sigma}_T}^T$.

Using (5.20) again, still setting $\sigma(t, S) \equiv 0$, but this time choosing for $\overline{\sigma}_t$ the constant volatility $\widehat{\sigma}_T$ yields:

$$P_{\mathrm{Market}}^T = P_{\widehat{\sigma}}^T = P_{\widehat{\sigma}=0}^T + e^{-rT} T \widehat{\sigma}_T^2$$

Inserting this expression of P_{Market}^T in (5.23) yields our final result:

$$\widehat{\sigma}_{\mathrm{VS},T} = \widehat{\sigma}_T$$

$\widehat{\sigma}_T$ and $\widehat{\sigma}_{\mathrm{VS},T}$ are identical and are simply related to the quadratic variation of $\ln S_t$ over $[0,T]$. VS forward variances ξ^T are then identical to log contract forward variances ζ^T and the instantaneous variance of S_t is the short end of the variance curve: $\overline{\sigma}_t = \sqrt{\xi_t^t}$. The joint dynamics of S_t and ξ_t is given by equations (4.36):

$$\begin{cases} dS_t &= \sqrt{\xi_t^t}\,S_t dW_t^S \\ d\xi_t^T &= \lambda_t^T dW_t^T \end{cases}$$

The case of cash-amount dividends

When there are cash-amount dividends the log contract no longer has constant dollar gamma. However, we show in Section 5.6.2 below that the log contract can be supplemented with European payoffs of intermediate maturities to generate a portfolio that, risk-managed at zero implied volatility, again satisfies condition (5.18).

In conclusion, in a diffusive setting:

• The VS volatility is the implied volatility of a portfolio of European options. This portfolio reduces to the log contract when there are no cash-amount dividends

• Any diffusive model calibrated to the vanilla smile yields the same value for $\widehat{\sigma}_{\mathrm{VS},T}$

5.3.2   In jump-diffusion models

Assume that, in addition to the diffusion in (5.19), S_t is allowed to abruptly jump at times generated by a Poisson process with constant intensity λ – we take zero interest rate and repo without loss of generality:

$$dS_t = \overline{\sigma}_t S_t dW_t^S + S_{t-}\left(JdN_t - \lambda \overline{J}dt\right) \tag{5.24}$$

where relative magnitudes J of successive jumps are iid random variables, $\overline{J} = E[J]$ and N_t is the counting process of the underlying Poisson process. J and N_t are assumed to be independent and $E[dN_t] = \lambda dt$. The drift in (5.24) is the compensator of the jump process; it ensures that the financing cost of S vanishes: $E[dS_t] = 0$.[5] We have for S_T:

$$S_T = S_0 e^{-\frac{1}{2}\int_0^T \overline{\sigma}_t^2 dt + \int_0^T \overline{\sigma}_t dW_t^S} e^{-\lambda \overline{J}T} \prod_{i=1}^{N_T}(1 + J_i)$$

where N_T is the (random) number of jumps occurring over $[0,T]$, whose probability distribution is $p_n = p(N_T = n) = e^{-\lambda T}\frac{(\lambda T)^n}{n!}$. In particular, $E[N_T] = \lambda T$. We get

---

[5]See Appendix A of Chapter 10, page 407, for the interpretation of the pricing equation in jump-diffusion models.

for the price of the log contract:

$$E[-2\ln S_T]$$

$$= -2\ln S_0 + 2\left(\lambda \overline{J}T + \frac{1}{2}E\left[\int_0^T \overline{\sigma}_t^2\,dt\right]\right) - 2\Sigma_{n=0}^\infty p_n E\left[\ln\left((1+J)^n\right)\right]$$

$$= -2\ln S_0 + 2\left(\lambda \overline{J}T + \frac{1}{2}E\left[\int_0^T \overline{\sigma}_t^2\,dt\right]\right) - 2\lambda T\overline{\ln(1+J)}$$

where we have used the fact that the J_i are iid and independent of N_t and $\Sigma_{n=0}^\infty np_n = \lambda T$. Inverting (5.10) we get the implied volatility of the log contract:

$$\widehat{\sigma}_T^2 = E\left[\frac{1}{T}\int_0^T \overline{\sigma}_t^2\,dt\right] - 2\lambda \overline{\ln(1+J) - J} \qquad (5.25)$$

Let us now turn to the VS payoff (5.7). In the limit of frequent observations:

$$\widehat{\sigma}_{VS,T}^2 = \frac{1}{T}E\left[\lim_{\Delta\to 0}\sum_{i=0}^{N-1}\ln^2\left(\frac{S_{i+1}}{S_i}\right)\right] = \frac{1}{T}\int_0^T E\left[\frac{(d\ln S_t)^2}{dt}\right]dt$$

We have:

$$d\ln S_t = -\left(\frac{\overline{\sigma}_t^2}{2} + \lambda\overline{J}\right)dt + \overline{\sigma}_t dW_t + \ln(1+J)\,dN_t$$

Squaring this expression and taking the expectation yields:

$$\frac{E[(d\ln S_t)^2]}{dt} = E[\overline{\sigma}_t^2] + \lambda\overline{\ln^2(1+J)} \qquad (5.26)$$

which gives, in the limit of frequent observations:

$$\widehat{\sigma}_{VS,T}^2 = E\left[\frac{1}{T}\int_0^T \overline{\sigma}_t^2\,dt\right] + \lambda\overline{\ln^2(1+J)} \qquad (5.27)$$

In the right-hand side of (5.26) the contribution of jumps to $E[(d\ln S_t)^2]/dt$ is weighted by λ, as for small dt there can be at most one jump, with probability λdt. If spot observations are not sufficiently frequent, more than one jump can occur during the interval Δt, with the result that the contribution of jumps is no longer linear in λ: in addition to $\overline{\sigma}\sqrt{\Delta t} \ll 1$, we also need $\lambda\Delta t \ll 1$.

Compare (5.25) and (5.27). $\widehat{\sigma}_T$ and $\widehat{\sigma}_{VS,T}$ are not equal anymore and their difference is given by:[6]

$$\widehat{\sigma}_{VS,T}^2 - \widehat{\sigma}_T^2 = \lambda\overline{\ln^2(1+J) + 2\ln(1+J) - 2J} \qquad (5.28)$$

[6]The difference between $\widehat{\sigma}_{VS,T}^2$ and $\widehat{\sigma}_T^2$ does not depend on T because in our example the jump process has constant intensity.

remember that we are using ξ^T to denote VS forward variances and ζ^T to denote log-contract forward variances. (5.28) implies that ξ_t, ζ_t are related though:

$$\xi_t^T - \zeta_t^T = \overline{\lambda \ln^2 (1 + J) + 2 \ln (1 + J) - 2J}$$

ζ_t, ξ_t are both driftless and the joint dynamics of S_t, ζ_t, ξ_t, is given by:

$$\begin{cases} dS_t = \overline{\sigma}_t S_t dW_t^S + S_{t-} \left(J dN_t - \lambda \overline{J} dt \right) \\ d\zeta_t^T = \lambda_t^T dW_t^T \\ d\xi_t^T = \lambda_t^T dW_t^T \end{cases}$$

where $\overline{\sigma}_t$ equals neither $\sqrt{\xi_t^t}$ nor $\sqrt{\zeta_t^t}$. We can use (5.27) to express $\overline{\sigma}_t$ as a function of ξ_t^t:

$$\overline{\sigma}_t = \sqrt{\xi_t^t - \lambda \overline{\ln^2 (1 + J)}}$$

The fact that $\overline{\sigma}_t$, $\sqrt{\xi_t^t}$ and $\sqrt{\zeta_t^t}$ are all different is typical of models for S_t that are not pure diffusions. We now estimate the order of magnitude of this difference and how it is related to the market smile used for calibration.

5.3.3   Difference of VS and log-contract implied volatilities

In the discussion that follows, we consider "pure" jump-diffusion/Lévy models, that is models with no dynamical variables besides S_t. This excludes mixtures of stochastic volatility and jump/Lévy processes, for example models where $\overline{\sigma}_t$ is a process correlated with S_t.

We will henceforth use "jump-diffusion" to denote jump/Lévy processes, that is processes with independent stationary increments for $\ln S_t$. In the context of the preceding section, this amounts to setting $\overline{\sigma}_t$ constant.

The dynamics in (5.24) serves as a basic prototype for the class of Lévy processes, but our conclusions have general relevance.

Let us expand the right-hand side of (5.28) in powers of J. The first non-vanishing contribution comes from J^3 and we get:

$$\hat{\sigma}_{\text{VS},T}^2 - \hat{\sigma}_T^2 \simeq -\frac{1}{3} \lambda \overline{J^3} \tag{5.29}$$

The fact that the first non-vanishing contribution is of order 3 in J is expected: at order 2 in J the effect of jumps is identical to that of a simple diffusion – see Appendix A of Chapter 10: whether the return was generated by a jump or by Brownian motion is immaterial.

Higher-order terms – $\lambda \overline{J^4}$, $\lambda \overline{J^5}$, etc. – contribute as well, but the order-three term is the leading contribution in the limit of small and frequent jumps. Indeed, when taking the limit $J \to 0$, λ should be increased so that the contribution of jumps to the quadratic variation of $\ln S$ – and the quadratic variation itself – stays fixed. Equation (5.26) shows that as $J \to 0$, for $\lambda \overline{J^2}$ to stay constant, λ has to scale like $\frac{1}{J^2}$. Terms $\lambda \overline{J^n}$ are then of order J^{n-2}: the largest contribution is generated by $n = 3$.

Calibrating jump parameters to the vanilla smile

In jump-diffusion models, jumps not only introduce a difference between $\widehat{\sigma}_T$ and $\widehat{\sigma}_{\text{VS},T}$; they also generate a smile. In Appendix A of Chapter 10 – see equation (10.26), page 413 – it is shown that in the limit $J \to 0$ the ATMF skew for maturity T, $\mathcal{S}_T$, is given, at order one in the skewness of $\ln S_T$, by:

$$\mathcal{S}_T \simeq \frac{\lambda \overline{J^3}}{6 \widehat{\sigma}_T^3 T} \tag{5.30}$$

which, using (5.29), gives the following approximation for the difference of $\widehat{\sigma}_T$ and $\widehat{\sigma}_{\text{VS},T}$ as a function of the ATMF skew for maturity T:

$$\widehat{\sigma}_{\text{VS},T}^2 - \widehat{\sigma}_T^2 \simeq -2 \widehat{\sigma}_T^3 \mathcal{S}_T T \tag{5.31}$$

Assuming the right-hand side is small:

$$\widehat{\sigma}_{\text{VS},T} \simeq \widehat{\sigma}_T \left(1 - \widehat{\sigma}_T \mathcal{S}_T T\right) \tag{5.32}$$

Thus, given a market smile, assuming a diffusion for S leads to:

$$\widehat{\sigma}_{\text{VS},T} = \widehat{\sigma}_T$$

while assuming a jump-diffusion process leads to (5.28), which for weak smiles is approximated by (5.32):

$$\widehat{\sigma}_{\text{VS},T} \simeq \widehat{\sigma}_T \left(1 - \widehat{\sigma}_T \mathcal{S}_T T\right)$$

Consider the case of a one-year maturity VS on an equity index. Typically, the difference of the implied volatilities for strikes 95% and 105% is of the order of 2 points of volatility, which gives $\mathcal{S}_T = -0.02/\ln(105/95) = -0.2$. Taking $\widehat{\sigma}_T = 20\%$, yields $\widehat{\sigma}_T \mathcal{S}_T T = -4\%$. Equation (5.32) then yields a difference of about one point of volatility between $\widehat{\sigma}_{\text{VS},T}$ and $\widehat{\sigma}_T$.

The size of this correction has prompted some to argue that replicating variance swaps with log contracts is inadequate and that variance swaps should be priced with jump/Lévy models calibrated on the market smile. Is this reasonable?

As we now show, the difference between $\widehat{\sigma}_{\text{VS},T}$ and $\widehat{\sigma}_T$ is due to the non-vanishing skewness of returns *at short time scales*. In the case of a jump-diffusion model, this skewness is inferred from the market smile of maturity T. Is this model-mediated relationship between *skewness* of short returns and *skew* of the T-maturity smile robust?

5.3.4   Impact of the skewness of daily returns – model-free

We now look at things in model-free fashion, with zero interest and repo, for simplicity.

Let r_i be the log-return of S over $[t_i, t_{i+1}]$: $r_i = \ln(\frac{S_{i+1}}{S_i})$, and imagine that we have sold a variance swap and are long a delta-hedged log contract. We assume that we are keeping this static position until $t = T$, risk-managing the VS at a fixed implied volatility $\hat{\sigma}_{VS,T}$ and the log contract at a fixed implied volatility $\hat{\sigma}_T$.

Our total P&L during Δt is the difference between the P&L of the delta-hedged log contract and the payoff of the VS over time interval $[t_i, t_{i+1}]$:

$$
\begin{aligned}
P\&L &= \left(Q^T(t_{i+1}, S_{i+1}) - Q^T(t_i, S_i)\right) - \frac{dQ^T}{dS}(t_i, S_i)(S_{i+1} - S_i) \qquad (5.33)\\
&\quad - (r_i^2 - \hat{\sigma}_{VS,T}^2 \Delta t)\\
&= \left(2(e^{r_i} - 1) - 2r_i - \hat{\sigma}_T^2 \Delta t\right) - \left(r_i^2 - \hat{\sigma}_{VS,T}^2 \Delta t\right) \qquad (5.34)\\
&= \left(2(e^{r_i} - 1) - 2r_i - r_i^2\right) - \left(\hat{\sigma}_T^2 - \hat{\sigma}_{VS,T}^2\right)\Delta t \qquad (5.35)
\end{aligned}
$$

where the first piece in the right-hand side of (5.33) is the P&L of the delta-hedged log contract over $[t_i, t_{i+1}]$. We have used the expressions of Q^T and $\frac{dQ^T}{dS}$ given in (5.10) and (5.11) to get (5.35).

Expand this P&L in powers of r_i. Up to order 2 in r_i, the payoff of the VS and the P&L of the delta-hedged log contract match. The first non-vanishing contribution comes from the order-three term and we get:

$$
P\&L \simeq \frac{r_i^3}{3} - \left(\hat{\sigma}_T^2 - \hat{\sigma}_{VS,T}^2\right)\Delta t \qquad (5.36)
$$

The combination of a static long position in a variance swap and a short position in a delta-hedged log contract generates at lowest order the daily P&L (5.36) which looks like a gamma/theta P&L except it involves r_i^3 rather than r_i^2.

Writing that the P&L in (5.36) vanishes on average yields the following relationship between the skewness of daily log-returns and the difference between $\hat{\sigma}_T$ and $\hat{\sigma}_{VS,T}$:

$$
\hat{\sigma}_{VS,T}^2 - \hat{\sigma}_T^2 \simeq -\frac{\langle r^3 \rangle}{3\Delta t} \simeq -\frac{s_{\Delta t}}{3}\hat{\sigma}_T^3 \sqrt{\Delta t} \qquad (5.37)
$$

where $s_{\Delta t}$ denotes the skewness of daily returns defined by: $s_{\Delta t} = \langle r^3 \rangle / \langle r^2 \rangle^{\frac{3}{2}}$. For the sake of relating $\langle r^3 \rangle$ to $s_{\Delta t}$ we have taken $\langle r^2 \rangle = \hat{\sigma}_T^2 \Delta t$; at order one in the difference $\hat{\sigma}_{VS,T} - \hat{\sigma}_T$ we could have equivalently used $\hat{\sigma}_{VS,T}^2 \Delta t$.

Assuming that the right-hand side is small, this results in the following adjustment for $\hat{\sigma}_{VS,T}$, at order one:

$$
\frac{\hat{\sigma}_{VS,T}}{\hat{\sigma}_T} - 1 \simeq -\frac{s_{\Delta t}}{6}\hat{\sigma}_T \sqrt{\Delta t} \qquad (5.38)
$$

The interpretation of the results in Sections 5.3.1 and 5.3.2 is now clear:

- In diffusive models $\langle r^3 \rangle$ scales like $\Delta t^{3/2}$. As $\Delta t \to 0$ the contribution of $\langle r^3 \rangle$ becomes negligible with respect to that of $\langle r^2 \rangle$, which scales like Δt: $\widehat{\sigma}_{\mathrm{VS},T} = \widehat{\sigma}_T$.

- In jump-diffusion models, the portion of $\langle r^3 \rangle$ that is generated by jumps is proportional to $\lambda \Delta t \overline{J^3}$, thus scales like Δt, just as $\langle r^2 \rangle$. Hence, $\widehat{\sigma}_T \neq \widehat{\sigma}_{\mathrm{VS},T}$. The implied value of the cubes of daily log-returns is non-vanishing and depends on jump parameters calibrated on the smile of maturity T.

5.3.5 Inferring the skewness of daily returns from market smiles?

Using (5.37) together with (5.31) yields:

$$-2\widehat{\sigma}_T^3 \mathcal{S}_T T \simeq -\frac{\widehat{s}_{\Delta t}}{3}\widehat{\sigma}_T^3 \sqrt{\Delta t}$$

which gives:

$$\widehat{s}_{\Delta t} \simeq 6\frac{\mathcal{S}_T T}{\sqrt{\Delta t}} \tag{5.39}$$

Consider the case of a one-year maturity and let us use the same level of ATMF skew as in the numerical example on page 161. Taking $\mathcal{S}_T = -0.2$ and $\Delta t = \frac{1}{252}$ gives $\widehat{s}_{\Delta t} \simeq -19$. This is a very large value, much larger than its historical average – see below.

This value of $\widehat{s}_{\Delta t}$ is derived from the smile of maturity T, the VS maturity, through equation (5.39). How is it that, out of a calibration to the market smile of maturity T, the jump-diffusion model is able to predict the value of the skewness of returns at short time scales? Is this prediction robust?

Expression (5.39) shows that, had we used a different value for T, we would have obtained a different estimate for $\widehat{s}_{\Delta t}$ – unless $\mathcal{S}_T$ scales like $\frac{1}{T}$.

It is a well-known property of jump-diffusion processes, that, for small jump amplitudes, the ATMF skew they generate scales like $\frac{1}{T}$ – see equation (5.30). Indeed these processes generate independent stationary increments for $\ln S_T$. Cumulants of $\ln S_T$ then scale linearly with T, which implies that the skewness s_T of $\ln S_T$ scales like $\frac{1}{\sqrt{T}}$.

As shown in Appendix B, perturbation of the lognormal Black-Scholes density at order one in the third-order cumulant yields identity (5.93) between the ATMF skew $\mathcal{S}_T$ and the skewness s_T for maturity T:

$$\mathcal{S}_T = \frac{s_T}{6\sqrt{T}} \tag{5.40}$$

A scaling of $s_T \propto \frac{1}{\sqrt{T}}$ then translates into the property that the ATMF skew scales like $\frac{1}{T}$, which is what equation (5.30) expresses.

The skew that a jump-diffusion model generates is a direct reflection of the skewness of increments of $\ln S_t$ at short time scales: for weak skews $\mathcal{S}_T \propto \frac{1}{T}$ and equation (5.39) supplies a value for the skewness of daily returns that does not depend on the particular value of T used.

The situation is very different in diffusive stochastic volatility models: the process for $\ln S_t$ is Gaussian at short time scales[7] and develops skewness at longer time scales by the fact that volatility is stochastic and correlated with S_t. The smile produced by a stochastic volatility model is *not* a reflection of the non-Gaussian character of returns at short time scales.

Thus, a jump-diffusion model yields a sizeable difference between $\widehat{\sigma}_T$ and $\widehat{\sigma}_{\text{VS},T}$ because it assumes a direct relationship between the ATMF skew of maturity T and $\widehat{s}_{\Delta t}$. This is a bold assumption, which, moreover, is not supported by market data as, typically, ATMF skews approximately scale like $\frac{1}{\sqrt{T}}$, rather than $\frac{1}{T}$.[8]

5.3.6 Preliminary conclusion

The difference between $\widehat{\sigma}_T$ and $\widehat{\sigma}_{\text{VS},T}$ depends on the non-Gaussian character of daily log-returns. Inferring the *implied* skewness of daily log-returns out of vanilla smiles through the filter of a calibrated model leads to a correction to $\widehat{\sigma}_{\text{VS},T}$ which is unreasonably model-dependent.

In fact the *implied* skewness of daily returns could only be accessible if the package consisting of a variance swap and the offsetting log contract was actually traded. While the *realized* skewness of daily returns is typically small – see below – the *implied* skewness that would then be backed out of the difference between $\widehat{\sigma}_T$ and $\widehat{\sigma}_{\text{VS},T}$ could be arbitrarily large.[9]

5.3.7 In reality

Imagine that there was no active VS market and we were asked to quote a VS – this would have been a typical situation in the late 90s. What would we have done?

---

[7]This statement does not stand in contradiction with the well-known property that the ATMF skew in stochastic volatility models does not vanish in the limit $T \to 0$. We show in Section 8.5 that, at order one in volatility of volatility, the *skewness* of short-maturity log-returns scales like $\sqrt{T}$, hence vanishes at short time scales: returns become Gaussian. The skew is given approximately by (5.40): because the *skewness* scales like $\sqrt{T}$, the *skew* tends to a finite limit. Generally, whether the short-maturity limit of the ATMF *skew* vanishes or not depends on whether the *skewness* vanishes faster or slower than $\sqrt{T}$ as $T \to 0$.

[8]In our discussion we have used the example of a jump-diffusion model with independent increments for $\ln S$: The skew $\mathcal{S}_T$ is then fully generated by the jump process only. In models that are mixtures of jump/Lévy process and stochastic volatility – for example jump-diffusion models where $\overline{\sigma}_t$ is stochastic and correlated with S_t, or subordinated Lévy processes where physical time t is replaced with the integral of a random positive diffusive process (see for example [26]) – $\mathcal{S}_T$ is a product of both jump and stochastic volatility portions of the model. Contrary to the jump-diffusion model that we have used so far, in such models the implied skewness of daily returns $\widehat{s}_{\Delta t}$ is not simply a function of the level of the market skew, as part of the latter is generated by stochastic volatility. We are exposed to the additional risk of letting the model determine how much of the vanilla smile is generated by the jump or Lévy component.

[9]So-called daily cliquets – strips of put options on daily index returns with far out-of-the-money strikes, say 80% – are a case in point. Their market prices can be drastically different than prices computed using historical densities of daily returns. See Chapter 10, page 391.

Typically we would have used a conservative estimate of the realized skewness to compute an adjustment to $\widehat{\sigma}_T$ using formula (5.38):[10]

$$\widehat{\sigma}_{\text{VS},T} \simeq \widehat{\sigma}_T \left(1 - \frac{s\Delta t}{6}\widehat{\sigma}_T\sqrt{\Delta t}\right)$$

For indexes, the *realized* skewness $s_{\Delta t}$ of daily log-returns, defined by: $s_{\Delta t} = \langle r^3 \rangle / \langle r^2 \rangle^{\frac{3}{2}}$, where we take $\langle r \rangle = 0$, is a (dimensionless) number of order 1.

Perhaps surprisingly, the skewness of daily returns of equity indexes is not always negative. Historical skewness is difficult to measure, as the skewness estimator is very sensitive to large returns. The skewness estimator applied to a historical sample that includes an inordinately large return will be swamped by the contribution of that one return. Whenever one computes realized skewness over sufficiently long periods of time that do not contain large returns, one finds a number with varying sign, of order 1.

Obviously, evaluating $s_{\Delta t}$ for the S&P 500 index over a historical sample that includes October 1987 will yield a large negative number, however its magnitude depends crucially on the size of the window used for its estimation, as this large number is generated by one single return. The notion that evaluating $s_{\Delta t}$ over a historical sample that includes a market crash – say, October 1987 – gives an estimate of $s_{\Delta t}$ that appropriately accounts for the possibility of crashes is thus a misguided idea. Whenever one wishes to adjust the price of a derivative for the impact of large market moves, it is much more reasonable to include an explicit stress-test impact in the derivative's price – see below (5.42) for the particular case of the variance swap.

Taking $s_{\Delta t} = -1$, $\widehat{\sigma}_T = 20\%$, $\Delta t = \frac{1}{252}$, and using (5.38) gives a relative adjustment $-\frac{s\Delta t}{6}\widehat{\sigma}_T\sqrt{\Delta t} = 0.2\%$.

It is in fact possible to assess the magnitude of the contribution of all orders – not just that of r^3 – by directly measuring on a historical sample the relative difference of the payoff of the VS and the P&L of the delta-hedged log contract (see the two expressions in (5.34)):

$$\frac{1}{2}\left(\frac{\langle r^2 \rangle}{\langle 2(e^r - 1) - 2r \rangle} - 1\right) \tag{5.41}$$

We have used the factor $1/2$ to convert a relative mismatch of variances into a relative mismatch of volatilities. Figure 5.1 shows the ratio (5.41) computed over 20 years of daily returns of the S&P 500 index, measured with a one-year sliding window.

---

[10]Because of the unavailability of out-of-the-money options needed in the replication of the log contract, we would also have charged an additional amount to cover for the cost of buying/selling options at then-prevailing market conditions whenever the spot moved.

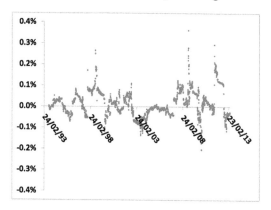

Figure 5.1: The ratio (5.41) evaluated for the S&P 500 index over 20 years, with a one-year sliding window.

As is apparent, ratio (5.41) is very noisy, even with a one-year estimator, but 0.2% is the right order of magnitude for the relative adjustment of $\hat{\sigma}_{VS,T}$ with respect to $\hat{\sigma}_T$.[11]

This represents a very minute correction: the level of realized skewness of daily returns does not warrant in practice a distinction between $\hat{\sigma}_T$ and $\hat{\sigma}_{VS,T}$.

On the modeling side, it is possible to build a stochastic volatility model that affords full control of the conditional distribution of daily returns, thereby allowing an assessment of the difference between $\hat{\sigma}_T$ and $\hat{\sigma}_{VS,T}$. This is done in Chapter 10 and the case of VSs is specifically covered in Section 10.2.4.

In case we would like to adjust $\hat{\sigma}_{VS,T}$ for the occurrence of large returns, for example to cover the cost of a stress-test P&L on a VS position, it is more reasonable to use formula (5.28). For example, a trading desk that was charged on an annual basis a fraction ε of its stress-test P&L would quote $\hat{\sigma}_{VS,T}$ according to:

$$\hat{\sigma}_{VS,T}^2 \;=\; \hat{\sigma}_T^2 + \varepsilon\big(\ln^2(1+J) + 2\ln(1+J) - 2J\big) \tag{5.42}$$

where J is the amplitude of the stress-test scenario – for equities J would be negative.

In the absence of a VS market we would then have chosen $\hat{\sigma}_{VS,T} = \hat{\sigma}_T$ with the possible addition of a reserve policy in the form of an adjustment given by (5.42).

It is important to note that adjustment (5.42) does not quantify the impact of large returns on a VS. It quantifies the impact of large returns on a position consisting of

---

[11]The largest values of ratio (5.41) are reached during the 1987 crash and lie outside Figure 5.1. The maximum is 2.5% (compare with the scale of the y axis), which is not a meaningful number, as it depends on the width of the window used for the estimator – here one year.

a VS together with the offsetting log contract, i.e. the portion of the impact of large returns on VSs *that is not already accounted for in vanilla option prices.*[12]

Assuming a monthly negative jump of 5% ($\varepsilon = 12$, $J = -5\%$), (5.42) yields for $\widehat{\sigma}_T = 20\%$ an adjustment $\widehat{\sigma}_{\text{VS},T} = \widehat{\sigma}_T + 0.13\%$.

5.4 Impact of strike discreteness

The previous section has been devoted to the assessment of the difference between the payoff of a VS and the P&L generated from delta-hedging a log contract. In reality, neither is the latter traded, nor can it be perfectly synthesized out of vanilla options, for the simple reason that only discrete strikes trade. The log contract is thus replaced with a piecewise affine profile.

Figure 5.2 shows the relative difference between the VS payoff and the P&L generated by delta-hedging an approximation of the log contract that uses discrete strikes K_i, such that the $\ln K_i$ are equally spaced. This difference is expressed as a relative adjustment factor on $\widehat{\sigma}_T$. For $\Delta \ln K \to 0$ this adjustment factor is the ratio (5.41), shown in Figure 5.1 for the case of a 1-year VS contract.

Each point in Figure 5.2 corresponds to the replication of a 1-year VS using S&P 500 daily closing quotes. The log contract is approximated by a strip of vanilla options with $\Delta \ln K = 5\%$ (resp. 1%) for the left-hand (resp. right-hand) graph, with $K_{\min} = 10\%S_0$, $K_{\max} = 500\%S_0$, where S_0 is the spot value at inception of the 1-year VS. We delta-hedge the vanilla portfolio at constant volatility.[13]

The right-hand graph in Figure 5.2 is similar to Figure 5.1: a strip of vanilla options with strikes spaced 1% apart provides an acceptable replication of the log contract, at least for a 1-year maturity.

The left-hand graph shows that with a coarser discretization, the replication of the VS payoff is much less accurate: a relative adjustment of $\widehat{\sigma}_T$ of 2% translates for $\widehat{\sigma}_T = 20\%$ in an adjustment of about plus or minus half a volatility point. Notice however, that, in contrast with the P&L impact of higher-order returns, the additional P&L generated by the imperfect replication of the log-contract payoff has no reason

---

[12]Interestingly, had the market standard for VS contracts featured standard returns: $\frac{S_{i+1}}{S_i} - 1$, rather than log-returns, the order-three term in expression (5.36) of the P&L would have been equal to $-\frac{2}{3}r^3$, rather than $\frac{1}{3}r^3$, where r is the log-return. At leading order, a short position in a VS contract combined with a long position in its delta-hedged vanilla-option hedge would generate positive, rather than negative P&L, for $J < 0$. Another advantage of using standard returns is that, even in the case of extreme bankruptcy ($S_{i+1} = 0$) the return remains well-defined.

[13]This constant volatility is taken equal to the realized volatility over the 1-year period that follows – a proxy for the actual market implied volatility of the vanilla portfolio, which we would use in reality. In practice the final P&L is not very sensitive to the actual implied volatility used for risk-managing the vanilla portfolio, especially if the log-contract profile is well approximated. Remember that the delta of the log contract in the Black-Scholes model does not depend on the implied volatility.

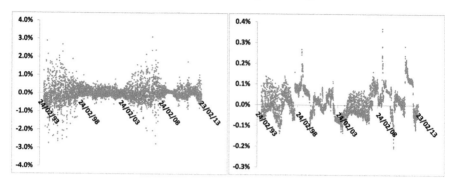

Figure 5.2: Mismatch of (a) the payoff of a 1-year VS, (b) the P&L generated by delta-hedging an approximation of the log-contract payoff of maturity 1 year using discrete strikes K_i with $\Delta \ln K = 5\%$ (left) and 1% (right). We have used S&P 500 daily quotes. This difference is expressed as a relative adjustment on the VS volatility $\widehat{\sigma}_T$.

to be biased and should be considered a noise – this is clearly seen in the left-hand graph of Figure 5.2.

5.5   Conclusion

- If log contracts were traded or, equivalently, if vanilla options of all strikes were available, both VS ($\widehat{\sigma}_{\mathrm{VS},T}$) and log contract ($\widehat{\sigma}_T$) implied volatilities would be market parameters.

  Their difference would supply a measure of the *implied* skewness of daily returns, which could be arbitrarily large, even though its *realized* counterpart is small. In such a situation VS and log-contract forward variances ξ^T and ζ^T would be different objects and the joint dynamics of S, ξ^T, ζ^T would read as:

  $$\begin{cases} dS_t &= \overline{\sigma}_t S_t dW_t^S \\ d\zeta_t^T &= \lambda_t^T dU_t^T \\ d\xi_t^T &= \psi_t^T dV_t^T \end{cases}$$

  $\overline{\sigma}_t, \ \zeta_t^T, \xi_t^T$ are all different.

- In the absence of liquid log contracts, we have shown that it makes no sense to price VSs using a jump-diffusion or Lévy model calibrated on the vanilla smile. By doing so, we use the model as a tool for inferring the skewness of returns at short time scales out of vanilla smiles; this is incongruous as there is no reason why these quantities should be related – except in the model.

Above all, there is no way of locking in this relationship – that is the difference between $\widehat{\sigma}_{\mathrm{VS},T}$ and $\widehat{\sigma}_T$ – by trading vanilla options.

- In practice VSs are much more liquid than far out-of-the-money vanilla options: while $\widehat{\sigma}_{\mathrm{VS},T}$ is a market parameter, $\widehat{\sigma}_T$ usually is not and depends on how one extrapolates implied volatilities for strikes that lie outside the liquid range.

  It is then common practice among index volatility market makers to choose this extrapolation such that replication of the log contract recovers the VS implied volatility – this enforces the equality $\widehat{\sigma}_T = \widehat{\sigma}_{\mathrm{VS},T}$.

- In the following chapters, we will thus make the assumption that the process for S_t is a diffusion, so that $\widehat{\sigma}_T = \widehat{\sigma}_{\mathrm{VS},T}$ and that the instantaneous volatility of S is given by the short end of the variance curve:

$$
\begin{cases}
dS_t &= \sqrt{\xi_t^t}\, S_t dW_t^S \\
d\xi_t^T &= \lambda_t^T dW_t^T
\end{cases}
\tag{5.43}
$$

One exception is Chapter 10 where we examine the impact of the conditional distribution of daily returns on derivative prices. We employ a model that gives us explicit control on the one-day smile and assess the difference $\widehat{\sigma}_{\mathrm{VS},T} - \widehat{\sigma}_T$.

- The two key properties expressed by (5.43) still hold in the presence of cash-amount dividends:

  - Forward VS variances are still driftless as the reasoning used in Section 5.1 still applies.
  - The instantaneous volatility of S is still given by the short end of the variance curve.

When $\widehat{\sigma}_{\mathrm{VS},T}$ and $\widehat{\sigma}_T$ are different

- For some very liquid indexes such as the S&P 500 or Euro Stoxx 50, far out-of-the-money puts are liquid. The spread between $\widehat{\sigma}_{\mathrm{VS},T}$ and $\widehat{\sigma}_T$ that one typically observes, even for short maturities, is not attributable to uncertainty about far out-of-money implied volatilities: the market does make a distinction between $\widehat{\sigma}_{\mathrm{VS},T}$ and $\widehat{\sigma}_T$.

  As discussed in Sections 5.3.7 and 5.4, there are valid reasons for this: (a) terms of order higher than 2 in the P&L of the delta-hedged log contract, whose implied value could be much larger than their realized value in Figure 5.1, (b) the fact that the log contract is in practice approximated by a portfolio of vanilla options with discrete strikes.

- In these cases, one can use expression (5.42) to dissociate $\widehat{\sigma}_{\text{VS},T}$ from $\widehat{\sigma}_T$, with J chosen and ε – possibly time-dependent – calibrated so that the market VS volatility $\widehat{\sigma}_{\text{VS},T}$ is recovered.

  This is not the same as using a jump model. Equation (5.42) expresses the spread between $\widehat{\sigma}_{\text{VS},T}$ and $\widehat{\sigma}_T$ as the *difference* in how a jump of magnitude J impacts a VS relative to a log contract. Vanilla options would still be priced with a *diffusive* model – say a local volatility or stochastic volatility model of type (5.43) calibrated to the vanilla smile – but whenever a spot-starting or forward-starting VS was priced, the adjustment specified in (5.42) would be used to generate $\widehat{\sigma}_{\text{VS},T}$ from $\widehat{\sigma}_T$.

- In a pricing library, realized variance – in the form of squared log-returns – would need to be identified as such so that it can be adjusted automatically. We still simulate SDE (5.43), except ξ_t^T has the status of a log-contract forward variance. Whenever the payoff at hand calls for observation of realized variance, adjustment (5.42) is applied automatically:

$$\ln^2\left(\frac{S_{i+1}}{S_i}\right) \;\to\; \ln^2\left(\frac{S_{i+1}}{S_i}\right) \;+\; (\lambda\Delta)\,\overline{\ln^2\left(1+J\right) + 2\ln\left(1+J\right) - 2J}$$

  It is also applied when the payoff calls for observation of implied realized variance – that is a forward VS variance ζ_t^T – for example in the case of a variance swaption:[14]

$$\zeta_t^T \;=\; \xi_t^T \;+\; \lambda\,\overline{\ln^2\left(1+J\right) + 2\ln\left(1+J\right) - 2J}$$

  This additive adjustment preserves the martingality of ζ_t^T. λ, J are chosen so as to match market values of $\widehat{\sigma}_{\text{VS},T} - \widehat{\sigma}_T$. λ is posssibly time-dependent, in which case it is replaced by $\frac{1}{T}\int_0^T \lambda_t dt$ in the formulas above. The exposure to the spread between $\widehat{\sigma}_{\text{VS},T}$ and $\widehat{\sigma}_T$ is easily monitored at the book level.

- For longer-dated VSs a large mismatch between $\widehat{\sigma}_{\text{VS},T}$ and $\widehat{\sigma}_T$ is more likely caused by a mispricing of the effect of interest-rate volatility – see Section 5.8 below – or an inappropriate dividend model.[15]

---

[14]In essence, this boils down to considering $\ln\left(S_{i+1}/S_i\right)^2$ as a short-period cliquet, and performing an ad-hoc adjustment for the forward-smile risk over interval $[t_i, t_{i+1}]$.

[15]Given market prices of vanilla options and the forward, one obtains different values for $\widehat{\sigma}_T$ depending upon whether one models dividends as proportional to the spot or as fixed cash amounts. $\widehat{\sigma}_T$ is higher when proportional dividends are used. There is an additional impact for index variance swaps, as the market convention mandates that daily returns be not stripped of the contribution of dividends. As shown in the next section this contribution to $\widehat{\sigma}_T$ is small, but can become unreasonably large if evaluated with an inappropriate dividend model – see Section 5.7 below.

5.6 Dividends

Dividends have two effects on the pricing of VSs: on the payoff itself and on its replication with vanilla options.

5.6.1 Impact on the VS payoff

Imagine a dividend d falls just after the close of day t_i: $S_{t_i^+} = S_{t_i^-} - d$.[16] The log-return over $[t_i, t_{i+1}]$ can then be written as:

$$\ln\left(\frac{S_{i+1}}{S_i}\right) = \ln\left(\frac{S_{i+1}}{S_i - d}\right) + \ln\left(\frac{S_i - d}{S_i}\right)$$

Let us assume that S follows a diffusion with instantaneous volatility $\bar{\sigma}$. Squaring this expression, taking its expectation over S_{i+1} and keeping terms up to order 2 in Δt yields:

$$E\left[\ln^2\left(\frac{S_{i+1}}{S_i}\right)\right] = \bar{\sigma}^2 \Delta t + \ln^2\left(\frac{S_i - d}{S_i}\right) + 2\left(r - q - \frac{\bar{\sigma}^2}{2}\right)\Delta t \ln\left(\frac{S_i - d}{S_i}\right) \quad (5.44)$$

Let us take the following typical values for a stock: $\bar{\sigma} = 30\%$, $\Delta t = \frac{1}{252}$, $\frac{d}{S_i} = 3\%$, $r - q = 3\%$: the order of magnitudes of the three terms above is, respectively: 3.10^{-4}, 10^{-3}, 2.10^{-6}: the drift of S over Δt is so small that the last term can be safely ignored. It turns out that the second term in the right-hand side of (5.44) can be discarded as well for the following reasons:

- For stocks, the usual convention of VS term sheets is to adjust the return over $[t_i, t_{i+1}]$ by the dividend amount d. The return used for the sake of computing realized variance is $\ln(\frac{S_{i+1}}{S_i - d})$.

- For indexes, the value of S_i is not adjusted for d, however an index is a basket of stocks: it jumps whenever a dividend is paid on one of its components by an amount equal to the dividend times the relative weight of that particular stock in the index. The yearly dividend yield of an index is the same order of magnitude as the yields of the components, except it is spread out over many dates in the year, corresponding to the dividend payment dates of the components.

  Take the example of the Euro Stoxx 50 index, with $n = 50$ dividend dates per year. Let us use a constant volatility $\bar{\sigma} = 20\%$ and assume that each of the n dividends is proportional to S, with a proportionality coefficient equal to q/n,

---

[16]The dividend is not paid to stockholders at that time. Rather, anyone who was not owning the share at t_i loses the right to the dividend payment – which occurs at a later date. The effect on S is however identical to that of a dividend payment occurring between t_i and t_{i+1}: S jumps by the value of the right to the dividend payment.

so that the yearly dividend yield q for the index is 3%. Keeping the first two terms in (5.44) yields for the realized volatility over one year:

$$\sigma_r = \sqrt{\bar{\sigma}^2 + n \ln^2 \left(1 - \frac{q}{n}\right)}$$

For large n, $n \ln^2 \left(1 - \frac{q}{n}\right)$ scales like $\frac{1}{n}$ and the contribution of dividends become negligible: using the numerical values above yields $\sigma_r = 20.005\%$.

The conclusion is that the impact of dividends on the VS payoff itself is negligible in practice.

5.6.2   Impact on the VS replication

Let us take zero interest rate and repo for simplicity and imagine that fixed cash dividends d_j fall at dates T_j. Consider a delta-hedged log contract at zero implied volatility. Its value is now given by

$$Q^T(t, S) = -2 \ln \left(S - \sum_{t < T_j < T} d_j\right)$$

The dollar gamma is given by:

$$\frac{S^2}{2} \frac{d^2 Q^T}{dS^2} = \frac{S^2}{\left(S - \sum_{t < T_j < T} d_j\right)^2} \tag{5.45}$$

Because of the presence of dividends with fixed cash amounts, it is no longer equal to one. It is however possible to assemble a portfolio that has constant dollar gamma by supplementing the log contract of maturity T with a set of European payoffs E^j of maturities T_j^-. Denote by T_N the last dividend date before T. At time $t = T_N^-$, the value of the log contract is $-2 \ln (S - d_N)$. To have a portfolio whose value is $-2 \ln S$, we need to go long an additional European payoff $E^N(S)$ of maturity T_N^- such that

$$-2 \ln (S - d_N) + E^N(S) = -2 \ln S$$

which yields:

$$E^N(S) = 2 \ln \left(\frac{S - d_N}{S}\right) \tag{5.46}$$

Working backward in time we can check that European payoffs E^j maturing at previous dates T_j^- have the same form as E^N: $E^j(S) = 2 \ln \left(\frac{S - d_j}{S}\right)$. In case interest rates are non-vanishing, the quantities of these intermediate payoffs are changed slightly, but keep the form (5.46).[17]

---

[17]Obviously, payoff $E^j(S)$ is only defined for $S > d_j$. The implied volatility surface used for pricing payoff E^j must ensure that this condition holds. In other words, the smile used as input must be such that the density of $S_{T_j^-}$ vanishes for $S_{T_j^-} < d_j$.

The conclusion is that $\hat{\sigma}_{VS}$ is no longer equal to $\hat{\sigma}_T$ but depends on smiles of intermediate maturities. The VS is replicated statically at second order in $\frac{\delta S}{S}$ by trading a log contract plus a series of European payoffs with maturities corresponding to dividend dates. Using (5.23), we have:

$$\hat{\sigma}^2_{VS,T} = \frac{e^{rT}}{T}\left(Q^T_{\text{market}} - Q^T_{\hat{\sigma}=0} + \sum_{T_j < T}\left(E^j_{\text{market}} - E^j_{\hat{\sigma}=0}\right)\right) \qquad (5.47)$$

For index VSs the contribution of dividends to the realized variance – examined in the previous section – has to be added by hand, in the form of additional $\ln^2\left(1 - \frac{d_j}{S}\right)$ payoffs – see section below.

5.7 Pricing variance swaps with a PDE

Indexes are large baskets; computing $\hat{\sigma}_{VS,T}$ with (5.47) entails replicating payoffs E_j for many maturities, corresponding to the (numerous) dividend dates of the components. It is more convenient to compute $\hat{\sigma}_{VS,T}$ by using a diffusive model calibrated on the market smile – we know from Section 5.3.1 and equation (5.23) that this yields the same value for $\hat{\sigma}_{VS,T}$ as (5.47).

As any diffusive model calibrated to the market smile will do, we can simply use a local volatility model. We have:

$$\hat{\sigma}^2_{VS,T} = \frac{1}{T}\int_0^T E[\sigma^2(t, S_t)]dt \qquad (5.48)$$

where $\sigma(t, S)$ is the local volatility function calibrated to the market smile, given by Dupire formula (2.3).[18] The expectation in the right-hand side of (5.48) is evaluated by solving the following backward PDE:

$$\frac{dU}{dt} + (r - q)S\frac{dU}{dS} + \frac{\sigma^2(t, S)}{2}S^2\frac{d^2U}{dS^2} = -\sigma^2(t, S) \qquad (5.49)$$

with terminal condition $U(t = T, S) = 0$ and the following matching condition at each dividend date: $U(T_j^-, S) = U(T_j^+, S - d_j)$. $\hat{\sigma}_{VS,T}$ is given by:

$$\hat{\sigma}_{VS,T} = \sqrt{\frac{U(0, S_0)}{T}} \qquad (5.50)$$

where S_0 is the spot value at $t = 0$.

---

[18]The reader can check that the derivation of (2.6) is still valid when there are dividends, however one has to ensure that prices of vanilla options maturing immediately before and after dividend dates obey appropriate matching conditions. For a fixed amount dividend d_j falling at T_j: $C(K, T_j^+) = C(K + d_j, T_j^-)$. If one prefers to work with implied volatilities directly, one has to make sure that the equivalent matching condition for implied volatilities holds. These are discussed in Section 2.3.1, page 34.

For indexes, the returns used for calculating the VS payout are not stripped of dividends. This extra contribution materializes as a discontinuity of U at dividend dates. For a dividend d_j falling at T_j this matching condition is – see (5.44):

$$U(T_j^-, S) = U(T_j^+, S - d_j(S)) + \ln^2\left(1 - \frac{d_j(S)}{S}\right) \qquad (5.51)$$

where $d_j(S)$ expresses that the dividend generally depends on the spot level. As argued in Section 5.6.1, this additional contribution of dividends to index VS levels should be small.

Using a pure cash amount dividend model may however result in a blatant overestimation. In fact, expression (5.51) requires $d_j(S) < S$, so $d_j(S)$ should be replaced with $\max\left(d_j(S), y_j^{\max} S\right)$ where $y_j^{\max}$ is the maximum yield allowed for dividend d_j, with $y_j^{\max} < 1$. If $y_j^{\max}$ is too large, the steep smiles of equity indexes may cause the contribution of payoff

$$\ln^2\left(1 - \max\left(\frac{d_j(S)}{S}, y_j^{\max}\right)\right)$$

to become unreasonably large. It is thus important to cap the effective yield of dividends when pricing index VSs.[19]

Adjustment for large returns

Imagine now that we would like to adjust the VS volatility for the impact of large returns – as in (5.42). This adjustment expresses the fact that while the VS can be perfectly hedged with vanilla options up to second order in $(S_{i+1} - S_i)$, higher-order terms impact the VS and the replicating vanilla portfolio differently. Expanding equation (5.42) in powers of the jump magnitude J and stopping at the lowest non-trivial order gives:[20]

$$\widehat{\sigma}_{VS,T}^2 = \widehat{\sigma}_T^2 + \varepsilon\left(\ln^2(1+J) + 2\ln(1+J) - 2J\right)$$
$$\simeq \widehat{\sigma}_T^2 - \frac{1}{3}\varepsilon J^3 \qquad (5.52)$$

where ε is the annualized probability of a jump. $\widehat{\sigma}_{VS,T}$ is still given by (5.50), except the PDE for U now reads:

$$\frac{dU}{dt} + (r - q)S\frac{dU}{dS} + \frac{\sigma^2(t,S)}{2}S^2\frac{d^2U}{dS^2} = -\left(\sigma^2(t,S) - \frac{1}{3}\varepsilon J^3\right) \qquad (5.53)$$

For constant J and ε the solution of (5.53) is simply $U(0, S_0) - \frac{1}{3}(\varepsilon T)J^3$, where $U(0, S_0)$ is the solution of (5.49).

---

[19] Capping the yield of dividends opens up a (small) can of worms: with dividends no longer an affine function of S, forwards become sensitive to volatility and the technique of Section 2.3.1 for calibrating the local volatility function no longer works exactly.
[20] Note the similarity with (5.36).

PDE (5.53) proves useful in situations when one needs to price weighted VSs consistently with VSs, in circumstances when VS market volatilities do not match the vanilla replication – presumably because VS market prices include an adjustment of type (5.52).

Weighted VSs are covered in Section 5.9 below. In weighted VSs, the realized variance is weighted by a function of the spot, $w(S)$.

Because weighted VSs can be replicated with vanilla options, they can be priced with PDE (5.49) where $\sigma^2(t, S)$ is simply replaced with $w(S)\sigma^2(t, S)$. What about adjustment (5.52) for higher-order terms?

Rather than assuming that J is constant, it is more reasonable to assume that the scale of a return occurring for a spot level S at time t is set by the local volatility $\sigma(t, S)$ – regardless of which component, be it Brownian motion or jump, generated that one return. We thus write:

$$\varepsilon J^3 \equiv -\mu \sigma^3(t, S)$$

where μ is a constant chosen so that market prices of VSs are recovered. The PDE for the weighted VS then reads:

$$\frac{dU}{dt} + (r - q) S \frac{dU}{dS} + \frac{\sigma^2(t, S)}{2} S^2 \frac{d^2U}{dS^2} = -w(S)\left(\sigma^2(t, S) + \frac{1}{3}\mu\sigma^3(t, S)\right)$$

where μ is calibrated to market VS quotes.

5.8 Interest-rate volatility

Assume that there are no cash-amount dividends. In a diffusive setting $\widehat{\sigma}_{\text{VS},T}$ is then equal to the implied volatility of the log contract.

As with any European payoff, the Black-Scholes implied volatility one backs out of a market price is in fact the integrated volatility of the forward $F_t^T = S_t e^{(r_t - q_t)(T-t)}$ rather than the integrated volatility of S_t, where r_t, q_t are the interest rate and repo prevailing at t for maturity T.

When interest rates are deterministic, the (lognormal) volatilities of F_t^T and S_t are identical and the Black-Scholes implied volatility is that of S_t. In the case of stochastic interest rates they are different – note that what matters is the volatility of the interest rate for the residual maturity.

We could still use $\widehat{\sigma}_{\text{VS},T} = \widehat{\sigma}_T$ if the VS contract paid the realized variance of the forward, i.e. the sum of $\ln^2\left(\frac{F_{i+1}^T}{F_i^T}\right)$, however this is not the case.

In what follows, we still use the notation $\widehat{\sigma}_T$ for the implied (Black-Scholes) volatility of the log contract, i.e. the VS volatility of the *forward*, and $\widehat{\sigma}_{\text{VS},T}$ for the VS volatility of the *spot*.

Let us assume that the instantaneous normal volatility at time t of the interest rate r_t of maturity T is constant, equal to σ_r. This dynamics is generated by the Ho&Lee model, which is a short rate model such that rates for all maturities have the same (normal) volatility, equal to that of the short rate:

$$dr_t = \left(\frac{df_{0,t}}{dt} + \sigma_r^2 t \right) dt + \sigma_r dW_t^r$$

where $f_{0,t}$ is the initial term structure of forward rates.

Denote by σ the (lognormal) volatility of S_t and ρ the correlation between S_t and r_t. We assume for simplicity zero repo. The instantaneous variance of F_t^T is given by:

$$
\begin{aligned}
E\left[\left(d \ln F_t^T\right)^2\right] &= E\left[\left(d \ln S_t + (T-t)\, dr_t\right)^2\right] \\
&= \left(\sigma^2 + 2\rho\,(T-t)\,\sigma\sigma_r + (T-t)^2\,\sigma_r^2\right) dt
\end{aligned}
$$

The integrated variance of the forward reads:

$$\frac{1}{T} \int_0^T E\left[\left(d \ln F_t^T\right)^2\right] = \frac{1}{T} \int_0^T \left(\hat{\sigma}_{\text{VS},T}^2 + 2\rho\,(T-t)\,\hat{\sigma}_{\text{VS},T}\sigma_r + (T-t)^2\,\sigma_r^2\right) dt$$

We thus get the following relationship between $\hat{\sigma}_T$ and $\hat{\sigma}_{\text{VS},T}$:

$$\hat{\sigma}_T^2 = \hat{\sigma}_{\text{VS},T}^2 + \rho\hat{\sigma}_{\text{VS},T}\sigma_r T + \frac{\sigma_r^2 T^2}{3} \qquad (5.54)$$

We leave it to the reader to invert (5.54) to get $\hat{\sigma}_{\text{VS},T}$ as a function of $\hat{\sigma}_T$. At order one in σ_r the adjustment reads:

$$\hat{\sigma}_{\text{VS},T} = \hat{\sigma}_T - \frac{\rho}{2}\sigma_r T \qquad (5.55)$$

Interest-rate volatility mostly affects long-maturity VSs. As an example take $\hat{\sigma}_T = 25\%$, an interest-rate volatility of 5 bps/day, a correlation equal to 50%, and a maturity of 5 years. (5.55) yields $\hat{\sigma}_{\text{VS},T} = 24\%$. The correction is about one point of volatility – this is not a small effect.

In the Ho&Lee model we have a constant σ_r, independent of $T - t$. We can of course use a time-dependent σ_r, calibrated on co-terminal swaptions, and the full term structure of VS volatilities rather than assuming a constant spot volatility.

5.9 Weighted variance swaps

We have so far concentrated on the standard VS, by far the most common variance instrument. Other variance payoffs exist, corresponding to other ways of weighting realized variance, as a function of the underlying. Their payoffs read:

$$\frac{1}{T-t}\sum_t^T w(S_i)\left(\ln^2\left(\frac{S_{i+1}}{S_i}\right) - \Delta t\,\hat{\sigma}^2\right)$$

$$= \frac{1}{T-t}\sum_t^T w(S_i)\ln^2\left(\frac{S_{i+1}}{S_i}\right) - \frac{\Delta t}{T-t}\hat{\sigma}^2\sum_t^T w(S_i) \qquad (5.56)$$

where Δt is 1 day and $\hat{\sigma}$ is the strike of the weighted VS, such that the latter is worth zero at inception.

As the second term in (5.56) is simply a string of European payoffs, we concentrate on the first one.

We now discuss the replication of these payoffs – we refer the reader to original work in [24], [25] and [65].

The standard VS is sensitive to contributions of order 3 in δS, and to the discretization of the replicating European profile, effects dicussed above, which impact weighted VSs in equal measure.

We focus here on the replication at order two in δS, paralleling the analysis in Section 5.2.

Can we find a European payoff $f(S)$ of maturity T such that, delta-hedging it at zero implied volatility generates as gamma P&L the desired variance payoff? Using the same notation as in (5.9), this condition reads:

$$\frac{1}{2}S^2\frac{d^2 P_{\hat{\sigma}=0}}{dS^2} = w(S)e^{-r(T-t)} \qquad (5.57)$$

$P_{\hat{\sigma}=0}$ is the price of European payoff f of maturity T, at zero implied volatility:

$$P_{\hat{\sigma}=0}(t, S) = e^{-r(T-t)}f\left(Se^{\mu(T-t)}\right)$$

where $\mu = r - q$. (5.57) translates into:

$$\frac{1}{2}S^2 f''\left(Se^{\mu(T-t)}\right)e^{2\mu(T-t)} = w(S)$$

or equivalently:

$$\frac{1}{2}S^2 f''(S) = w\left(Se^{-\mu(T-t)}\right) \qquad (5.58)$$

which must be obeyed $\forall t, \forall S$.

The standard VS corresponds to $w \equiv 1$. Taking $f(S) = -2\ln S$ indeed takes care of (5.58).

For other weighting schemes, (5.58) cannot hold, owing to the dependence of the right-hand side on t.

Let us then *decide* that $\mu = 0$: we risk-manage a long position in European payoff f (a) at zero implied volatility, (b) with $q = r$. $P_0 \equiv P_{\hat{\sigma}=0}^{\mu=0}$ solves the Black-Scholes PDE with $q = r$, $\sigma = 0$:

$$\frac{dP_0}{dt} = rP_{\hat{\sigma}=0}^{\mu=0}, \quad P_0(t = T, S) = f(S) \tag{5.59}$$

The P&L during δt of a long position in payoff f, delta-hedged, reads:

$$\begin{aligned} P\&L &= \left(P_0\left(t + \delta t, S + \delta S\right) - P_0\left(t, S\right)\right) - \frac{dP_0}{dS}\left(\delta S - (r - q)\,S\delta t\right) - rP_0\delta t \\ &= e^{-r(T-t)}w(S)\left(\frac{\delta S}{S}\right)^2 + S\frac{dP_0}{dS}(r - q)\delta t \tag{5.60} \end{aligned}$$

where we have expanded the P&L at order two in δS, using (5.59) and the property that $\frac{1}{2}S^2\frac{d^2 P_0}{dS^2} = w(S)e^{-r(T-t)}$.

The first portion in (5.60) is exactly what we need to replicate our weighted VS at order two in δS.[21]

The second portion can be canceled by selling at inception a quantity $(r - q)\delta t$ of European options of maturity t with payoff $S\frac{dP_0}{dS}(t, S) = e^{-r(T-t)}S\frac{df}{dS}$.

In conclusion, a weighted VS of maturity T with weight $w(S)$ is replicated at order two in δS with:

- a European payoff $f(S)$ of maturity T with f such that $\frac{d^2 f}{dS^2} = 2\frac{w(S)}{S^2}$, delta-hedged at zero implied volatility, and with $q = r$

- a continuous density $(r - q)$ of (unhedged) European options of maturities τ spanning $[0, T]$ whose payoffs are $e^{-r(T-\tau)}S\frac{df}{dS}$

We now review some examples of weighted VSs.

Gamma swap: $w(S) = S$

The replicating European payoff of maturity T is $f(S) = 2S \ln S$ and the intermediate European payoffs are log contracts.

In case $r - q = 0$, for example if the underlying is a future, the replicating portfolio only consists of the final payoff $2S \ln S$. Strike $\hat{\sigma}$ of the gamma swap then equals the implied volatility of this payoff, $\hat{\sigma}_{S \ln S}$, which is given explicitly as a function of vanilla implied volatilities by:

$$\hat{\sigma}_{S \ln S}^2 = \int_{-\infty}^{+\infty} dy\, \frac{e^{-\frac{y^2}{2}}}{\sqrt{2\pi}} \hat{\sigma}_{K(y)T}^2$$

$$y(K) = \frac{\ln\left(\frac{F_T}{K}\right)}{\hat{\sigma}_{KT}\sqrt{T}} + \frac{\hat{\sigma}_{KT}\sqrt{T}}{2}$$

---

[21]Factor $e^{-r(T-t)}$ is appropriate, as the gamma P&L is generated at time t while the VS payoff is delivered at T.

This formula is derived on page 143, in Section 4.3.1 of Chapter 4.

The $S \ln S$ contract is replicated with a density $\frac{2}{K}$ of vanilla options, as opposed to a density $\frac{2}{K^2}$ for the standard VS, thus lessening the contribution of expensive low-strike vanilla options, in case of a strong skew.

In exchange for relinquishing part of the realized variance for low spot values, we get a lower strike: $\hat{\sigma}_{S \ln S} \leq \hat{\sigma}_{VS}$.

Arithmetic variance swap $w(S) = S^2$

The payoff of the arithmetic VS is: $\Sigma_i (S_{i+1} - S_i)^2$.

It is not traded, though it enjoys a unique property among weighted VSs: it is exactly replicable, even in the presence of large returns.

Indeed, the replicating European payoff of maturity T is a parabola. The expansion of the carry P&L at order two in δS in (5.60) is exact, as there are no higher-order terms.

Corridor variance swap: $w(S) = 1_{S \in [L,H]}$

In corridor variance swaps $w(S)$ is an indicator function. They are very popular, either as corridors, or as down-VSs ($w(S) = 1_{S \in [0,H]}$), or up-VSs ($w(S) = 1_{S \in [L,\infty]}$). Their replication has been first studied by Keith Lewis and Peter Carr in [24].

The terminal replicating European payoff is a truncated log contract, synthesized with a density $\frac{2}{K^2}$ of vanilla options of strikes $K \in [L, H]$. $f(S)$ is given by:

$$
\begin{array}{ll}
-2 \ln S & S \in [L,\ H] \\
-2 \ln H - \frac{2}{H}(S - H) & S \geq H \\
-2 \ln L - \frac{2}{L}(S - L) & S \leq L
\end{array}
$$

The intermediate European payoffs are simple combinations of zero-coupon bonds and vanilla options struck at L and H.

How should returns that cross a barrier be treated? Consider for example a situation with $S_i < H$ and $S_{i+1} > H$.

For this particular return, our long, delta-hedged, position in payoff f generates the following P&L – taking zero interest rate for simplicity:

$$
\begin{aligned}
P\&L &= \left(f(S_{i+1}) - f(S_i)\right) - \frac{df}{dS}(S_i)(S_{i+1} - S_i) \\
&= -2 \ln H - \frac{2}{H}(S_{i+1} - H) + 2 \ln S_i + \frac{2}{S_i}(S_{i+1} - S_i)
\end{aligned}
$$

This is not equal to $\left(\frac{S_{i+1} - S_i}{S_i}\right)^2$, which is what the corridor VS payoff would prescribe, at order two in $(S_{i+1} - S_i)$. Expanding our P&L at order two in $(S_i - H)$ and $(S_{i+1} - H)$ yields:

$$
P\&L = \frac{(S_{i+1} - S_i)^2}{H^2} - \frac{(S_{i+1} - H)^2}{H^2}
$$

We leave it to the reader to work out the other three cases. Unfortunately, none of the provisions typically found in term sheets of corridor VSs matches the P&L that our replication strategy generates.

This mismatch has to be estimated separately and factored in the strike of the corridor VS.

Appendix A – timer options

Let us take zero repo and interest rate and consider a short position in a European option of maturity T, delta-hedged in the Black-Scholes model with a fixed implied volatility $\widehat{\sigma}$.

We assume for now that our option's payoff is convex so that $\frac{d^2 P_{\widehat{\sigma}}}{dS^2} \geq 0$. Our final P&L at order one in δt and two in δS is given by expression (1.9), page 9. In the continuous limit:

$$P\&L = -\int_0^T \frac{S_t^2}{2} \frac{d^2 P_{\widehat{\sigma}}}{dS^2}(t, S_t)(\sigma_t^2 - \widehat{\sigma}^2)dt \tag{5.61}$$

where σ_t is the instantaneous realized volatility and $P_{\widehat{\sigma}}$ is the Black-Scholes expression for the option's price.

In case σ_t lies consistently above/below $\widehat{\sigma}$ we will lose/make money. It can happen though that while the realized volatility over the option's maturity, $\sqrt{\frac{1}{T} \int_0^T \sigma_t^2 dt}$, is lower than $\widehat{\sigma}$, $P\&L < 0$ as in expression (5.61) the difference $(\sigma_t^2 - \widehat{\sigma}^2)$ is weighted by the option's dollar gamma, which varies with t and S.

Only for VSs, whose dollar gamma is constant, does the final P&L only depend on the difference between realized volatility over $[0, T]$ and implied volatility.

Rather than keeping $\widehat{\sigma}$ fixed, consider adjusting it in real time so as to absorb, over each interval between two delta rehedges, the gamma/theta P&L.[22] Denote by $\widehat{\sigma}_t$ the implied volatility we use at time t. During δt, our total P&L, including the mark-to-market P&L from remarking our option position at $t + \delta t$ with the implied volatility $\widehat{\sigma}_t + \delta\widehat{\sigma}$, is:

$$P\&L = -\frac{S_t^2}{2} \frac{d^2 P_{\widehat{\sigma}}}{dS^2}(\sigma_t^2 - \widehat{\sigma}_t^2)\delta t - \frac{dP_{\widehat{\sigma}}}{d\widehat{\sigma}}\delta\widehat{\sigma} \tag{5.62}$$

with $\delta\widehat{\sigma}$ chosen so that $P\&L = 0$.

We now pause to derive an ancillary result relating vega to the dollar gamma in the Black-Scholes model.

---

[22]This introduction draws from a presentation given by Bruno Dupire at the 2007 Global Derivatives conference.

A.1 Vega/gamma relationship in the Black-Scholes model

Denote by V the vega: $V = \frac{dP_{\widehat{\sigma}}}{d\widehat{\sigma}}$, where $P_{\widehat{\sigma}}$ is the Black-Scholes price with implied volatility $\widehat{\sigma}$. Taking the derivative of the Black-Scholes equation (1.4) with respect to $\widehat{\sigma}$ yields:

$$\frac{dV}{dt} + (r-q)S\frac{dV}{dS} + \frac{\widehat{\sigma}^2}{2}S^2\frac{d^2V}{dS^2} - rV = -\widehat{\sigma}S^2\frac{d^2P_{\widehat{\sigma}}}{dS^2} \qquad (5.63)$$

At maturity, $V(t=T,S) = 0$, $\forall S$. V is thus only generated by the source term in (5.63):

$$V(t,S) = \widehat{\sigma}\int_t^T E_{t,S}\left[e^{-r(\tau-t)}S_\tau^2\frac{d^2P_{\widehat{\sigma}}}{dS^2}(\tau,S_\tau)\right]d\tau \qquad (5.64)$$

Setting $x = \ln S$, the Black-Scholes equation reads:

$$\frac{dP_{\widehat{\sigma}}}{dt} + \left(r-q-\frac{\widehat{\sigma}^2}{2}\right)\frac{dP_{\widehat{\sigma}}}{dx} + \frac{\widehat{\sigma}^2}{2}\frac{d^2P_{\widehat{\sigma}}}{dx^2} = rP_{\widehat{\sigma}} \qquad (5.65)$$

Take the derivative of (5.65) n times with respect to x. Since neither r, nor q, nor $\widehat{\sigma}$ depend on x, $\frac{d^nP_{\widehat{\sigma}}}{dx^n}$ solves the same PDE as $P_{\widehat{\sigma}}$. We thus have:

$$\frac{d^nP_{\widehat{\sigma}}}{dx^n}(t,x) = E_{t,x}\left[e^{-r(T-t)}\frac{d^nP_{\widehat{\sigma}}}{dx^n}(T,x_T)\right]$$

$e^{-r(T-t)}\frac{d^nP_{\widehat{\sigma}}}{d\ln S^n}$ is thus a martingale.

Set $n=1$: $\frac{dP_{\widehat{\sigma}}}{d\ln S} = S\frac{dP_{\widehat{\sigma}}}{dS}$ – we get the result that the discounted dollar delta is a martingale.

Now set $n=2$: $\frac{d^2P_{\widehat{\sigma}}}{d\ln S^2} = S^2\frac{d^2P_{\widehat{\sigma}}}{dS^2} + S\frac{dP_{\widehat{\sigma}}}{dS}$. Combining this with the result for $n=1$ implies that the discounted dollar gamma is a martingale as well.

More generally $e^{-r(T-t)}S^n\frac{d^nP_{\widehat{\sigma}}}{dS^n}$ is a martingale for all n; this is also true in the Black-Scholes model with deterministic time-dependent volatility.

Thus, the expectation in (5.64) is simply equal to the dollar gamma evaluated at time t for spot S. (5.64) then becomes:

$$V(t,S) = \widehat{\sigma}S^2\frac{d^2P_{\widehat{\sigma}}}{dS^2}\int_t^T d\tau$$

Thus

$$\frac{dP_{\widehat{\sigma}}}{d\widehat{\sigma}} = S^2\frac{d^2P_{\widehat{\sigma}}}{dS^2}\widehat{\sigma}(T-t) \qquad (5.66)$$

Going back to (5.62), expressing now $\frac{dP_{\widehat{\sigma}}}{d\widehat{\sigma}}$ in terms of $S^2\frac{d^2P_{\widehat{\sigma}}}{dS^2}$, the condition $P\&L = 0$ translates into:

$$\frac{1}{2}(\sigma_t^2 - \widehat{\sigma}_t^2)\delta t + (T-t)\widehat{\sigma}_t\delta\widehat{\sigma} = 0$$

Denote by Q_t the quadratic variation of $\ln S$ realized since $t = 0$: $Q_t = \int_0^t \sigma_\tau^2 d\tau$. Replacing $\sigma_t^2 \delta t$ with δQ_t, we get, at order one in δt:[23]

$$\delta\big(\widehat{\sigma}_t^2(T - t)\big) = -\delta Q_t$$

$\widehat{\sigma}_t$ is thus given by:

$$\widehat{\sigma}_t^2 = \frac{\widehat{\sigma}_{t=0}^2 T - Q_t}{T - t} \tag{5.67}$$

Imagine that, over $[0, t]$, the integrated realized volatility matches the initial implied volatility, thus $Q_t = \widehat{\sigma}_{t=0}^2 t$. If we had delta-hedged our option in standard fashion, our P&L at time t would be given by (5.61) with $T \equiv t$. Even though the average realized volatility matches the implied volatility, there is no reason why our theta/gamma P&L would vanish.

In our situation (5.67) shows instead that, if $Q_t = \widehat{\sigma}_{t=0}^2 t$ then $\widehat{\sigma}_t = \widehat{\sigma}_{t=0}$: our option is valued at t with an implied volatility that is equal to its initial value, and we have generated exactly zero carry P&L.

In case $Q_t > \widehat{\sigma}_{t=0}^2 t$ then $\widehat{\sigma}_t < \widehat{\sigma}_{t=0}$: the negative gamma/theta P&L is offset by a positive mark-to-market vega P&L.

We then risk-manage our option by adjusting $\widehat{\sigma}_t$ according to (5.67), generating zero P&L. Two things can happen, according to whether the realized volatility over $[0, T]$ exceeds $\widehat{\sigma}_{t=0}$ or not:

- $Q_T < \widehat{\sigma}_{t=0}^2 T$: from (5.67), $\widehat{\sigma}_T$ is infinite. In the Black-Scholes model, however, the price of a European option is not a separate function of $\widehat{\sigma}$ and $(T - t)$, but a function of $Q = \widehat{\sigma}^2(T - t)$. We will thus use the notation P_Q rather than $P_{\widehat{\sigma}}$. From (5.67), $Q = \widehat{\sigma}_{t=0}^2 T - Q_T$: there is some time value left in our option. Since at T we pay the intrinsic value to the client – given by $P_{Q=0}(T, S_T)$ – we make a net positive P&L given by:

$$P\&L = P_{Q=(\widehat{\sigma}_{t=0}^2 T - Q_T)}(T, S_T) - P_{Q=0}(T, S_T)$$

  We have assumed that our option has convex payoff, thus $P\&L \geq 0$.

- $Q_T > \widehat{\sigma}_{t=0}^2 T$: Q_t is a process that starts from zero at $t = 0$ and increases: there exists an intermediate time τ such that $Q_\tau = \widehat{\sigma}_{t=0}^2 T$. From (5.67), at time τ, $\widehat{\sigma}_\tau = 0$. Moreover, for $t > \tau$, if we kept using (5.67) we would have $\widehat{\sigma}_t^2 < 0$. At $t = \tau$ we thus stop adjusting $\widehat{\sigma}$ and delta-hedge our option over $[\tau, T]$ using $\widehat{\sigma}_\tau$, i.e. zero implied volatility. Our net P&L is then given by an expression similar to (5.61), except it is restricted to the interval $[\tau, T]$ and there is no theta contribution:

$$P\&L = -\int_\tau^T \frac{S_u^2}{2} \frac{d^2 P_{\widehat{\sigma}=0}}{dS^2}(u, S_u)\sigma_u^2 dt$$

  Since our option's payoff is convex, this P&L is negative.

---

[23]In our context $\widehat{\sigma}_t$ is a stochastic process that does not have a diffusive term – it only has a drift. In an expansion at order one in δt, we only keep terms of order one in $\delta \widehat{\sigma}_t$.

The conclusion is that, unlike what happens with the standard delta-hedging process, with our hedging scheme, the sign of our final P&L is exactly that of the difference between realized and implied volatility or, equivalently, between realized quadratic variation Q_T and $\hat{\sigma}^2_{t=0}T$. The magnitude of the P&L is random.

A.2   Model-independent payoffs based on quadratic variation

With the hedging scheme of the previous section, no P&L is generated until we reach maturity unless Q_t reaches our quadratic variation "budget" $\hat{\sigma}^2_{t=0}T$. Our P&L at T depends on the remaining quadratic variation "budget" $\hat{\sigma}^2_{t=0}T - Q_T$. This suggests we could create an exactly hedgeable claim by delivering the payout at a random maturity defined as the time τ when Q_τ reaches a pre-specified value $\mathcal{Q}$.

A *timer* option is such an option; it pays a payoff $f(S_\tau)$ at time τ such that $Q_\tau = \mathcal{Q}$, where $\mathcal{Q}$, called the quadratic variation budget, is specified in the term sheet, in lieu of maturity.[24] From our analysis above, the price of this option – for zero interest rate and repo – is thus simply $P(S, Q, \mathcal{Q}) \equiv \mathcal{P}_{BS}(S, Q; \mathcal{Q})$ where $\mathcal{P}_{BS}$ is given by:

$$\mathcal{P}_{BS}(S, Q; \mathcal{Q}) = E\left[f\left(Se^{-\frac{\mathcal{Q}-Q}{2}+\sqrt{\mathcal{Q}-Q}Z}\right)\right] \qquad (5.68)$$

where Z is a standard normal variable. It is the Black-Scholes price, calculated with an effective volatility equal to 1 and an effective maturity equal to $\mathcal{Q} - Q$.

The price of a timer option is thus model-independent, as by following the delta-hedging outlined above, we replicate exactly the option's payoff, no matter what the realized volatility is. Physical time does not enter the pricing function.

Timer option prices do not depend on market implied volatilities – this makes timer options attractive for underlyings that lack a liquid options market. Carrying a naked gamma/theta position, in the case of a standard option, is too risky, even though we may have selected a conservative implied volatility. In the timer version, the carry P&L – see below – is smaller. This enables trades that would not be contemplated otherwise.

Timer options effectively started trading around 2007, but the idea of replacing physical time with quadratic variation and designing payoffs that are not sensitive to volatility assumptions well predates timer options – see Avi Bick's 1995 article [14].

In practice, quadratic variation is measured using log returns of daily closes S_i:

$$Q_{i+1} = Q_i + \ln^2\left(\frac{S_{i+1}}{S_i}\right)$$

---

[24]While Société Générale has marketed these options under the name of *timer* options they have also been traded under the name of *mileage* options.

Are there other model-independent payoffs involving the quadratic variation? Write the price of such an option as $P(t, S, Q)$. The P&L during δt of a short, delta-hedged position reads, at order one in δt and two in δS:

$$P\&L = -\frac{dP}{dt}\delta t - \frac{S^2}{2}\frac{d^2 P}{dS^2}\frac{\delta S^2}{S^2} - \frac{dP}{dQ}\delta Q \tag{5.69a}$$

$$= -\frac{dP}{dt}\delta t - \left[\frac{S^2}{2}\frac{d^2 P}{dS^2} + \frac{dP}{dQ}\right]\frac{\delta S^2}{S^2} \tag{5.69b}$$

where, by definition, $\delta Q = \frac{\delta S^2}{S^2}$. This P&L vanishes if the following conditions hold:

$$\frac{dP}{dt} = 0 \tag{5.70a}$$

$$\frac{S^2}{2}\frac{d^2 P}{dS^2} + \frac{dP}{dQ} = 0 \tag{5.70b}$$

(5.70a) expresses that P does not depend on physical time. (5.70b) is identical to the Black-Scholes equation, except time is replaced with quadratic variation; we only need to supplement (5.70b) with the terminal profile $P(S, \mathcal{Q})$, where $\mathcal{Q}$ is the quadratic variation budget.

Thus, in addition to the "European" timer option discussed above whose price is given by (5.68), many familiar payoffs have a *timer* counterpart.

We can define a timer barrier option that pays $f(S_\tau)$ when Q_τ reaches $\mathcal{Q}$ unless S hits a barrier B in which case the option pays a rebate $R(Q_\tau)$. This type of payoff will come in handy when we construct an upper bound for prices of options on realized variance – see Section 7.6.10. The barrier can also be made a function of Q.

Path-dependent variables can be used except they are not allowed to involve physical time.[25]

Quadratic variation does not need to accrue uniformly; we can choose to weight realized variance by a function of S: $\delta Q_t = \mu(S)\sigma_t^2 \delta t$; for example realized variance is not counted whenever S lies above or below a given threshold. (5.70b) is then replaced with:

$$\frac{S^2}{2}\frac{d^2 P}{dS^2} + \mu(S)\frac{dP}{dQ} = 0$$

Selecting $\mu(S) = S^2$ corresponds to accruing the quadratic variation of S, rather than $\ln S$.

---

[25]In timer options we make a stochastic time change from physical time to quadratic variation: $t \to Q_t = \int_0^t \sigma_u^2 du$ and from S_t to $S_Q^\star$ defined by: $S_{Q_t}^\star = S_t$; $S_Q^\star$ is lognormally distributed with a quadratic variation of $\ln S^\star$ equal to Q.

Consider a path-dependent variable f that is a function of the path of S_t over $[0, T]$, which we denote by $[S_t]_0^T$. A timer option whose payoff involves f remains model-independent only if the condition: $f([S_t]_0^T) = f([S_Q^\star]_0^{Q_T})$ holds.

While the (continuously sampled) $\min_t S_t$ and $\max_t S_t$ satisfy this condition, the Asian average $M_t = \frac{1}{t}\int_0^t S_u du$ does not.

We can also define multi-asset timer options. For example define Q as the quadratic variation of $\ln\left(\frac{S_2}{S_1}\right)$. An option paying $S_1 f\left(\frac{S_2}{S_1}\right)$ when Q hits $\mathcal{Q}$ is model-independent.

This is easily shown by using S_1 as numeraire: the value of all assets, including the value of our timer option as well as S_2 are expressed in units of S_1. Q is then simply the quadratic variation of S_2 expressed in units of S_1 – in these new units this option becomes a standard timer option on S_2.

It can be shown that these are the only model-independent options that are functions of S_1, S_2, Q.

Finally, model-independent payoffs whose prices do not depend on physical time can also be created using $\min_\tau S_\tau$ and $\max_\tau S_\tau$ rather than quadratic variation – see [31].

A.3 How model-independent are timer options?

Our analysis above applies to the ideal situation of real-time delta-hedging, and for a continuous process for S – hence the expansion at order two in δS and order one in δt for δQ_t. What about the real case?

In standard options delta-hedging offsets the directional position on S; our P&L starts with terms of order two in δS whose contribution, as illustrated in Section 1.2, is not small.

One order is gained with timer options: condition (5.70b) ensures that the δS^2 term in the gamma P&L is offset by a corresponding change of the quadratic variation. Our P&L now starts with terms of order three.

Consider a "European" timer option, that is with no path-dependence. $P(Q, S)$ solves equation (5.70b). In typical term sheets of timer options, quadratic variation is defined as the sum of squared *log-returns*: $\delta Q = \ln\left(1 + \frac{\delta S}{S}\right)^2$. Expanding $P(S, Q)$ at order 3 in δS, setting $\delta Q = \frac{\delta S^2}{S^2} - \frac{\delta S^3}{S^3}$ and using (5.70b):

$$
\begin{aligned}
P\&L &= -\frac{S^2}{2}\frac{d^2 P}{dS^2}\frac{\delta S^2}{S^2} - \frac{dP}{dQ}\delta Q - \frac{S^3}{6}\frac{d^3 P}{dS^3}\frac{\delta S^3}{S^3} - \frac{d^2 P}{dQdS}\delta Q\delta S \\
&= \left(\frac{S^2}{2}\frac{d^2 P}{dS^2} + \frac{S^3}{3}\frac{d^3 P}{dS^3}\right)\frac{\delta S^3}{S^3}
\end{aligned}
\tag{5.71}
$$

Consider a timer call option. For spot values near the strike, the prefactor in (5.71) is dominated by $\frac{S^2}{2}\frac{d^2 P}{dS^2}$: $P\&L \simeq \frac{S^2}{2}\frac{d^2 P}{dS^2}\frac{\delta S^3}{S^3}$, to be compared with the gamma P&L of the standard, non-timer, option: $-\frac{S^2}{2}\frac{d^2 P}{dS^2}\frac{\delta S^2}{S^2}$. The prefactors are identical; the P&L in the timer version is smaller by a factor $\frac{\delta S}{S}$.

Timer options are thus less risky than their non-timer counterparts but are in practice not fully model-independent.[26] Mathematically, "model independent" means

---

[26] Timer options on single stocks – which can experience large overnight drawdowns – are particularly risky.

model-independent as long as the process for S_t is a continuous semimartingale. Practically, the meaning of "model-independent" is that the P&L vanishes up to order two in δS.

We have assumed so far vanishing interest rate and repo. The essence of timer options is that quadratic variation replaces physical time. Financing costs/benefits, however, are paid/received *prorata temporis*: physical time re-enters the picture. Taking into account interest rate and repo, the P&L in (5.69) now reads:

$$
\begin{aligned}
P\&L &= -\frac{dP}{dt}\delta t - \frac{S^2}{2}\frac{d^2P}{dS^2}\frac{\delta S^2}{S^2} - \frac{dP}{dQ}\delta Q + rP\delta t - (r-q)S\frac{dP}{dS}\delta t \\
&= -\left[\frac{dP}{dt} - rP + (r-q)S\frac{dP}{dS}\right]\delta t - \left[\frac{S^2}{2}\frac{d^2P}{dS^2} + \frac{dP}{dQ}\right]\frac{\delta S^2}{S^2}
\end{aligned}
$$

The conditions ensuring "model-independence" are now:

$$
\frac{S^2}{2}\frac{d^2P}{dS^2} + \frac{dP}{dQ} = 0 \tag{5.72a}
$$

$$
\frac{dP}{dt} - rP + (r-q)S\frac{dP}{dS} = 0 \tag{5.72b}
$$

(5.72b) implies that $P(t,S,Q)$ has the following form:

$$
P(t,S,Q) = e^{rt}p\big(Se^{-(r-q)t},Q\big)
$$

(5.72a) yields the following condition for $p(x,Q)$:

$$
\frac{x^2}{2}\frac{d^2p}{dx^2} + \frac{dp}{dQ} = 0
$$

Setting the terminal condition $p(x,Q = \mathcal{Q}) = f(x)$ then fully determines p. At time τ, when Q_τ reaches $\mathcal{Q}$, we pay the amount $e^{r\tau}f(S_\tau e^{-(r-q)\tau})$ to the client. Commercially, this is less attractive than simply paying $f(S_\tau)$.

Select a (distant) maturity T and redefine f as: $f(x) \to e^{-rT}f(xe^{(r-q)T})$. The payoff at τ is then $e^{-r(T-\tau)}f(S_\tau e^{(r-q)(T-\tau)})$, which is equivalent to settling the payoff $f(F_\tau^T)$ at T, where F_τ^T is the forward at time τ, spot S_τ, for maturity T. $P(t,S,Q)$ is given by:

$$
P(t,S,Q) = e^{-r(T-t)}\mathcal{P}_{BS}(Se^{(r-q)(T-t)},Q;\mathcal{Q}) \tag{5.73}
$$

where $\mathcal{P}_{BS}$ is defined in (5.68). This specification of a timer option remains in fact model-independent when interest rates are stochastic.

Thus, for non-vanishing interest rate and repo, model-independent payoffs still exist, but become somewhat convoluted.

Equivalently, a standard timer option now acquires a spurious sensitivity to realized – or implied – volatility, as this volatility determines the duration over which financing costs are paid/received. For example, in the Black-Scholes model, with an implied volatility $\hat{\sigma}$, the price of a standard vanilla timer option becomes:

$$P(t, S, Q) = e^{-r\frac{Q-Q}{\hat{\sigma}^2}}\mathcal{P}_{BS}(Se^{(r-q)\frac{Q-Q}{\hat{\sigma}^2}}, Q; Q)$$

P explicitly depends on $\hat{\sigma}$ thus is not model-independent anymore.[27]

Beside a reserve that covers third-order terms in the P&L and an adjustment to account for interest rate, repo and dividends, two additional corrections to the model-independent price are needed.

The final quadratic variation is always larger than the allotted budget Q as, usually, the expiry of the timer option is defined as the first day when Q_τ exceeds Q: this overshoot needs to be factored in the price.

Also, term sheets of timer options specify a maximum maturity $T_{\max}$, typically $T_{\max} = 2\frac{Q}{\hat{\sigma}^2}$ where $\hat{\sigma}$ is a reference volatility. The corresponding price adjustment is very model-dependent.

A.4 Leveraged ETFs

Consider a leveraged ETF (leveraged exchange traded fund [LETF]). The fund's strategy consists in investing in a security S, with a fixed leverage β. Ignoring borrowing costs, we would expect the performance of our investment I, over a given time horizon, to be β times the performance of S: $\ln(I_T/I_0) \simeq \beta \ln(S_T/S_0)$.

At time t we need to hold $\beta\frac{I_t}{S_t}$ units of S. $\frac{I_t}{S_t}$ will keep changing – except in the uninteresting case $\beta = 1$. We thus need to rebalance our position – in the case of LETFs on a daily basis. This rebalancing is analogous to the readjustment of the delta of an option position and likely exposes us to realized volatility.

Let I_t be the fund net asset value (NAV) at time t and S_t the security it invests in. As the fund manager, we borrow the amount βI_t, which we invest in S, while accruing interest on the amount I_t. Over $[t, T+\delta t]$, our return $r_I = \frac{I_{t+\delta t}}{I_t} - 1$ is given by:

$$r_I = \beta(r_S - (r-q)\delta t) + r\delta t$$

where $r_S = \frac{S_{t+\delta t}}{S_t} - 1$. Expanding at order two in $\delta\ln S$ and one in δt the log-return $\delta\ln I = \ln(1 + r_I)$ is given by:

$$\begin{aligned}
\delta\ln I &= \ln(1 + \beta(e^{\delta\ln S} - 1 - (r-q)\delta t) + r\delta t) \\
&= \beta\delta\ln S + (r - \beta(r-q))\delta t - \frac{\beta(\beta-1)}{2}\delta\ln S^2
\end{aligned}$$

---

[27]Because it is generated by rate and repo sensitivities, the vega of a timer option is very unlike that of its non-timer counterpart.

Keeping terms up to $\delta \ln S^2$:

$$\delta \ln I = \beta \delta \ln S + (r - \beta(r - q)) \delta t - \frac{\beta(\beta - 1)}{2} \delta Q$$

where $\delta Q = \frac{\delta S^2}{S^2}$. This yields:

$$I_t = I(t, S_t, Q_t) \tag{5.74}$$

$$I(t, S, Q) = I_0 e^{rt} \left(\frac{S}{S_0 e^{(r-q)t}} \right)^{\beta} e^{-\frac{\beta(\beta-1)}{2} Q} \tag{5.75}$$

I_t is the result of a pure delta strategy: (5.74) shows that the fund's NAV replicates (up to second order in δS) the (exotic) payoff $I(t, S_t, Q_t)$ that involves both S_t and its realized variance – starting from NAV I_0 at $t = 0$.[28] The reader can check that $I(t, S, Q)$ indeed satisfies conditions (5.72): the LETF is a perpetual contract, with no quadratic variation budget Q.

Set a maturity T: the payoff I_T at T can be generated out of an initial investment I_0. Equivalently the payoff $\frac{I_T}{I_0} - e^{rT}$ at T can be synthesized with zero initial cash. Multiply by e^{-rT} and take the limit $\beta \to 0$:

$$\lim_{\beta \to 0} \frac{1}{\beta} \left(e^{-rT} \frac{I_T}{I_0} - 1 \right) = \ln \left(\frac{S_T}{S_0 e^{(r-q)T}} \right) + \frac{Q_T}{2} \tag{5.76}$$

The package in the right-hand side of (5.76) is replicated with zero initial cash. Equivalently, the value at $t = 0$ of a payoff of maturity T that delivers either $-\ln \left(\frac{S_T}{S_0 e^{(r-q)T}} \right)$ or $\frac{Q_T}{2}$ is equal – hence the replication strategy of the VS once again.

Finally, options on LETFs exist as well. Unlike LETFs, LETF options are highly model-dependent.

Appendix B – perturbation of the lognormal distribution

In the Black-Scholes model implied volatilities are flat. A smile appears whenever the distribution of S is not lognormal, such as in stochastic volatility or jump/Lévy models.

These models collapse onto the Black-Scholes model when a given parameter – say volatility of volatility or jump size – vanishes. It is thus possible to carry out an

---

[28]For typical values of β, such as $\beta = 2$ or $\beta = -2$, the contribution from realized variance impacts negatively the performance of the ETF, more so for negative values of β – an aspect of LETF trading that some investors seem to have overlooked.

expansion in powers of this parameter, around the Black-Scholes case. In Chapter 8 we go through such an expansion for general forward variance models.

Here we consider the general case of a distribution of S_T which we assume to be slightly non-lognormal. Our aim is to derive the expansion of European option prices at order one in the parameters that quantify the non-lognormality of S_T – the cumulants of $\ln S_T$.

While this idea is not new (see for example [5]), it is important to ensure that, at the chosen order in the expansion, some quantities stay fixed. The constraint that the forward $F_T = E[S_T]$ should be unchanged is typically enforced.

In the context of forward variance models, forward variances ξ_0^τ are underlyings in their own right whose values should be left unchanged as well.

We thus require that prices of log contracts – hence VS volatilities – be unaffected in the expansion.

Let us start with a lognormal density for S_T. The density ρ_0 of $z = \ln(\frac{S_T}{F_T})$, where F_T is the forward for maturity T, is given by:

$$\rho_0(z) = \frac{1}{\sqrt{2\pi\Sigma^2}} e^{-\frac{(z-\mu)^2}{2\Sigma^2}}$$

where μ, Σ are the average and standard deviation of the unperturbed lognormal density whose volatility we denote by $\hat{\sigma}_0$; namely:

$$\Sigma = \hat{\sigma}_0\sqrt{T}$$

$$\mu = -\frac{\hat{\sigma}_0^2 T}{2} = -\frac{\Sigma^2}{2}$$

Let us denote by $L(q)$ the logarithm of the characteristic function of a given density ρ:

$$L(q) = \ln\left(\int_{-\infty}^{+\infty} e^{-qz}\rho(z)\,dz\right) \qquad (5.77)$$

$L(q)$ is called the cumulant-generating function. For the normal density ρ_0, $L_0(q)$ reads:

$$L_0(q) = -\mu q + \frac{\Sigma^2}{2}q^2 = \frac{\Sigma^2}{2}\left(q + q^2\right) \qquad (5.78)$$

$L_0(q)$ is a polynomial of order 2 – this is a distinguishing feature of normal distributions. For a general density ρ, cumulants κ_n are defined as the coefficients of the Taylor expansion of $L(q)$ around $q = 0$ – when it exists:

$$L(q) = \sum_{n=1}^{\infty} \frac{(-1)^n}{n!}\kappa_n q^n$$

One can check by taking derivatives of $L(q)$ in (5.77) and evaluating them at $q = 0$ that the first and second cumulants are related respectively to the mean and variance of ρ:

$$\kappa_1 = \overline{z}, \quad \kappa_2 = \overline{(z - \overline{z})^2}$$

where $\overline{f}$ denotes $E[f]$. Likewise, the third and fourth cumulants are related to centered moments of ρ:

$$\kappa_3 = \overline{(z - \overline{z})^3}, \quad \kappa_4 = \overline{(z - \overline{z})^4} - 3\overline{(z - \overline{z})^2}^2$$

Since cumulants κ_n for $n > 2$ vanish for a normal distribution, it is natural to consider a small perturbation $\delta\kappa_n$ of the cumulants of $L_0(q)$ for $n \geq 3$:

$$L(q) = L_0(q) + \sum_{n=3}^{\infty} \frac{(-1)^n}{n!} \delta\kappa_n q^n \tag{5.79}$$

We now derive the perturbed density $\rho = \rho_0 + \delta\rho$ at order one in the $\delta\kappa_n$.

B.1 Perturbing the cumulant-generating function

While we perturb ρ_0 to generate a smile for implied volatilities, we wish to keep the forward unchanged. By definition of z:

$$E[S_T] = F_T E[e^z] = F_T e^{L(-1)}$$

which imposes the constraint $L(-1) = 0$. Inspection of (5.78) shows that $L_0(p)$ obviously obeys this condition.

Imagine that only one $\delta\kappa_n$ – say $\delta\kappa_3$ – is non-vanishing. Then the constraint $L(-1) = 0$ cannot be accommodated unless we shift cumulants of order 1 and 2 by an amount proportional to $\delta\kappa_3$. We then rewrite (5.79) as:

$$L(q) = L_0(q) - \delta\kappa_1 q + \delta\kappa_2 \frac{q^2}{2} + \sum_{n=3}^{\infty} \frac{(-1)^n}{n!} \delta\kappa_n q^n$$

Several choices are possible:

- Take $\delta\kappa_1 \neq 0$, $\delta\kappa_2 = 0$: this is the choice typically made in the literature, thereby translating the distribution of z by an amount $\delta\kappa_1$ given by $\delta\kappa_1 = -\sum_{n=3}^{\infty} \frac{1}{n!}\delta\kappa_n$ and leaving the standard deviation of z unchanged.[29] This results in the following expression for $L(q)$:

$$L(q) = L_0(q) + \sum_{n=3}^{\infty} \frac{\delta\kappa_n}{n!}\left((-1)^n q^n + q\right)$$

[29]Choosing not to alter κ_2 has little financial motivation since, in the presence of a smile, the standard deviation of $\ln S_T$ is not related simply to implied volatilities of vanilla options.

- Take $\delta\kappa_1 = 0$, $\delta\kappa_2 = -2\sum_{n=3}^{\infty} \frac{1}{n!}\delta\kappa_n$. This yields:

$$L(q) = L_0(q) + \sum_{n=3}^{\infty} \frac{\delta\kappa_n}{n!}\left((-1)^n q^n - q^2\right) \tag{5.80}$$

- Take both $\delta\kappa_1 \neq 0$, $\delta\kappa_2 \neq 0$:

$$L(q) = L_0(q) + \sum_{n=3}^{\infty} \frac{\delta\kappa_n}{n!}\left((-1)^n q^n - q^2 + \theta_n\left(q + q^2\right)\right) \tag{5.81}$$

where the θ_n are arbitrary.

B.2 Choosing a normalization and generating a density

In diffusive models the VS volatility for maturity T is equal to the implied volatility of the log contract and is given by:

$$\hat{\sigma}_{\mathrm{VS},T}^2 T = \hat{\sigma}_T^2 T = E[-2\ln(S_T/F_T)]$$

From the definition (5.77) of the cumulant-generating function, we get:

$$\hat{\sigma}_T^2 = \frac{2}{T}\frac{dL}{dq}\bigg|_{q=0} = -\frac{2}{T}\kappa_1 \tag{5.82}$$

Thus, keeping κ_1 unchanged guarantees that the log-contract implied volatility is unchanged. For diffusive models, this guarantees that $\hat{\sigma}_{\mathrm{VS},T}$ is also unchanged.

This is a very desirable feature for stochastic volatility models: forward VS variances are underlyings whose initial values should be left unchanged in a perturbation of the model's parameters. We then choose expression (5.80) for $L(q)$.

It is a classical result that a general perturbation in the cumulants of $L_0(q)$:

$$L(q) = L_0(q) + \sum_{n=1}^{\infty} \frac{\delta\kappa_n}{n!}q^n$$

translates at order one in the $\delta\kappa_n$ into the following perturbation of the density:

$$\rho(z) = \rho_0(z) + \delta\rho(z)$$

$$\delta\rho(z) = \sum_{n=1}^{\infty} \frac{\delta\kappa_n}{\Sigma^n\sqrt{n!}}H_n\left(\frac{z-\mu}{\Sigma}\right)\rho_0(z) \tag{5.83}$$

where the H_n are a family of orthogonal polynomials – the Hermite polynomials – defined by:

$$H_n(z) = \frac{(-1)^n}{\sqrt{n!}}e^{\frac{z^2}{2}}\frac{d^n}{dz^n}\left(e^{-\frac{z^2}{2}}\right) \tag{5.84}$$

with the following properties:

$$\int_{-\infty}^{\infty} \frac{1}{\sqrt{2\pi}} e^{-\frac{z^2}{2}} H_n(z) H_m(z) \, dz = \delta_{nm}$$

$$\int_{-\infty}^{\infty} \frac{1}{\sqrt{2\pi}} e^{-\frac{z^2}{2}} H_n(z) e^{-qz} dz = \frac{(-1)^n q^n}{\sqrt{n!}} e^{\frac{q^2}{2}}$$

Using (5.84), (5.83) can be rewritten in a simpler form – this is the Gram-Charlier formula:

$$\delta\rho(z) = \sum_{n=1}^{\infty} \frac{\delta\kappa_n}{n!} (-1)^n \frac{d^n \rho_0(z)}{dz^n}$$

Choosing now normalization (5.80) results in the following expression of $\delta\rho$ at order one in $\delta\kappa_n$, $n \geq 3$:

$$\delta\rho(z) = \sum_{n=3}^{\infty} \frac{\delta\kappa_n}{n!} \left((-1)^n \frac{d^n}{dz^n} - \frac{d^2}{dz^2} \right) \rho_0(z) \qquad (5.85)$$

Normalization (5.81) would have yielded:

$$\delta\rho(z) = \sum_{n=3}^{\infty} \frac{\delta\kappa_n}{n!} \left((-1)^n \frac{d^n}{dz^n} - \frac{d^2}{dz^2} + \theta_n \left(\frac{d^2}{dz^2} + \frac{d}{dz} \right) \right) \rho_0(z) \qquad (5.86)$$

The perturbed density is $\rho = \rho_0 + \delta\rho$.

B.3   Impact on vanilla option prices and implied volatilities

Denote by δP the perturbation in the price of a vanilla option generated by $\delta\rho$. Its payoff – for a call - is given by:

$$(S_T - K)^+ = K \left(\frac{S_T}{K} - 1 \right)^+ = K f \left(\frac{F_T e^z}{K} \right)$$

where $f(x) = (x-1)^+$.

Starting from the expression of the price P_0 in the Black-Scholes model:

$$P_0 = e^{-rT} \int_{-\infty}^{\infty} \rho_0(z) K f \left(\frac{F_T e^z}{K} \right) dz = e^{-rT} \int_{-\infty}^{\infty} \rho_0(u - \ln F_T) K f \left(\frac{e^u}{K} \right) du$$

and taking the derivative with respect to $\ln S$ – remembering that $F_T = S e^{(r-q)T}$ – yields:

$$\frac{d^n P_0}{d \ln S^n} = \frac{d^n P_0}{d \ln F_T^n} = e^{-rT} \int_{-\infty}^{\infty} (-1)^n \frac{d^n \rho_0}{dz^n} (z) K f \left(\frac{F_T e^z}{K} \right) dz \qquad (5.87)$$

The perturbation $\delta\rho$ in (5.85) then results in a price variation δP given by:

$$\delta P = \sum_{n=3}^{\infty} \frac{\delta\kappa_n}{n!} \left(\frac{d^n}{d\ln S^n} - \frac{d^2}{d\ln S^2} \right) P_0 \qquad (5.88)$$

If we do not require that the implied volatility of the log contract be unchanged and use formula (5.81) for $\delta\rho$ we get:

$$\delta P = \sum_{n=3}^{\infty} \frac{\delta\kappa_n}{n!} \left[\left(\frac{d^n}{d\ln S^n} - \frac{d^2}{d\ln S^2} \right) + \theta_n \left(\frac{d^2}{d\ln S^2} - \frac{d}{d\ln S} \right) \right] P_0 \qquad (5.89)$$

To obtain the perturbation of implied volatilities $\delta\widehat{\sigma}$ at order one in the $\delta\kappa_n$ simply divide δP by the option's vega $\frac{dP_0}{d\widehat{\sigma}}$, which is related to its dollar gamma through equation (5.66):

$$\frac{dP_0}{d\widehat{\sigma}} = S^2 \frac{d^2 P_0}{dS^2} \widehat{\sigma}T = \left(\frac{d^2 P_0}{d\ln S^2} - \frac{dP_0}{d\ln S} \right) \widehat{\sigma}T$$

Equation (5.89) translates into:

$$\delta\widehat{\sigma} = \frac{1}{\widehat{\sigma}_0 T} \sum_{n=3}^{\infty} \frac{\frac{\delta\kappa_n}{n!} \left(\frac{d^n P_0}{d\ln S^n} - \frac{d^2 P_0}{d\ln S^2} \right)}{\frac{d^2 P_0}{d\ln S^2} - \frac{dP_0}{d\ln S}} + \frac{1}{\widehat{\sigma}_0 T} \sum_{n=3}^{\infty} \frac{\delta\kappa_n}{n!} \theta_n \qquad (5.90)$$

As (5.90) makes it plain, choosing $\theta_n \neq 0$ has the effect of simply adding a constant shift to $\delta\widehat{\sigma}$, independent of the option's strike.

B.4 The ATMF skew

We now derive the expression of the ATMF skew $\mathcal{S}_T = \frac{d\widehat{\sigma}_{KT}}{d\ln K}\Big|_{F_T}$ at order one in the $\delta\kappa_n$. Because the second piece in the right-hand side of (5.90) generates a uniform shift of implied volatilities, it does not contribute to $\mathcal{S}_T$, hence $\mathcal{S}_T$ does not depend on the θ_n.

Starting from (5.90), using (5.87) to express derivatives of P_0 in terms of derivatives of ρ_0, and using the Black-Scholes expression of the gamma yields, after some tedious algebra, the following result:

$$\mathcal{S}_T = \frac{1}{\sqrt{T}} \sum_{n=3}^{\infty} \frac{\delta\kappa_n}{n!} \frac{\int_0^{\infty} \left((-1)^n \left(\frac{1}{2} \frac{d^n \rho_0}{dz^n} + \frac{d^{n+1}\rho_0}{dz^{n+1}} \right) - \left(\frac{1}{2} \frac{d^2 \rho_0}{dz^2} + \frac{d^3 \rho_0}{dz^3} \right) \right) (e^z - 1) dz}{\Sigma\rho_0(0)}$$

$$(5.91)$$

Remember that Σ is not the volatility, but the unperturbed standard deviation of $\ln S_T$: $\Sigma = \widehat{\sigma}_0 \sqrt{T}$.

Using formula (5.91) for $\mathcal{S}_T$ at order one in the $\delta\kappa_n$ is advantageous in situations when it is easier to approximately compute cumulants than carry out an expansion in powers of a given parameter. Let us concentrate on the contribution of $\delta\kappa_3$ and $\delta\kappa_4$. They are usually expressed in terms of the skewness s and kurtosis κ of $\delta\rho$:

$$\delta\kappa_3 = s\Sigma^3, \; \delta\kappa_4 = \kappa\Sigma^4$$

Straightforward, though tedious, calculation of the numerator in (5.91) yields:

$$\mathcal{S}_T = \frac{1}{\sqrt{T}} \left(\frac{s}{6} + \frac{\kappa}{12}\widehat{\sigma}_0\sqrt{T} + \cdots \right) \qquad (5.92)$$

This recovers the result in [5].

If we only consider the contribution of the third-order cumulant, δP is given by:

$$\delta P = \frac{\delta\kappa_3}{6} \left[\left(\frac{d^3}{d\ln S^3} - \frac{d^2}{d\ln S^2} \right) + \theta_3 \left(\frac{d^2}{d\ln S^2} - \frac{d}{d\ln S} \right) \right] P_0$$

with $\theta_3 = 0$ in case we keep the log contract implied volatility unchanged. (5.92) supplies the following simple relationship relating the ATMF *skew* to the *skewness* of $\ln S_T$:

$$\mathcal{S}_T = \frac{s}{6\sqrt{T}} \qquad (5.93)$$

We have focused on the ATMF skew, but could have derived from (5.90) an expression for the ATMF implied volatility as well.

While approximations of the ATMF volatility at order one in $\delta\kappa_3$ and $\delta\kappa_4$ are usually insufficiently accurate for practical use, relationship (5.93) is remarkably robust – presumably because it does not involve any volatility reference level, and only depends on skewness – a dimensionless number.

Chapter's digest

5.1 Variance swap forward variances

▶ Variance swap volatilities are strikes of variance swap contracts. $\widehat{\sigma}_{VS,T}(t)$ is such that, at inception, a contract delivering at T the following payoff is worth zero.

$$\frac{1}{T-t}\sum_{i=0}^{N-1}\ln^2\left(\frac{S_{i+1}}{S_i}\right) - \widehat{\sigma}^2_{VS,T}(t)$$

Forward VS variances ξ_t^T are defined as:

$$\xi_t^T = \frac{d}{dT}\left((T-t)\,\widehat{\sigma}^2_{VS,T}(t)\right)$$

They are positive and driftless.

❧ ❧ ❧ ❧ ❧

5.2 Relationship of variance swaps to log contracts

▶ Up to second order in $S_{i+1} - S_i$, VSs can be replicated by delta-hedging a static position in European payoff $-2\ln S_T$, called the log contract. In the absence of cash-amount dividends, log-contract, hence VS, implied volatilities are given by formula (5.17):

$$\widehat{\sigma}^2_{VS,T} = \int_{-\infty}^{+\infty} dy\, \frac{e^{-\frac{y^2}{2}}}{\sqrt{2\pi}}\widehat{\sigma}^2_{K(y)T}$$

$$y(K) = \frac{\ln\left(\frac{F_T}{K}\right)}{\widehat{\sigma}_{KT}\sqrt{T}} - \frac{\widehat{\sigma}_{KT}\sqrt{T}}{2}$$

❧ ❧ ❧ ❧ ❧

5.3 Impact of large returns

▶ The property that a delta-hedged log contract replicates the payoff of a VS holds at second order in $S_{i+1} - S_i$.

▶ In a diffusive model, the VS volatility $\widehat{\sigma}_{VS,T}$ and the log-contract implied volatility $\widehat{\sigma}_T$ match: any diffusive model calibrated to a given market smile yields the same value for $\widehat{\sigma}_{VS,T}$.

▶ In a jump-diffusion model, the difference between $\widehat{\sigma}_{VS,T}$ and $\widehat{\sigma}_T$ is given by: (5.28):

$$\widehat{\sigma}^2_{VS,T} - \widehat{\sigma}^2_T = \lambda\ln^2(1+J) + 2\ln(1+J) - 2J$$

This is expressed, as a function of the ATMF skew of the smile for maturity T, $\mathcal{S}_T$, at order one in $\mathcal{S}_T$, through (5.32):

$$\widehat{\sigma}_{\mathrm{VS},T} \simeq \widehat{\sigma}_T \left(1 - \widehat{\sigma}_T \mathcal{S}_T T\right)$$

The difference between $\widehat{\sigma}_{\mathrm{VS},T}$ and $\widehat{\sigma}_T$ is independent on T only if $\mathcal{S}_T$ decays as $\frac{1}{T}$.

▶ In model-free fashion the difference between $\widehat{\sigma}_{\mathrm{VS},T}$ and $\widehat{\sigma}_T$ is given by (5.38), which involves the skewness $s_{\Delta t}$ of daily returns.

$$\frac{\widehat{\sigma}_{\mathrm{VS},T}}{\widehat{\sigma}_T} - 1 \simeq -\frac{s_{\Delta t}}{6}\widehat{\sigma}_T \sqrt{\Delta t}$$

▶ Inferring the skewness of returns at short time scales from market smiles is very model-dependent and unreasonable.

▶ The realized skewness of daily returns is such that the adjustment it warrants for $\widehat{\sigma}_{\mathrm{VS},T}$ is minute. The implied value of this adjustment, however, could be arbitrarily large.

▶ Are there variance payoffs that can be exactly replicated, even for large returns? There is only one, whose payoff is:

$$\sum_i (S_{i+1} - S_i)^2$$

It is replicated by delta-hedging a parabolic profile.

❧ ❧ ❧ ❧ ❧

5.4 Impact of strike discreteness

▶ The fact that, in practice, only discrete strikes – rather than continuous ones – can be traded, further adds to the imperfection of the replication of the VS.

❧ ❧ ❧ ❧ ❧

5.5 Conclusion

▶ In practice one can take $\widehat{\sigma}_{\mathrm{VS},T}$ and $\widehat{\sigma}_T$ to be equal, in effect setting $\widehat{\sigma}_T = \widehat{\sigma}_{\mathrm{VS},T}$. VS and log-contract forward variances are identical objects. The model is simulated according to SDE (5.43):

$$\begin{cases} dS_t &= \sqrt{\xi_t^t} S_t dW_t^S \\ d\xi_t^T &= \lambda_t^T dW_t^T \end{cases}$$

▶ For very liquid indexes, for which $\widehat{\sigma}_{VS,T}$ and $\widehat{\sigma}_T$ are both observable, a practical solution is to stay within a diffusive model, driven by SDE (5.43). The spread between $\widehat{\sigma}_{VS,T}$ and $\widehat{\sigma}_T$ is taken into account by adjusting the realized variance according to:

$$\ln^2\left(\frac{S_{i+1}}{S_i}\right) \;\to\; \ln^2\left(\frac{S_{i+1}}{S_i}\right) + (\lambda\Delta)\,\overline{\ln^2\left(1+J\right) + 2\ln\left(1+J\right) - 2J}$$

and adjusting the implied realized variance – or the forward VS variance – according to:

$$\zeta_t^T \;=\; \xi_t^T \;+\; \lambda\overline{\ln^2\left(1+J\right) + 2\ln\left(1+J\right) - 2J}$$

when the payoff calls for observation of implied VS volatilities. λ, J are chosen to match market values for the spread between $\widehat{\sigma}_{VS,T}$ and $\widehat{\sigma}_T$. The term structure of this spread is captured by making λ time-dependent.

❧ ❧ ❧ ❧ ❧

5.6 Dividends

▶ The impact of dividends on the VS payoff itself is zero for stocks, whose returns are corrected for the dividend impact, and minute for indexes.

▶ Fixed cash-amount dividends impact the replication of VSs. The log contract is supplemented with additional European payoffs with maturities matching the dividend schedule.

❧ ❧ ❧ ❧ ❧

5.7 Pricing variance swaps with a PDE

▶ VS volatilities for indexes are most easily calculated by solving PDE (5.49):

$$\frac{dU}{dt} + (r-q)\,S\frac{dU}{dS} + \frac{\sigma^2\,(t,S)}{2}S^2\frac{d^2U}{dS^2} \;=\; -\sigma^2\,(t,S)$$

▶ The adjustment for large returns is performed by adding an extra term – one solves PDE (5.53):

$$\frac{dU}{dt} + (r-q)\,S\frac{dU}{dS} + \frac{\sigma^2\,(t,S)}{2}S^2\frac{d^2U}{dS^2} \;=\; -\left(\sigma^2\,(t,S) - \frac{1}{3}\varepsilon J^3\right)$$

❧ ❧ ❧ ❧ ❧

5.8 Interest-rate volatility

▶ $\widehat{\sigma}_T$ is the implied volatility of the log-contract, a European payoff. Thus it really is the implied volatility of the forward for maturity T, F_t^T. VSs, on the other hand, pay the realized variance of S_t. Interest-rate volatility creates a difference between realized volatilities of S_t and F_t^T – at order one in interest-rate volatility, the resulting adjustment for $\widehat{\sigma}_{VS,T}$ is given in (5.55):

$$\widehat{\sigma}_{VS,T} \;=\; \widehat{\sigma}_T - \frac{\rho}{2}\sigma_r T$$

❧❧ ❧❧ ❧❧ ❧❧ ❧❧

5.9 Weighted variance swaps

▶ In weighted VSs, realized variance is weighted with a function of the spot $w(S)$. At order two in δS, weighted VSs can be replicated exactly. Standard examples of weighted VSs include the Gamma swap, the arithmetic swap, for which replication is exact, and corridor variance swaps. The latter require an additional adjustment to their strike to take into account barrier crossings.

❧❧ ❧❧ ❧❧ ❧❧ ❧❧

Appendix A – timer options

▶ Timer options expire when a given quadratic-variation budget $\mathcal{Q}$ is exhausted. The price of a timer option is a function of t, S and the current quadratic variation Q, measured using daily log-returns of the underlying stock or index:

$$Q_{i+1} - Q_i = \ln^2\left(\frac{S_{i+1}}{S_i}\right)$$

▶ For vanishing interest rate and repo, no dividends, and if the process for S_t is a diffusion, timer options are model-independent. They are replicated by a plain delta strategy and their prices are given by a Black-Scholes formula with an effective volatility equal to 1 and an effective maturity equal to $\mathcal{Q} - Q$. Physical time disappears.

▶ In practice, timer options are not exactly model-independent. While order-two contributions in δS vanish, higher-order terms contribute to the carry P&L. Moreover, the presence of non-vanishing interest rate and repo as well as dividends reintroduces the dependence on physical time. Two additional effects need to be priced-in: the overshoot in realized quadratic variation with respect to the budget, and the provision of a maximum maturity in the term sheet.

▶ Leveraged ETFs are another breed of quadratic-variation-based payoffs that are model-independent, in a diffusive setting, as they are replicated by a plain delta strategy. Starting from a value I_0, the NAV I_t at time t of the ETF is $I(t, S_t, Q_t)$ where $I(t, S, Q)$ is given by (5.75):

$$I(t, S, Q) = I_0 e^{rt} \left(\frac{S}{S_0 e^{(r-q)t}}\right)^\beta e^{-\frac{\beta(\beta-1)}{2}Q}$$

where β is the ETF's leverage.

❧❧ ❧❧ ❧❧ ❧❧ ❧❧

Appendix B – perturbation of the lognormal distribution

▶ The smile of vanilla options for maturity T is generated by the non-lognormality of the distribution of S_T. Many types of models can be collapsed onto the Black-Scholes by setting a parameter to zero – stochastic volatility and jump-diffusion models are two examples. It is useful to have an expansion of implied volatilities at order one in such parameter.

▶ The non-lognormality of the distributon of S_T is quantified by the cumulants κ_n of the distribution of $\ln S_T$. For a lognormal distribution of S_T, $\kappa_n = 0$, $\forall n \geq 3$. We carry out an expansion of implied volatilities in the κ_n, $n \geq 3$, at order one.

▶ As we perturb the cumulant-generating function, we require that prices of log contracts be unaffected, so that VS volatilities, in diffusive models, be unchanged.

▶ The perturbation of implied volatilities around a lognormal density of implied volatility $\widehat{\sigma}_0$ is given, at order one in the cumulants, by: (5.90):

$$
\delta\widehat{\sigma} = \frac{1}{\widehat{\sigma}_0 T} \sum_{n=3}^{\infty} \frac{\frac{\delta\kappa_n}{n!}\left(\frac{d^n P_0}{d\ln S^n} - \frac{d^2 P_0}{d\ln S^2}\right)}{\frac{d^2 P_0}{d\ln S^2} - \frac{dP_0}{d\ln S}} + \frac{1}{\widehat{\sigma}_0 T} \sum_{n=3}^{\infty} \frac{\delta\kappa_n}{n!}\theta_n
$$

where $\theta_n = 0$ if we demand that log-contract implied volatilities stay unchanged.

▶ At order one in the third cumulant – expressed in terms of the skewness s of $\ln S_T$ through $\delta\kappa_3 = s^3(\widehat{\sigma}_0\sqrt{T})^3$ – the ATMF skew is given by (5.93):

$$
\mathcal{S}_T = \frac{s}{6\sqrt{T}}
$$

Chapter 6

An example of one-factor dynamics: the Heston model

In the next chapter we cover models built on a specification of the dynamics of forward VS variances that can be calibrated to a term structure of VS volatilities, or a term-structure of vanilla implied volatilities of an arbitrary moneyness.

We now briefly pause to consider the Heston model, which is not a forward variance model, since in its native form it can be calibrated to the VS volatility of one single maturity.

It is instructive to assess the Heston model and its capabilities in the framework of forward variances. This exercise will help us introduce suitability criteria which we then apply to the forward variance models covered in Chapter 7. Moreover, the Heston model is an archetypal example among first-generation stochastic volatility models, that is models written in terms of the instantaneous variance V_t.

Unless stated otherwise, forward VS variances ξ_t^T will henceforth be called simply "forward variances" and we use $\hat{\sigma}_T$ interchangeably for VS or log-contract implied volatilities.

6.1 The Heston model

The Heston model [60] is a first-generation model; it owes its popularity to the fact that, being an affine model, the Laplace transform of its moment-generating function for $\ln S$ is analytically known. Numerical inversion of this transform then yields vanilla option prices – when there are no dividends in fixed cash amounts.

The analytics of the Heston model is abundantly covered in the literature – see for example [48], [69]. Rather, we concentrate on the joint spot/volatility dynamics that this model generates. We assume for simplicity zero interest rate and repo.

The Heston model is a diffusive one-factor model – instead of forward variances, the instantaneous variance is modeled, according to the following SDEs:

$$\begin{cases} dS_t = \sqrt{V_t}S_t dW_t \\ dV_t = -k(V_t - V^0)dt + \sigma\sqrt{V_t}dZ_t \end{cases} \quad (6.1)$$

V_t is the instantaneous variance – with respect to our previous notation:

$$V_t = \xi_t^t = \overline{\sigma}_t^2$$

σ is commonly called "volatility of volatility", though it is not a lognormal volatility as it has dimension time^{-1}. The Brownian motions W_t, Z_t are correlated, with correlation ρ. V^0 is a constant.

6.2 Forward variances in the Heston model

Forward variances ξ_t^T are defined by:

$$\xi_t^T = E_t[V_T]$$

Taking the expectation of the second equation in (6.1) and using the compact notation $\overline{V}_u = E_t[V_u]$ yields:

$$d\overline{V}_T = -k(\overline{V}_T - V^0)dT \tag{6.2}$$

whose solution is

$$\overline{V}_T = V^0 + e^{-k(T-t)}(V_t - V^0)$$

which, using our notation reads:

$$\xi_t^T = V^0 + e^{-k(T-t)}\left(\xi_t^t - V^0\right) \tag{6.3}$$

VS volatilities $\widehat{\sigma}_t^T$ are given by:

$$\widehat{\sigma}_T^2(t) = \frac{1}{T-t}\int_t^T \xi_t^\tau d\tau = V^0 + \frac{1-e^{-k(T-t)}}{k(T-t)}(V_t - V^0) \tag{6.4}$$

Differentiating (6.3) gives:

$$d\xi_t^T = \sigma e^{-k(T-t)}\sqrt{\xi_t^t}\, dZ_t$$

ξ_t^T is driftless – as it should. In the framework of forward variances the Heston dynamics (6.1) reads:

$$\begin{cases} dS_t = \sqrt{\xi_t^t}S_t dW_t \\ d\xi_t^T = \sigma e^{-k(T-t)}\sqrt{\xi_t^t}dZ_t \end{cases} \tag{6.5}$$

The Heston model is thus a one-factor model for forward variances where the instantaneous volatility of all forward variances ξ_t^T is proportional to the instantaneous volatility $\overline{\sigma}_t = \sqrt{\xi_t^t}$. It is a Markov-functional model for forward variances, as ξ_t^T is a *function* of ξ_t^t, given by (6.3).

Still, the Heston model should not be considered as a particular version of the forward variance models of Chapter 7; a one-dimensional Markov representation exists *only* if the initial values $\xi_{t=0}^T$ of forward variances satisfy condition (6.2):

$$\frac{d\xi_0^T}{dT} = -k\left(\xi_0^T - V^0\right)dT$$

It is not able to accommodate general term structure of VS volatilities.

Before analyzing further the dynamics of the Heston model, let us discuss the issue of the drift of V_t in first-generation models.

6.3   Drift of V_t in first-generation stochastic volatility models

The traditional approach to these models typically found in papers and textbooks can be summarized as follows:

- Start with historical dynamics of the instantaneous variance:

$$dV_t = \mu\left(t, S, V, p\right)dt + \alpha dZ_t$$

where p are model parameters – such as k, V^0 in the Heston model.

- In risk-neutral dynamics, drift of V_t is altered by "market price of risk" λ, which is an arbitrary function of t, S, V:

$$dV_t = \left(\mu\left(t, S, V, p\right) + \lambda\left(t, S, V\right)\right)dt + \alpha dZ_t$$

- A few lines down the road, jettison "market price of risk" and conveniently decide that risk-neutral drift has same functional form as historical drift, except parameters now have stars:

$$dV_t = \mu\left(t, S, V, p^\star\right)dt + \alpha dZ_t$$

- Calibrate (starred) parameters on vanilla smile.

Discussions surrounding the "market price of risk" and the uneasiness generated by its *a priori* arbitrary form and hasty disposal are pointless – the "market price of risk" is a nonentity.

V_t itself is an artificial object: for different t, V_t represents a different forward variance. Its drift is then only a reflection of the term-structure of forward variances. Differentiating the identity $V_t = \xi_t^t$ and using the following dynamics for ξ_t^T:

$$d\xi_t^T = \lambda_t^T dZ_t^T$$

yields

$$dV_t = \left. \frac{d\xi_t^T}{dT} \right|_{T=t} dt + \lambda_t^t dZ_t^t$$

The drift of the instantaneous variance is thus simply the slope at time t of the short end of the variance curve. One can check that taking the derivative at ξ_t^T with respect to T in (6.3) indeed yields the drift of V_t in the Heston model.

6.4 Term structure of volatilities of volatilities in the Heston model

Using (6.4) we get the following SDE for $\hat{\sigma}_T$:

$$d\hat{\sigma}_T = \bullet\, dt + \frac{\sigma}{2} \frac{1 - e^{-k(T-t)}}{k\,(T-t)} \frac{\hat{\sigma}_t}{\hat{\sigma}_T} dZ_t \tag{6.6}$$

where $\hat{\sigma}_t = \sqrt{V_t}$. Let us examine the term structure of the volatility of VS volatilities, that is the dependence of the volatility of $\hat{\sigma}_T$ on T. We have the two following limiting regimes:

$$T - t \ll \frac{1}{k} \qquad d\hat{\sigma}_T \simeq \bullet\, dt + \frac{\sigma}{2}\left(1 - \frac{k\,(T-t)}{2}\right)\frac{\hat{\sigma}_t}{\hat{\sigma}_T} dZ_t \tag{6.7}$$

$$T - t \gg \frac{1}{k} \qquad d\hat{\sigma}_T \simeq \bullet\, dt + \frac{\sigma}{2}\frac{1}{k\,(T-t)}\frac{\hat{\sigma}_t}{\hat{\sigma}_T} dZ_t \tag{6.8}$$

Thus, for long maturities, the instantaneous volatility of $\hat{\sigma}_T$ decays like $\frac{1}{T-t}$. For a flat term structure of VS volatilities, (6.6) implies that the instantaneous lognormal volatility of $\hat{\sigma}_T$ is:

$$\mathrm{vol}(\hat{\sigma}_T) \propto \frac{1 - e^{-k(T-t)}}{k\,(T-t)} \tag{6.9}$$

Figure 6.1 shows the lognormal volatilities of VS implied volatilities of the Euro Stoxx 50 index, compared to expression (6.9), suitably rescaled, along with a power-law fit.

As Figure 6.1 shows, volatilities of VS volatilities are typically larger than volatilities of the underlying itself. This is not due to the fact that we are using VS volatilities – volatilities of ATM volatilities are similar.

As is apparent, while realized levels of volatilities of VS volatilities are consistent with a power law dependence on maturity, they cannot be captured by the dynamics of the Heston model over a wide range of maturities. The two values of k used in Figure 6.1, 2 and 0.4, have been chosen so as to best match the short and long end of the term structure of the volatilities of VS volatilities in the graph.

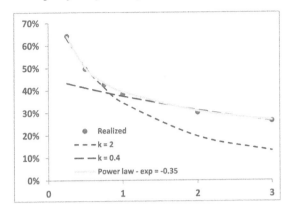

Figure 6.1: Volatility of VS volatilities of the Euro Stoxx 50 index as a function of maturity (years), evaluated on the period [2005, 2010] (dots), along with (a) volatilities of VS volatilities in the Heston model given by (6.9) for two different values of k (dotted lines), (b) a power-law fit $\propto T^{-0.35}$.

6.5   Smile of volatility of volatility

In diffusive models, for short maturities, the ATMF implied volatility is approximately equal to the instantaneous volatility, which is equal to $\sqrt{V}$. From (6.1) we get:

$$d\sqrt{V} \;=\; \bullet\, dt + \frac{\sigma}{2}dZ_t$$

Thus, in the Heston model, short ATMF implied volatilities are approximately normal, rather than lognormal, with a normal volatility equal to $\frac{\sigma}{2}$. Figure 6.2 shows the 3-month ATMF implied volatility as well as its realized (lognormal) volatility computed over a 6-month sliding window.

Inspection of the scales of the left-hand and right-hand axes again confirms that volatilities of short-dated volatilities are larger than volatilities themselves. Also, Figure 6.2 shows that high levels of volatility tend to coincide with high levels of volatility of volatility. In this respect, implied volatilities are in fact more than lognormal: their dynamics seems to be of the type:

$$d\hat{\sigma}_{\text{ATM}} \;=\; \bullet\, dt + \hat{\sigma}_{\text{ATM}}^{\gamma}dZ_t$$

with $\gamma > 1$. This should be compared with the value $\gamma = 0$ that the Heston model generates for short maturities.

What about implied volatilities for longer maturities? Equation (6.8) shows that the instantaneous volatility of long-dated VS volatilities decays like $\frac{1}{T-t}$. Equation

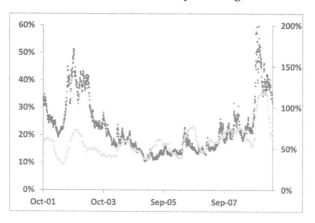

Figure 6.2: The 3-month ATMF implied volatility of the Euro Stoxx 50 index (darker dots, left-hand axis) together with its (lognormal) volatility, evaluated with a six-month sliding window (lighter dots, right-hand axis) from October 2001 to May 2009.

(6.4) implies that, since V_t is positive, $\widehat{\sigma}_T(t)$ has a floor:

$$\widehat{\sigma}_T(t) \geq \widehat{\sigma}_T^{\min}(t) = \sqrt{V^0}\sqrt{1 - \frac{1 - e^{-k(T-t)}}{k(T-t)}}$$

Figure 6.3 shows $\frac{\widehat{\sigma}_T^{\min}(t)}{\sqrt{V^0}}$ as a function of $T - t$; note that, from (6.4), $\sqrt{V^0}$ is the level of VS volatility for long-dated maturities, calibrated at $t = 0$. Forward VS volatilities are thus floored at a fraction of the initial long-run VS volatility level: The consequence for the smile of volatility of volatility is that volatilities of VS volatilities vanish as VS volatilities come near the floor. As is apparent in Figure 6.3, this floor on $\widehat{\sigma}_T(t)$ is not a minor effect.

6.6   ATMF skew in the Heston model

We now turn our attention to the dependence of the ATMF skew on maturity and volatility level. We obtain an approximation of the ATMF skew at order one in the volatility of volatility σ.

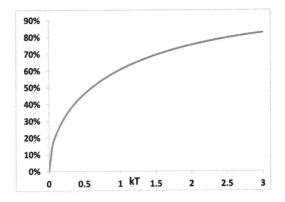

Figure 6.3: $\frac{\hat{\sigma}_T^{\min}(t)}{\sqrt{V^0}}$ as a function of $k(T-t)$.

6.6.1 The smile at order one in volatility of volatility

While we provide in Chapter 8 a general expression of the ATMF skew at order one in the volatility of volatility for general stochastic volatility models, we now carry out a derivation for the specific case of the Heston model.

The Heston model is homogeneous, thus vanilla implied volatilities are not a separate function of spot and strike, but of the ratio of the strike to the forward of the option's maturity. To reduce bookkeeping we now set interest rate and repo to zero and reinstate them once we have our final formulas.

From SDEs (6.1) we get the following equation for the price $P(t, S, V)$ of a European option:

$$\frac{dP}{dt} - k\left(V - V^0\right)\frac{dP}{dV} + \frac{V}{2}S^2\frac{d^2P}{dS^2} + \frac{\sigma^2}{2}V\frac{d^2P}{dV^2} + \rho\sigma VS\frac{d^2P}{dSdV} = 0 \quad (6.10)$$

with the terminal condition at maturity $P(T, S, V) = f(S)$, where f is the option's payoff. Denote by $P^0(t, S, V)$ the solution of (6.10) with $\sigma = 0$:

$$\frac{dP^0}{dt} - k\left(V - V^0\right)\frac{dP^0}{dV} + \frac{V}{2}S^2\frac{d^2P^0}{dS^2} = 0 \quad (6.11)$$

with terminal condition $P^0(T, S, V) = f(S)$, where f is the option's payoff.

To gain some intuition on the solution of (6.11) let us go back to the stochastic representation (6.1). For $\sigma = 0$, V is deterministic: the SDE for V_t in (6.1) becomes an ODE, identical to (6.2). Given the value V_t at time t, the value of V_τ at a later time τ is given by

$$V_\tau(V_t) = V^0 + (V_t - V^0)e^{-k(\tau-t)} \quad (6.12)$$

For $\sigma = 0$, the Heston model becomes a lognormal model with deterministic time-dependent volatility $\sigma(\tau)$ given by:

$$\sigma^2(\tau) = V_\tau(V_t)$$

The implied volatility at time t for maturity T is given by:

$$\widehat{\sigma}_T^2(t, V) = \frac{1}{T-t} \int_t^T V_\tau(V) \, d\tau = V^0 + (V - V^0) \frac{1 - e^{-k(T-t)}}{k(T-t)} \quad (6.13)$$

where V is the instantaneous variance at time t. P^0 is thus simply given by the Black-Scholes formula evaluated with implied volatility $\widehat{\sigma}_T(t, V)$:

$$P^0(t, S, V) = P_{BS}(t, S, \widehat{\sigma}_T(t, V))$$

Let us expand P:

$$P = P^0 + \delta P$$

where δP is of order one in σ and let us insert this expression in equation (6.10), keeping terms up to order one in σ. Using the fact that P^0 obeys (6.11) we are left with the following equation for δP:

$$\frac{d\delta P}{dt} - k(V - V^0) \frac{d\delta P}{dV} + \frac{V}{2} S^2 \frac{d^2 \delta P}{dS^2} = -\rho \sigma V S \frac{d^2 P^0}{dS dV} \quad (6.14)$$

with the terminal condition $\delta P(T, S, V) = 0$. δP is generated by the source term in the right-hand side. The second derivative of P_0 with respect to V does not appear, as it is multiplied by σ^2 – at order 1 in σ only the mixed derivative of P_0 with respect to S, V contributes.

The solution of (6.14) at time 0 is given by:

$$\delta P(t, S, V) = \mathop{E_t^0}_{\substack{S_t = S \\ V_t = V}} \left[\int_t^T \rho \sigma V_\tau S_\tau \left. \frac{d^2 P^0}{dS dV} \right|_{\tau, S_\tau, V_\tau} d\tau \right]$$

where the subscript 0 indicates that the expectation is taken with respect to the dynamics (6.1) with $\sigma = 0$, that is in a Black-Scholes model with deterministic time-dependent volatility $\sigma(\tau)$.

In the Black-Scholes model the vega of a European option is related to its dollar gamma – see (5.66), page 181:

$$\frac{dP_{BS}}{d\widehat{\sigma}} = S^2 \frac{d^2 P_{BS}}{dS^2} \widehat{\sigma}(T - \tau)$$

We then have:

$$\frac{dP_0}{dV} = S^2 \frac{d^2 P^0}{dS^2} \widehat{\sigma}_T(\tau, V)(T - \tau) \frac{d\widehat{\sigma}_T(\tau, V)}{dV}$$

$$= \frac{1 - e^{-k(T-\tau)}}{2k} S^2 \frac{d^2 P^0}{dS^2}$$

where we have used expression (6.13) for $\widehat{\sigma}_T(\tau, V)$. δP now reads:

$$\delta P(t, S, V) = \mathop{E_t^0}_{\substack{S_t = S \\ V_t = V}} \left[\int_t^T \frac{\rho \sigma}{2} V_\tau(V_t) \frac{1 - e^{-k(T-\tau)}}{k} S_\tau \frac{d}{dS} S^2 \left. \frac{d^2 P^0}{dS^2} \right|_{\tau, S_\tau, V_\tau} d\tau \right]$$

where V_τ is given by expression (6.12) as a function of V_t, the instantaneous variance at time 0.

We now take $t = 0$ and simply denote by V the instantaneous variance at $t = 0$. We have shown in Appendix A of Chapter 5 that, in the Black-Scholes model with deterministic time-dependent volatility, $e^{-r(\tau-t)} \frac{d^n P^0}{d \ln S^n}(\tau, S_\tau)$ is a martingale. Using:

$$S \frac{d}{dS} S^2 \frac{d^2}{dS^2} = \frac{d^3}{d \ln S^3} - \frac{d^2}{d \ln S^2}$$

we get:

$$\delta P = \frac{\rho\sigma}{2} \left[\int_0^T V_\tau(V) \frac{1 - e^{-k(T-\tau)}}{k} \, d\tau \right] \left(\frac{d^3 P^0}{d \ln S^3} - \frac{d^2 P^0}{d \ln S^2} \right)_{t=0,S,V} \tag{6.15}$$

where $V_\tau(V)$ is given by:

$$V_\tau(V) = V^0 + (V - V^0)e^{-k\tau}$$

Compare (6.15) with expression (5.88), page 193: δP has the same expression as a function of the second and third order derivatives of P^0 with respect to $\ln S$: perturbing at order one in the volatility of volatility amounts to perturbing at order one in the third-order cumulant with fixed forward variances.[1]

This is not surprising. Indeed, the ODE for $E[V_t]$ in the Heston model – see equation (6.2) – does not involve σ. Consequently, the perturbation in powers of σ leaves forward variances unchanged at all orders.

The interpretation of the integral over τ in the prefactor is not straightforward at this stage. It will become clear when we carry out the derivation for general stochastic volatility models – see Section 8.6.

The expansion of the implied volatility $\hat\sigma_{KT}$ at order one in σ is: $\hat\sigma_{KT} = \hat\sigma_T(0, V)$ $+\delta\hat\sigma_{KT}$ where $\delta\hat\sigma_{KT}$ is given by:

$$\delta\hat\sigma_{KT} = \left(\frac{dP_{BS}}{d\hat\sigma} \right)^{-1} \delta P$$

$$= \frac{1}{\hat\sigma_T(0,V)\,T} \left[\int_0^T \frac{\rho\sigma}{2} V_\tau(V) \frac{1 - e^{-k(T-\tau)}}{k} \, d\tau \right] \frac{\frac{d^3 P^0}{d \ln S^3} - \frac{d^2 P^0}{d \ln S^2}}{\frac{d^2 P^0}{d \ln S^2} - \frac{dP^0}{d \ln S}}$$

$$= \frac{1}{\hat\sigma_T(0,V)\,T} \left[\int_0^T \frac{\rho\sigma}{2} V_\tau(V) \frac{1 - e^{-k(T-\tau)}}{k} \, d\tau \right] \frac{d}{d \ln S} \ln \left(\frac{d^2 P^0}{d \ln S^2} - \frac{dP^0}{d \ln S} \right)$$

---

[1]The mistrustful reader is encouraged to compute κ_3 at order one in σ – an easy task as the characteristic function of $\ln S$ is analytically known in the Heston model – to verify that one indeed recovers (6.15) from (5.88).

where we have expressed vega in terms of gamma using relationship (5.66). Using the formula for the Black-Scholes dollar gamma, we get:

$$\delta\hat{\sigma}_{KT} = \frac{1}{\hat{\sigma}_T^3 T^2}\left[\frac{\rho\sigma}{2}\int_0^T V_\tau\left(V\right)\frac{1-e^{-k(T-\tau)}}{k}\,d\tau\right]\left(\frac{\hat{\sigma}_T^2 T}{2}+\ln\frac{K}{S}\right)$$

We now reinstate interest rate and repo. As mentioned above, we only need to replace $\frac{K}{S}$ with $\frac{K}{F_T}$, where F_T is the forward for maturity T:

$$\delta\hat{\sigma}_{KT} = \frac{1}{\hat{\sigma}_T^3 T^2}\left[\frac{\rho\sigma}{2}\int_0^T V_\tau\left(V\right)\frac{1-e^{-k(T-\tau)}}{k}\,d\tau\right]\left(\frac{\hat{\sigma}_T^2 T}{2}+\ln\frac{K}{F_T}\right) \qquad (6.16)$$

We have used the shorthand notation $\hat{\sigma}_T = \hat{\sigma}_T\left(0,V\right) = \sqrt{V^0+(V-V^0)\frac{1-e^{-kT}}{kT}}$. The notation $\hat{\sigma}_T$ is appropriate, as $\hat{\sigma}_T$ is the VS volatility both in the order-zero and order-one expansion in σ. Denoting the ATMF skew by $\mathcal{S}_T$ and the ATMF volatility by $\hat{\sigma}_{F_T T}$, (6.16) gives:

$$\hat{\sigma}_{F_T T} = \hat{\sigma}_T\left(1+\frac{\hat{\sigma}_T T}{2}\mathcal{S}_T\right) \qquad (6.17a)$$

$$\mathcal{S}_T = \left.\frac{d\hat{\sigma}_{KT}}{d\ln K}\right|_{F_T} = \frac{1}{\hat{\sigma}_T^3 T^2}\frac{\rho\sigma}{2}\int_0^T V_\tau\frac{1-e^{-k(T-\tau)}}{k}\,d\tau \qquad (6.17b)$$

Short maturities
Let us take the limit $kT \ll 1$. Formula (6.16) translates at lowest order in kT in the following expressions for $\hat{\sigma}_{F_T T}$ and $\mathcal{S}_T$:

$$\hat{\sigma}_{F_T T} = \sqrt{V}\left(1+\frac{\rho\sigma T}{8}\right) \qquad (6.18a)$$

$$\mathcal{S}_T = \frac{\rho\sigma}{4\sqrt{V}} = \frac{\rho\sigma}{4\hat{\sigma}_{F_T T}} \qquad (6.18b)$$

where the second equality in (6.18b) is correct at order 1 in σ.

Long maturities
We now take the limit $kT \gg 1$. Keeping only terms of order 1 in $1/kT$ we get:

$$\hat{\sigma}_{F_T T} = \sqrt{V^0}\left(1+\frac{\rho\sigma}{4k}\right)+\frac{1}{2kT}\left(\frac{V-V^0}{\sqrt{V^0}}+\frac{\rho\sigma}{4k}\frac{V-3V^0}{\sqrt{V^0}}\right) \qquad (6.19a)$$

$$\mathcal{S}_T = \frac{\rho\sigma}{2\sqrt{V^0}}\frac{1}{kT} \qquad (6.19b)$$

6.6.2 Example

Consider the example of a smile for a three-month maturity generated with the following parameters, whose values are typical of index smiles: $V^0 = 0.04$, $k = 1$, $\sigma = 0.6$, $\rho = -80\%$.

Figure 6.4 shows $\hat{\sigma}_{KT}$ as a function of $\ln\left(\frac{K}{F_T}\right)$ for three different values of V: 1%, 4%, 16%, corresponding approximately to ATMF volatilities around 10%, 20%, and 40%, respectively.

The reason why we choose to vary V while keeping other parameters constant is that V is the only state variable of the Heston model. Figure 6.4 provides an illustration of future smiles that the Heston model generates at time T_1 for maturity $T_2 = T_1 + 3$ months – as a function of V_{T_1}.

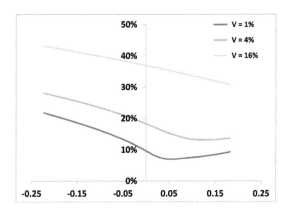

Figure 6.4: Implied volatilities $\hat{\sigma}_{KT}$ as a function of $\ln\left(\frac{K}{F_T}\right)$ for a maturity $T = 3$ months, in the Heston model, generated with the following parameters: $V^0 = 0.04$, $k = 1$, $\sigma = 0.6$, $\rho = -80\%$, for three different values of V.

The quality of approximations (6.17a, 6.17b) for these parameter values is assessed in Tables 6.1 and 6.2.

V	1%	4%	16%
$\mathcal{S}_T^{\text{real}}$	−8.1	−6.0	−3.1
$\mathcal{S}_T^{\text{approx}}$	−8.8	−5.5	−3.0

Table 6.1: ATMF skew in the Heston model for a three-month maturity, for different values of V with $V^0 = 0.04$, $k = 1$, $\sigma = 0.6$, $\rho = -80\%$, computed either exactly: $\mathcal{S}_T^{\text{real}}$ or using approximation (6.17b): $\mathcal{S}_T^{\text{approx}}$. Results have been multiplied by 10 to represent approximately the difference in volatility points of implied volatilities for strikes $0.95F_T$ and $1.05F_T$.

V	1%	4%	16%
$\widehat{\sigma}_T$	11.6	20	38.2
$(\widehat{\sigma}_{F_TT} - \widehat{\sigma}_T)_{\text{real}}$	-1.9	-1.7	-1.2
$(\widehat{\sigma}_{F_TT} - \widehat{\sigma}_T)_{\text{approx}}$	-0.1	-0.3	-0.5

Table 6.2: $\widehat{\sigma}_T$ and the difference $\widehat{\sigma}_{F_TT} - \widehat{\sigma}_T$, both in volatility points, in the Heston model for a three-month maturity, for different values of V with $V^0 = 0.04$, $k = 1$, $\sigma = 0.6$, $\rho = -80\%$, computed either exactly (real) or using approximation (6.17a) (approx).

The ATMF skew is acceptably captured by an approximation at order one in σ: the maximum relative error is about 10%.

In contrast, the difference $\widehat{\sigma}_{F_TT} - \widehat{\sigma}_T$, is poorly estimated by (6.17a). Though both $\mathcal{S}_T$ and $\widehat{\sigma}_{F_TT} - \widehat{\sigma}_T$ are of order one in σ, the approximation for $\mathcal{S}_T$ is more robust.

This is generally observed in stochastic volatility models for equities: one typically needs to carry out the expansion of the ATMF volatility at order two in volatility of volatility to reach acceptable accuracy – more on this in Section 8.2.

6.6.3 Term structure of the ATMF skew

Equations (6.18b), (6.19b) show that while the ATMF skew tends to a constant for $T \to 0$, it decays like $\frac{1}{T}$ for long maturities, at order one in σ. The $\frac{1}{T}$ decay for long maturities is expected: because V is mean-reverting, for maturities T such that $T \gg \frac{1}{k}$, the distribution of returns of $\ln S$ over long periods becomes independent on the initial value of V, hence returns of $\ln S$ over long time scales become independent. The third-order cumulant κ_3 of $\ln S$ then scales like T, the skewness s of $\ln S$ scales like $\frac{1}{\sqrt{T}}$ and (5.92) then implies that $\mathcal{S}_T \propto \frac{1}{T}$.

In the special case when $V = V^0$, the term structure of VS volatilities is flat, V_t is constant and (6.16) takes the following simple form:

$$\left.\frac{d\widehat{\sigma}_{KT}}{d\ln K}\right|_{F_T} = \frac{\rho\sigma}{2\sqrt{V^0}}\frac{kT + e^{-kT} - 1}{(kT)^2} \tag{6.20}$$

Figure 6.5 shows an example of the term structure of the ATMF skew for the Euro Stoxx 50 and S&P 500 indexes, together with a power-law fit and a best fit using formula (6.20). The maturity dependence of the market skew indeed exhibits a power-law-like behavior, which cannot be captured by the Heston model for both short and long maturities: the Heston model is a one-factor model, with an embedded time scale $1/k$.

The issue here is not only about whether we are or aren't able to fit the vanilla smile. Rather, when risk-managing cliquets we may need to carry a naked forward smile position: it is then necessary to assess whether the model is able to generate *future* smiles – that is vanilla smiles at future dates – that are comparable to historically observed vanilla smiles – see the discussion below.

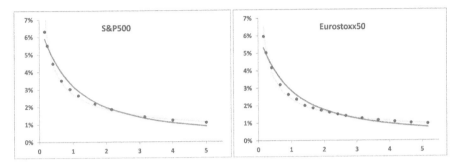

Figure 6.5: The ATMF skew as a function of maturity for the Euro Stoxx 50 and S&P 500 indexes (dots), observed on October 22, 2010, expressed as the difference in volatility points of the implied volatilities of strikes $0.95F_T$ and $1.05F_T$. A best fit using a power law with exponent 0.55 (lighter line) as well as a best fit using formula (6.20) ($k = 2.9$) (darker line) are shown as well.

6.6.4   Relationship between ATMF volatility and skew

For short maturities, equation (6.18b) shows that, in the Heston model the ATMF skew is inversely proportional to the ATMF implied volatility – this is also evidenced in Figure 6.6. Is this dependence observed in reality?

Figure 6.6 displays the ATM volatility together with the ATM skew for a 3-month maturity, for the Euro Stoxx 50 index.[2] We can see that while skew and volatility seem to behave fairly independently, they are, if anything, positively correlated rather than negatively. It seems unreasonable to hard-wire an inverse dependence of the ATMF skew to the ATMF volatility in our model.

6.7   Discussion

The above analysis has highlighted some discrepancies between the spot/volatility dynamics generated by the Heston model on one hand, and observed in reality on the other hand – see also [8]. What makes the Heston model unsuitable for handling exotics however, rather than its inability to reproduce *exactly* the observed historical dynamics, is the lack of flexibility it affords.

---

[2]We have used the ATM volatility and skew for simplicity – using ATMF data would have yielded a similar graph.

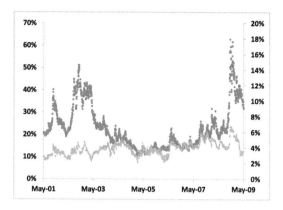

Figure 6.6: The ATM skew (lighter dots, right-hand axis) as the difference of the implied volatilities of the 95% and 105% strikes, and the ATM volatility (darker dots, left-hand axis) of the Euro Stoxx 50 index, for a 3-month maturity.

Indeed, from a trading perspective, one may choose to use parameter levels different than their historical averages and structural dependencies other than what is historically observed, even for parameters that have no implied counterpart. For example, even though the historical realized correlation between two quantities may be negligibly small, we will use bid/offer levels that are different than zero, depending on the size and sign of the option's sensitivity to correlation.

Likewise, imagine that in reality the short ATMF skew was indeed approximately inversely proportional to the short ATMF volatility, thus in line with the behavior generated in the Heston model. Still, when selling an exotic option that has positive forward ATMF volatility/skew cross-gamma: $\frac{d^2 P}{d\hat\sigma_{F_T T} dS_T} > 0$, we may choose to conservatively price with a model that generates positive or vanishing covariance between ATMF volatility and skew – rather than negative.

Making the level of the short ATMF skew independent on the level of short ATMF volatility is easy: we only need to replace the SDE for V_t in (6.1) with:

$$dV_t \;=\; \bullet\, dt \;+\; \nu V_t dZ_t$$

This would also have the advantage of making the short ATMF volatility lognormal rather than normal.

Other deficiencies of the Heston model are structural. For example, the scaling with maturity of the volatility of volatility in (6.6) and of the ATMF skew in (6.20) is intimately related to the fact that the Heston model is a one-factor model. Some have advocated making the parameters of the Heston model time-dependent so as to alter the dependence of $\mathcal{S}_T$ on T and achieve accurate calibration of the vanilla smile.

Making V^0 time-dependent so as to best match the term structure of VS volatilities is appropriate, as forward variances can be hedged by trading variance swaps.[3] Making σ and ρ time-dependent is more questionable.

Indeed, consider the example of a call-spread cliquet paying at time T_2 the payoff $\left(\frac{S_{T_2}}{S_{T_1}}-95\%\right)^+ - \left(\frac{S_{T_2}}{S_{T_1}}-105\%\right)^+$, where $T_2-T_1 \ll \frac{1}{k}$. This call spread has negligible sensitivity to volatility but is very sensitive to the ATMF skew observed at T_1 for maturity T_2. The order-one expansion in σ in (6.18b) yields the following expression for the ATMF skew at T_1 for maturity T_2:

$$\mathcal{S}_{T_2}\left(T_1\right) = \frac{\rho\left(T_1\right)\sigma}{4\sqrt{V_{T_1}}}$$

where we have allowed ρ to be time-dependent so as to match the term structure of the market *vanilla* ATMF skew. By pricing the cliquet with the Heston model thus calibrated we are betting on a given level of *forward* skew $\mathcal{S}_{T_2}\left(T_1\right)$, derived from calibration to today's term structure of the ATMF skew of vanilla options. Worse, the model will actually generate hedge ratios for our forward-start call spread on vanilla options of maturities T_1 and T_2.

This is not justified, as it is not possible to hedge forward-skew risk by trading vanilla options – we refer the reader to the illuminating experiment in Section 3.1.7. In case we are unable to take an offsetting position in a different exotic that has comparable forward-smile risk and have to keep the cliquet and its forward skew exposure on our book, it is more reasonable to *select* a conservative level of forward skew we are comfortable with, rather than having it dictated by calibration to market prices of instruments that are incapable of hedging it. The same goes for volatility-of-volatility risk.

Only if we are able to trade cliquets of varying maturities is it licit to make σ or ρ time-dependent, so as to match the term structure of the implied forward skew.

Fundamentally, the problem with the Heston model lies not so much with the model itself, its lack of flexibility, or even its peculiar idiosyncrasies, but with its usage (and its users): which practical pricing or hedging issue naturally calls for SDE (6.5)?

---

[3]Exact calibration is not guaranteed as $V_0\left(t\right)$ has to remain positive.

Chapter's digest

▶ In terms of forward variances, the SDE of the Heston model reads:

$$
\begin{cases}
dS_t &= \sqrt{\xi_t^t} S_t dW_t \\
d\xi_t^T &= \sigma e^{-k(T-t)} \sqrt{\xi_t^t} dZ_t
\end{cases}
$$

with the following constraint on the initial values of forward variances:

$$
\frac{d\xi_0^T}{dT} = -k\left(\xi_0^T - V^0\right) dT
$$

VS volatilities are given by:

$$
\widehat{\sigma}_T^2(t) = V^0 + \frac{1 - e^{-k(T-t)}}{k(T-t)}(V_t - V^0)
$$

thus are floored.

▶ The drift of the instantaneous variance V_t in stochastic volatility models has nothing to do with the "market price of risk". It is related to the initial slope of the variance curve:

$$
dV_t = \left.\frac{d\xi_t^T}{dT}\right|_{T=t} dt + \lambda_t^t dZ_t^t
$$

▶ For a flat term-structure of VS volatilities, at order one in volatility of volatility, the ATMF skew of the Heston model is given by:

$$
\mathcal{S}_T = \frac{\rho\sigma}{2\sqrt{V^0}} \frac{kT + e^{-kT} - 1}{(kT)^2}
$$

and the (lognormal) volatility of the VS volatility of maturity T, $\widehat{\sigma}_T$, has the form:

$$
\text{vol}(\widehat{\sigma}_T) \propto \frac{1 - e^{-kT}}{kT}
$$

▶ At order one in volatility of volatility, the ATMF skew of the Heston model for short maturities is given by:

$$
\mathcal{S}_T = \frac{\rho\sigma}{4\sqrt{V}}
$$

and for long maturities:

$$
\mathcal{S}_T = \frac{\rho\sigma}{2\sqrt{V^0}} \frac{1}{kT}
$$

It decays as $\frac{1}{T}$.

Chapter 7

Forward variance models

We catch up to where we left off, at the end of Chapter 4. We examine diffusive stochastic volatility models built on the dynamics of continuous forward VS variances – they are exactly calibrated to an initial term structure of VS volatilities, by construction. They can alternatively be calibrated to a term structure of implied volatilities of other payoffs, for example ATMF vanilla options, or power payoffs.

We concentrate on the control of the term structure of volatility of volatility, the term structure of the ATMF skew, the smile of volatility of volatility and cover options on realized variance and VIX instruments.

The last section deals with discrete forward variance models, a type of stochastic volatility model that is particularly suited to the analysis of exotic option risks in terms of volatility of volatility, spot/volatility covariance and future skew.

7.1 Pricing equation

The Heston model – studied in the preceding chapter – is an elementary attempt at designing a model such that implied volatilities are not frozen anymore and have their own dynamics. This is done by specifying an SDE for the instantaneous variance V_t, a non-physical object. It has then been our task to extract the dynamics of implied volatilities that this SDE gives rise to.

In this chapter we model implied volatilities directly. From the discussion in Section 4.3.6, page 148, and Section 5.5, page 168, it is clear that the easiest objects to model are VS forward variances ξ^T.

The ξ^T will be our state variables, in addition to S, and we will design models so that we have a direct handle on the volatilities of VS volatilities – instantaneous or discrete, forward or spot-starting. The price of an option in such a model is given by:

$$P(t, S, \xi)$$

where ξ is the variance curve.

Consider a short position in an option of maturity T – first unhedged. Our P&L during δt is

$$-\left[P(t + \delta t, S + \delta S, \xi + \delta \xi) - (1 + r\delta t) P(t, S, \xi)\right]$$

The delta hedge consists of $\frac{dP}{dS}$ shares and $\frac{\delta P}{\delta \xi^u}$ forward VS contracts of maturity u, for all u in $[t,\ T]$, where $\frac{\delta P}{\delta \xi^u}$ is a functional derivative, since ξ^u is a function of u. Our total P&L during δt, including now our delta- and vega-hedge, is:

$$
P\&L \;=\; -\left[P(t+\delta t, S+\delta S, \xi+\delta\xi) - (1+r\delta t)\,P(t,S,\xi) \right]
$$
$$
+ \frac{dP}{dS}\left(\delta S - (r-q)S\delta t\right) + \int_t^T \frac{\delta P}{\delta\xi^u}\delta\xi^u
$$

We remind the reader of the fact that (forward) VSs – which provide exact delta hedges for the ξ^u – can be entered into at zero cost. The ξ^u have no financing cost – thus zero risk-neutral drift – hence the simple form of the contribution of $\delta\xi^u$ to the P&L.

Expanding at order one in δt and two in δS and $\delta\xi^u$:[1]

$$
P\&L \;=\; -\frac{dP}{dt}\delta t + rP\delta t - (r-q)S\frac{dP}{dS}\delta t
$$
$$
-\frac{S^2}{2}\frac{d^2P}{dS^2}\frac{\delta S^2}{S^2} - \frac12\int_t^T du \int_t^T du' \frac{d^2P}{\delta\xi^u\delta\xi^{u'}}\delta\xi^u\delta\xi^{u'} - \int_t^T du\, S\frac{d^2P}{dS\delta\xi^u}\frac{\delta S}{S}\delta\xi^u
$$
$$
\tag{7.1}
$$

Using the same criteria that led us to the Black-Scholes equation in Section 1.1, we specify break-even levels for the random second-order terms in the P&L. Denote by $\sigma(t)$ the instantaneous break-even volatility of S_t and by $\nu\,(t,u,u')$ and $\mu\,(t,u)$ the instantaneous break-even covariances for, respectively, $\delta\xi^u$, $\delta\xi^{u'}$ and $\delta\xi^u$, $\frac{\delta S}{S}$:

$$
\mu\,(t,u)\,\delta t \;=\; \left\langle \frac{\delta S}{S}\delta\xi^u \right\rangle_t
\tag{7.2a}
$$
$$
\nu\,(t,u,u')\,\delta t \;=\; \left\langle \delta\xi^u\delta\xi^{u'} \right\rangle_t
\tag{7.2b}
$$

Obviously μ is only defined for $u \geq t$ and ν for $u \geq t$, $u' \geq t$.

Break-even variances and covariances can be at most a function of our state variables S and ξ, unless we introduce additional degrees of freedom in the model. We thus write: $\sigma(t,S,\xi)$, $\nu\,(t,u,u',S,\xi)$, $\mu\,(t,u,S,\xi)$. In the models we will work with in the sequel, ν and μ do not depend on S, so let us write our covariance functions more simply as: $\nu\,(t,u,u',\xi)$, $\mu\,(t,u,\xi)$.

We would like our carry P&L at order one in δt, two in $\delta S, \delta\xi$ to read:

$$
P\&L \;=\; -\frac{S^2}{2}\frac{d^2P}{dS^2}\left(\frac{\delta S^2}{S^2} - \sigma(t,S,\xi)^2\delta t\right)
\tag{7.3a}
$$
$$
-\frac12\int_t^T du \int_t^T du' \frac{d^2P}{\delta\xi^u\delta\xi^{u'}}\left(\delta\xi^u\delta\xi^{u'} - \nu\,(t,u,u',\xi)\,\delta t\right)
\tag{7.3b}
$$
$$
-\int_t^T du\, S\frac{d^2P}{dS\delta\xi^u}\left(\frac{\delta S}{S}\delta\xi^u - \mu\,(t,u,\xi)\,\delta t\right)
\tag{7.3c}
$$

---

[1] We use economical notations for what should really read: $\frac{\delta^2 P}{\delta\xi^u\delta\xi^{u'}}$ and $\frac{d\delta P}{dS\delta\xi^u}$.

The gamma P&L in (7.3a) can be offset by trading a short-maturity VS. So that no free theta is generated, we must impose that the break-even volatility at time t is equal to the instantaneous VS volatility: $\sigma(t, S, \xi)^2 = (\hat{\sigma}_t^t)^2 = \xi^t$.

Identifying the δt terms in (7.1) and (7.3) supplies us with the pricing equation in our model:

$$
\frac{dP}{dt} + (r - q) S \frac{dP}{dS} + \frac{\xi^t}{2} S^2 \frac{d^2 P}{dS^2}
$$
$$
+ \frac{1}{2} \int_t^T du \int_t^T du' \nu(t, u, u', \xi) \frac{d^2 P}{\delta\xi^u \delta\xi^{u'}} + \int_t^T du\, \mu(t, u, \xi) S \frac{d^2 P}{dS \delta\xi^u} = rP
\tag{7.4}
$$

with the terminal condition $P(t, S, \xi, t = T) = g(S)$ where g is the option's payoff, in the case of a European option.

Generally, g can depend on the full path of S_t for a path-dependent option, or the full path of S_t and of the ξ_t^τ if the payoff involves observations of VS volatilities.[2]

The probabilistic interpretation of (7.4) is that P is given – in the case of a European payoff – by:

$$
P = E\big[g(S_T) \mid S_t = S, \ \xi_t^u = \xi^u\big]
$$

with the following SDEs for S_t and ξ_t^u:

$$
\begin{cases}
dS_t &= (r - q)S_t dt + \sqrt{\xi_t^t} S_t dW_t^S \\
d\xi_t^u &= \lambda_t^u dW_t^u
\end{cases}
$$

with λ_t^u and correlations between W_t^S and W_t^u such that:

$$
\lim_{dt \to 0} \frac{1}{dt} E_t\big[d\ln S_t d\xi_t^u\big] = \sqrt{\xi_t^t} \lambda_t^u \frac{1}{dt} E_t\big[dW_t^S dW_t^u\big] = \mu(t, u, \xi)
\tag{7.5}
$$

$$
\lim_{dt \to 0} \frac{1}{dt} E_t\big[d\xi_t^u d\xi_t^{u'}\big] = \lambda_t^u \lambda_t^{u'} \frac{1}{dt} E_t\big[dW_t^u dW_t^{u'}\big] = \nu(t, u, u', \xi)
\tag{7.6}
$$

7.2 A Markov representation

While equation (7.4) is general, it is an infinite-dimensional equation that is not solvable unless the ξ^u possess a Markov-functional representation. Failing that, in a Monte Carlo simulation the (infinitely many) ξ^u need to be evolved individually. This is not possible, unless one resorts to an approximation. If instead a Markov-functional representation exists, the ξ^u can be expressed as a function of a small set of state variables.

---

[2]Consider for example VS swaptions.

In addition to this technical condition, we also require that the dynamics of forward variances be financially motivated.[3]

Consider a forward variance ξ^T, where T is an arbitrary date, and let us start with a lognormal dynamics for ξ^T – Figure 6.2 in Section 6.5 suggests that this assumption is a reasonable starting point. How should the volatility of ξ^T_t depend on t and T?

If there existed a market of options on ξ^T with maturities ranging from t to T, the volatility risk of ξ^T could be hedged away and the volatility of ξ^T would be derived from market implied volatilities. However, volatility of volatility is only traded in very special forms, for example through options on realized variance, through VIX futures and options, or cliquets. We have already considered cliquets beforehand and will analyze in detail other instruments further on, and characterize the type of volatility-of-volatility risk they are sensitive to.

In general we will have no choice but to carry a position on the realized volatility of ξ^T and thus will need to make assumptions that we will depend on. It is then reasonable to make the assumption of time-homogeneity: the volatility of ξ^T only depends on $T-t$, with a dependence that is adjustable so that, for example, volatilities of spot-starting VS volatilities $\hat{\sigma}_T$ in the model can be made to match their historical counterparts. Let us then write:

$$d\xi^T_t = \omega\,(T - t)\,\xi^T_t\,dW^T_t \tag{7.7}$$

The solution of this SDE is

$$\ln(\xi^T_t) = \ln(\xi^T_0) - \frac{1}{2}\int_0^t \omega^2\,(T - \tau)\,d\tau + \int_0^t \omega\,(T - \tau)\,dW^T_\tau \tag{7.8}$$

Imagine the same Brownian motion W_t drives the dynamics of all ξ^T. Equation (7.8) expresses $\ln \xi^T$ as a weighted average of increments of W_t, with a weight $\omega\,(T - t)$ that depends on T, hence is specific to forward variance ξ^T. Even though a single Brownian motion drives our model, simulation of the forward variance curve at time t requires knowledge of the full path of W_t, as weights for different forward variances are different – all of the ξ^T have to be simulated individually.

However, if ω is of the form:

$$\omega\,(u) = \omega e^{-ku} \tag{7.9}$$

$$\int_0^t \omega\,(T - \tau)\,dW_\tau = e^{-kT}\int_0^t e^{k\tau}\,dW_\tau$$

The dependence on T factors out and knowledge of one quantity – $\int_0^t e^{k\tau}dW_\tau$ – allows the construction of the full variance curve at time t: a Markov-functional representation exists.[4]

---

[3]See [15] for a characterization of the conditions on the volatility structure of a futures curve such that the resulting dynamics admits a finite-dimensional Markov representation. The ξ^T are indeed akin to a futures curve, as they are driftless.

[4]In [20] Hans Buehler studies Markov representations of the variance curve of the type: $\xi^T_t = G\,(\mathbf{X}_t, T - t)$ where $\mathbf{X}_t$ is a vector diffusive process, and provides a few examples of $(\mathbf{X}_t, G)$ couples

Choosing an exponentially decaying volatility function is equivalent to driving the dynamics of forward variances with one Ornstein-Ühlenbeck (OU) process X_t:

$$dX_t = -kX_t dt + dW_t, \ X_0 = 0$$

X_t and its variance are given by:

$$X_t = \int_0^t e^{-k(t-\tau)} dW_\tau \qquad E[X_t^2] = \frac{1 - e^{-2kt}}{2k}$$

With $\omega(u)$ of the form (7.9), the solution of SDE (7.7) reads:

$$\xi_t^T = \xi_0^T \exp\left(\omega e^{-k(T-t)} X_t - \frac{\omega^2}{2} e^{-2k(T-t)} E[X_t^2]\right) \qquad (7.10)$$

ω is the lognormal volatility of $\xi_t^{T=t}$, a foward variance with vanishing maturity.

7.3 N-factor models

Let us use N Brownian motions and write the SDE of ξ_t^T as:

$$d\xi_t^T = \omega \alpha_w \xi_t^T \sum_i w_i e^{-k_i(T-t)} dW_t^i \qquad (7.11)$$

where α_w is a normalizing factor such that the instantaneous lognormal volatility of $\xi_t^{T=t}$ is ω. Volatilities of volatilities are more natural objects than volatilities of variances. We thus introduce the lognormal volatility ν of a VS volatility of vanishing maturity, which is the square root of ξ_t^t. Its instantaneous volatility is half that of ξ_t^t. We have:

$$\omega = 2\nu \qquad (7.12a)$$

$$\alpha_w = \frac{1}{\sqrt{\displaystyle\sum_{ij} w_i w_j \rho_{ij}}} \qquad (7.12b)$$

The solution of (7.11) is given by:

$$\xi_t^T = \xi_0^T \exp\left(\omega \Sigma_i w_i e^{-k_i(T-t)} X_t^i - \frac{\omega^2}{2} \Sigma_{ij} w_i w_j e^{-(k_i+k_j)(T-t)} E[X_t^i X_t^j]\right)$$
$$(7.13)$$

where the N OU processes X^i are defined by:

$$dX_t^i = -k_i X_t^i dt + dW_t^i, \ X_{t=0}^i = 0 \qquad (7.14)$$

---

that ensure that ξ_t^T are martingales. This amounts to enforcing a parametric representation of the variance curve – arbitrary VS term structures cannot be accommodated. He later relaxes this constraint by setting $\xi_t^T = \xi_0^T G\left(\mathbf{X}_t, T - t\right)$.

7.3.1 Simulating the N-factor model

Because it is driven by Ornstein-Ühlenbeck processes the N-factor model is easily and exactly simulable. We start at $t = 0$ with:

$$X_0^i = 0, \quad E[X_0^i X_0^j] = 0$$

Imagine we have X_t^i, and $E[X_t^i X_t^j]$ at time $t = \tau_n$ and we need to generate them at time $\tau_{n+1} = \tau_n + \delta\tau$. The solution of (7.14) at time t is given by:

$$X_t^i = e^{-k_i t} X_0^i + \int_0^t e^{-k_i(t-u)} dW_u^i$$

$X_{\tau_{n+1}}^i$ thus reads:

$$
\begin{aligned}
X_{\tau_{n+1}}^i &= e^{-k_i \tau_{n+1}} X_0^i + \int_0^{\tau_{n+1}} e^{-k_i(\tau_{n+1}-u)} dW_u^i \\
&= e^{-k_i \delta\tau} X_{\tau_n}^i + \int_{\tau_n}^{\tau_{n+1}} e^{-k_i(\tau_{n+1}-u)} dW_u^i
\end{aligned}
$$

Introducing the Gaussian random variable $\delta X^i = \int_{\tau_n}^{\tau_{n+1}} e^{-k_i(\tau_{n+1}-u)} dW_u^i$, with zero mean, $X_{\tau_{n+1}}^i$ is generated from $X_{\tau_n}^i$ through:

$$X_{\tau_{n+1}}^i = e^{-k_i \delta\tau} X_{\tau_n}^i + \delta X^i \qquad (7.15)$$

Using this expression for $X_{\tau_{n+1}}^i$ and taking expectations:

$$E[X_{\tau_{n+1}}^i X_{\tau_{n+1}}^j] = e^{-(k_i+k_j)\delta\tau} E[X_{\tau_n}^i X_{\tau_n}^j] + E[\delta X^i \delta X^j] \qquad (7.16)$$

To generate the Gaussian random variables δX^i we only need their covariance matrix, which is given by:

$$E\left[\delta X^i \delta X^j\right] = \rho_{ij} \frac{1 - e^{-(k_i+k_j)\delta\tau}}{k_i + k_j} \qquad (7.17)$$

where ρ_{ij} is the correlation of Brownian motions W_t^i and W_t^j.

Thus, in case we do not need to simulate the spot process – for example if we are dealing with payoffs on realized or implied variance – no time stepping is required: the X_t^i are generated exactly for times t at which instantaneous or VS variances are needed, as mandated by the derivative's term sheet.

The stochastic volatility degrees of freedom of the N-factor lognormal model are easily and exactly simulated; this is a very attractive feature, especially when compared with the Heston model.[5]

---

[5]It is a well-known fact that mere simulation of the (one-factor) process V_t in the Heston model is excessively arduous, especially for large volatilities of volatilities. This one issue has contributed its fair share of papers to the mathematical finance literature.

In addition, as shown in Section 7.7 further below, we can easily relax the lognormality of forward variances while preserving the Markov-functional property of the model and retaining the capability of exactly simulating the dynamics of the variance curve.

Note that, unlike the X_t^i, the "convexity terms" $E[X_t^i X_t^j]$ are non-random, hence do not need to be simulated by the time-stepping process in (7.16); they can simply be computed in advance for times t of interest.

Simulating the spot process

Over the interval $[\tau_n, \tau_{n+1}]$ the process for $\ln S$ is discretized as:

$$\delta \ln S = \left(r - q - \frac{\xi_t^t}{2} \right) \delta t + \sqrt{\xi_t^t} \delta W^S$$

where δW^S is a Gaussian random variable of variance δt. The covariance of δW^S and δX^i can be computed at once using their expressions:

$$\delta X^i = \int_{\tau_n}^{\tau_{n+1}} e^{-k_i(\tau_{n+1}-u)} dW_u^i \qquad \delta W^S = \int_{\tau_n}^{\tau_{n+1}} dW_u^S$$

We get:

$$E \left[\delta W^S \delta X^i \right] = \rho_{iS} \frac{1 - e^{-k_i \delta \tau}}{k_i} \tag{7.18}$$

where ρ_{iS} is the correlation between W^i and W^S.

Using expressions (7.17) and (7.18) for the various covariances, Gaussian random variables δW^S and δX^i are easily generated.

In case we are only interested in obtaining the vanilla smile generated by our model, there are more efficient techniques than simulating S_t and evaluating vanilla payoffs – we refer the reader to Appendix A of Chapter 8, page 336.

7.3.2 Volatilities and correlations of variances

The instantaneous volatility of ξ_t^T is, from SDE (7.11):

$$w(T - t) = (2\nu)\alpha_w \sqrt{\Sigma_{ij} w_i w_j \rho_{ij} e^{-(k_i+k_j)(T-t)}} \tag{7.19}$$

and the instantaneous correlation of two forward variances $\xi^T, \xi^{T'}$ is given by:

$$\rho_t(\xi^T, \xi^{T'}) = \frac{\Sigma_{ij} w_i w_j \rho_{ij} e^{-\left(k_i(T-t)+k_j(T'-t)\right)}}{\sqrt{\Sigma_{ij} w_i w_j \rho_{ij} e^{-(k_i+k_j)(T-t)}} \sqrt{\Sigma_{ij} w_i w_j \rho_{ij} e^{-(k_i+k_j)(T'-t)}}} \tag{7.20}$$

Consider the VS volatility for maturity T, $\hat{\sigma}_T(t)$: $\hat{\sigma}_T^2(t) = \frac{1}{T-t} \int_t^T \xi_t^\tau d\tau$. The dynamics of $\hat{\sigma}_T(t)$ is given by:

$$d\hat{\sigma}_T = \nu \alpha_w \frac{1}{\hat{\sigma}_T} \sum_i w_i \left(\frac{1}{T-t} \int_t^T \xi_t^\tau e^{-k_i(\tau-t)} d\tau \right) dW_t^i + \bullet \, dt \tag{7.21}$$

We now introduce the notation $\nu_T(t)$ for the instantaneous lognormal volatility of $\widehat{\sigma}_T$ at time t. $\nu_T(t)$ is given by:

$$
\begin{cases}
\nu_T(t) = \nu\alpha_w \sqrt{\sum_{ij} w_i w_j \rho_{ij}\, f_i(t,T)\, f_j(t,T)} \\[2em]
f_i(t,T) = \dfrac{\int_t^T \xi_t^\tau e^{-k_i(\tau-t)}\,d\tau}{\int_t^T \xi_t^\tau\,d\tau}
\end{cases}
\tag{7.22}
$$

The instantaneous volatility of a very short VS volatility is ν:

$$
\nu_t(t) = \nu
$$

As is clear from (7.22) ν is a global scale factor for volatilities of volatilities.

What about volatilities of forward VS volatilities? Consider two dates T_1, T_2 with $t \le T_1 \le T_2$ and define the foward VS volatility $\widehat{\sigma}_{T_1 T_2}$ as:

$$
\widehat{\sigma}_{T_1 T_2}(t) = \sqrt{\frac{1}{T_2-T_1}\int_{T_1}^{T_2}\xi_t^\tau\,d\tau}
$$

The instantaneous volatility $\nu_{T_1 T_2}(t)$ of $\widehat{\sigma}_{T_1 T_2}$ is given by:

$$
\begin{cases}
\nu_{T_1 T_2}(t) = \nu\alpha_w \sqrt{\sum_{ij} w_i w_j \rho_{ij}\, f_i(t,T_1,T_2)\, f_j(t,T_1,T_2)} \\[2em]
f_i(t,T_1,T_2) = \dfrac{\int_{T_1}^{T_2} \xi_t^\tau e^{-k_i(\tau-t)}\,d\tau}{\int_{T_1}^{T_2} \xi_t^\tau\,d\tau}
\end{cases}
\tag{7.23}
$$

Flat term structure of VS volatilities

In the case of a flat term structure of VS volatilities, ξ_t^τ does not depend on τ. The integral in $f_i(t,T)$ in (7.22) can be evaluated analytically and we get the following simple formula for the instantaneous volatility of $\widehat{\sigma}_T$ at time t:

$$
\nu_T(t) = \nu\alpha_w \sqrt{\sum_{ij} w_i w_j \rho_{ij} I\big(k_i(T-t)\big) I\big(k_j(T-t)\big)}
\tag{7.24}
$$

where

$$
I(x) = \frac{1-e^{-x}}{x}
\tag{7.25}
$$

Likewise, the instantaneous volatility $\nu_{T_1 T_2}(t)$ of the *forward* VS volatility $\widehat{\sigma}_{T_1 T_2}$ is given by:

$$
\nu_{T_1 T_2}(t) = \nu\alpha_w \sqrt{\sum_{ij} w_i w_j \rho_{ij} I\big(k_i(T_2-T_1)\big) I\big(k_j(T_2-T_1)\big) e^{-(k_i+k_j)(T_1-t)}}
$$

$$
\tag{7.26}
$$

As is clear from (7.24), whenever the VS term structure is flat at time t, $\nu_T(t)$ and $\nu_{T_1 T_2}(t)$ are respectively functions of $T - t$ and $T_1 - t, T_2 - t$ only: the model is time-homogeneous.

In what follows, for the sake of setting model parameters, we will frequently use this situation as a reference case.

7.3.3 Vega-hedging in finite-dimensional models

Imagine we use N OU processes. In a lognormal model for forward variances we then have a Markov-functional representation: all forward variances can be written as *functions* of the N OU processes X^i and time. One might argue that we should delta-hedge – in our case vega-hedge – forward variance risk using N variance swaps of different maturities only, so as to neutralize sensitivities with respect to the N factors X^i. In the case of a one-factor model we could pick a particular maturity and our delta hedge would consist of one variance swap only.

However the function of a delta – in our case vega – strategy is to immunize our position at order one against all deformations $\delta \xi^T$ of the variance curve – not only those allowed by the covariance structure of the model. Only if the deltas $\frac{dP}{d\xi^T}$ are traded are we then able to materialize during δt the usual gamma/theta P&L with break-even levels specified by the covariance functions μ and ν in the pricing equation (7.4).

From SDE (7.11) for ξ_t^T and SDE:

$$dS_t = (r - q)S_t dt + \sqrt{\xi_t^t} S_t dW_t^S$$

for S_t, we get the spot/variance and variance/variance covariance functions in the N-factor model:

$$\mu(t, u, \xi) = \omega \alpha_w \sqrt{\xi_t^t \xi_t^u} \sum_i \rho_{SX^i} w_i e^{-k_i(u-t)} \tag{7.27a}$$

$$\nu(t, u, u', \xi) = \omega^2 \alpha_w^2 \xi_t^u \xi_t^{u'} \sum_{ij} \rho_{ij} w_i w_j e^{-k_i(u-t)} e^{-k_j(u'-t)} \tag{7.27b}$$

where ρ_{ij} is the correlation of W^i and W^j and ρ_{SX^i} the correlation of W^i and W^S.

Thus, with regard to deltas, the deformation modes of the variance curve generated by the N processes have no special significance. Model factors simply set the structure and rank of the break-even covariance matrix of the gamma/theta P&L of a hedged position. We refer the reader to the discussion of a similar issue – the delta in the local volatility model – in Section 2.7.8, page 77.

It is important to stress that calculation of deltas is not connected in any way to the covariance structure of the hedging instruments in the model at hand.

7.4 A two-factor model

How many OU processes should we use? How should we select their time scales $1/k_i$? Let us start with a one-factor model:

$$d\xi_t^T = (2\nu) e^{-k(T-t)} \xi_t^T dW_t$$

From (7.24) the instantaneous volatility of $\hat{\sigma}_T$ in the case of a flat term structure of VS volatilities at time t is:

$$\nu_T(t) = \nu I\big(k(T-t)\big) = \nu \frac{1 - e^{-k(T-t)}}{k(T-t)}$$

Observe that this expression is identical to formula (6.9) in the Heston model – we know from our study in Chapter 6 that one factor does not offer sufficient flexibility with regard to the dynamics of forward variances.

We now try with two OU processes X^1 and X^2. Denote by k_1, k_2 their mean-reversion constants, and by ρ_{12} the correlation between the Brownian motions driving X^1 and X^2. We introduce the mixing parameter $\theta \in [0,1]$ and denote by α_θ the normalization constant previously noted α_w in (7.12) – recall that the instantaneous lognormal volatility of the instantaneous *variance* ξ_t^t is equal to 2ν:

$$d\xi_t^T = (2\nu)\xi_t^T \alpha_\theta \left((1-\theta) e^{-k_1(T-t)} dW_t^1 + \theta e^{-k_2(T-t)} dW_t^2 \right) \qquad (7.28)$$

$$\alpha_\theta = 1/\sqrt{(1-\theta)^2 + \theta^2 + 2\rho_{12}\theta(1-\theta)} \qquad (7.29)$$

We introduce processes x_t^T defined by:

$$x_t^T = \alpha_\theta \left[(1-\theta) e^{-k_1(T-t)} X_t^1 + \theta e^{-k_2(T-t)} X_t^2 \right] \qquad (7.30)$$

where X_t^1, X_t^2 are OU processes:

$$\begin{cases} dX_t^1 = -k_1 X_t^1 dt + dW_t^1, & X_0^1 = 0 \\ dX_t^2 = -k_2 X_t^2 dt + dW_t^2, & X_0^2 = 0 \end{cases}$$

x_t^T is a driftless Gaussian process:

$$dx_t^T = \alpha_\theta \left[(1-\theta) e^{-k_1(T-t)} dW_t^1 + \theta e^{-k_2(T-t)} dW_t^2 \right]$$

whose quadratic variation is given by:

$$\big\langle (dx_t^T)^2 \big\rangle = \eta^2 (T-t) dt \qquad (7.31a)$$

$$\eta(u) = \alpha_\theta \sqrt{(1-\theta)^2 e^{-2k_1 u} + \theta^2 e^{-2k_2 u} + 2\rho_{12}\theta(1-\theta) e^{-(k_1+k_2)u}} \qquad (7.31b)$$

By definition of α_θ, $\eta(0) = 1$.

SDE (7.28) now simply reads:

$$d\xi_t^T = (2\nu)\xi_t^T \, dx_t^T \tag{7.32}$$

Its solution is:

$$\xi_t^T = \xi_0^T f^T\left(t, x_t^T\right) \tag{7.33}$$

$$f^T(t, x) = e^{\omega x - \frac{\omega^2}{2}\chi(t,T)} \tag{7.34}$$

where $\omega = 2\nu$ and $\chi(t, T)$ is given by:

$$
\begin{aligned}
\chi(t,T) &= \int_{T-t}^{T} \eta^2(u) \, du \tag{7.35}\\
&= \alpha_\theta^2 \Bigg[(1-\theta)^2 e^{-2k_1(T-t)} \frac{1 - e^{-2k_1 t}}{2k_1} + \theta^2 e^{-2k_2(T-t)} \frac{1 - e^{-2k_2 t}}{2k_2} \\
&\quad + 2\theta(1-\theta)\rho_{12} e^{-(k_1+k_2)(T-t)} \frac{1 - e^{-(k_1+k_2)t}}{k_1 + k_2} \Bigg]
\end{aligned}
$$

(7.33) expresses the property that ξ_t^T has a Markov representation as a function of x_t^T – a Gaussian process. We have a Markov-functional model for ξ_t^T. The reason for introducing x_t^T will become clear further below when we consider VIX futures.

Presently, the mapping function f is just an exponential, thus forward variances are lognormally distributed, but we will use other forms for f in Section 7.7.1.

We take $k_1 > k_2$ without loss of generality and call X^1 the short factor and X^2 the long factor. From (7.21):

$$
\frac{d\hat{\sigma}_T}{\hat{\sigma}_T} = \nu\alpha_\theta \left((1-\theta) \frac{\int_t^T \xi_t^\tau e^{-k_1(\tau-t)} d\tau}{\int_t^T \xi_t^\tau d\tau} dW_t^1 + \theta \frac{\int_t^T \xi_t^\tau e^{-k_2(\tau-t)} d\tau}{\int_t^T \xi_t^\tau d\tau} dW_t^2 \right) + \bullet \, dt \tag{7.36}
$$

$$
= \nu\alpha_\theta \left((1-\theta) A_1 dW_t^1 + \theta A_2 dW_t^2 \right) + \bullet \, dt \tag{7.37}
$$

with A_i given by:

$$
A_i = \frac{\int_t^T \xi_t^\tau e^{-k_i(\tau-t)} d\tau}{\int_t^T \xi_t^\tau d\tau} \tag{7.38}
$$

The instantaneous volatility of a VS volatility $\nu_T(t)$ is given by:

$$
\nu_T(t) = \nu\alpha_\theta \sqrt{(1-\theta)^2 A_1^2 + \theta^2 A_2^2 + 2\rho_{12}\theta(1-\theta) A_1 A_2} \tag{7.39}
$$

For a flat term-structure of VS volatilities:

$$
A_i = I\big(k_i(T-t)\big) = \frac{1 - e^{-k_i(T-t)}}{k_i(T-t)}
$$

7.4.1   Term structure of volatilities of volatilities

How flexible is a two-factor model? As illustrated in Figure 6.1, for equity indexes, volatilities of VS volatilities usually display a power-law dependence on maturity, with an exponent that typically lies between 0.3 and 0.6.

In the sequel we will make frequent use of the following time-homogeneous benchmark form for $\nu_T(t)$:

$$\nu_T^B(t) \;=\; \sigma_0 \left(\frac{\tau_0}{T-t} \right)^{\alpha} \tag{7.40}$$

where τ_0 is a reference maturity and σ_0 is the volatility of $\hat{\sigma}_{t+\tau_0}(t)$. Typically we will take $\alpha = 0.4$, $\tau_0 = 3$ months and $\sigma_0 = 100\%$. Figure 6.1 shows that the realized volatility of a 3-month VS volatility is around 60% for the Euro Stoxx 50 index. Implied levels for σ_0 derived from prices of options on realized variance are about twice as large – hence our choice for σ_0.

Figure 7.1 shows $\nu_T^B(t)$, as well as expression (7.39) for $\nu_T(t)$ generated by a two-factor model for a flat term structure of VS volatilities, at $t = 0$. We have chosen three different sets of parameters, differentiated by the value of the correlation between processes X^1 and X^2. We have used $\rho_{12} = -70\%,\ 0,\ 70\%$ and have selected the remaining parameters ν, θ, k_1, k_2 so as to best match our benchmark (7.40) for maturities from one month to 5 years.

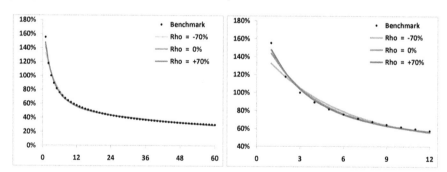

Figure 7.1: The left-hand graph displays the term structure of instantaneous volatilities at $t = 0$ of VS volatilities $\nu_T(t)$ (y axis) as a function of T (x axis, in months) generated by the benchmark form (7.40) as well as the two-factor model, with the different sets of parameters listed in Table 7.1. The right-hand graph focuses on maturities less than 1 year.

As is clear from Figure 7.1, the two-factor model is able to capture a power-law dependence for volatilities of volatilities over a wide range of maturities – similarly good agreement is achieved for other values of α. Moreover, for a given α, many different sets of parameters exist that provide an equally acceptable fit to our benchmark $\nu_T^B(t)$. Table 7.1 displays the parameters used.

	ν	θ	k_1	k_2	ρ_{12}
Set I	150%	0.312	2.63	0.42	-70%
Set II	174%	0.245	5.35	0.28	0%
Set III	186%	0.230	7.54	0.24	70%

Table 7.1: Three sets of parameters matching $\nu_T^B(t)$ in (7.40) with $\sigma_0 = 100\%$, $\tau_0 = 0.25$, $\alpha = 0.4$, for maturities up to 5 years. The resulting term structures of volatility of volatility are shown in Figure 7.1. ν is the instantaneous (lognormal) volatility of a VS volatility of vanishing maturity.

Notice how the time scales of the OU processes $1/k_1$, $1/k_2$ are clearly separated, thus generating a volatility-of-volatility term structure that cannot be captured in a one-factor model. Figure 7.1 demonstrates that very similar term structures of instantaneous volatilities of spot-starting VS volatilities are obtained in sets I, II, III, employing very different time scales.

7.4.2 Volatilities and correlations of forward variances

What distinguishes parameters sets corresponding to the same value of α? VS variances are equally weighted baskets of forward variances: $\hat{\sigma}_T^2 = \frac{1}{T} \int_0^T \xi^\tau d\tau$. It is instructive to look at the volatilities and correlations of the forward variances themselves. Figure 7.2 shows the instantaneous volatilities at $t = 0$ of forward variances (not volatilities) using the three sets in Table 7.1, while Figure 7.3 displays $\rho(\xi_t^T, \xi_t^{T'})$. We have used expressions (7.19) and (7.20), specialized to the case of the two-factor model.

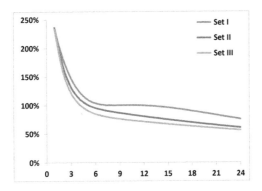

Figure 7.2: Instantaneous volatilities of forward variances ξ_t^T at $t = 0$ as a function of T (in months), using parameter sets in Table 7.1.

While volatilities of forward variances are higher in Set I than in Set III, the opposite is true of correlations. This is natural since the volatilities of VS variances –

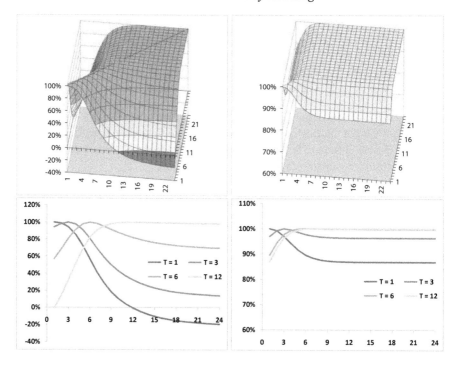

Figure 7.3: Top: $\rho(\xi_t^T, \xi_t^{T'})$ at $t = 0$ as a function of T, T' (in months) in Set I (left) and Set III (right). Bottom: slices of $\rho(\xi_t^T, \xi_t^{T'})$ for $T' = 1$ month, 3 months, 6 months, 12 months, in Set I (left) and Set III (right).

which are baskets of forward variances – are almost identical in both sets as they have been calibrated to the same benchmark: the higher the volatilities of the basket components, the lower their correlations. Observe at the top of Figure 7.3 how $\rho(\xi_t^T, \xi_t^{T'})$ becomes almost constant, equal to one, for T, T' larger than a given threshold, especially for Set III.

Inspection of expression (7.20) for $\rho(\xi_t^T, \xi_t^{T'})$ shows that correlations are unchanged if all k_i are shifted by the same constant: the relevant time scales for correlations are not the $\frac{1}{k_i}$, but the quantities $\frac{1}{k_i - k_j}$. In a two-factor model $\rho(\xi_t^T, \xi_t^{T'})$ is thus only a function of $k_1 - k_2$: the correlation structure has a single time scale $\frac{1}{k_1 - k_2}$.

In Set III, the values of k_1, k_2 are, respectively 7.54 and 0.24. For $T - t \gg \frac{1}{k_1 - k_2} = 1.64$ months, the contribution of the short factor is negligible and the ξ_t^T behave as in a one-factor model, with 100% correlations among themselves. Their correlations with variances $\xi_t^{T'}$ with $T' - t \gg \frac{1}{k_1 - k_2}$ do not depend on T anymore – this is clearly seen in the slices of $\rho(\xi_t^T, \xi_t^{T'})$ for $T - t = 6$ months and $T - t = 12$

months in the right-hand graph at the bottom of Figure 7.3. Also note that, while long-dated variances are driven by process X^2, short-dated variances are driven by the linear combination $\theta X^1 + (1 - \theta) X^2$: even with $\rho_{12} = 0$, there is a fair amount of correlation between short- and long-dated variances; in this respect Set II is more akin to Set III than to Set I.

7.4.3   Smile of VS volatilities

We have chosen to model instantaneous forward variances ξ^T as lognormal processes, based on historical evidence that VS volatilities are lognormal rather than normal – see Figure 6.2. VS *variances* $\widehat{\sigma}_T^2$, which are baskets of the ξ^T, will not be exactly lognormal and neither will VS *volatilities* $\widehat{\sigma}_T$. Their non-lognormality can be assessed by pricing variance swaptions, that is options to enter at T_1 into a long position in a VS of maturity T_2 with a strike K. The payoff of such a VS at T_2 is $\left(\sigma_r^2 - K\right)$, where σ_r is the realized volatility over $[T_1, T_2]$.

The option is exercised at T_1 only if the forward VS volatility $\widehat{\sigma}_{T_1 T_2}$ observed at T_1 is larger than $\sqrt{K}$: we exercise the swaption and sell a VS struck at the market implied VS volatility $\widehat{\sigma}_{T_1 T_2} (T_1)$. The payout of this strategy at T_2 is:[6]

$$\left(\sigma_r^2 - K\right) - \left(\sigma_r^2 - \widehat{\sigma}_{T_1 T_2}^2 (T_1)\right) \;=\; \widehat{\sigma}_{T_1 T_2}^2 (T_1) - K$$

The underlying of the VS swaption is thus the *forward* VS volatility $\widehat{\sigma}_{T_1 T_2}$ and the VS swaption is a call option of maturity T_1 on its square.[7] Expressing $\widehat{\sigma}_{T_1 T_2}^2 (T_1)$ as a function of forward variances observed at T_1, the swaption payoff reads:

$$\left(\frac{1}{T_2 - T_1} \int_{T_1}^{T_2} \xi_{T_1}^u \, du - K\right)^{+}$$

Figure 7.4 shows the smile of variance swaptions with $T_1 = 3$ months and $T_2 = 6$ months, in Set I and Set III, in the case of a flat term structure of VS volatilities. Implied volatilities for $\widehat{\sigma}_{T_1 T_2}^2 (T_1)$ have been computed by simply inverting the Black-Scholes formula as $\widehat{\sigma}_{T_1 T_2}^2$ is driftless. In the two-factor model $\widehat{\sigma}_{T_1 T_2}(T_1)$ is a function of two Gaussian variables $X_{T_1}^1, X_{T_1}^2$: variance swaptions are simply priced by two-dimensional quadrature.[8]

While the term structure of VS volatilities is flat at $t = 0$, it is not at future dates. For the sake of computing the instantaneous volatility of $\widehat{\sigma}_{T_1 T_2}$ at time t, let us make the approximation that the VS term structure at time t is flat. The instantaneous volatility of $\widehat{\sigma}_{T_1 T_2}$ at t is then given by $\nu_{T_1 T_2} (t)$ in (7.26). The instantaneous volatility of $\widehat{\sigma}_{T_1 T_2}^2 (t)$ is twice as large. We then get the following strike-independent approximation of the implied volatility $2\widehat{\nu}_{T_1 T_2}(T_1)$ by integrating the square of $\nu_{T_1 T_2} (t)$ in

---

[6]See the footnote on page 151 for the normalization of VS payoffs in actual VS term sheets.

[7]Note the similarity with cliquets – see Section 1.3.2.

[8]Since call and put payoffs are not smooth functions, one should employ for best performance a Gaussian quadrature with abscissas and weights determined for the one-sided Gaussian density.

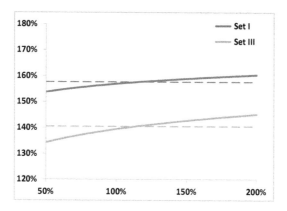

Figure 7.4: Implied volatilities of variance swaptions – that is of $\widehat{\sigma}^2_{T_1 T_2}(T_1)$ – as a function of *volatility* moneyness: $\sqrt{K}/\widehat{\sigma}_{T_1 T_2}(t=0)$ with $T_1 = 3$ months, $T_2 = 6$ months in sets I and III. Dotted lines correspond to the strike-independent level $2\widehat{\nu}_{T_1 T_2}(T_1)$, where the integral in (7.41) has been computed numerically.

(7.26) over $[0, T_1]$.

$$2\widehat{\nu}_{T_1 T_2}(T_1) = 2\sqrt{\frac{1}{T_1}\int_0^{T_1}\nu^2_{T_1 T_2}(t)\,dt} \qquad (7.41)$$

$2\widehat{\nu}_{T_1 T_2}(T_1)$ appears in Figure 7.4 as a dashed line.

Figure 7.4 displays the weak positively sloping smile that is typical of baskets of lognormal underlyings: while not exactly lognormal, VS volatilities are close to lognormal and approximation (7.41) is fairly accurate.

We will introduce further on an extension of the model that allows for full control of the smile of forward variances.

7.4.4   Non-constant term structure of VS volatilities

We now consider the effect of the shape of the term structure of VS volatilities on their instantaneous volatilities. Figure 7.5 shows the instantaneous volatilities at $t = 0$ of VS volatilities $\widehat{\sigma}_T$ as a function of T for a positively sloping, a negatively sloping, and a flat term structure of VS volatilities.

Consider formula (7.22) for $\nu_T(t)$. Volatilities of VS volatilities are larger for negatively sloping $\widehat{\sigma}_T$: this can be understood by noting that in our model short-dated instantaneous variances ξ^T have larger volatilities than longer-dated ones. A negatively sloping term structure of VS volatilities implies that the initial values of these shorter-dated variances are larger than those of longer-dated variances. This increases their relative weight in the expression of $\widehat{\sigma}^2_T$, thus increasing ν_T – see expression (7.22) for $\nu_T(t)$.

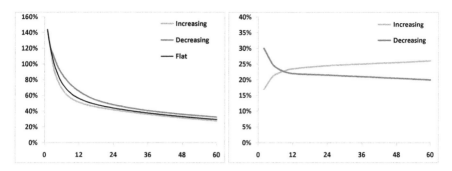

Figure 7.5: Left: instantaneous volatilities at $t = 0$ of VS volatilities ν_T as a function of T (months) computed in Set II for three different term structures of VS volatilities. Right: increasing and decreasing term structures of VS volatilities used in left-hand graph.

Note that simply multiplying all $\hat\sigma_T$ by the same constant leaves volatilities of volatilities unchanged, as the ξ^T are lognormal.

The smiles of VS volatilities shown in Figure 7.4, computed for a flat VS curve then have an additional dependence on the slope of the term structure of VS volatilities.

7.4.5 Conclusion

A two-factor model provides sufficient control on volatilities of forward variances so that the benchmark form $\nu_T^B(t)$ in (7.40) can be matched over a wide range of maturities. Furthermore, very similar term structures of volatilities of spot-starting VS volatilities $\hat\sigma_T$ can be obtained using different sets of parameters, which allows for separation of volatilities of (a) *spot-starting* VS volatilities and (b) *forward* VS volatilities.

While instantaneous volatilities of *spot-starting* VS volatilities are identical in these different sets, instantaneous volatilities and correlations of forward variances are different, hence volatilities of *forward* VS volatilities differ. This is clearly seen in the case of the 3-months in 3-months swaption implied volatilities in Figure 7.4: we get values around 160% using Set I and 140% using Set III – this disconnection of volatilities of *spot-starting* and *forward-starting* VS volatilities cannot be achieved within a one-factor model.

The correlation structure of forward variances in the two-factor model is rather poor as it is determined by a single time scale $\frac{1}{k_1-k_2}$; making it richer would be the primary motivation for introducing a third factor. Finally, while not exactly lognormal, VS volatilities are almost lognormal – Figure 7.5 highlights the fact that their volatilities will depend somewhat on the term structure of VS volatilities.

The question of which parameter set to use can only be settled on a case-by-case basis by analyzing the nature of the volatility-of-volatility risk of the payoff at hand.

In case the underlyings of our option are *spot-starting* volatilities, all sets corresponding to the same value of α will yield very similar prices. If instead *forward* volatilities are the real underlyings, we will need to discriminate among sets generating equal levels of volatilities of *spot-starting* volatilities, yet different levels of volatilities and correlations of *forward* volatilities.

7.5   Calibration – the vanilla smile

What do vanilla smiles look like in the two-factor model? This question should be asked jointly with another question: which financial observables should it be calibrated to, and which instruments used as hedges?

The natural building blocks of forward variance models are (a) the spot and (b) forward variances – or VS volatilities.

Calibrating a model amounts to *deciding* which (vanilla) instruments our exotic option price is a function of, along with the spot. The consequence is these instruments are our hedge instruments.

Calibrating our forward variance model to VSs implies these are the hedge instruments we use.

Alternatively, we can calibrate the ξ_0^t so that the term structure of ATMF or ATM volatilities – or implied volatilities for an arbitrary moneyness – is recovered.[9] Using the corresponding vanilla options as hedges – together with the spot – leads to a well-defined gamma/theta carry P&L for a hedged position that involves covariances of implied volatilities of the calibrated instruments with each other and with the spot.[10]

In practice, lognormal volatilities of ATMF volatilities in the two-factor model are not much different than those of VS volatilities – thus formula (7.39) can be used to set model parameters so that desired levels of volatilities of ATMF volatilities are obtained.

Forward variance models can thus equivalently be viewed as genuine market models for the spot and the term structure of implied volatilities for a given moneyness, with the capability – as illustrated by the example of the two-factor model – of accommodating an exogenously specified dynamics for this one-dimensional set of instruments.

Once we have calibrated a term-structure of implied volatilities, say VS or ATMF, we can select the parameters of the model so as to achieve a best-fit of the whole

---

[9]Efficient techniques for generating vanilla smiles are surveyed in Appendix A of Chapter 8.

[10]Derivation of this practically relevant result is very similar to how we derive the carry P&L in the local volatility model – see Section 2.7 of Chapter 2, page 66 – or in local-stochastic volatility models – see Section 12.3 of Chapter 12, page 463.

smile. This calibration is of a very different nature than calibration to the VS/ATMF volatility term structure. In the latter case the "calibration" process amounts to inputing the values of the model's underliers, that is ATMF or VS volatilities – this can hardly be called a calibration.

Inferring model *parameters* from market prices of vanilla options is a quite different matter. These parameters are then used to price more exotic structures. The resulting hedge ratios may be meaningless as they are likely to reflect structural relationships that are model-specific rather than express genuine matching of risks of a congruent nature.[11]

Characterizing the smiles generated by forward variance models and their determinants is then an important issue which is dealt with in detail in Chapter 8.

Smiles generated by the two-factor model are discussed in Section 8.7, page 326.

Efficient techniques for generating vanilla smiles in stochastic volatility models are surveyed in Appendix A of Chapter 8 – see page 336.

The relationship of the vanilla smile to the dynamics of implied volatilities is another important subject that is covered in Chapter 9 – the special case of the two-factor model is examined in Section 9.7, page 363.

7.6 Options on realized variance

Typically, options on realized variance comprise call and put payoffs on the realized variance of an equity underlying – usually an index – with the same convention for realized volatility as that of variance swaps. The payoff of a call on realized variance is:

$$\frac{1}{2\widehat{\sigma}_{\text{ref}}}\left(\sigma_r^2\left(T\right)-\widehat{\sigma}^2\right)^+ \quad \text{with} \quad \sigma_r^2\left(T\right)=\frac{252}{N}\sum_{i=0}^{N-1}\ln^2\left(\frac{S_{i+1}}{S_i}\right)$$

where $\widehat{\sigma}$ is the (volatility) strike, $\sigma_r\left(T\right)$ is the realized volatility over the option's maturity, and $\widehat{\sigma}_{\text{ref}}$ is usually chosen equal to the VS volatility $\widehat{\sigma}_T$ for the option's maturity. The reason for this normalization is that, for an at-the-money option, $\widehat{\sigma}=\widehat{\sigma}_T$ and at order one in $\sigma_r\left(T\right)-\widehat{\sigma}_T$, the payoff of the ATM call on variance is simply $\left(\sigma_r\left(T\right)-\widehat{\sigma}_T\right)^+$.

Volatility swaps, whose payoff is:

$$\sigma_r\left(T\right)-\widehat{\sigma}$$

trade as well; calls and puts on realized volatility trade occasionally.

---

[11]Think for example of volatility of volatility in the Heston model. This parameter drives both the curvature of the vanilla smile near the money and the volatility of VS volatilities, which affects prices of options on realized variance.

A call option on realized variance can be promptly priced in any stochastic volatility model: this produces a number that acquires the status of a price once we have identified the main risks of our option, the instruments that can be used to hedge them, and appropriate carry levels for the residual risks. To this end we now develop a simple model (SM) that is sufficiently robust that it can be used practically. The path we follow is similar to how one derives a well-known approximation for Asian options – see footnote 14 below.[12]

7.6.1   A simple model (SM)

We drop for now the $\frac{1}{2\sigma_{\text{ref}}}$ factor without loss of generality and focus on the payoff $(\sigma_r^2 - \widehat{\sigma}^2)^+$ – we will also take vanishing interest rate and repo for simplicity. Let us use the notation Q_t to denote the quadratic variation defined as the sum of daily squared log-returns over the interval $[0, t]$:

$$Q_t = \sum_0^t \ln^2\left(\frac{S_{i+1}}{S_i}\right) = t\sigma_r^2(t)$$

where $\sigma_r(t)$ is the realized volatility over $[0, t]$. The option's payoff is $\left(\frac{1}{T}Q_T - \widehat{\sigma}^2\right)^+$. This suggests that the underlying of our option is Q_t – can it be replicated?

The most natural candidate for hedging Q is a VS: consider entering a VS contract of maturity T at time t, struck at an implied volatility $\widehat{\sigma}_T(t)$, whose payoff at T is:

$$\frac{1}{T-t}\sum r_i^2 - \widehat{\sigma}_T^2(t) \quad , \quad r_i = \ln\left(\frac{S_{i+1}}{S_i}\right)$$

where the sum runs over all returns over the interval $[t, T]$.

Imagine entering this VS contract at time t – at no cost – and consider its value at some later time t'. While at t the market value of the realized quadratic variation over $[t, T]$ was $(T - t)\widehat{\sigma}_T^2(t)$, at t', the quadratic variation over $[t, t']$ has already been realized and is equal to $Q_{t'} - Q_t$, while the market value of the quadratic variation over $[t', T]$ is $(T - t')\widehat{\sigma}_T^2(t')$. The P&L over $[t, t']$ of our VS position is thus:

$$\begin{aligned}
P\&L &= \frac{1}{T-t}\left[(Q_{t'} - Q_t) + (T - t')\widehat{\sigma}_T^2(t')\right] - \widehat{\sigma}_T^2(t) \\
&= \frac{1}{T-t}\left(Q_{t'} + (T - t')\widehat{\sigma}_T^2(t')\right) - \frac{1}{T-t}\left(Q_t + (T - t)\widehat{\sigma}_T^2(t)\right) \\
&= \frac{T}{T-t}\left[\frac{Q_{t'} + (T - t')\widehat{\sigma}_T^2(t')}{T} - \frac{Q_t + (T - t)\widehat{\sigma}_T^2(t)}{T}\right] \quad (7.42)
\end{aligned}$$

The above equation expresses the property that taking at time t a position in $\frac{T-t}{T}$ variance swaps of maturity T perfectly hedges the "underlying" U defined by:

$$U_t = \frac{Q_t + (T - t)\widehat{\sigma}_T^2(t)}{T} \quad (7.43)$$

---

[12] A presentation based on this same idea was given by Zhenyu Duanmu in 2004 – see [38].

over any finite interval $[t, t']$ as, from (7.42), the P&L of such a position is simply $P\&L = (U_{t'} - U_t)$.

The conclusion is that, while Q_t itself is not hedgeable, the package consisting of $Q_t + (T - t)\,\hat{\sigma}_T^2\,(t)$ can be exactly replicated: U_t is a legitimate underlying. Moreover the pricing drift of U is zero, since taking a VS position involves no cash outlay. This is consistent with the expression of U in terms of forward variances:

$$U_t = \frac{1}{T}\left(\int_0^t \xi_\tau^\tau d\tau + \int_t^T \xi_t^\tau d\tau\right) \tag{7.44}$$

where the first piece in the right-hand side of (7.44) corresponds to past observations. As the ξ^τ are driftless, so is U. From definition (7.43), the values of U at $t = 0$ and $t = T$ are, respectively:

$$U_0 = \hat{\sigma}_T^2\,(0) \qquad U_T = \frac{1}{T}Q_T$$

The payoff of the call on realized variance is then simply:

$$\left(U_T - \hat{\sigma}^2\right)^+ \qquad .$$

and the price P of this option can be expressed as the expectation of the payoff $\left(U_T - \hat{\sigma}^2\right)^+$ under a dynamics of U_t that is driftless, with the initial condition $U_0 = \hat{\sigma}_T^2\,(0)$.

A dynamics for U_t
From (7.43) we have:

$$dU_t = \frac{T-t}{T}\underbrace{d\left(\hat{\sigma}_T^2\,(t)\right)}_{\text{Diffusive portion}} \tag{7.45}$$

where we are only keeping the diffusive portion of $d\left(\hat{\sigma}_T^2\,(t)\right)$ as U_t is driftless. Let us assume that $\hat{\sigma}_T\,(t)$ is lognormal with deterministic volatility $\nu_T\,(t)$: $\hat{\sigma}_T^2\,(t)$ is lognormal with volatility $2\nu_T\,(t)$. We get the following SDE for U_t:

$$dU_t = 2\frac{T-t}{T}\nu_T\,(t)\,\hat{\sigma}_T^2\,(t)\,dW_t$$

which we rewrite as:

$$\frac{dU_t}{U_t} = 2R_t\frac{T-t}{T}\nu_T\,(t)\,dW_t \tag{7.46}$$

with R_t given by:

$$R_t = \frac{\hat{\sigma}_T^2\,(t)}{U_t} = \frac{T\hat{\sigma}_T^2\,(t)}{Q_t + (T-t)\,\hat{\sigma}_T^2\,(t)} = \frac{T\hat{\sigma}_T^2\,(t)}{t\sigma_r^2\,(t) + (T-t)\,\hat{\sigma}_T^2\,(t)} \tag{7.47}$$

SDE (7.46) is problematic as the dynamics of U_t is not autonomous – it involves $\hat{\sigma}_T^2(t)$ through R_t. But for this prefactor, the dynamics of U_t would be lognormal. The quadratic variation of $\ln U$ over $[t, T]$ is:

$$\int_t^T 4R_\tau^2 \left(\frac{T-\tau}{T}\right)^2 \nu_T^2(\tau)\, d\tau \tag{7.48}$$

R_τ is the ratio of two integrated variances, thus is a number of order one.

We now make an approximation: let us replace R_τ with a constant – we approximate R_τ by its value at time t, the pricing date:

$$R_\tau \simeq R_t$$

SDE (7.46) now becomes:

$$\frac{dU_\tau}{U_\tau} = 2R_t \frac{T-\tau}{T} \nu_T(\tau)\, dW_\tau$$

where t is now the pricing date. U is then lognormal and the option price $P(t, U)$ is given by a Black-Scholes formula evaluated with an effective volatility σ_{eff}:

$$P(t, U) = P_{\text{BS}}(t, U, \sigma_{\text{eff}}, T) \tag{7.49a}$$

$$\sigma_{\text{eff}}^2 = \frac{1}{T-t} \int_t^T 4R_t^2 \left(\frac{T-\tau}{T}\right)^2 \nu_T^2(\tau)\, d\tau \tag{7.49b}$$

The number of VS contracts of maturity T needed at time t to hedge our option is:[13]

$$\frac{T-t}{T} \frac{dP}{dU}$$

where the ratio $\frac{T-t}{T}$ expresses the fact that it takes $\frac{T-t}{T}$ VS contracts of maturity T at time t to replicate U.

At inception $U_{t=0} = \hat{\sigma}_T^2(0)$, so that $R_{t=0} = 1$. The option premium is then given by:[14]

---

[13] Actual VS term-sheets specify the VS payoff as $\frac{1}{2\hat{\sigma}_T}(\sigma_\tau^2 - \hat{\sigma}_T^2)$. We are not using the $1/(2\hat{\sigma}_T)$ prefactor for now.

[14] A similar route can be followed to derive a well-known approximation for the price of an Asian option in the Black-Scholes model, that is an option that pays at time T a European payoff on $\frac{1}{T}\int_0^T S_t dt$. Assume zero interest rate/repo for simplicity and call M_t the running average at time t: $M_t = \frac{1}{t}\int_0^t S_\tau d\tau$. While M_t cannot be exactly hedged by taking a position on S, the package U_t defined by $U_t = \frac{tM_t + (T-t)S_t}{T}$ is exactly replicated over $[t, t+dt]$ by trading $\frac{T-t}{T}$ units of S: U_t is driftless. The initial and final values of U are: $U_0 = S_0$, $U_T = M_T$. The SDE for U_T is: $\frac{dU_t}{U_t} = R_t \frac{T-t}{T}\sigma dW_t$ with $R_t = \frac{S_t}{U_t}$, where σ is the volatility of S. Let us make the approximation $\frac{S_t}{U_t} = 1$. U_t is then lognormal with an effective volatility over $[0, T]$ given by: $\sigma_{\text{eff}}^2 = \frac{1}{T}\int_0^T \left(\frac{T-t}{T}\right)^2 \sigma^2 dt = \frac{\sigma^2}{3}$. The option price is then simply given by the Black-Scholes formula $P_{\text{BS}}(t=0, S_0, \sigma_{\text{eff}}, T)$, with $\sigma_{\text{eff}} = \frac{\sigma}{\sqrt{3}}$.

$$P(t = 0) = P_{BS}\left(t = 0, \widehat{\sigma}_T^2(0), \sigma_{\text{eff}}, T\right) \tag{7.50a}$$

$$\sigma_{\text{eff}}^2 = \frac{1}{T} \int_0^T 4 \left(\frac{T - \tau}{T}\right)^2 \nu_T^2(\tau)\, d\tau \tag{7.50b}$$

These are the basic equations of the SM.

7.6.2 Preliminary conclusion

So far, our main results are, assuming the approximation $R_\tau \equiv R_t$ is satisfactory – this will be checked further on:

- Options on realized variance are simply hedged by dynamically trading variance swaps of the option's residual maturity.

- Their value does not depend on the detailed dynamics of instantaneous forward variances – it only depends on the instantaneous volatility $\nu_T(t)$ of $\widehat{\sigma}_T(t)$, for all times $t \in [0, T]$: the only pricing ingredient is the curve $\nu_T(t)$, such as those in Figure 7.1.

- The SM is a legitimate, arbitrage-free, model. The only approximation we are making is in the volatility of U_t – hence in the value of σ_{eff}. We have replaced R_τ with its value at the pricing date, R_t so that U_t is lognormal.

In the following section we will assess the accuracy of the SM using various forms for $\nu_T(t)$.

7.6.3 Examples

Our conclusion suggests that options on realized variance are in fact simple instruments as their only volatility-of-volatility risk is the exposure to the realized volatility over $[0, T]$ of the VS volatility of maturity T.

We now check this prediction by pricing an at-the-money call on realized variance of maturity 6 months and 1 year in the two-factor model and in the SM.

$\nu_T(\tau)$ in formula (7.50b) for σ_{eff} is the instantaneous volatility of the VS volatility for maturity T observed at t, $\widehat{\sigma}_T(t)$. In our benchmark (7.40), $\nu_T(\tau)$ is specified directly:

$$\nu_T^B(t) = \sigma_0 \left(\frac{\tau_0}{T - t}\right)^\alpha \tag{7.51}$$

In the two-factor model, instead, the model generates its own dynamics for $\widehat{\sigma}_T(t)$. Set II parameters are such that, for a flat VS term structure, instantaneous volatilities of VS volatilities at $t = 0$ – given by (7.39) – best match the benchmark form (7.51). What about instantaneous volatilities of VS volatilities at *future* times t – which enter expression (7.50b) for σ_{eff}?

In expression (7.50b) for σ_{eff} we use the following expression for $\nu_T(t)$, which we denote $\nu_T^0(t)$:

$$\nu_T^0(t) = \nu \alpha_\theta \sqrt{(1-\theta)^2 f_1^0(t,T)^2 + \theta^2 f_2^0(t,T)^2 + 2\rho_{12}\theta(1-\theta) f_1^0(t,T) f_2^0(t,T)}$$
(7.52)

with $f_i^0(t,T)$ given by:

$$f_i^0(t,T) = \frac{\int_t^T \xi_0^\tau e^{-k_i(\tau-t)} d\tau}{\int_t^T \xi_0^\tau d\tau}$$

The reader can check – see expression (7.22) – that the actual expression of the instantaneous volatility of $\hat\sigma_T(t)$ at time t, $\nu_T(t)$, specialized to the two-factor model, is identical to (7.52), except, $f_i(t,T)$ is used in place of $f_i^0(t,T)$:

$$f_i(t,T) = \frac{\int_t^T \xi_t^\tau e^{-k_i(\tau-t)} d\tau}{\int_t^T \xi_t^\tau d\tau}$$

Using the time-deterministic form (7.52) instead of $\nu_T(t)$ – a process – amounts to making the assumption that at time t forward variances ξ_t^τ are equal to their initial values ξ_0^τ. For a flat term structure of VS volatilities, $\nu_T^0(t) = \nu_T(t)$ as given by expression (7.39), page 227.

Even though the VS term structure is flat at $t = 0$, its shape at future times in the two-factor model will not be flat: the instantaneous volatility of $\hat\sigma_T(t)$ will depend on the VS term structure prevailing at time t and may differ from $\nu_T^0(t)$.

How good is the SM with σ_{eff} in (7.50) computed with $\nu_T^0(t)$, compared with prices produced by a Monte Carlo simulation of the two-factor model?

Option prices appear in Table 7.2. We have used a flat initial term structure of VS volatilities at 20% and have followed the standard market practice for the normalization of the call payoff:

$$\frac{1}{2\hat\sigma_T}\left(\sigma_r^2 - \hat\sigma^2\right)^+$$

Table 7.2 also lists prices produced by (7.50), with $\nu_T(t) = \nu_T^B(t)$. In the latter case σ_{eff} is given by:

$$\sigma_{\text{eff}} = \frac{2\sigma_0}{\sqrt{3-2\alpha}}\left(\frac{T_0}{T}\right)^\alpha$$
(7.53)

Observe how close prices generated by the SM with $\nu_T(t) = \nu_T^0(t)$ are to exact prices. Also note how prices generated with parameter sets I, II, III almost match, and how they almost match prices produced by the SM with $\nu_T(t) \simeq \nu_T^B(t)$. This confirms that, indeed, the primary risk of options on realized volatility is the exposure to the realized volatility of the VS volatility for the residual maturity.

		Set I	Set II	Set III	Benchmark
6 months	exact	2.97%	2.96%	2.94%	
	SM	2.93%	2.88%	2.86%	2.82%
1 year	exact	3.13%	3.08%	3.06%	
	SM	3.02%	2.99%	2.98%	3.01%

Table 7.2: Prices of an ATM call option on realized variance computed in the two-factor model with parameter sets in Table 7.1, page 229, either in a Monte Carlo simulation (exact), or using the SM with $\nu_T^0(t)$ given by (7.52) (SM). Prices in the last column are computed using the SM with $\nu_T(t)$ given by benchmark (7.51) with $\sigma_0 = 100\%$, $\tau_0 = 0.25$, $\alpha = 0.4$. The term structure of VS volatilities is flat at 20%.

Finally, notice how prices very weakly depend on maturity. This is due to the fact that the value of α in our benchmark, 0.4, is close to 0.5. For $\alpha = 0.5$, σ_{eff} in (7.53) is inversely proportional to $\sqrt{T}$, $\sigma_{\text{eff}}\sqrt{T}$ is constant and the price given by (7.50a) does not depend on T.

7.6.4 Accounting for the term structure of VS volatilities

So far, in the derivation of the SM, we have made the assumption $R_\tau \equiv R_t$. At inception $R_0 = 1$. However, (7.47) shows that R_τ will be different than 1 whenever the realized volatility over $[0, \tau]$, $\sigma_r(\tau)$, is different than $\hat{\sigma}_T(\tau)$.

Prices in Table 7.2 have been computed with flat VS volatilities equal to 20%. Within the two-factor model, while $\sigma_r(\tau)$ and $\hat{\sigma}_T(\tau)$ are random, the expectation of their squares is equal to $20\%^2$, thus the expectations of the numerator and denominator of R_τ within the model are equal, and approximation $R_\tau \equiv R_0 = 1$ is appropriate. The good agreement of exact and SM prices in Table 7.2 shows that the fluctuation of R_τ around 1 has little impact on the option's premium.

Obviously, from a pricing point of view, the assumption $R_\tau \equiv 1$ will need to be corrected for non-constant term structures of VS volatilities. The dependence on the term structure of VS volatilities that the option price then acquires will need to be hedged, however.

In fact, regardless of the shape of the term structure of VS volatilities – and even for flat ones – the realized volatility, hence Q_t, will in practice be whatever it wants to be and realized values of R_τ will be substantially different than 1.

For example, imagine that, in reality, realized volatilities are systematically lower than implied VS volatilities – which is usually the case for indexes: $Q_t/t < \hat{\sigma}_T^2(t)$. R_t will be systematically larger than 1 and (7.46) indicates that the realized volatility of U will be larger than the level we have priced. If we have sold a call on realized variance, our daily gamma/theta P&L on U will be negative, thus we will lose money steadily even though the instantaneous realized volatility at time τ of $\hat{\sigma}_T$ matches our pricing level $\nu_T(\tau)$. Can we hedge this exposure of R_τ to the realized volatility of S_t?

As we will see shortly, the issues of (a) accounting for the term structure of VS volatilities in the option price, and (b) hedging the exposure of R_τ to the realized volatility of the underlying are connected.

Let us first amend (7.49b) to take into account the term structure of VS volatilities. For general term structures the expectations at time t of the numerator and denominator of R_τ in (7.47) are:

$$E_t[U_\tau] = U_t \equiv \frac{Q_t + (T-t)\,\widehat{\sigma}_T^2(t)}{T}$$

$$E_t[\widehat{\sigma}_T^2(\tau)] = \frac{1}{T-\tau}\int_\tau^T \xi_t^u u = \widehat{\sigma}_{\tau T}^2(t)$$

where $\widehat{\sigma}_{\tau T}(t)$ is the forward VS volatility at time t for the interval $[\tau, T]$. We now make the approximation:

$$R_\tau \simeq \frac{E_t[\widehat{\sigma}_T^2(\tau)]}{E_t[U_\tau]} = \frac{T\widehat{\sigma}_{\tau T}^2(t)}{Q_t + (T-t)\,\widehat{\sigma}_T^2(t)} \qquad (7.54)$$

Formula (7.49b) for σ_{eff} now becomes:

$$\sigma_{\text{eff}}^2 = \frac{4}{T-t}\int_t^T \left(\frac{T-\tau}{T}\right)^2 \left(\frac{T\widehat{\sigma}_{\tau T}^2(t)}{Q_t + (T-t)\,\widehat{\sigma}_T^2(t)}\right)^2 \nu_T^2(\tau)\,d\tau \qquad (7.55)$$

At time $t=0, Q_0 = 0$, σ_{eff} and the option's price are given by:

$$\sigma_{\text{eff}}^2 = \frac{4}{T}\int_0^T \left(\frac{T-\tau}{T}\right)^2 \left(\frac{\widehat{\sigma}_{\tau T}^2(0)}{\widehat{\sigma}_T^2(0)}\right)^2 \nu_T^2(\tau)\,d\tau \qquad (7.56a)$$

$$P(t=0) = P_{\text{BS}}\big(t=0, \widehat{\sigma}_T^2(0), \sigma_{\text{eff}}, T\big) \qquad (7.56b)$$

In what follows we use this new expression for σ_{eff} instead of (7.50b).

σ_{eff} now depends on the full term structure of VS volatilities up to T, with the consequence that the option price acquires an exposure to intermediate VS volatilities which it is necessary to hedge.

7.6.5   Vega and gamma hedges

The continuous density $\lambda(\tau)$ of intermediate VSs is given by the functional derivative of P with respect to $\widehat{\sigma}_\tau^2$. Writing $\widehat{\sigma}_{\tau T}^2(t)$ as:

$$\widehat{\sigma}_{\tau T}^2(t) = \frac{(T-t)\,\widehat{\sigma}_T^2(t) - (\tau-t)\,\widehat{\sigma}_\tau^2(t)}{T-\tau}$$

(7.55) can be rewritten as:

$$\sigma_{\text{eff}}^2 = \frac{4}{T-t}\int_t^T \left(\frac{(T-t)\,\widehat{\sigma}_T^2(t) - (\tau-t)\,\widehat{\sigma}_\tau^2(t)}{Q_t + (T-t)\,\widehat{\sigma}_T^2(t)}\right)^2 \nu_T^2(\tau)\,d\tau \qquad (7.57)$$

$\lambda(\tau)$ is given by:

$$\lambda(\tau) = \frac{\delta P}{\delta \widehat{\sigma}_\tau^2}\bigg|_t$$

$$= -8 \frac{dP}{d\sigma_{\text{eff}}^2} \frac{(\tau - t)(T - \tau)}{(T - t)T} \frac{T\widehat{\sigma}_{\tau T}^2(t)}{\left(Q_t + (T - t)\,\widehat{\sigma}_T^2(t)\right)^2} \nu_T^2(\tau) \qquad (7.58)$$

Differentiating σ_{eff}^2 with respect to $\widehat{\sigma}_T^2$ also produces a discrete quantity of variance swaps of maturity T which we denote by μ_T:

$$\mu_T = \frac{dP}{d\sigma_{\text{eff}}^2} \frac{d\sigma_{\text{eff}}^2}{d\widehat{\sigma}_T^2}$$

These come in addition to the VS of maturity T that offsets $\frac{dP}{dU}$.

While we trade these intermediate VSs as well as an additional VS of maturity T to hedge the mark-to-market P&L generated by the dependence of R_τ – hence of σ_{eff} – on intermediate VS volatilities, these VSs will also generate spurious gamma/theta P&Ls.

These P&Ls are in fact offsetting similar P&Ls from the short option position. To see that this is indeed the case, rewrite expression (7.57) of σ_{eff}^2 as:

$$\sigma_{\text{eff}}^2 = \frac{4}{T - t} \int_t^T \left(\frac{\left(Q_t + (T - t)\,\widehat{\sigma}_T^2(t)\right) - \left(Q_t + (\tau - t)\,\widehat{\sigma}_\tau^2(t)\right)}{Q_t + (T - t)\,\widehat{\sigma}_T^2(t)} \right)^2 \nu_T^2(\tau)\,d\tau$$
$$(7.59)$$

Recall that a position in a VS of maturity T replicates the package $Q_t + (T - t)\,\widehat{\sigma}_T^2(t)$ exactly – see equation (7.42). As is manifest in equation (7.59) $\widehat{\sigma}_\tau^2(t)$ does not appear alone, but associated with Q_t in such a way that the VS position that hedges the exposure to $\widehat{\sigma}_\tau^2(t)$ actually hedges $Q_t + (\tau - t)\,\widehat{\sigma}_\tau^2(t)$ – the same goes for the exposure to $\widehat{\sigma}_T^2(t)$. In other words, the extra VS position that hedges the sensitivity of σ_{eff}^2 to the VS term structure also hedges the sensitivity of R_τ to Q_t.

The conclusion is that in the SM, the gamma/theta P&L generated by the VS position that hedges the exposure of σ_{eff}^2 to the term structure of VS volatilities exactly offsets the gamma/theta P&L generated by the dependence of σ_{eff}^2 on the quadratic variation Q_t and VS volatilities $\widehat{\sigma}_\tau$ for $\tau \in [0, T]$.

Thus our vega hedge also functions as a gamma/theta hedge. We will need to check whether this property, obtained in the SM with the help of approximation (7.54), holds more generally.

The continuous density $\lambda(\tau)$ and discrete quantity μ_T of hedging VSs are related by a simple equation. Expression (7.56a) shows that σ_{eff} is unchanged if all VS volatilities are rescaled uniformly – thus for a small change $\delta\widehat{\sigma}_T^2 = \varepsilon\widehat{\sigma}_T^2$, at order one in ε, $\delta\sigma_{\text{eff}}^2 = 0$:

$$\int_0^T \lambda(\tau)\,\delta\widehat{\sigma}_\tau^2 d\tau + \mu_T\delta\widehat{\sigma}_T^2 = 0$$

which implies that

$$\int_0^T \lambda(\tau)\,\widehat{\sigma}_\tau^2 d\tau + \mu_T \widehat{\sigma}_T^2 = 0$$

For the case of a flat term structure of VS volatilities, this simplifies to:

$$\int_0^T \lambda(\tau)\,d\tau + \mu_T = 0 \tag{7.60}$$

The aggregate vega of the VS position that hedges σ_{eff} thus vanishes: the total vega of the VS hedge reduces to the vega of the VS position of maturity T that hedges U.

7.6.6 Examples

We now use expression (7.56a) for σ_{eff}^2 to study the nature of the VS hedge, concentrating first on the case of a flat term structure of VS volatilities. The pricing date is January 1, 2010 and the option's maturity is January 1, 2011. We have used market conventions: the payoff at T of the ATM call on realized variance and the payoff at τ of a VS of maturity $\tau \in [0, T]$ are, respectively:

$$\frac{1}{2\widehat{\sigma}_T}\left(\frac{Q_T}{T} - \widehat{\sigma}^2\right)^+ , \qquad \frac{1}{2\widehat{\sigma}_\tau}\left(\frac{Q_\tau}{\tau} - \widehat{\sigma}_\tau^2\right) \tag{7.61}$$

The vega of a VS contract is thus 1.

With the benchmark

We first use the SM with our benchmark form for $\nu_T^B(t)$. We use a flat term structure of VS volatilities: while the *price* computed using either expression (7.56a) or expression (7.50b) for σ_{eff} is identical, the VS *hedge* is different. The VS hedge, together with the dollar gammas generated by the hedging VSs are reported in Table 7.3.

In practice, VSs do not trade for all maturities, but for maturities corresponding to monthly or quarterly expiries of listed options: we have chosen monthly expiries, including an unrealistic one-day maturity. For the sake of computing hedge ratios for the intermediate maturities τ_i in Table 7.3 we have used a simple piecewise affine interpolation of $\tau\widehat{\sigma}_\tau^2$ over each interval $[\tau_i, \tau_{i+1}]$. This ensures that as $\widehat{\sigma}_{\tau_i}$ is shifted, only forward variances ξ_u for $u \in [\tau_{i-1}, \tau_{i+1}]$ vary – in particular our variance option will have no vega on VS volatilities of maturities longer than the option's maturity.[15]

The rightmost column in Table 7.3 lists the dollar gammas generated by each intermediate VS. As expressed by equation (5.9) the dollar gamma of a VS – without the $\frac{1}{2\widehat{\sigma}_\tau}$ prefactor – is $2e^{-r(T-t)}$. Using the market convention in (7.61) and zero interest rate, the dollar gamma of a VS of maturity τ is given by:

$$S^2\frac{d^2\text{VS}}{dS^2} = \frac{1}{\widehat{\sigma}_\tau\tau} \tag{7.62}$$

---

[15]This is not guaranteed if less rustic interpolation schemes, such as splines, are used.

Maturities	SM - benchmark	
	VS hedge	VS dollar gamma
02-Jan-10	-0.01%	-13%
31-Jan-10	-0.4%	-25%
02-Mar-10	-0.8%	-25%
02-Apr-10	-1.2%	-25%
02-May-10	-1.6%	-25%
02-Jun-10	-2.0%	-24%
02-Jul-10	-2.2%	-22%
01-Aug-10	-2.6%	-22%
01-Sep-10	-2.8%	-21%
01-Oct-10	-3.0%	-20%
01-Nov-10	-3.1%	-18%
01-Dec-10	-2.9%	-16%
01-Jan-11	87.7%	438%
Aggregate	**65.1%**	**182%**

Single maturity		
01-Jan-11	65.1%	325%

Table 7.3: VS vega hedges and the corresponding dollar gammas for an ATM call
on realized variance of maturity 1 year starting on January 1, 2010, computed in
the SM with $\nu_T(t)$ given by the benchmark form (7.51) with $\sigma_0 = 100\%$, $\tau_0 =$
0.25, $\alpha = 0.4$, as well as aggregate vega and dollar gamma. The term structure of
VS volatilities is flat at 20%. The bottom line shows the same results when only the
VS volatility for the option's maturity is used for pricing and hedging.

Table 7.3 shows that a call option on realized variance has negative sensitivities
to VS volatilities for intermediate maturities. These sensitivities are identical for a
call and a put struck at the same strike, as they are proportional to $\frac{dP}{d\sigma_{\text{eff}}}$ – see (7.58).[16]

Recall that the VS hedge for the option's maturity consists of two pieces: one
piece hedges U and acts as a delta; the other hedges σ_{eff} and acts as a vega: as is
apparent from the bottom of Table 7.3, the former is the largest contributor to the
vega of the aggregate VS hedge.

The bottom of Table 7.3 also shows the vega and dollar gamma of the VS hedge
obtained if we discard intermediate maturities and price our option using only the
VS volatility for the option's maturity, in which case σ_{eff} is given by (7.50b). Notice
how the aggregate vegas are identical in both situations – which is expected from
(7.60), which expresses that for a flat VS term structure the aggregate vega of the
VSs that hedge σ_{eff} vanishes.

---

[16]Besides, a long position in a call combined with a short position in a put with matching strikes is a
forward, which is perfectly hedged with a VS of maturity T and has zero sensitivity to intermediate VS
volatilities.

Notice though how the aggregate gammas are very different, depending upon whether we use the full term structure of VS volatilities or not. Indeed, short-maturity VSs, despite their small vega, do contribute large dollar gammas[17] – this is clear from formula (7.62).

In practice VSs for such short maturities do not trade: a *long* position on a call or put on realized variance generates a *short* gamma position that must be offset with vanilla options.

Checking the gamma in the SM

What about the gamma of the option on realized variance? We have proved in Section 7.6.4 that in the SM the vega hedge also functions as a perfect gamma hedge. In the SM, the option's dollar gamma can then be computed by summing the dollar gammas of the hedging VSs. From Table 7.3 the dollar gamma thus computed is 182%.

This is only applicable to the SM. There exists however a model-independent expression of the dollar gamma for models such that the dynamics of forward variances is independent on the spot level. This is the case for the forward variance models considered so far.

Imagine that S is shifted by δS: the quadratic variation is shifted from Q to $Q + \left(\frac{\delta S}{S}\right)^2$. The dollar gamma of our option is then given by:

$$S^2 \frac{d^2 P}{dS^2} = 2 \frac{dP}{dQ}$$

We have:

$$\frac{dP}{dQ} = \frac{P(Q_0 + \delta Q) - P(Q_0)}{\delta Q}$$

for δQ small where $Q_0 = 0$ is the quadratic variation at $t = 0$, and δQ is a small increment of Q. Quadratic variation is additive thus, all things being equal, a small perturbation of Q_0 generated by a change of the initial spot value results in the same perturbation in Q_T.[18] Using convention (7.61) for the variance option payoff and denoting by $\hat{\sigma}$ its (volatility) strike, at first order in δQ:

$$P(Q_0 + \delta Q, \hat{\sigma}) = \frac{1}{2\hat{\sigma}T} E\left[\left(\frac{Q_T + \delta Q}{T} - \hat{\sigma}^2\right)^+\right]$$

$$= \frac{1}{2\hat{\sigma}T} E\left[\left(\frac{Q_T}{T} - \left(\hat{\sigma}^2 - \frac{\delta Q}{T}\right)\right)^+\right] = P\left(Q_0, \hat{\sigma} - \frac{\delta Q}{2\hat{\sigma}T}\right)$$

The dollar gamma of the call option is thus related to the sensitivity to its (volatility) strike through:

$$S^2 \frac{d^2 P}{dS^2} = -\frac{1}{\hat{\sigma}T} \frac{dP}{d\hat{\sigma}} \qquad (7.63)$$

---

[17]This is apparent in the contribution of the one-day maturity VS to the aggregate dollar gamma – this was the reason for including this unrealistic VS maturity in the hedge portfolio.

[18]This is not the case in local volatility or mixed local/stochastic volatility models: a change in the initial spot value affects both the values and volatilities of forward variances: Q_T is not simply shifted by δQ.

Let us stress that while (7.63) holds for the forward variance models we are studying, it does not hold in local volatility or mixed local-stochastic volatility models – more generally models such that covariances of forward variances depend on the spot level.

Using (7.63) with P given by the SM yields a dollar gamma of 175% for our 1-year ATM call. The fact that the value of 182% obtained – in the framework of the SM – by summing the dollar gammas of the VS hedge is close to this is another check of the validity of approximation (7.54) for R_τ in the SM.

With the two-factor model

We now use the two-factor model with Set II parameters, and compare exact VS hedges with those given by the SM with $\nu_T(t) = \nu_T^0(t)$. Results appear in Table 7.4. Hedge ratios computed either in a Monte Carlo simulation of the two-factor model or by using the SM are very similar.

Maturities	VS hedge		VS dollar gamma	
	Exact	SM	Exact	SM
02-Jan-10	-0.02%	-0.01%	-31%	-27%
31-Jan-10	-0.9%	-0.8%	-55%	-46%
02-Mar-10	-1.2%	-1.0%	-37%	-30%
02-Apr-10	-1.2%	-1.5%	-23%	-30%
02-May-10	-1.0%	-1.2%	-14%	-18%
02-Jun-10	-0.8%	-1.1%	-9%	-13%
02-Jul-10	-0.7%	-1.0%	-7%	-10%
01-Aug-10	-0.5%	-0.7%	-5%	-6%
01-Sep-10	-0.4%	-0.3%	-3%	-2%
01-Oct-10	-0.1%	-0.1%	0%	-1%
01-Nov-10	0.5%	1.3%	3%	8%
01-Dec-10	1.9%	2.4%	10%	13%
01-Jan-11	68.2%	68.8%	341%	344%
Aggregate	63.8%	64.9%	169%	183%

Single maturity				
01-Jan-11	63.8%	64.9%	319%	325%

Table 7.4: VS vega hedge for an ATM call on realized variance of maturity 1 year starting on January 1, 2010, computed with (a) a Monte Carlo simulation of the two-factor model with Set II parameters (Exact); (b) the SM with $\nu_T^0(t)$ given by (7.52) – as well as the corresponding dollar gammas. The term structure of VS volatilities is flat at 20%. The bottom line shows the same results when only the VS volatility for the option's maturity is used for pricing and hedging.

Comparison of Table 7.3 and Table 7.4 shows, however, that VS hedge ratios in the two-factor model or in the benchmark are different, even though the aggregate vegas are almost exactly equal: the VS hedge is distributed differently. Why is this?

Both the second column of Table 7.3 and the third column of Table 7.4 are computed using the SM except in the first case expression (7.56a) of σ_{eff}^2 makes use of $\nu_T^B(t)$, while in the latter case $\nu_T^0(t)$ in (7.52) is used.

Set II parameters are such that, with the initial flat term structure of VS volatilities, $\nu_T^0(t) \simeq \nu_T^B(t)$. ν_T^B is by construction independent on the term structure of VS volatilities.

Unlike ν_T^B, $\nu_T^0(t)$ depends on this term structure – see expression (7.52). In the two-factor model, the instantaneous volatility of $\hat{\sigma}_T(t)$ depends on the term structure of VS volatilities – this is illustrated in Figure 7.5, page 233.

This sensitivity of the instantaneous volatility of $\hat{\sigma}_T(t)$ to the VS term structure is reflected in an additional sensitivity of the option on realized variance to the VS term structure. This accounts for the difference of the VS hedges in Tables 7.3 and 7.4.

The reason why aggregate vegas in (a) the second column of Table 7.3, (b) the third column of Table 7.4 almost exactly match is that we are using a flat term structure of VS volatilities. Indeed, in a uniform rescaling of VS volatilities $\nu_T^0(t)$ is unchanged, as can be checked on expression (7.52). With a flat term structure, a rescaling of VS volatilities is akin to a uniform shift. Thus, in a uniform shift of VS volatilities, $\nu_T^0(t)$ is unchanged, which implies that the sum of the additional VS hedges contributed by the sensitivity of $\nu_T^0(t)$ to the VS term structure must vanish.

Turning now to the dollar gamma, from Table 7.4 the aggregate dollar gamma of the VS hedge is 183%, when the exact dollar gamma, computed through the sensitivity to the option's strike is 166%. The SM property that the vega hedge of options on realized variance functions as a gamma hedge still holds approximately.

Conclusion

We have pinpointed an additional element of model-dependence of options on realized variance: the dependence of volatilities of volatilities on the slope – not the level – of the VS term structure.

Depending on which assumption is made, one obtains different VS hedges – column 2 in Table 7.3 or column 3 in Table 7.4 – even though the initial price is identical. Both hedges are acceptable as they are generated by legitimate models – indeed the benchmark can be mimicked in the two-factor model, at the expense of time-homogeneity.[19] Making volatilities of VS volatilities independent on the term structure is a simpler assumption to work with and a book of options on realized variance can be risk-managed using the SM.

On the other hand the dependence generated by the two-factor model is not implausible – see Figure 7.5, page 233. In the two-factor model, volatilities of volatilities are larger for decreasing term structures of VS volatilities, which is what one would indeed expect. Also, the two-factor model can be used to risk-manage options

---

[19]One only needs to: (a) make it a one-factor model ($\theta = 0$) and set $k_1 = 0$, (b) make ν in (7.28) time-dependent: $\nu \to \nu(t) = \sigma_0 \left(\frac{T_0}{T-t}\right)^\alpha$. Instantaneous volatilities of $\hat{\sigma}_T(t)$ are then exactly equal to $\nu_T^B(t)$ and are independent on the term structure of VS volatilities. Because the maturity T is hard-wired in the model, the latter is not time-homogeneous anymore.

on both spot-starting and forward-starting variance, as well as other payoffs that have volatility-of-volatility risk.

It is then more reasonable to pick *one* parameter set for the two-factor model that matches the benchmark with given σ_0, τ_0, α, and use it to risk-manage all options in a book.

7.6.7 Non-flat VS volatilities

Here we focus on the case of sloping VS term structures. In the two-factor model, for a given parametrization, instantaneous volatilities of VS volatilities depend on the VS term structure.

To be able to measure the "intrinsic" sensitivity of the option on realized variance to the VS term structure, we now recalibrate parameters ν, θ, k_1, k_2 of the two-factor model so that, for each term structure at hand, $\nu_T^0(t) \simeq \nu_T^B(t)$, where $\nu_T^0(t)$ is given by (7.52). For $\nu_T^0(t)$, we take $\sigma_0 = 100\%$, $\tau_0 = 0.25$, $\alpha = 0.4$.

Table 7.5 lists prices of an ATM call on realized variance of maturities 6 months and 1 year in the two-factor model, for two term structures of VS volatilities – respectively decreasing and increasing – shown in Figure 7.6.

	1 year		6 months		Parameters				
	exact	approx	exact	approx	ν	θ	k_1	k_2	ρ_{12}
Decreasing	3.46%	3.32%	4.28%	4.11%	2.27	0.30	12.8	1.06	0%
Increasing	4.46%	4.36%	3.03%	2.95%	2.22	0.29	12.0	0.96	0%

Table 7.5: Prices of an ATM call on realized variance computed in the two-factor model for the case of a decreasing (resp. increasing) term structure of VS volatilities, either in a Monte Carlo simulation (exact) or in the approximate model with $\nu_T^0(t)$ given by (7.52) (approx). Parameter values appear in right-hand side of table.

For the 1-year maturity case, the option's strike $\hat{\sigma}_T$ is 25% whereas in the 6-month case $\hat{\sigma}_T = 31\%$ (resp. $\hat{\sigma}_T = 19\%$) for the decreasing (resp. increasing) term structure of VS volatilities.

As Table 7.5 shows, exact and approximate prices are very close, considering the unrealistically steep term structures we have used. This again confirms that the volatility of U_t is the main contributor to the value of the option on realized variance.

For the 1-year maturity case, the option's strike is $\hat{\sigma}_T = 25\%$, for both increasing and decreasing term structures. Observe the sizeable impact of the term structure of VS volatilities on the option's price – of the order of 1%.

7.6.8 Accounting for the discrete nature of returns

In the model developed so far, the time value of an option on realized variance is generated by the volatility of $\hat{\sigma}_T(t)$, which is the only source of randomness.

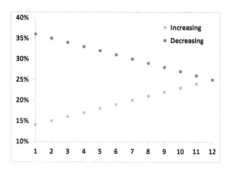

Figure 7.6: Term structures of VS volatilities at $t = 0$ used for generating prices in Table 7.5. Maturities are expressed in months.

Setting $\nu_T(t) \equiv 0$ in expression (7.56a) leads to $\sigma_{\text{eff}} = 0$: the price of an ATM call on realized variance vanishes.

However, the volatility of U_t in (7.46) vanishes when $\nu_T(t) \equiv 0$ only because we have chosen a continuous-time framework: the quadratic variation Q_t accrues continuously in our model. In realized variance payoffs, however, Q_t is expressed as the sum of squared daily log-returns up to time t.

In the Black-Scholes model with constant volatility, for example, using the standard definition for quadratic variation yields:

$$Q_t = \lim_{\Delta t \to 0} \sum_{0}^{t-1} \ln^2\left(\frac{S_{i+1}}{S_i}\right) = \sigma^2 t$$

which implies that Q_t is deterministic and the price of an ATM call on realized variance vanishes. Consider instead the following definition for Q_t:

$$Q_t = \sum_{0}^{t-1} \ln^2\left(\frac{S_{i+1}}{S_i}\right)$$

Q_t is the standard estimator of realized variance over $[0, t]$. The log-return $r = \ln\left(\frac{S_{i+1}}{S_i}\right)$ is given by:

$$r = \sigma\sqrt{\Delta t}Z - \frac{\sigma^2 \Delta t}{2} \tag{7.64a}$$

$$\simeq \sigma\sqrt{\Delta t}Z \tag{7.64b}$$

where Z is a gaussian random variable with unit variance and Δt is the interval between two observations of S, i.e. 1 day.

Volatility levels of typical equity underlyings – $\sigma\sqrt{\Delta t}$ are of the order of 1% to 2% – are such that $\sigma\sqrt{\Delta t} \ll 1$, justifying the approximation in (7.64b). While the expectation of Q_t is still (approximately) $\sigma^2 t$, Q_t is a finite sum of squared random variables and has non-vanishing variance. Even in a Black-Scholes model with constant volatility, an ATM call on variance acquires a non-zero value. This value is generated by the intrinsic variance of the volatility estimator we are using.

Let us then switch to discrete time and consider the evolution of U_t over the time interval $[t, t + \Delta t]$. Let r be the log-return of S_t over $[t, t + \Delta t]$. The variation of U_t during Δt reads:

$$
\begin{aligned}
U_{t+\Delta t} - U_t &= \frac{(Q_t + r^2) + (T - t - \Delta t)\,\widehat{\sigma}_T^2\,(t + \Delta t)}{T} - \frac{Q_t + (T - t)\,\widehat{\sigma}_T^2\,(t)}{T} \\
&= \frac{T - t - \Delta t}{T}\left(\widehat{\sigma}_T^2\,(t + \Delta t) - \widehat{\sigma}_T^2\,(t)\right) + \frac{1}{T}\left(r^2 - \widehat{\sigma}_T^2\,(t)\,\Delta t\right)
\end{aligned}
\tag{7.65}
$$

The first piece in (7.65) is the discrete-time counterpart of the $\frac{T-t}{T}d(\widehat{\sigma}_T^2)$ term in (7.45), which is the only contribution to dU_t in a continuous-time setting.

Let us assume that the volatility of $\widehat{\sigma}_T^2$ vanishes so that this contribution vanishes and let us concentrate on the contribution of the second piece to the variance of $U_{t+\Delta t} - U_t$.

Write the daily return over Δt as:

$$
r = \sigma_t\sqrt{\Delta t}Z
\tag{7.66}
$$

where Z is a (possibly non-gaussian) random variable with unit variance, and σ_t is the instantaneous (discrete) volatility over $[t, t + \Delta t]$. Let us make the approximation that the term structure of the VS curve at t is flat: $\sigma_t = \widehat{\sigma}_T(t)$. The second piece in (7.65) then reads:

$$
\frac{\widehat{\sigma}_T^2\,(t)\,\Delta t}{T}\left(Z^2 - 1\right)
$$

where, just as in (7.64b), we have discarded terms of higher order in $\sigma\sqrt{\Delta t}$. Computing now the variance of $\frac{U_{t+\Delta t}}{U_t}$ contributed by the second term in (7.65) we get:

$$
\begin{aligned}
E\left[\left(\frac{U_{t+\Delta t}}{U_t} - 1\right)^2\right] &= \frac{1}{U_t^2}\left(\frac{\widehat{\sigma}_T^2\Delta t}{T}\right)^2 E\left[\left(Z^2 - 1\right)^2\right] \\
&= \left(\frac{\widehat{\sigma}_T^2}{U_t}\right)^2\left(\frac{\Delta t}{T}\right)^2(2 + \kappa)
\end{aligned}
\tag{7.67}
$$

where κ is the (excess) kurtosis of Z: $\kappa = E[Z^4] - 3$. Let us assume that the term structure of VS volatilities is flat, so that $E[U_t] = \widehat{\sigma}_T^2$ and let us replace U_t with its expectation. The prefactor in (7.67) is then simply equal to 1. This yields:

$$
E\left[\left(\frac{U_{t+\Delta t}}{U_t} - 1\right)^2\right] \simeq \left(\frac{\Delta t}{T}\right)^2(2 + \kappa)
\tag{7.68}
$$

We now revert to the continuous-time framework. The quadratic variation of U_t over $[0, T]$ now acquires an extra contribution and reads :

$$\int_0^T 4R_\tau^2 \left(\frac{T-\tau}{T}\right)^2 \nu_T^2(\tau)\, d\tau \ + \ N\left(\frac{\Delta t}{T}\right)^2 (2+\kappa) \qquad (7.69)$$

where $N = \frac{T}{\Delta t}$ is the number of returns in the interval $[0, T]$. Dividing now (7.69) by T provides the following amended expression for σ_{eff}^2 which supersedes (7.56a):

$$\sigma_{\text{eff}}^2 \ = \ \frac{1}{T}\int_0^T 4\left(\frac{T-\tau}{T}\right)^2 \left(\frac{\widehat{\sigma}_{\tau T}^2(0)}{\widehat{\sigma}_T^2(0)}\right)^2 \nu_T^2(\tau)\, d\tau \ + \ \frac{2+\kappa}{NT} \qquad (7.70)$$

where N is the number of returns over $[0, T]$: $N = \frac{T}{\Delta t}$. The second piece in (7.70) is generated by the intrinsic variance of the variance estimator itself: as expected, its relative contribution to σ_{eff}^2 is largest for short maturities.

Which value should we pick for κ? κ is the *conditional* kurtosis of daily log-returns, that is the kurtosis of Z in (7.66). It is the portion of the *unconditional* kurtosis – the kurtosis of r – that is not generated by fluctuations of σ_t – i.e. the scale – of daily returns. In practice, as already mentioned in Section 1.2.2, measuring the *unconditional* kurtosis is already tricky; 5 is a typical level for equity underlyings.

Sorting out which portion of κ is attributable to fluctuations of σ_t – a quantity that is not directly observable – or to the intrinsic kurtosis of scaled log-returns (Z) is even more challenging. In numerical examples below we have used $\kappa = 2$. The case of conditional lognormal daily returns corresponds to $\kappa = 0$.

Setting $\kappa = -2$ in formula (7.70) suppresses the contribution of the kurtosis of daily returns to σ_{eff} altogether.

Figure 7.7 shows prices of an ATM call option on realized variance for maturities 1 month to 1 year. We have used Set II parameters, flat VS volatilities equal to 20% and have used a ratio of 21 returns per month. We compare approximate prices obtained with expression (7.70) for σ_{eff} with $\kappa = 2$ or $\kappa = -2$ with (exact) prices computed in a Monte Carlo simulation of the two-factor model. While the dynamics of forward variances is generated in standard fashion, we draw daily returns with a Student distribution with $\nu = 7$ degrees of freedom, so that their conditional kurtosis is equal to 2.[20]

Figure 7.7 highlights the fact that the discrete nature of returns is mostly apparent for short-maturity options – say 3 months or less. Observe how prices computed in our approximate model with σ_{eff} given by (7.70) are in good agreement with prices computed in a Monte Carlo simulation of the full-blown forward variance model –

---

[20] The density of the Student distribution is given in (10.1), page 392. Recall that the kurtosis of the Student distribution is $\frac{6}{\nu-4}$, and that only moments of order less than ν exist. Using a Student distribution for the log-return has the consequence that the expectation of the return itself (the forward, equal to $\exp(r)$) is infinite. Here we are only concerned with the estimation of moments of r up to order 4. We refer the reader to Chapter 10 for an example of stochastic volatility model with Student-distributed returns.

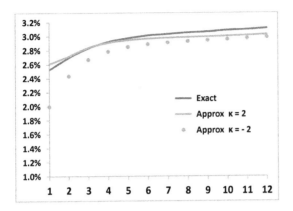

Figure 7.7: Price of an ATM call on realized variance as a function of maturity, from 1 month to 1 year, computed either in a Monte Carlo simulation of the forward variance model (exact), or in the approximate model with σ_{eff} given by (7.70) with $\kappa = 2$ and $\kappa = -2$.

with Student-like daily returns for S_t. The slight negative bias of approximate prices with respect to exact prices is also seen in Table 7.5.

7.6.9 Conclusion

- The primary risk of options on realized variance is the exposure to the realized volatility of $\hat{\sigma}_T(t)$, the implied volatility for the option's maturity. This risk is well captured with an approximate model for the dynamics of $\hat{\sigma}_T(t)$: an option on realized variance is delta-hedged with VSs of the option's maturity. Forward variance models generating different volatilities and correlations of forward variances but the same volatility for $\hat{\sigma}_T(t)$ yield very similar prices.

- Additional VSs for intermediate maturities are needed to hedge the residual exposure to the VS term structure and to properly gamma-hedge the option on realized variance against the realized volatility of the underlying. The approximate model does not price the gamma costs arising from dynamical trading of these intermediate VSs. Comparison of approximate and exact prices suggests however that these costs are small.

- Quadratic variation does not accrue continuously but is expressed as a discrete sum of squared daily log-returns: this results in a specific contribution of the intrinsic variance of the volatility estimator to the option's price.

7.6.10   What about the vanilla smile? Lower and upper bounds

Nowhere in our analysis have we calibrated our model to the vanilla smile – except for determining VS volatilities. This is natural as we are only using VSs as hedges. Can the information in the vanilla smile be somehow used? In what measure do market prices of vanilla options restrict prices of options on realized variance?

We know that the density of S_T can be extracted from the vanilla smile of maturity T – see equation (2.8), page 29. If we now assume that the process for S_t is a diffusion – which in practitioner terms means that the expansion of the daily P&L of a delta-hedged position at order two in $\frac{\delta S}{S}$ is adequate – much more information can be extracted from the vanilla smile.

Indeed, by delta-hedging European options we are able to create payoffs that involve the realized variance weighted by the dollar gamma, a function of t and S. The VS is an example of such a payoff – with a constant weight – but more complex payoffs involving S_t and increments of the quadratic variation Q_t can be synthesized.

What about call and put payoffs on Q_T? These are not replicable by a combination of a static position in a European payoff and a delta strategy on S. Given a particular market smile, can we nevertheless derive model-independent bounds for the prices of calls and puts on realized variance?

Model-independent lower and upper bounds for prices of options on realized variance are most naturally determined by solving the stochastic control problem touched upon in Appendix A of Chapter 2, in the context of the Lagrangian Uncertain Volatility Model. The higher bound $\overline{\mathcal{P}}(F)$ is obtained as the highest price generated by processes for the instantaneous volatility σ_t such that vanilla option prices are recovered:

$$\overline{\mathcal{P}}(F) = \max_{\substack{\sigma_t \in [\sigma_{\min},\sigma_{\max}] \\ E_\sigma[O_i] = \mathcal{P}_{\text{Market}}(O_i)}} E_\sigma[F]$$

where we set $\sigma_{\min} = 0$, $\sigma_{\max} = +\infty$. F is the payoff of the option on realized variance, and the O_i are payoffs of vanilla options.

Likewise, the model-independent lower bound $\underline{\mathcal{P}}(F)$ is obtained by minimizing $E_\sigma[F]$. In the literature, this problem has been tackled differently, making use of particular solutions of the Skorokhod embedding problem – see [42] as well as [62], [28], [34].

The $\underline{\mathcal{P}}(F)$, $\overline{\mathcal{P}}(F)$ bounds thus derived are sharp: in case they are violated, an arbitrage strategy consisting in a static position in an option on realized variance plus a European payoff together with a dynamic delta strategy nets a positive P&L. Both the European payoff and the delta strategy have to be determined numerically. The delta strategy consists in delta-hedging the European payoff, either in timer-option fashion ($\sigma = \sigma_{\max} = \infty$), or at zero volatility ($\sigma = \sigma_{\min} = 0$).[21]

---

[21]For options that involve the quadratic variation Q in addition to S, equation (2.130), page 88, is replaced – with zero rates and repo – with:

To give the reader an intuition for how these delta-hedging regimes come into play, we now go through the derivation of a heuristic lower bound for a call on realized variance proposed by Bruno Dupire[22] and a heuristic upper bound, proposed by Peter Carr and Roger Lee and published in [28].[23]

Both are non-optimal bounds in the sense that the lower/upper bound lie lower/higher than $\underline{P}(F)/\overline{P}(F)$.

A lower bound

Assume zero interest rate and repo and consider a short position in a European option of maturity T with convex payoff $f(S)$. Let us delta-hedge this option in timer option manner with a quadratic variation budget given by $Q = \hat{\sigma}^2 T$, where $\hat{\sigma}$ is the (volatility) strike of the option on realized variance.[24]. The option's value is given by: $\mathcal{P}^f_{BS}(S_t, Q_t; Q)$, where $\mathcal{P}^f_{BS}$ is given by expression (5.68), page 183, and Q_t is the realized quadratic variation since $t = 0$.

If $Q_T < \hat{\sigma}^2 T$, there remains at maturity some residual quadratic variation budget and we make a positive P&L equal to:

$$P\&L^{Q_T<Q}_{\text{Final}} = \mathcal{P}^f_{BS}(S_T, Q_T; Q) - \mathcal{P}^f_{BS}(S_T, Q_T; Q_T) \qquad (7.71)$$

If instead we exhaust our budget at time $\tau < T$: $Q_\tau = Q$, we delta-hedge our option at zero implied volatility on $[\tau, T]$ and our final P&L is:

$$P\&L^{Q_T>Q}_{\text{Final}} = -\frac{1}{2}\int_\tau^T S_t^2 \left.\frac{d^2 f}{dS^2}\right|_{S_t} \bar{\sigma}_t^2 dt \qquad (7.72)$$

where $\bar{\sigma}_t$ is the instantaneous realized volatility.

Let us now assume that the convex profile f is such that $\frac{S^2}{2}\frac{d^2 f}{dS^2}$ either vanishes or is equal to a positive constant, Γ: $\frac{S^2}{2}\frac{d^2 f}{dS^2} = \theta(S)\Gamma$ and $\Gamma \geq 0$, with $\theta(S) = 0$ or 1. Then:

$$P\&L^{Q_T>Q}_{\text{Final}} \geq -\Gamma\int_\tau^T \bar{\sigma}_t^2 dt \qquad (7.73)$$

$$\geq -\Gamma(Q_T - Q_\tau) = -\Gamma(Q_T - Q) \qquad (7.74)$$

$$\frac{dP}{dt} + \max_{\sigma=\sigma_{\min},\sigma_{\max}} \sigma^2\left(\frac{S^2}{2}\frac{d^2 P}{dS^2} + \frac{dP}{dQ}\right) = 0$$

with $\sigma_{\min} = 0$, $\sigma_{\max} = +\infty$. In this equation P is the price of payoff $F - \Sigma\lambda_i O_i$ where F is the exotic option's payoff – in our case a call on realized variance – and the O_i are vanilla options. Because $\sigma_{\min} = 0$, $\sigma_{\max} = +\infty$, the solution of this PDE is then such that (a) $\frac{dP}{dt} = 0$; (b) in regions of S where the max is obtained with $\sigma = \sigma_{\max} = +\infty$, we have $\frac{S^2}{2}\frac{d^2 P}{dS^2} + \frac{dP}{dQ} = 0$. These are exactly conditions (5.70), page 184, that characterize timer options. I thank Pierre Henry-Labordère for pointing this out to me.

[22]Presented at the 2005 Global Derivatives conference.

[23]In what follows all P&Ls are computed at order one in δt and order two in δS; provided this is adequate, the sub- and super-replicating strategies derived below indeed work. Mathematically, they hold for processes of S_t that are continuous semimartingales. Practically, they hold as long as contributions of order three and higher in δS to our daily P&L can be ignored.

[24]See Section 5.9 on timer options.

Consider a call option on realized variance with strike $\hat{\sigma}$, whose payoff is

$$\frac{1}{2\hat{\sigma}_T}\left(\frac{Q_T}{T} - \hat{\sigma}^2\right)^+ = \frac{1}{2\hat{\sigma}_T T}(Q_T - Q)^+$$

where $\hat{\sigma}_T$ is the VS volatility for maturity T. Denote by $C^{\hat{\sigma}}_{\text{Mkt}}$ its market price and let us buy $2\hat{\sigma}_T T\Gamma$ units of it.

In case $Q_T > Q$, this option's payout more than offsets the negative P&L $P\&L^{Q_T>Q}_{\text{Final}}$: our total final P&L is positive.

If instead $Q_T < Q$, the option on realized variance expires worthless, and our final total P&L, given by $P\&L^{Q_T<Q}_{\text{Final}}$ is positive as well.

Thus our short position in the European payoff $f(S)$ combined with a long position in $2\hat{\sigma}_T T\Gamma$ options on realized variance of strike $\hat{\sigma}$ nets a profit in all cases. It must then be that the cost for entering this position is positive. Setting up this position requires a cash amount equal to the market price of the realized variance call minus the difference between market and model prices of the European payoff at $t = 0$, which is equal to:

$$P^f_{\text{Mkt}} - P^f_{BS}(S_0, 0; Q)$$

The condition that the cash amount needed at $t = 0$ be positive then reads:

$$(2\hat{\sigma}_T T\Gamma)C^{\hat{\sigma}}_{\text{Mkt}} - \left(P^f_{\text{Mkt}} - P^f_{BS}(S_0, 0; Q)\right) \geq 0$$

This yields the following lower bound for $C^{\hat{\sigma}}_{\text{Mkt}}$:

$$C^{\hat{\sigma}}_{\text{Mkt}} \geq \frac{1}{2\hat{\sigma}_T T}\frac{1}{\Gamma}\left(P^f_{\text{Mkt}} - P^f_{BS}(S_0, 0; \hat{\sigma}^2 T)\right) \qquad (7.75)$$

where we have replaced Q with $\hat{\sigma}^2 T$. This condition holds for any European payoff f such that $\frac{S^2}{2}\frac{d^2 f}{dS^2}$ either vanishes or is equal to Γ. Which payoff should we pick, so that the lower bound for $C^{\hat{\sigma}}_{\text{Mkt}}$ provided by the right-hand side of (7.75) is highest?

f can be replicated with cash, forwards and vanilla options of all strikes, of which only the latter contribute to the right-hand side of (7.75). From formula (3.6) for the replication of European payoffs, the density of vanilla options of strike K in the replicating portfolio is equal to the second derivative of the payoff, thus is equal to $2\theta(K)\frac{\Gamma}{K^2}$. The right-hand side of (7.75) can thus be written as:

$$\frac{1}{2\hat{\sigma}_T T}\frac{1}{\Gamma}\int_0^\infty 2\theta(K)\frac{\Gamma}{K^2}\left(P^K_{\text{Mkt}} - P^K_{BS}(S_0, 0; \hat{\sigma}^2 T)\right)dK \qquad (7.76)$$

where P^K_{Mkt} is the market price for a vanilla option of strike K and P^K_{BS} is its Black-Scholes price. There is no need to distinguish between calls and puts, since by call-put parity $P^K_{\text{Mkt}} - P^K_{BS}$ is identical for a call or a put struck at the same strike.

Let us introduce the implied volatility for strike K, $\widehat{\sigma}_K$. By definition of $\widehat{\sigma}_K$: $P^K_{\text{Mkt}} = \mathcal{P}^K_{BS}(S_0, 0; \widehat{\sigma}^2_K T)$. (7.76) is thus equal to:

$$\frac{1}{2\widehat{\sigma}_T T} \int_0^\infty \frac{2}{K^2} \theta(K) \Big(\mathcal{P}^K_{BS}(S_0, 0; \widehat{\sigma}^2_K T) - \mathcal{P}^K_{BS}(S_0, 0; \widehat{\sigma}^2 T) \Big) dK$$

The highest value for this expression is obtained by setting $\theta(K) = 1$ for strikes such that $\widehat{\sigma}_K > \widehat{\sigma}$ and $\theta(K) = 0$ otherwise. We thus get our final expression for the (sub-optimal) lower bound :

$$C^{\widehat{\sigma}}_{\text{Mkt}} \geq \frac{1}{2\widehat{\sigma}_T T} \int_0^\infty 1_{\widehat{\sigma}_K > \widehat{\sigma}} \frac{2}{K^2} \Big(\mathcal{P}^K_{BS}(S_0, 0; \widehat{\sigma}^2_K T) - \mathcal{P}^K_{BS}(S_0, 0; \widehat{\sigma}^2 T) \Big) dK$$
$$(7.77)$$

This idea can be extended to the case of a call on forward-starting variance – see [28].

Imagine that implied volatilities all lie above the (volatility) strike of the call on realized variance: $1_{\widehat{\sigma}_K > \widehat{\sigma}} = 1 \ \forall K$. This implies also that $\widehat{\sigma}_T \geq \widehat{\sigma}$: the call on realized variance is in the money. We then recognize in the right-hand side of (7.77) the difference between the market price for the VS for maturity T – see equation (5.16), page 154 – and its price for a flat smile equal to $\widehat{\sigma}$. This yields:

$$C^{\widehat{\sigma}}_{\text{Mkt}} \geq \frac{1}{2\widehat{\sigma}_T} \big(\widehat{\sigma}^2_T - \widehat{\sigma}^2 \big)$$

which expresses that, with zero rates, the price of a call option is larger than its intrinsic value.

Likewise, if $1_{\widehat{\sigma}_K > \widehat{\sigma}} = 0 \ \forall K$, $\widehat{\sigma}_T \leq \widehat{\sigma}$: the call on realized variance is out of the money, and (7.77) again expresses that $C^{\widehat{\sigma}}_{\text{Mkt}}$ lies above the intrinsic value, in this case zero.

An upper bound
We present here one version of Peter Carr and Roger Lee's super-replicating strategy – the upper bound in [28] is sharper.

We work with zero interest rate and repo and temporarily omit the $\frac{1}{2\widehat{\sigma}_T T}$ prefactor in the payoff of the call on realized variance:

$$(Q_T - \Sigma)^+$$

with $\Sigma = \widehat{\sigma}^2 T$ where $\widehat{\sigma}$ is the strike of the option on realized variance. Pick two barriers L and H such that $L \leq S_0 \leq H$ where S_0 is the initial spot level. It is possible to generate model-independently the payoff $(Q_\tau - \Sigma)^+$ where τ is defined as the time when S_τ first hits L or H – see Section 5.9 on timer options.

This payoff is the timer equivalent of a perpetual barrier option that pays a rebate $R(t) = (t - \Sigma)^+$ when S first hits L or H.

Denote by $\mathcal{B}_{LH}(S_0, Q; \Sigma)$ the price of this option – as a timer option price it does not depend on t. $\mathcal{B}_{LH}$ solves the following PDE:

$$\frac{S^2}{2}\frac{d^2\mathcal{B}_{LH}}{dS^2} + \frac{d\mathcal{B}_{LH}}{dQ} = 0 \tag{7.78}$$

with boundary conditions: $\mathcal{B}_{LH}(L, Q; \Sigma) = \mathcal{B}_{LH}(H, Q; \Sigma) = (Q - \Sigma)^+$

- If $\tau > T$, at $t = T$ we have:

$$
\begin{aligned}
\mathcal{B}_{LH}(S_T, Q_T; \Sigma) &= E_T[(Q_\tau - \Sigma)^+] = E_T[(Q_\tau - Q_T + Q_T - \Sigma)^+] \\
&\geq (Q_T - \Sigma)^+ \tag{7.79}
\end{aligned}
$$

where E_T is a shorthand notation for E_{T,S_T,Q_T}, an expectation taken with respect to PDE (7.78), and we have used the property that the quadratic variation increases with time: $Q_\tau - Q_T \geq 0$. Thus, if $\tau > T$, a long position in the timer barrier option that pays $(Q_\tau - \Sigma)^+$ super-replicates the payoff of the option on realized variance of maturity T.

- What if $\tau \leq T$? If S hits either L or H at time $\tau < T$ our timer option pays us $(Q_\tau - \Sigma)^+$. Since our aim is to super-replicate payoff $(Q_T - \Sigma)^+$, we need to trade an additional instrument of maturity T that generates $(Q_T - Q_\tau)$. At second order in δS, delta-hedging the profile $-2\ln S$ at zero implied volatility generates the realized quadratic variation:

$$d(-2\ln S_t) + \frac{2}{S_t}dS_t = dQ_t$$

The P&L from the costless delta strategy $\frac{2}{S_t}dS_t$, combined with a long position in an option that pays $-2\ln S_T$ at T, generates the quadratic variation up to T. Since we only need to generate the quadratic variation starting at τ, when S hits either L or H, let us adjust $-2\ln S$ by an affine function of S that preserves the convexity of $-2\ln S$ but ensures that the resulting payoff profile vanishes for $S = L$ and $S = H$. The resulting payoff $f_{LH}(S)$ is:

$$f_{LH}(S) = -2\ln S + 2\left(\frac{\ln H - \ln L}{H - L}(S - L) + \ln L\right)$$

$f_{LH}(S) \leq 0$ for $S \in [L, H]$ and is positive otherwise. We have:

$$
\begin{aligned}
Q_T - Q_\tau &= (f_{LH}(S_T) - f_{LH}(S_\tau)) - \int_\tau^T \left.\frac{df_{LH}}{dS}\right|_{S_t} dS_t \tag{7.80a} \\
&= f_{LH}(S_T) - \int_\tau^T \left.\frac{df_{LH}}{dS}\right|_{S_t} dS_t \tag{7.80b}
\end{aligned}
$$

since by construction $f_{LH}(S_\tau) = 0$. By buying the timer option discussed above as well as the European option that pays $f_{LH}(S_T)$ we generate at T

the amount: $(Q_\tau - \Sigma)^+ + (Q_T - Q_\tau)$. It is easy to check that this again super-replicates the call on variance:

$$(Q_\tau - \Sigma)^+ + (Q_T - Q_\tau) \geq (Q_T - \Sigma)^+$$

Our super-replicating portfolio thus comprises, in addition to the barrier timer option, a European option of maturity T that pays $f_{LH}(S_T)$. In the case $\tau > T$, however, the value of our portfolio at T is $\mathcal{B}_{LH}(S_T, Q_T; \Sigma) + f_{LH}(S_T)$. Since $f_{LH}(S_T) \leq 0$ for $S \in [L, H]$, to ensure that the super-replication in (7.79) still works, we just need to replace $f_{LH}(S_T)$ with $g_{LH}(S_T)$ given by:

$$g_{LH}(S) = \max(f_{LH}(S), 0)$$

Because $g_{LH} \geq f_{LH}$, the super-replication still holds for $\tau \leq T$, provided we trade the delta $\frac{df_{LH}}{dS}$ over $[\tau, T]$.

In conclusion, reverting to the usual normalization for the payoff of calls on realized variance: $\frac{1}{2\widehat{\sigma}_T T}(Q_T - \widehat{\sigma}^2 T)$, whose price is $C_{\text{Mkt}}^{\widehat{\sigma}}$, we have:

$$C_{\text{Mkt}}^{\widehat{\sigma}} \leq \frac{1}{2\widehat{\sigma}_T T}\left(\mathcal{B}_{LH}(S_0, 0; \widehat{\sigma}^2 T) + \mathcal{G}_{LH}(S_0, T)\right) \tag{7.81}$$

where $\mathcal{G}_{LH}(S_0, T)$ is the market price of the option that pays $g_{LH}(S_T)$ at T. As it is a European payoff, it can be replicated with vanilla options: $\mathcal{G}_{LH}(S_0, T)$ only depends on the vanilla smile for maturity T.

$\mathcal{B}_{LH}(S_0, 0; \widehat{\sigma}^2 T)$ instead, only depends on L and H. The best higher bound is thus obtained for the (L, H) couple that minimizes the right-hand side of (7.81).

The reasoning for the upper bound can be extended to calls on forward-starting variance – see [28].

As a sanity check, take $L = H = S_0$. Then $\mathcal{B}_{LH}(S_0, 0; \widehat{\sigma}^2 T) = 0$ and $g_{LH}(S)$ is given by:

$$g_{LH}(S) = f_{LH}(S)$$
$$= -2\ln\left(\frac{S}{S_0}\right) + \frac{2}{S_0}(S - S_0)$$

We know from (5.12) that, with zero rate and repo, the market price of an option that pays $-2\ln\left(\frac{S_T}{S_0}\right)$ is $\sigma_T^2 T$. Thus, $\mathcal{G}_{LH}(S_0, T) = \sigma_T^2 T$ and (7.81) expresses the fact that the price of a call option is bounded above by the forward. This higher bound is not indecent; it is the limit of a call price when volatility is taken to infinity.

Conclusion
Information in the vanilla smile can be used to bound prices of options on realized variance. Using these bounds as bid/offer levels leads to prices that are, in practice, too conservative.[25]

---

[25]These bounds can be interpreted as a measure of model risk – in the absence of jumps, that is, in practitioner terms, when only terms of order up to two in δS are considered in our daily P&L. When

7.6.11 Options on forward realized variance

Consider two dates T_1, T_2. An option on forward realized variance pays at T_2 a vanilla payoff on the variance realized over the interval $[T_1, T_2]$. This type of option is attractive when the term structure of VS volatilities is such that forward VS implied volatilities appear to be much lower/higher than reasonable estimates of future realized volatility.

At $t = T_1$ the option is simply a spot-starting option on realized variance, which we have just extensively analyzed. Its price at T_1 is a function of VS volatilities $\widehat{\sigma}_T(T_1), T \in [T_1, T_2]$ given, for example in the SM, by expressions (7.56), but where we sit at $t = T_1$ rather than at $t = 0$:

$$\sigma_{\text{eff}}^2 = \frac{4}{T_2 - T_1} \int_{T_1}^{T_2} \left(\frac{T_2 - \tau}{T_2 - T_1} \right)^2 \left(\frac{\widehat{\sigma}_{\tau T_2}^2 (T_1)}{\widehat{\sigma}_{T_2}^2 (T_1)} \right)^2 \nu_{T_2}^2 (\tau) \, d\tau$$

$$P(t = T_1) = P_{\text{BS}}\left(t = T_1, \widehat{\sigma}_{T_2}^2 (T_1), \sigma_{\text{eff}}, T_2 \right) \qquad (7.82)$$

The value of our option at T_1 is then a particular form of a swaption payoff. The price at $t = 0$ of an option on forward realized variance is thus the price of a particular variance swaption of maturity T_1. It depends on assumptions about the volatility of forward variances $\xi_t^\tau, \tau \in [T_1, T_2]$, over the interval $t \in [0, T_1]$.

Consider replacing σ_{eff} with 0 in (7.82); the value at T_1 of our option is then simply the call – or put – payoff applied to $\widehat{\sigma}_{T_2}^2 (T_1)$. This is the payoff of a standard VS swaption, that is the option to enter at T_1 a VS contract of maturity T_2.[26] An option on forward realized variance is thus more expensive than the corresponding VS swaption.[27]

Spot-starting options on realized variance can be economically priced and risk-managed in the SM as the only ingredient is the curve $\nu_T(\tau), \tau \in [0, T]$: the instantaneous volatility of the VS volatility for the residual maturity. Options on forward realized variance, on the other hand, require a specification of volatilities of *forward* VS volatilities. These cannot be backed out of the curve $\nu_T(\tau), \tau \in [0, T]$. An option on forward realized variance can then only be priced in a full-blown model.

In the two-factor model we can choose parameters such that, while volatilities of *spot-starting* VS volatilities are identical, volatilities of *forward* VS volatilities are different. Sets I, II, III in Table 7.1, page 229, have this property. They are chosen so as to match the benchmark $\nu_T^B(t)$ in (7.40) with $\sigma_0 = 100\%$, $\tau_0 = 0.25$ $\alpha = 0.4$.

---

higher-order terms are taken into account – which is the case if jumps are allowed – even the bounds on a simple VS become sizeably spaced; see for example [63]. By construction, the processes that generate these bounds are worst-case scenarios and can hardly be considered realistic.

[26]Variance swaption smiles in the two-factor model are shown in Section 7.4.3.

[27]This is because $\widehat{\sigma}_{T_2}^2 (T_1)$, which is the underlying of the VS swaption, is the expectation at $t = T_1$ of σ_r^2, where σ_r is the realized volatility over $[T_1, T_2]$. We then have, in model-independent fashion: $E_{T_1}[(\sigma_r^2 - K)^+] \geq (E[\sigma_r^2] - K)^+ = (\widehat{\sigma}_{T_2}^2 (T_1) - K)^+$.

Table 7.6 lists prices of a 6 month-in-6 month ATM call option on forward realized variance in Sets I, II, III, together with (a) prices of spot-starting options with the same maturity from Table 7.2, page 241, (b) prices of 6 month-in-6 month variance swaptions. We can see that (a) options on forward realized variance are indeed more expensive than both spot-starting options and swaptions, (b) the price in Set I is appreciably higher than that in Set III, even though prices of spot-starting options are almost identical.

The difference between the second and third line of Table 7.6 is a measure of the additional time value contributed by the volatility of forward volatilities ξ_t^T during the interval $[T_1, T_2]$.

	Set I	Set II	Set III
Spot-starting	2.97%	2.96%	2.94%
Foward-starting	4.25%	4.09%	3.94%
Swaption	3.12%	2.90%	2.69%

Table 7.6: Prices of an ATM call option on forward realized variance (top) and ATM call VS swaption (bottom) with $T_1 = 6$ months, $T_2 = 1$ year, computed in the two-factor model with parameter sets in Table 7.1, page 229.
Prices for a spot-starting option of the same maturity, from Table 7.2, are shown for reference. The term structure of VS volatilities is flat at 20%.

7.7 VIX futures and options

We discuss here the application of the two-factor model to VIX instruments. This will allow us to relax the assumption of lognormality of forward variances.

VIX futures trade on the CBOE and expire on the morning of the Wednesday that is exactly 30 days prior to the monthly expiration dates of listed S&P 500 options. The settlement value of the expiring VIX future is the 30-day log-contract implied volatility, computed using market prices of listed S&P 500 options.[28] Ignoring the effect of fixed cash-amount dividends, the log-contract implied volatility at time t for maturity T is given by expression (5.12):

$$\widehat{\sigma}_T^2(t) = \frac{e^{r(T-t)}}{T-t} \left(Q_{\mathrm{mkt},t}^T - Q_{\widehat{\sigma}=0,t}^T \right)$$

where $Q_{\mathrm{mkt},t}^T$ is the market price at time t of the payoff $-2\ln S_T$ and $Q_{\widehat{\sigma}=0,t}^T$ is the price of the same payoff using vanishing volatility. We now set $t = T_i$, the expiration

---

[28]The settlement value of the VIX future is in fact one hundred times this VS volatility – we ignore this multiplier in what follows.

date of the ith future, and $T - t = \Delta = 30$ days, and use the notation $\widehat{\sigma}_{\mathrm{VIX},T_i}$ for the settlement value of the VIX future expiring at T_i. As indicated in Section 5.2, the payoff $-2\ln S_T$ is replicated by a continuous density $\frac{2}{K^2}$ of vanilla options, which yields:

$$\widehat{\sigma}^2_{\mathrm{VIX},T_i}(T_i) = \frac{2e^{r\Delta}}{\Delta}\left(\int_0^{F_{T_i}^{T_i+\Delta}} P_{\mathrm{mkt},T_i}^{K,T_i+\Delta}\frac{dK}{K^2} + \int_{F_{T_i}^{T_i+\Delta}}^{\infty} C_{\mathrm{mkt},T_i}^{K,T_i+\Delta}\frac{dK}{K^2} \right) \qquad (7.83)$$

where $P_{\mathrm{mkt},t}^{KT}$ (resp. $C_{\mathrm{mkt},t}^{KT}$) is the (discounted) market price at time t of a put (resp. call) option of strike K, maturity T on the S&P 500 and $F_{T_i}^{T_i+\Delta}$ is the forward of the S&P 500 index for maturity $T_i + \Delta$ observed at T_i. In the formula actually used by the CBOE, the integrals above are discretized using the trapezoidal rule and the integration does not run from 0 to ∞ but is cut off for small and large strikes wherever a zero bid price is encountered for two consecutive strikes.[29] The resulting approximation is quite good, as strikes of listed S&P 500 options are closely spaced and out-of-the-money strikes are reasonably liquid, typically from 50% to 125%.

Note that $\widehat{\sigma}_{\mathrm{VIX},T_i}$ does not share the same convention as VS payoffs for annualizing volatility: in (7.83) $\frac{1}{\Delta}$ is used instead of $\frac{252}{N}$. Depending on the exact number of trading days for the 30-day period at hand, this difference in conventions may translate into a difference of the order of one point of volatility. Further below, we will use the same $\frac{1}{\Delta}$ convention when comparing VIX and log-contract – or VS – quotes.

Exchange-traded options on VIX futures exist as well, with the same expiration dates as the underlying VIX futures. Figure 7.8 shows the values of the 5 VIX futures for expiries July through December, observed on January 14, 2011, as well as the smiles of the associated options – see Figure 7.9 below for the expiry dates of the VIX futures.[30]

The settlement value of a VIX future is a log-contract volatility, not a VS volatility. However, for ease of exposition, since forward VS variances and forward log-contract variances are both driftless and can be modeled identically – see the discussion in Section 5.5, page 168 – we make no distinction in what follows between $\widehat{\sigma}_T$ and $\widehat{\sigma}_{\mathrm{VS},T}$. $\widehat{\sigma}_T = \widehat{\sigma}_{\mathrm{VS},T}$ and forward log-contract and VS variances are identical, equal to ξ_t^T.

The distinction between $\widehat{\sigma}_T$ and $\widehat{\sigma}_{\mathrm{VS},T}$ becomes relevant again in the discussion of the arbitrage between S&P 500 VSs and VIX instruments, in Section 7.7.4.

---

[29]We refer the reader to *www.cboe.com/micro/vix/vixwhite.pdf* for the exact procedure for computing $\widehat{\sigma}_{\mathrm{VIX},T}$.

[30]The VIX itself is a number published in real time: $\widehat{\sigma}_{\mathrm{VIX},t}$ is the 30-day VS volatility obtained through an interpolation of VS volatilities for two consecutive expiration dates of S&P 500 listed options, computed with the same methodology as for the settlement value of VIX futures. While the VIX index has become a popular indicator of real-time market temperature we do not consider it in what follows as – just like the temperature in New York City – it cannot be traded.

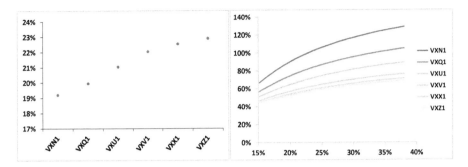

Figure 7.8: Left: VIX futures as of June 14, 2011 for expiries ranging from July (VXN1) to December (VXZ1). Right: Smiles of VIX futures.

7.7.1 Modeling VIX smiles in the two-factor model

Let F_t^i be the VIX future for expiry T_i, observed at time t. At $t = T_i$, F_t^i is equal to the VIX index, that is the 30-day VS volatility observed at T_i:

$$F_{t=T_i}^i = \widehat{\sigma}_{\text{VIX},T_i}(T_i) \tag{7.84}$$

where we denote by $\widehat{\sigma}_{\text{VIX},T_i}$ the forward VS volatility corresponding to the VIX future expiring at T_i – it is defined for $t < T_i$:

$$\widehat{\sigma}_{\text{VIX},T_i}^2(t) = \frac{1}{\Delta} \int_{T_i}^{T_i+\Delta} \xi_t^T \, dT = E_t[(F_{T_i}^i)^2] \tag{7.85}$$

F_t^i is the value of a future, hence it is driftless. For $t < T_i$, F_t^i is thus given by:

$$F_t^i = E[F_{T_i}^i] \tag{7.86}$$

So far, we have modeled instantaneous forward variances as lognormal processes. While the resulting discrete forward variances are not exactly lognormal, they are close to lognormal – see for example swaption smiles in Figure 7.4. As is apparent in Figure 7.8, implied volatilities of VIX futures exhibit substantial smiles: we need to relax the lognormality of instantaneous forward variances, while preserving a Markov representation, if possible.

Let us start with the two-factor model specified in Section 7.4, page 226. Forward variances ξ_t^T are given by (7.33) and (7.34):

$$\xi_t^T = \xi_0^T f^T(t, x_t^T)$$
$$f^T(t, x) = e^{\omega x - \frac{\omega^2}{2} \chi(t,T)}$$

In the basic version of the two-factor model, the mapping function is an exponential, hence ξ_t^T is lognormal.

A Markov-functional model

More general forms for $f(t, x)$ can be considered. The condition that ξ_t^T is driftless translates into the following PDE for $f^T(t, x)$:

$$\frac{df^T}{dt} + \frac{\eta^2 (T - t)}{2} \frac{d^2 f^T}{dx^2} = 0 \qquad (7.87)$$

where $\eta^2 (T - t)$ is the instantaneous variance of x_t^T. η is given in (7.31b):

$$\eta(u) = \alpha_\theta \sqrt{(1 - \theta)^2 e^{-2k_1 u} + \theta^2 e^{-2k_2 u} + 2\rho_{12}\theta (1 - \theta) e^{-(k_1 + k_2)u}}$$

Once a terminal condition at time $t = T$, $f^T(T, x)$ is chosen, the mapping function $f^T(t, x)$ is generated for dates $t < T$ by solving (7.87). Any solution of (7.87) needs to be suitably normalized so that the initial value of the forward variance ξ_t^T is recovered at $t = 0$, for $x = 0$ – which is another way of saying that $E[\xi_t^T] = \xi_0^T$, $\forall t$:

$$f^T(0, 0) = 1$$

The dynamics of ξ^T reads:

$$\frac{d\xi_t^T}{\xi_t^T} = \frac{d \ln f^T}{dx} (t, x_t^T) \, dx_t^T$$

The instantaneous volatility of ξ^T is thus given by:

$$\text{vol}(\xi_t^T) = \eta(T - t) \left| \frac{d \ln f^T}{dx} \right|_{x = f^{T-1}(\xi_t^T, t)} \qquad (7.88)$$

$f^{T^{-1}}$ is well-defined only if f^T is monotonic; this is verified in the parametric model used in the section that follows. The right-hand side of (7.88) is a function of ξ_t^T and t, thus what we have is a local volatility model for ξ_t^T.

A simple parametrization

Any linear combination of exponential solutions (7.34) solves (7.87) – using positive weights ensures that $f^T(t, x) \geq 0$, for all t, x. With such an ansatz for $f^T(t, x)$ smiles for ξ^T are positively sloping, as for large values of x the linear combination is dominated by terms with large values of ω. This is not a serious limitation as VIX smiles for liquid strikes are usually positively sloping. Let us then use just two exponentials – this ansatz was proposed in [10]. We introduce the volatility-of-volatility smile parameters γ_T, β_T, ω_T and set:

$$f^T(t, x) = (1 - \gamma_T) e^{\omega_T x - \frac{\omega_T^2}{2} \chi(t, T)} + \gamma_T e^{\beta_T \omega_T x - \frac{(\beta_T \omega_T)^2}{2} \chi(t, T)} \qquad (7.89)$$

$\gamma_T \in [0, 1]$ is a mixing parameter. For $\gamma_T = 0$ or $\gamma_T = 1$, ξ^T is lognormal. β_T is also chosen in $[0, 1]$ – for $\beta_T = 0$, ξ^T is simply a displaced lognormal. The instantaneous volatility of ξ_T is given by (7.88). Setting $t = 0$, $x = 0$ yields:

$$\text{vol}(\xi_t^T)\big|_{t=0} = \omega_T \left((1 - \gamma_T) + \beta_T \gamma_T \right) \eta(T)$$

Let us introduce the dimensionless parameter ζ^T and write ω_T as:

$$\omega_T = \frac{2\nu}{(1 - \gamma_T) + \beta_T \gamma_T} \zeta_T \tag{7.90}$$

The instantaneous volatility of ξ_t^T at $t = 0$ is then given by:

$$\mathrm{vol}(\xi_t^T)\big|_{t=0} = 2\nu\zeta_T\eta(T)$$

ζ_T is thus simply a scale factor for the instantaneous volatility of ξ_t^T at $t = 0$. We now use parameters $\gamma^T, \beta^T, \zeta^T$ to adjust the smile of VIX futures, for a given set of parameters $\nu, \theta, k_1, k_2, \rho$.

Calibration of VIX futures and options

We first choose a set of parameters $\nu, \theta, k_1, k_2, \rho$ to generate the underlying basic dynamics of our model – specifically one of the three sets in Table 7.1.

Then for each expiry T_i we determine forward variances ξ_0^T for $T \in [T_i, T_i + \Delta]$ as well as parameters $\gamma^T, \beta^T, \zeta^T$ so that market prices of (a) the VIX future F_t^i expiring at T_i and (b) VIX options maturing at T_i are matched.[31] Mathematically our model needs to ensure that:

$$E_t\left[\widehat{\sigma}_{\mathrm{VIX},T_i}(T_i)\right] = F_t^i$$
$$E_t\left[\left(\widehat{\sigma}_{\mathrm{VIX},T_i}(T_i) - K\right)^+\right] = \mathcal{C}_t^{Ki}, \quad E_t\left[\left(K - \widehat{\sigma}_{\mathrm{VIX},T_i}(T_i)\right)^+\right] = \mathcal{P}_t^{Ki}$$

where $\mathcal{C}_t^{Ki}, \mathcal{P}_t^{Ki}$ are, respectively, *undiscounted* market prices of call and put options with strike K, maturity T_i on the VIX index, and $\widehat{\sigma}_{\mathrm{VIX},T_i}$ is given by (7.85) as a function of forward variances. We assume that within each interval $[T_i, T_i + \Delta]$ forward variances at $t = 0$ are flat and denote them by ξ_0^i. Similarly we use constant values for $\gamma^T, \beta^T, \zeta^T$ over $[T_i, T_i + \Delta]$ – which we denote by $\gamma^i, \beta^i, \zeta^i$.

The forward variance for interval $[T_i, T_{i+1}]$ that underlies the VIX future expiring at T_i is thus a function of t, X_t^1, X_t^2 given by:

$$\widehat{\sigma}^2_{\mathrm{VIX},T_i}\left(t, X_t^1, X_t^2\right) = \xi_0^i \frac{1}{\Delta} \int_{T_i}^{T_{i+1}} f^\tau(t, x_t^\tau)\, d\tau \tag{7.91}$$

and both the VIX future and prices of options on $\widehat{\sigma}_{\mathrm{VIX},T_i}(T_i)$ are obtained through a two-dimensional Gaussian quadrature on $(X_{T_i}^1, X_{T_i}^2)$.

Figure 7.9 shows the values of parameters $\gamma^i, \beta^i, \zeta^i$ calibrated to the market smiles of VIX futures as of June 14, 2011 for strikes in the interval $[15\%, 40\%]$. In this strike range the difference between market and calibrated implied volatilities lies well within the bid/offer spread which is about 3 points of volatility – the smiles

---

[31]Note that there is no guarantee that the variance curve thus determined matches SP500 log-contract implied volatilities – more on this in Section 7.7.4 below.

in Figure 7.8 are in fact those generated by our model with Set II parameters.[32] Calibration is performed by least-squares minimization with a sufficiently large weight on the future itself that it is exactly calibrated.[33]

It should be mentioned that for very large strikes – that are much less liquid – market implied volatilities fall off, a feature that our parametrization is unable to capture.

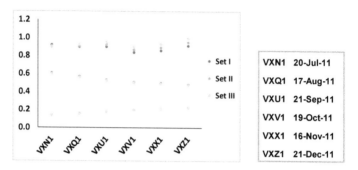

Figure 7.9: Left: values of $\gamma^i, \beta^i, \zeta^i$ calibrated on market smiles of VIX futures on June 14, 2011, using parameter sets in Table 7.1 – bottom: β^i, middle: γ^i, top: ζ^i. Right: expiries of VIX futures. The corresponding futures and their smiles appear in Figure 7.8.

Calibration has been performed using the three sets of parameters $\nu, \theta, k_1, k_2, \rho$ listed in Table 7.1, page 229. Figure 7.9 also lists the expiry dates of the corresponding VIX futures.[34] Calibrated values of $\gamma^i, \beta^i, \zeta^i$ for the six futures considered are similar; moreover they hardly depend on which set is used: only ζ is appreciably larger in Set III than in Sets I and II. This is expected. Sets I, II, III generate almost identical term

---

[32]Contrary to what is claimed at times, simultaneous jumps in the underlying and its volatility are not needed to generate market-compatible VIX smiles. While in this section we focus exclusively on VIX futures, that is, on the dynamics of forward variances, we provide in Section 8.7.2 an example of parametrization for the correlations of S_t with X_t^1, X_t^2 that generates a term structure of the ATMF skew that agrees with market smiles of vanilla options on S_t. It is thus possible to approximately match both the S&P 500 smile and VIX smiles without resorting to simultaneous jumps – provided the mismatch between VS volatilities either derived from the S&P 500 VS market or derived from the VIX market is small – see the discussion in Section 7.7.4 below.

[33]The reader may wonder why, rather than calibrating ξ^i along with $\gamma^i, \beta^i, \zeta^i$, we do not determine ξ^i using equation (7.98) in Section 7.7.4 below and then calibrate $\gamma^i, \beta^i, \zeta^i$ on F_t^i and the smile of VIX options maturing at T_i. Imagine that our model is able to perfectly calibrate VIX futures and options; then the (calibrated) value of ξ^i will obey equation (7.98). This will not be the case, however, if market and model smiles differ significantly outside the strike range used for calibration. In the latter case, using (7.98) will generate a value for ξ^i that may make it impossible to find values of $\gamma^i, \beta^i, \zeta^i$ such that F_t^i *and* market prices of near-the-money VIX options are recovered.

[34]Because expiry dates of VIX futures are not spaced 30 days apart, intervals $[T_i, T_i+\Delta]$ for consecutive futures may overlap – see the case of VXU1 and VXV1 for example. γ, β, ζ are then assumed constant over each interval $[T_i, \min(T_i + \Delta, T_{i+1})]$.

structures for volatilities of spot-starting VS volatilities, however, as manifested in Figure 7.2, volatilities of forward volatilities are lowest in Set III – this is compensated for by an increase in ζ.

Calibration is equally satisfactory for Sets I, II, III. Which set should one choose? Different sets for parameters $\nu, \theta, k_1, k_2, \rho$ generate different distributions of the realized variance of F_t^i, as well as different correlation structures for the F_t^i. Define σ_t as the square root of the expected instantaneous variance of F_t^i:

$$\sigma_t^2 = E_0 \left[\frac{\langle (d \ln F_t^i)^2 \rangle}{dt} \right] \tag{7.92}$$

In our two-factor Markov-functional model, F_t^i is a function of X_t^1, X_t^2, given by numerical evaluation of the expectation in (7.86). σ_t is thus simply given by:

$$\sigma_t = \sqrt{E\left[\alpha_{X^1}^2 + \alpha_{X^2}^2 + 2\rho_{12}\alpha_{X^1}\alpha_{X^2}\right]}$$

where $\alpha_{X^1}, \alpha_{X^2}$ are the sensitivities of $\ln F_t^i$ with respect to X_t^1, X_t^2 and the expectation is taken over X_t^1, X_t^2 – this is efficiently evaluated through a two-dimensional Gaussian quadrature.

Figure 7.10 displays σ_t for the VIX future expiring on December 21, 2011, for dates t ranging from June 16, 2011 to the future's expiry.

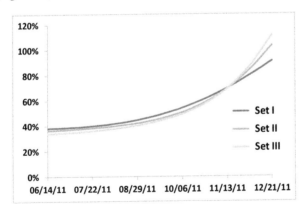

Figure 7.10: σ_t – as defined in (7.92) – for VIX future VXZ1, for dates ranging from June 14 to its expiry date, in Sets I, II, III.

While the integrated value of σ_t^2 is identical in Sets I, II, III, the distribution is different.[35]

---

[35]By definition, σ_t^2 is the expected instantaneous realized variance of F_t^i. The integral of σ_t^2 over $[0, T_i]$ is thus the implied variance of the log contract payoff: $-2\ln(F_{T_i}^i)$, which only depends on the smile of $F_{T_i}^i$ – the VIX smile for maturity T_i.

Controlling the term structure of the instantaneous volatilities of VIX futures is then the criterion for choosing one particular set for $\nu, \theta, k_1, k_2, \rho$.

While curves in Figure 7.10 are different, they are not terribly different; the reader may think that the range of volatility distributions of forward volatilities spanned by the two-factor model is limited.

That this is not the case is illustrated in the right-hand graph of Figure 7.13 below, where we use a one-factor model for simplicity and vary k_1 – still maintaining calibration to VIX smiles. Notice how different the volatility distributions are.

In Figure 7.10 parameters of Sets I, II, III have been used. They have the property that they generate approximately the same volatilities for spot-starting volatilities of all maturities. This additional constraint accounts for the narrower range of volatility distributions in Figure 7.10 than in Figure 7.13.

Characterizing the dynamics in the model

What about the dynamics of VIX futures in our model? The variance curve observed at time t is a function of $\left(X_t^1, X_t^2\right)$, however each instantaneous forward variance ξ^T is a *function* of x_t^T. We have a one-factor Markov-functional model for each *instantaneous* forward variance ξ^T – i.e. a local volatility model[36] – whose local volatility function is given by (7.88).

Processes x_t^T for $T \in [T_i, \ T_i + \Delta]$ are different linear combinations of X_t^1, X_t^2, thus, literally F_t^i is a function of $\left(X_t^1, X_t^2\right)$; however since typically $\Delta \ll \frac{1}{k_1}, \frac{1}{k_2}$, F_t^i can practically be considered a function of the single quantity $x_t^{T_i}$. We thus have essentially a multi-asset local volatility model for VIX futures.

7.7.2 Simulating VIX futures in the two-factor model

In the parametric model specified by (7.89), page 264, continuous forward variances ξ_t^T are modeled as a function of x_t^T. Forward variances $\widehat{\sigma}_{\mathrm{VIX},T_i}^2(t)$ are explicitly known for all t.

Consider VIX future F^i. At $t = T_i$, $F_{T_i}^i = \sqrt{\widehat{\sigma}_{\mathrm{VIX},T_i}^2(T_i)}$: the values of VIX futures *at their settlement dates* are readily available – this is sufficient for pricing VIX options.

Consider however a path-dependent payoff that depends on F_t^i for $t < T_i$, for example an option on a VIX ETF or ETN – see Section 7.7.3 below. Pricing such an option requires simulation of F_t^i at dates $t < T_i$. In the continuous forward variance model specified by (7.89), $F_{t=T_i}^i$ is a function of $X_{T_i}^1$ and $X_{T_i}^2$ given by (7.84). F_t^i is given by:

$$F_t^i = E_t\left[F_{T_i}^i\left(X_{T_i}^1, X_{T_i}^2\right)\right] \tag{7.93}$$

---

[36]See Section 2.10.

While $\hat{\sigma}^2_{\text{VIX},T_i}(t)$ is readily available, F_t^i for $t < T_i$ has to be computed by two-dimensional quadrature on $X_{T_i}^1, X_{T_i}^2$.[37]

In case F_t^i is needed for many dates – say on a daily basis – it is preferable to use a discrete forward variance model of the type discussed in Section 7.8.2 below. In these models VIX futures – rather than forward variances – are modeled directly.

7.7.3 Options on VIX ETFs/ETNs

VIX ETFs or ETNs typically maintain a rolling position in VIX futures.[38] Denoting by X_t the value of the ETF:

$$\frac{dX_t}{X_t} = rdt + \sum_{i,\, T_i > t} w_t^i \frac{dF_t^i}{F_t^i} \tag{7.94}$$

where the ETF's allocation strategy is expressed in the weight w_t^i. The VXX, one of the most popular ETNs, maintains a long position in the first and second nearby futures, so that the weighted duration of both futures is approximately 30 days. It would be most natural to set $w_t^i \equiv w(T_i - t)$ with $w(\tau)$ given by:

$$\begin{cases} w(\tau) = \frac{\tau}{\Delta} & \tau \in [0, \Delta] \\ w(\tau) = 2 - \frac{\tau}{\Delta} & \tau \in [\Delta, 2\Delta] \\ w(\tau) = 0 & \tau > 2\Delta \end{cases} \tag{7.95}$$

where Δ is the interval between two VIX futures' expiries. $w(\tau)$ appears in Figure 7.11. This allocation strategy results in the following dynamics for the VXX:

$$\frac{dX_t}{X_t} = rdt + \left[w(T^{1\text{st}} - t)\frac{dF_t^{1\text{st}}}{F_t^{1\text{st}}} + \left(1 - w(T^{1\text{st}} - t)\right)\frac{dF_t^{2\text{nd}}}{F_t^{2\text{nd}}} \right]$$

where $T^{1\text{st}}(t)$ is the expiry of the first nearby future $F_t^{1\text{st}}$, and $T^{2\text{nd}}(t)$ that of the following future.

The VXX prospectus states that it is the *number* of futures that is proportional to τ and $\Delta - \tau$, rather than the *notional* invested in both futures:[39]

$$w_{1\text{st}} = \frac{(T^{1\text{st}} - t)\, F_t^{1\text{st}}}{(T^{1\text{st}} - t)\, F_t^{1\text{st}} + (\Delta - (T^{1\text{st}} - t))\, F_t^{2\text{nd}}}$$

$$w_{2\text{nd}} = 1 - w_{1\text{st}}$$

---

[37]This is achieved efficiently by first finding the linear combination of $X_{T_i}^1$ and $X_{T_i}^2$ that accounts for the bulk of the variance of $F_{T_i}^i$. The resulting quadrature is then almost one-dimensional, especially since Δ is small. The chosen algorithm should ensure that F_t^i is a martingale so that self-financing strategies that invest in VIX futures – such as VIX ETNs – have the correct forward.

[38]ETF stands for "exchange traded fund" – it is a fund whose shares trade much like stocks. ETN stands for "exchange traded note". It is very similar to an ETF except there are no segregated assets backing the ETN: the holder of an ETN bears the credit risk of the note's issuer. In theory (7.94) the drift of X_t should be supplemented with the credit spread of the issuer. Market appetite for borrowing the ETN – this is the case for the VXX – may however be such that the repo is large enough that it more than offsets the credit spread.

[39]See the prospectus of the VXX ETN at *www.ipathetn.com/static/pdf/vix-prospectus.pdf*

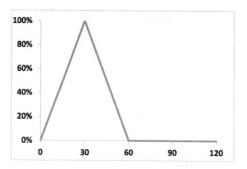

Figure 7.11: $w(\tau)$ for the VXX ETN, as a function of τ (days). We have made the simple assumption that VIX expiries are spaced 30 days apart.

If $F_t^{1st} = F_t^{2nd}$ the weights become identical to those given by expression (7.95) for $w(\tau)$, which is the convention we use in what follows for the sake of simplicity.

The VXX smiles in Figure 7.13 below are computed in a Monte Carlo simulation that uses the proper convention – using either convention generates very similar prices, unless the term structure of VIX futures is unreasonably steep.

Consider now options on the VXX, which are listed. Can VXX options be priced off VIX smiles?

VIX implied volatilities quantify the realized volatility of VIX futures up to their expiry date. In contrast, SDE (7.94) for the VXX and the profile of $w(\tau)$ in Figure 7.11 show that returns of the VXX are a weighted average of returns of two VIX futures that always have less than two months to expiry. Thus two natural questions arise:

- Assuming we are calibrated to VIX smiles, what is the impact of the distribution of the volatility of VIX futures?

- What is the impact of correlation between VIX futures?

Volatilities of VIX futures

In what follows we use the parametric model of Section 7.8.2 where VIX futures are modeled directly. We wish to assess the effect of different distributions of the realized volatility of each future throughout its life. To this end we use a one-factor model ($\theta = 0$) so that the instantaneous correlation of VIX futures is constant, equal to 100%, and vary the value of k_1 while remaining calibrated to VIX smiles.

We use VIX market data as of June 8, 2012; the calibrated smiles along with the levels of VIX futures appear in Figure 7.12.[40]

---

[40]Note that, as we vary k_1, calibration to VIX smiles remains identical. Indeed, as we vary k_1, the variance of $x_{T_i}^{T_i}$ changes. However, calibration makes up for this through a change in ζ_i so that the variance of $w_i x_{T_i}^{T_i}$ is unchanged. From (7.104), $F_{T_i}^i$ is a function of the Gaussian variable $w_i x_{T_i}^{T_i}$: if its variance is unchanged as we vary k_1, so is the density of $F_{T_i}^i$. As a result, calibrated values of γ_i, β_i do not depend on k_1 and the VIX smiles generated by the model do not depend on k_1 either.

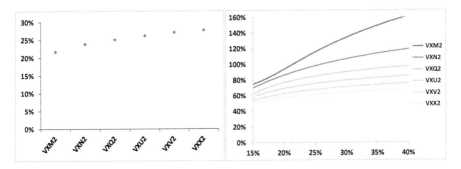

Figure 7.12: Left: VIX futures as of June 8, 2012 for expiries ranging from June 20 (VXM2) to November 21 (VXX2). Right: Smiles of VIX futures.

Figure 7.13 shows the VXX smiles generated by different values of k_1 for the listed maturity of December 21, 2012 – together with the market smile.[41] It also shows the instantaneous volatility of VIX future VXV2, which expires on October 17, 2012. This instantaneous volatility, σ_t, is obtained as the square root of the expectation of the instantaneous variance: $\sigma_t^2 = E[\left(\frac{dF_t^i}{F_t^i}\right)^2]$. This expectation is easily computed by Gaussian quadrature on $x_t^{T_i}$.

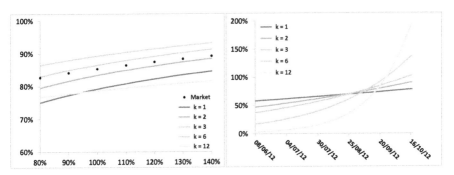

Figure 7.13: Left: VXX smiles as of June 8, 2012 for the December 21st maturity, for different values of k_1, compared to the market smile. Right: instantaneous volatility of the VIX future expiring on October 17, 2012 (VXV2) for different values of k_1.

Consider the case $k_1 = 1$. For this (low) value of k_1 the instantaneous volatilities of VIX futures have little term structure and are distributed more or less evenly over their lives (see Figure 7.13 for the case of the VXV2 future, whose ATM im-

---

[41]Listed options on the VXX are American. However, because of the low level of interest rates, the fact that the VXX pays no dividends and the high volatility of the VXX, they can in practice be considered as European for a wide range of strikes around the money.

plied volatility is 70%) with no particular concentration on the final two months before expiry. The resulting VXX implied volatilities are lower than market implied volatilities, which is not surprising.

Conversely, consider the case $k_1 = 12$. As the graph in the right-hand side of Figure 7.13 shows, the volatility of VIX futures is now concentrated right before expiry. However, as is clear from Figure 7.11, this is where their weight in the VXX vanishes: again we expect low VXX implied volatilities, which the left-hand graph in Figure 7.13 confirms.

One can see that the highest implied volatilities are obtained for $k_1 \simeq 6$, with $k_1 = 2$, $k_1 = 3$ generating VXX smiles that most closely approximate the market VXX smile.[42]

Correlations of VIX futures

We have thus far used 100% correlation for simplicity. The realized correlation of the first two nearby futures typically ranges from 85% to 100%. How does correlation impact VXX implied volatilities?

Let us make the simple assumption that the volatilities of the two nearby futures F^i and F^{i+1} are identical and constant, equal to σ, and that their correlation is constant, equal to ρ. The instantaneous volatility of the VXX, $\sigma_X(t)$, is given for $t \in [T_{i-1}, T_i]$ by:

$$\sigma_X^2(t) = \sigma^2 \left(w^2(T_i - t) + w^2(T_{i+1} - t) + 2\rho w(T_i - t)w(T_{i+1} - t) \right)$$

where $w(\tau)$ appears in Figure 7.11. If weights $w(T_i - t)$ and $w(T_{i+1} - t)$ were constant, equal to $\frac{1}{2}$, we would get the standard formula for the volatility of an equally weighted basket: $\sigma_X^2 = \frac{1+\rho}{2}\sigma^2$.

In our case the product $w(T_i - t)w(T_{i+1} - t)$ vanishes both for $t = T_{i-1}$ and for $t = T_i$: the effect of the cross term will be comparatively smaller. Integrating $\sigma_X^2(t)$ over $[T_{i-1}, T_i]$:

$$\widehat{\sigma}_X^2 = \frac{1}{\Delta} \int_{T_{i-1}}^{T_i} \sigma_X^2(t)dt = \frac{2+\rho}{3}\sigma^2$$

Consider the two cases $\rho_{\min} = 80\%$, $\rho_{\max} = 100\%$. The ratio of implied volatilities $\frac{\widehat{\sigma}_X^{\max}}{\widehat{\sigma}_X^{\min}}$ is then equal to $\sqrt{\frac{2+\rho_{\max}}{2+\rho_{\min}}} = 103.5\%$: a difference of correlation of 20 points gives rise to a variation of VXX implied volatilities of a few points only.

Conclusion

In conclusion, VIX smiles do not provide enough information for confining prices of options on VIX ETF(N)s sufficiently. The example of the VXX demonstrates that, even though consistency with VIX smiles is enforced, implied volatilities of the VXX are very dependent on assumptions about how the volatility of each VIX future

---

[42]For $k_1 = 3$ the volatility-of-volatility adjustment factors ζ_i become (serendipitously) essentially identical for all VIX futures, thus making the model time-homogeneous.

is distributed throughout its life, and also depend moderately on the correlation structure of VIX futures.

Thus VXX options are not redundant instruments – they supply information on the volatility distribution of VIX futures.[43]

7.7.4 Consistency of S&P 500 and VIX smiles

From market prices of VIX futures and options we can derive forward log-contract volatilities for the S&P 500 index. The square of the forward volatility for time interval $[T_i, T_i + \Delta]$ is given by:

$$\widehat{\sigma}^2_{T_i, T_{i+\Delta}}(t) = E_t\left[\left(F^i_{T_i}\right)^2\right] \tag{7.96}$$

We know from Section 3.1.3 that any European payoff on $F^i_{T_i}$ can be replicated by a static position consisting of cash, forwards (or futures) and vanilla options on $F^i_{T_i}$. The decomposition in (3.6) applied to the function $f(S) = S^2$ reads:

$$S^2 = S^2_* + 2S_* (S - S_*) + 2 \int_0^{S_*} (K - S)^+ dK + 2 \int_{S_*}^\infty (S - K)^+ dK \tag{7.97}$$

where S_* is arbitrary. We now apply this identity to $S = F^i_{T_i}$ with $S_* = F^i_t$ and translate this equality of payoffs in an equality of prices.

Adding up the (undiscounted) prices of the different components in the right-hand side of (7.97) yields the following consistency condition relating S&P 500 forward volatilities to market prices of VIX futures and options:

$$\widehat{\sigma}^2_{T_i, T_{i+\Delta}}(t) = \left(F^i_t\right)^2 + 2 \int_0^{F^i_t} \mathcal{P}^{Ki}_t dK + 2 \int_{F^i_t}^\infty \mathcal{C}^{Ki}_t dK \tag{7.98}$$

where $\mathcal{P}^{Ki}_t$ (resp. $\mathcal{C}^{Ki}_t$) are undiscounted market prices of put (resp. call) options on the VIX of maturity T_i. We have used the fact that the price of a payoff linear in $\left(F^i_T - F^i_t\right)$ – the second piece in (7.97) – vanishes. In contrast to the replication of the log contract, the densities of calls and puts on the VIX are constant, so that $\widehat{\sigma}^2_{T_i, T_{i+\Delta}}(t)$ has little dependence on the exact cutoff used in the integrals in the above expression.

Log-contract versus VS volatility

Because of the definition of the settlement value of VIX futures, $\widehat{\sigma}_{T_i, T_{i+\Delta}}$ as defined in (7.96) is a log-contract volatility. (7.98) thus expresses an identity between market prices of VIX instruments and of log-contracts, which, unlike VSs, do not trade.

---

[43]Note the similarity with the issue of pricing interest rate swaptions in a model calibrated on LIBOR caps/floors.

We will thus assume that the difference between $\widehat{\sigma}_T$ and $\widehat{\sigma}_{\mathrm{VS},T}$ is small, so that we can use VSs *in lieu* of log contracts.[44]

Figure 7.14 shows F_t^i, $\widehat{\sigma}_{T_i,T_{i+\Delta}}(t)$ as given by replication on the VIX market through (7.98) and $\widehat{\sigma}_{T_i,T_{i+\Delta}}(t)$ as generated by interpolation of market quotes of S&P 500 VS implied volatilities observed on January 14, 2011.

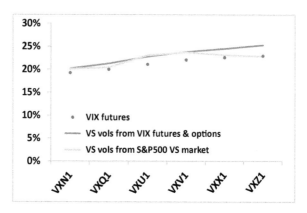

Figure 7.14: VIX futures, VS volatilities as generated by (7.98) and VS volatilities as derived from the S&P 500 VS market, all in 365 convention.

The contribution from the time values of VIX calls and puts in (7.98) is of the order of one to two points of volatility. Notice how VS volatilities derived either from the S&P 500 VS market or from the VIX market do not coincide. This suggests an arbitrage strategy: imagine that a particular forward VS volatility as derived from the VIX market through (7.98) lies higher than its counterpart derived from the regular S&P 500 VS market. We sell VIX futures and options in the proportions expressed by (7.98) and buy a *forward* $[T_i, T_i + \Delta]$ VS. At $t = T_i$, upon settlement of VIX futures and options we unwind the regular VS position – which by then is a spot-starting VS – at an implied volatility equal to the settlement value of the expiring VIX future: the P&L of this strategy is $\widehat{\sigma}^2_{T_i,T_{i+\Delta}}(t)_{\mathrm{VIX\ mkt}} - \widehat{\sigma}^2_{T_i,T_{i+\Delta}}(t)_{\mathrm{VS\ mkt}}$. Practically, however, setting up this strategy is not as straightforward:

- Short-maturity forward variance swaps on the S&P 500 index are not liquid – they are built by combining a long position in a VS of maturity $T_i + \Delta$ with a short position in a VS of maturity T_i. Even though we may be charged a bid/offer spread on one leg only, the resulting spread for the $[T_i, T_i + \Delta]$ *forward* VS volatility will be sizeable. Moreover, S&P 500 VS contracts only trade for maturities corresponding to the expiries of listed S&P 500 options –

---

[44]As discussed in Section 5.5, there are valid reasons why $\widehat{\sigma}_T$ and $\widehat{\sigma}_{\mathrm{VS},T}$ could be different. Typically, because sellers of VSs would lose on large drawdowns of the underlying index, we expect that $\widehat{\sigma}_{\mathrm{VS},T} > \widehat{\sigma}_T$. It so happens that usually – see below – equivalent log-contract volatilities derived from the VIX market lie higher than $\widehat{\sigma}_{\mathrm{VS},T}$. In case $\widehat{\sigma}_{\mathrm{VS},T} > \widehat{\sigma}_T$, the discrepancy with the S&P 500 market is even stronger.

the third Friday of each month. While $T_i + \Delta$ is an S&P 500 expiry by definition of the VIX index, T_i is not as it falls 30 days before, on a Wednesday. Forward VS volatilities that can practically be traded on the VS market are either two days shorter or five days longer than the corresponding VS volatilities synthesized from the VIX market. Considering that a 30-day interval comprises approximately 20 returns, carrying an open gamma position on two – or three[45] – squared returns entails appreciable risk. It is then preferable to group together packages corresponding to several adjacent VIX futures – say three – so that bid/offer costs and risks are mitigated.

- VIX options are only available for discrete strikes. The continuous portfolio of VIX call and put options in (7.98) is in practice replaced by a discrete portfolio. Quantities of VIX futures and options are then determined so as to achieve the most expensive sub-replication of the $\left(F^i_{T_i}\right)^2$ payoff – when the forward VS volatility derived from the VIX market is higher than that of the S&P 500 VS market – or the cheapest super-replication of $\left(F^i_{T_i}\right)^2$ in the opposite case.[46]

- At $t = T_i$ we need to unwind the forward VS position at an implied VS volatility that is exactly equal to $\hat{\sigma}_{\mathrm{VIX},T_i}$. This is achieved by selling vanilla options – which will subsequently be delta-hedged until $T_i + \Delta$ – on the S&P 500 in exactly the same quantities and for the same prices used in the calculation of the settlement of F^i by the CBOE. This is possible as VIX futures settle at the open of the S&P 500 options' market: orders can be placed and are executed at the open at the same prices that are used for the calculation of $F^i_{T_i}$. Still we know from Section 5.3.7 that the package consisting of a VS together with its offsetting vanilla replication is not risk-free. It is thus best to unwind the arbitrage position before T_i, should an opportunity arise.

Violations of (7.98) can then only be arbitraged by S&P 500 volatility market makers.

In the author's experience, arbitrage opportunities that used to arise involved, most often than not, VIX-synthesized VS volatilities lying higher than their S&P 500 VS counterparts. As of the time of writing, these opportunities seem to hardly occur anymore.

Are there other structural connections between S&P 500 and VIX smiles that could be practically arbitraged when violated? For examples are implied volatilities of (a) S&P 500 put options, (b) VIX call options related?

In [36] Stefano de Marco and Pierre Henry-Labordère consider the sub- and super-replication of VIX options using VIX futures, the S&P 500 index and S&P 500 options. They derive optimality conditions, which can be solved numerically,

---

[45]There are three business days from Friday to Wednesday.

[46]The most expensive sub-replicating and cheapest super-replicating portfolios are determined with the simplex algorithm. In the latter case, because of the convexity of the parabola, super-replication only holds for a limited range of values of $F^i_{T_i}$.

and also obtain analytical non-optimal upper and lower bounds; with the upper bound optimal under a condition involving the S&P 500 smiles for maturities T_i and T_{i+1}. These bounds are widely spaced, but more interestingly it does not seem that there is much more information to be extracted from the S&P 500 smile, other than log-contract implied volatilities.

See also reference [77], where Andrew Papanicolaou derives a bound on the moment-generating function of a squared VIX future from prices of moments of the S&P 500 index.

7.7.5 Correlation structure of VIX futures

How do correlations of VIX futures generated by the two-factor model compare with those observed in reality?

Consider the first 7 futures $F_t^i, i = 1 \ldots 7$. Figure 7.15 shows instantaneous correlations $\rho(F_t^1, F_t^i)$ (left-hand graph) and $\rho(F_t^i, F_t^{i+1})$ (right-hand graph) in the two-factor model, for an observation date t that lies 15 days before the expiry of the first VIX future; this represents an "average" correlation level. The dots correspond to historical correlations evaluated from February 16, 2010 to February 15, 2012 – from which we have excluded roll dates.

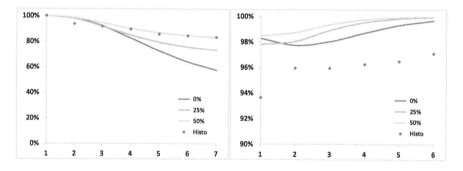

Figure 7.15: Left: correlations $\rho(F_t^1, F_t^i)$ of the first 7 VIX futures with the first future in the two-factor model for different values of the factor/factor correlation, and as observed in reality (dots), as a function of i. Right: correlations $\rho(F_t^i, F_t^{i+1})$ of contiguous futures.

We have used three parameter sets for the two-factor model, all calibrated to the benchmark form (7.40) for $\nu_T^B(t)$, page 228, with $\sigma_0 = 100\%$, $\tau_0 = 3$ months, $\alpha = 0.4$, characterized by different levels of correlation ρ_{12} between the two factors: $\rho_{12} = 0\%, 25\%, 50\%$. The set with $\rho = 0\%$ is Set II in Table 7.1.

It is apparent that correlations of the first future with other futures are acceptably captured with $\rho_{12} = 50\%$. The right-hand side graph highlights however that correlations of adjacent futures are then systematically higher in the two-factor model than in reality. In the two-factor model, VIX futures with long expiries have almost 100% correlation.

This is due to the fact, already pointed out in Section 7.4.2, page 229, that in the two-factor model correlations between forward variances involve one time scale only: $\frac{1}{k_1 - k_2}$.

We cannot realistically expect to achieve a good fit of historical correlations of VIX futures by employing one single time scale. Regaining some flexibility with respect to the correlation structure would be a valid motivation for introducing a third factor in the model.

7.7.6 Impact on smiles of options on realized variance

Consider an option on realized variance starting on June 14, 2011 and maturing in January 20, 2012, that is 30 days after the expiry of the VIXZ1 future, and let us price it using Set II parameters and the corresponding volatility-of-volatility smile parameters $\gamma^i, \beta^i, \zeta^i$ in Figure 7.9.

Figure 7.16 shows the implied volatility of this option as a function of *volatility moneyness*, computed by inverting the Black-Scholes formula with the (driftless) underlying equal to the square of the VS volatility for the option's maturity. These results are generated by a Monte Carlo simulation of the two-factor model with conditionally Gaussian returns for $\ln S$. Two other sets of implied volatilities are plotted as well – see caption.

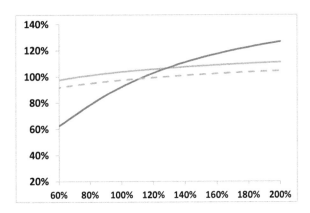

Figure 7.16: Implied volatilities of options on realized variance. The inception date is June 14, 2011 and the maturity is January 20, 2012. Implied volatilities of the realized variance are shown as a function of *volatility* moneyness using Set II parameters and three configurations for $\gamma^i, \beta^i, \zeta^i$: dark line: values in Figure 7.9; light line: same but setting $\gamma^i = 0$; dotted line: setting $\gamma^i = 0, \zeta^i = 1$.

Comparison of the dark and light curves in Figure 7.16 shows that the impact of the smile of forward variances is by no means small. The continuous light curve shows the implied volatilities as generated by Set II parameters with $\gamma^i = 0$: it incorporates the volatility-of-volatility adjustment factors ζ^i as derived from calibration on the VIX smiles. The dotted line is obtained using Set II parameters only: $\gamma^i = 0, \zeta^i = 1$. Comparison of the light continuous and dotted lines indeed confirms that ζ is a simple adjustment factor for volatilities of forward variances: implied volatilities are altered approximately uniformly.

While Figure 7.16 is an indication of the effect of the smile of forward variances on the price of an option on realized variance, the size of this effect is highly model-dependent.

Calibration on the smiles of VIX options sets the value of the expected integrated variance of VIX futures, $\int_t^{T_i} \sigma_u^2 du$, but not its distribution. Furthermore, the correlation of VIX futures is left undetermined, yet it is instrumental in determining the volatility of VS volatilities for the option's maturity, which is the main determinant of its price.

7.7.7   Impact on the vanilla smile

The effect of VIX smiles on the vanilla smile of the underlying itself is covered in Section 8.10.

7.8   Discrete forward variance models

So far in this chapter we have modeled instantaneous forward variances ξ_t^T. We parametrize the two-factor model so that it generates:

- the desired term structure of volatilities of volatilities,

- the desired term structure of ATMF skew – this is discussed in Chapter 8 – or the desired level of future ATMF skew, if we have forward-start options in mind.

As we vary volatility-of-volatility parameter ν, we alter volatilities of volatilities in the model, but also the ATMF skew of the vanilla smile, and also future ATMF skews.

Likewise, changing spot/volatility correlations alters both the ATMF skew of the vanilla smile, and the skew of forward-starting options. The fact that we do not have independent handles on:

- the vanilla smile,

- the smile of forward-start options,

- volatilities of volatilities,

is a typical shortcoming of continuous forward variance models, a particularly worrisome one when one needs to risk-manage complex path-dependent options that are subject to these three types of risk.

In what follows, we present models that allow for separation of these risks and permit an assessment of their individual impact on exotic option prices; they were first presented in [9].[47]

We refer the reader to the discussion of the risks of forward-start options in Section 3.1.6 of Chapter 3, page 111, if she/he has not read it yet.

Most exotic payoffs that have forward-smile risk involve returns for a set time scale, for example monthly or quarterly returns. One typical example is an accumulator: take 12 monthly returns of the S&P 500 index, cap them individually at, say 3%, then sum them up and pay this sum, floored at zero, as an annual coupon.

A discrete forward variance model is tied to a particular schedule. The latter is defined in the term sheet of the exotic option at hand, thus dates T_i in the schedule are typically uniformly spaced, say by a month or a quarter. Nonetheless, we assume an arbitrary schedule in what follows.

Our aim is to separately control the future smiles over each individual time interval $[T_i, T_{i+1}]$ and also the vanilla smile, in addition to the term structure of volatilities of volatilities.

The model is built in two stages:

- first define a dynamics for discrete forward variances over intervals $[T_i, T_{i+1}]$,

- then specify a dynamics for S_t over each interval.

A benefit of discrete models is that VIX futures can be modeled directly, rather than forward variances – see Section 7.8.2 below. Our first step can be replaced with:

- first define a dynamics for VIX futures.

7.8.1 Modeling discrete forward variances

Let ξ_t^i be the discrete forward VS variance for interval $[T_i, T_{i+1}]$. It is similar to the continuous forward variances employed so far, except it is defined as a finite difference rather than a derivative:

$$\xi_t^i \equiv \widehat{\sigma}_{T_i,T_{i+1}}^2(t) = \frac{(T_{i+1} - t)\widehat{\sigma}_{T_{i+1}}^2(t) - (T_i - t)\widehat{\sigma}_{T_i}^2(t)}{T_{i+1} - T_i}$$

---

[47]There cannot be complete disconnection between the vanilla smile and future smiles. Consider vanilla smiles for maturities T_1 and $T_2 > T_1$ and the corresponding densities $\rho_1(S_1)$, $\rho(S_2)$. Given ρ_1 and ρ_2, the transition density $\rho_{12}(S_2|S_1, \bullet)$, which determines future smiles generated by the model – where $\bullet$ stands for state variables other than S – cannot be chosen arbitrarily. It has to comply with the Chapman-Kolmogorov condition:

$$\rho_2(S_2) = \int E[\rho_{12}(S_2|S_1, \bullet)|S_1]\rho_1(S_1)\,dS_1, \quad \forall S_2$$

Still, the presence of the $\bullet$ state variables (X_t^1, X_t^2, in the two-factor model) affords considerable freedom in selecting ρ_{12}. We refer the reader to Section 3.1.7 of Chapter 3, page 113, for examples of how loosely cliquet prices are constrained by the vanilla smile.

where $\hat{\sigma}_T(t)$ is the VS volatility for maturity T, and $\hat{\sigma}_{T_i,T_{i+1}}(t)$ the forward VS volatility for interval $[T_i, T_{i+1}]$, observed at t. In the diffusive models we work with, implied volatilities of VSs and log-contracts are identical, thus $\hat{\sigma}_{T_i,T_{i+1}}(t)$ is also the implied volatility at t of the payoff that pays $\ln(\frac{S_{T_{i+1}}}{S_{T_i}})$ at T_{i+1}.

As with continuous variance models, just because we use forward variances as basic building blocks does not mean we necessarily use VSs as hedge instruments. Our model can be calibrated to a term structure of implied volatilities for a given moneyness for maturities T_i – for example ATMF volatilities.

The corresponding vanilla options are then our hedge instruments, along with the spot, and the carry P&L of a hedged position is of the genuine gamma/theta form. We refer the reader to the discussion in Section 7.5 in the context of continuous models, whose conclusions apply to discrete models as well.

Just as their continuous counterparts, the ξ_T^i are driftless.[48] We can thus recycle the two-factor model and, mimicking (7.28), write the SDE of ξ_t^i as:

$$d\xi_t^i = (2\nu_i)\xi_t^i \, \alpha_{\theta_i} \left((1-\theta_i) \, e^{-k_1(T_i-t)} dW_t^1 + \theta_i e^{-k_2(T_i-t)} dW_t^2 \right) \qquad (7.99)$$

$$\alpha_{\theta_i} = 1/\sqrt{(1-\theta_i)^2 + \theta_i^2 + 2\rho_{12}\theta_i(1-\theta_i)}$$

where index i for parameters θ and ν keeps track of the forward variance ξ^i they apply to.

While θ and ν depend on i, we use the same values for k_1, k_2, ρ_{12} for all intervals – otherwise we lose the two-dimensional Markov representation of the ξ_t^i. We also use the same values as in the continuous model, so that discrete and continuous versions of the model can be mapped onto another – see below.

The solution of (7.99) reads:

$$\xi_t^i = \xi_0^i e^{\omega_i x_t^{T_i} - \frac{\omega_i^2}{2}\chi(t,T_i)} \qquad (7.100)$$

$$x_t^{T_i} = \alpha_{\theta_i} \left[(1-\theta_i) \, e^{-k_1(T_i-t)} X_t^1 + \theta_i e^{-k_2(T_i-t)} X_t^2 \right]$$

with $\omega_i = 2\nu_i$. The driftless processes $x_t^{T_i}$ are defined in (7.30) and $\chi(t,T)$ is defined in (7.35), page 227.

Mapping a continuous to a discrete model

The spacing between two successive dates T_i, T_{i+1} is specific to each exotic payoff – it is different for different payoffs. Still, risks of the same nature should be priced at the same level accross the book, for example volatility-of-volatility risk. We thus need to parametrize our discrete model so that some model features remain unchanged with respect to its continuous counterpart.

---

[48]The exposure to ξ^i can be delta-hedged by going long T_{i+1} VSs of maturity T_{i+1} and short T_i VSs of maturity T_i, with no cash borrowing or lending involved.

Consider the forward volatility $\hat{\sigma}_{T_i,T_{i+1}}(t)$. It is given, in the discrete and continuous model, respectively, by:

$$\begin{cases} \hat{\sigma}_{T_i,T_{i+1}}(t) = \sqrt{\xi_t^i} \\ \hat{\sigma}_{T_i,T_{i+1}}(t) = \sqrt{\frac{1}{T_{i+1}-T_i} \int_{T_i}^{T_{i+1}} \xi_t^\tau d\tau} \end{cases}$$

The SDE of forward variance $\hat{\sigma}_{T_i,T_{i+1}}^2(t)$ is given, in both models, by:

$$\frac{d\hat{\sigma}_{T_i,T_{i+1}}^2}{\hat{\sigma}_{T_i,T_{i+1}}^2} = 2\nu_i\alpha_{\theta_i}\left((1-\theta_i)e^{-k_1(T_i-t)}dW_t^1 + \theta_i e^{-k_2(T_i-t)}dW_t^2\right) \quad (7.101a)$$

$$\frac{d\hat{\sigma}_{T_i,T_{i+1}}^2}{\hat{\sigma}_{T_i,T_{i+1}}^2} = 2\nu\alpha_\theta\left((1-\theta)A_i^1(t)e^{-k_1(T_i-t)}dW_t^1 + \theta A_i^2(t)e^{-k_2(T_i-t)}dW_t^2\right)$$

$$(7.101b)$$

where $A_i^n(t)$ reads:

$$A_i^n(t) = \frac{\int_{T_i}^{T_{i+1}} \xi_t^\tau e^{-k_n(\tau-T_i)}d\tau}{\int_{T_i}^{T_{i+1}} \xi_t^\tau d\tau}$$

(7.101a) stems from (7.99) directly while (7.101b) is adapted from the corresponding expression (7.36), page 227, for a spot-starting VS volatility.

The two SDEs in (7.101) cannot be identical in both models, for all t, for all configurations of the forward variance curve ξ_t^τ, but let us demand that they coincide for all t, for forward variances ξ_t^τ equal to their initial values ξ_0^τ, that is with A_i^n equal to $A_i^n(0)$.

A quick glance at the right-hand sides of (7.101a) and (7.101b) shows this is possible only if k_1, k_2, ρ are identical in both models, where ρ is the correlation between W_t^1 and W_t^2.

The conditions on ν_i, θ_i read:

$$\nu_i\alpha_{\theta_i}(1-\theta_i) = \nu\alpha_\theta(1-\theta)A_i^1(0)$$
$$\nu_i\alpha_{\theta_i}\theta_i = \nu\alpha_\theta\theta A_i^2(0)$$

This yields:

$$\begin{cases} \theta_i = \frac{\theta A_i^2(0)}{\theta A_i^2(0)+(1-\theta)A_i^1(0)} \\ \nu_i = \nu\frac{\alpha_\theta\theta}{\alpha_{\theta_i}\theta_i}A_i^2(0) \end{cases} \quad (7.102)$$

Provided θ_i and ν_i are given by (7.102), the dynamics of forward volatilities $\hat{\sigma}_{T_i,T_{i+1}}(t)$ is identical in the discrete and continuous versions of the model – for forward variances ξ_t^τ equal to their initial values ξ_0^τ. Consequently:

• instantaneous volatilities of VS volatilities of maturities T_i

- instantaneous correlations of forward VS volatilities $\widehat{\sigma}_{T_i, T_{i+1}}$ and $\widehat{\sigma}_{T_j T_{j+1}}$

are identical as well, in both models, for forward variances ξ_t^T equal to their initial values ξ_0^T.

Our criterion for mapping consists in requiring that the SDEs in (7.101) match $\forall t$ for $\xi_t^T = \xi_0^T$. We now give an illustration of the fact that, indeed, the dynamics of the ξ^i in both discrete and continuous models is very similar.

An example
Consider the case of a flat term structure of VS volatilities, equal to 20%. In this case

$$A_i^n(0) = \frac{1 - e^{-k_n (T_{i+1} - T_i)}}{k_n (T_{i+1} - T_i)}$$

θ_i and ν_i then only depend on $T_{i+1} - T_i$. They are shown in Figure 7.17.

Figure 7.17 also shows prices of a VS ATM swaption, whose payoff is

$$\frac{1}{2\widehat{\sigma}_{T, T+\Delta}(0)} \left(\widehat{\sigma}_{T, T+\Delta}^2(T) - \widehat{\sigma}_{T, T+\Delta}^2(0) \right)^+$$

for various values of Δ.

The expiry T of the swaption is 1 year, and the discrete model is mapped for each value of Δ using (7.102) with $T_i = T$, $T_{i+1} = T + \Delta$.

Parameters of the continuous model appear in Table 7.7. They are chosen so as to generate a term-structure of VS volatilities that closely fits benchmark (7.51), page 239, with $\alpha = 0.6$, $\tau_0 = 3$ months, $\sigma_0 = 125\%$.

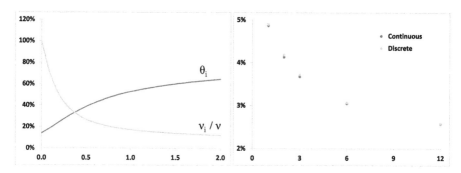

Figure 7.17: Left: θ_i and $\frac{\nu_i}{\nu}$ as a function of Δ (years). Right: prices of a 1-year ATM VS swaption in both continuous and discrete models, as a function of the maturity of the underlying VS volatility (months).

Prices in both discrete and continuous models are very similar, thus illustrating that with mapping (7.102), both models generate very similar dynamics for discrete forward VS variances.

ν	θ	k_1	k_2	ρ_{12}
310%	0.139	8.59	0.47	0%

Table 7.7: Parameters of the continuous two-factor forward variance model.

Smiling discrete forward variances

With the dynamics in (7.99), forward volatilities $\widehat{\sigma}_{T_i,T_{i+1}}$ are lognormal, thus VS swaption smiles are flat.[49] For the sake of generating upward-sloping smiles, we can use the same ansatz as in (7.89), page 264, and replace (7.100) with:

$$\xi_t^i = \xi_0^i \left((1 - \gamma_i) e^{\omega_i x_t^{T_i} - \frac{\omega_i^2}{2} \chi(t,T_i)} + \gamma_i e^{\beta_i \omega_i x_t^{T_i} - \frac{(\beta_i \omega_i)^2}{2} \chi(t,T_i)} \right)$$

with the normalization in (7.90):

$$\omega_i = \frac{2\nu_i}{(1 - \gamma_i) + \beta_i \gamma_i} \zeta_i$$

7.8.2 Direct modeling of VIX futures

VIX instruments are introduced in Section 7.7, in the context of the two-factor (continuous) forward variance model.

VIX futures F_t^i at dates $t < T_i$ are accessible in the continuous model, at the cost of a two-dimensional quadrature – this is explained in Section 7.7.2. For payoffs requiring frequent observations of VIX futures, using a model where VIX futures – rather than variances – are modeled directly is preferable.

We now discuss this particular breed of discrete forward variance models – they are Markov-functional models for VIX futures.

Denote by T_i the expiry of future F_t^i. At expiry, a VIX future is equal to the log-contract implied volatility for maturity $T_i + \Delta$, where $\Delta = 30$ days. Borrowing the notations of Section 7.7.1:

$$F_{T_i}^i = \widehat{\sigma}_{\mathrm{VIX},T_i}(T_i) = \widehat{\sigma}_{T_i,T_i+\Delta}(T_i)$$

VIX futures expire on Wednesdays, 30 days before the expiry of listed S&P 500 options, thus two consecutive VIX expiries are spaced either (a) 28 days or (b) 35 days apart. In case (a), $T_i + \Delta > T_{i+1}$: the forward VS volatilities $\widehat{\sigma}_{T_i,T_i+\Delta}$ and $\widehat{\sigma}_{T_{i+1},T_{i+1}+\Delta}$ that underlie, respectively, futures F^i and F^{i+1} overlap. In case (b), $T_i + \Delta < T_{i+1}$, and there is no overlap. We assume that $T_{i+1} - T_i = \Delta$.[50]

---

[49] In the continuous version of the model, instantaneous forward variances ξ_t^T are lognormal, but discrete forward variances, hence forward VS volatilities, are not. Figure 7.4, page 232, highlights the resulting slight positive skew of VS swaptions.

[50] Practically, in case (b), we set $\widehat{\sigma}_{T_i,T_{i+1}}(T_i) \equiv \widehat{\sigma}_{T_i,T_i+\Delta}(T_i)$, whereas in case (a), when both underlying forward variances overlap, we set: $\widehat{\sigma}_{T_i+\Delta,T_{i+1}+\Delta}(T_{i+1}) \equiv$

$$\sqrt{\frac{\Delta \widehat{\sigma}_{T_{i+1},T_{i+1}+\Delta}^2 (T_{i+1}) - (T_i + \Delta - T_{i+1}) \widehat{\sigma}_{T_i,T_i+\Delta}^2 (T_i)}{T_{i+1} - T_i}}.$$

A parametric form

Since VIX futures are driftless – just like forward variances ξ^T – we can use the discrete two-factor model directly on VIX futures and, mirroring (7.99) write:

$$dF_t^i = \nu_i F_t^i \, \alpha_{\theta_i} \left((1 - \theta_i) \, e^{-k_1(T_i - t)} dW_t^1 + \theta_i e^{-k_2(T_i - t)} dW_t^2 \right) \qquad (7.103)$$

$$\alpha_{\theta_i} = 1 / \sqrt{(1 - \theta_i)^2 + \theta_i^2 + 2\rho_{12}\theta_i (1 - \theta_i)}$$

where we use ν_i instead of $2\nu_i$ since F_t^i is a volatility rather than a variance.

The solution of (7.103) is given by:

$$F_t^i = F_0^i e^{\omega_i x_t^{T_i} - \frac{\omega_i^2}{2} \chi(t,T_i)}$$

$$x_t^{T_i} = \alpha_{\theta_i} \left[(1 - \theta_i) \, e^{-k_1(T_i - t)} X_t^1 + \theta_i e^{-k_2(T_i - t)} X_t^2 \right]$$

with $\omega_i = \nu_i$.

Upward-sloping smiles can be generated by using the same ansatz as in (7.89):

$$F_t^i = F_0^i \left((1 - \gamma_i) \, e^{\omega_i x_t^{T_i} - \frac{\omega_i^2}{2} \chi(t,T_i)} + \gamma_i e^{\beta_i \omega_i x_t^{T_i} - \frac{(\beta_i \omega_i)^2}{2} \chi(t,T_i)} \right) \qquad (7.104)$$

with:

$$\omega_i = \frac{\nu_i}{(1 - \gamma_i) + \beta_i \gamma_i} \zeta_i$$

VIX smiles generated with this parametrization are very similar to those generated by the equivalent parametrization for continuous forward variances in Section 7.7.1.

A non-parametric form

While ansatz (7.104) is adequate for capturing VIX smiles, as a parametric form it only allows for certain types of smile shapes. We now build a model that can be calibrated to arbitrary VIX smiles, as long as they are non-arbitrageable; it is an example of the Markov-functional models discussed in Section 2.10.

Reasoning as in Section 7.7.1, let us write F_t^i as:

$$F_t^i = F_0^i f^i(t, x_t^{T_i}) \qquad (7.105)$$

Note that (7.104) is but a particular form of (7.105), with f^i the sum of two exponentials.

The mapping function $f^i(t, x)$ has to be such that (a) at $t = T_i$, $F_{t=T_i}^i$ is distributed so that the corresponding VIX smile observed at $t = 0$ is recovered, (b) F_t^i is driftless.

Condition (b) implies that f^i obeys PDE (7.87):

$$\frac{df^i}{dt} + \frac{\eta_i^2 (T_i - t)}{2} \frac{d^2 f^i}{dx^2} = 0 \qquad (7.106)$$

where $\eta_i^2 \, (T_i - t)$ is the instantaneous variance of $x_t^{T_i}$. η_i is given in (7.31b):

$$\eta_i \, (u) \;=\; \alpha_{\theta_i} \sqrt{(1 - \theta_i)^2 \, e^{-2k_1 u} + \theta_i^2 e^{-2k_2 u} + 2\rho_{12}\theta_i \, (1 - \theta_i) \, e^{-(k_1 + k_2)u}}$$

Once the terminal profile $f^i \, (T_i, x)$ is specified, solving (7.106) produces $f^i \, (t, x)$ for all $t \leq T_i$.

The terminal condition for f^i, $f^i \, (T_i, x)$ must be such that the mapping $x_{T_i}^{T_i} \rightarrow f^i(T_i, x_{T_i}^{T_i})$ generates the VIX market smile for maturity T_i.

Consider a VIX level K and denote by $\mathcal{D}^K$ the undiscounted market price of a digital option of strike K, maturity T_i that pays 1 if $F_{T_i}^i < K$ and zero otherwise. $\mathcal{D}^K$ is straightforwardly derived from the vanilla smile of future F^i, as a digital is essentially a narrow put spread: $\mathcal{D}^K = \frac{dP^K}{dK}$. We have:

$$\mathcal{D}^K \;=\; P\big(F_{T_i}^i < K\big) \;=\; P\big(f^i(T_i, x_{T_i}^{T_i}) < k\big)$$
$$=\; \mathcal{N}_i\big(f^{i^{-1}}(T_i, k)\big)$$

where k is the moneyness: $k = K/F_0$ and $\mathcal{N}_i$ is the cumulative distribution function of the centered Gaussian random variable $x_{T_i}^{T_i}$, whose variance is known in closed form.

This yields $f^{i^{-1}} \, (T_i, k) = \mathcal{N}_i^{-1} \, (\mathcal{D}^K)$. By choosing a large number of values of moneyness k we determine $f^{i^{-1}}$. Since $\mathcal{D}^K$ is an increasing function of K, so is $f^{i^{-1}}$. $f^{i^{-1}}$ is monotonic thus f^i is well-defined and monotonic as well.[51]

Characterizing the dynamics of VIX futures

What kind of dynamics does (7.105) generate for F_t^i? Making use of (7.106):

$$\frac{dF_t^i}{F_t^i} \;=\; \frac{d \ln f^i}{dx}\big(t, x_t^{T_i}\big) dx_t^{T_i}$$

The instantaneous volatility of F^i is thus given by:

$$\mathrm{vol}(F^i) \;=\; \eta \, (T_i - t) \left| \frac{d \ln f^i}{dx} \right|_{x = f^{i^{-1}}(t, F_t^i)} \tag{7.107}$$

---

[51]Do we have an assurance that $f^i \, (0, 0) = 1$ – so that we get indeed the right initial value for the VIX future? This is equivalent to $E[f^i(T_i, x_{T_i}^{T_i})] = 1$, that is the mappping function f^i ensures the forward of $F_{T_i}^i$ is correctly priced. This is not guaranteed: even though the finite set of digitals $\mathcal{D}^K$ we have used to build f^i is correctly priced, their integral – which is equal to the forward – depends on how we interpolate/extrapolate f^{-1} or f. We thus may need to uniformly rescale f^i to make sure $f^i \, (0, 0) = 1$.

$f^{i^{-1}}$ is well-defined only if f^i is monotonic, which is the case.[52] The right-hand side of (7.107) is a function of F_t^i and t, thus what we have is really a local volatility model for F_t^i.

What about forward VS volatilities?

The benefit of modeling VIX futures directly is that they are readily accessible in a simulation. Forward VS volatilities $\widehat{\sigma}_{T_i, T_i+\Delta}$ are directly accessible at T_i since, by definition of the settlement value of VIX futures:

$$F_{T_i}^i = \widehat{\sigma}_{T_i, T_i+\Delta}(T_i)$$

What if we also need $\widehat{\sigma}_{T_i, T_i+\Delta}(t)$ for $t \leq T_i$?

VIX futures are given by:

$$F_{T_i}^i = F_0^i f^i(t, x_t^{T_i})$$

Moreover, forward *variances* $\widehat{\sigma}_{T_i, T_i+\Delta}^2$ are driftless. We can thus represent $\widehat{\sigma}_{T_i, T_i+\Delta}$ as:

$$\widehat{\sigma}_{T_i, T_i+\Delta}(t) = F_0^i \sqrt{g^i(t, x_t^{T_i})}$$

The terminal condition of g^i is:

$$g^i(T_i, x) = f^i(T_i, x)^2$$

and $g^i(t, x)$ for $t \leq T_i$ is obtained by solving PDE (7.106):

$$\frac{dg^i}{dt} + \frac{\eta_i^2 (T_i - t)}{2} \frac{d^2 g^i}{dx^2} = 0$$

In conclusion, we have a model calibrated to VIX smiles where all VIX futures and the corresponding forward VS volatilities are easily generated. We only need to simulate two Ornstein-Uhlenbeck processes: X_t^1 and X_t^2.[53]

---

[52] The mapping function at T_i, $f^i(T_i, x)$ is monotonic, as it is derived from market prices of digital options of maturity T_i – no-arbitrage requires monotonicity of digital option prices with respect to their strike. Next, (7.106) implies that if $f^i(T_i, x)$ is monotonic, so is $f^i(t, x)$. Indeed, take the derivative of (7.106) with respect to x: $\frac{df^i}{dx}(t, x)$ solves the same PDE as f^i. It can thus be written as an expectation: $\frac{df^i}{dx}(t, x) = E_t[\frac{df^i}{dx}(T_i, x_{T_i}^{T_i}) \mid x_t^{T_i} = x]$. $\frac{df^i}{dx}(T_i, x) \geq 0 \; \forall x$ then implies $\frac{df^i}{dx}(t, x) \geq 0 \; \forall x, \forall t$.

[53] The ease with which we build multi-asset, multi-factor Markov-functional models may look suspicious to readers with a fixed income background. In fixed income, multi-factor Markov-functional models are notoriously difficult to build, because determination of the final mapping function of an underlying – swap or LIBOR rate – involves an annuity ratio that depends on the mapping function of a different, contiguous, asset. In our context, it is as if prices of European options of maturity T_i on VIX future F^i no longer read $E\left[h(f^i(T_i, x_{T_i}^{T_i}))\right]$, but $E\left[h(f^i(T_i, x_{T_i}^{T_i})) f^{i+1}(T_i, x_{T_i}^{T_{i+1}})\right]$. In the multi-factor case $x_t^{T_i}$ and $x_t^{T_{i+1}}$ are different processes, hence the simple calibration procedure outlined in Section 7.8.2 would no longer work. Fortunately, unlike swap or LIBOR rates, forward variances or VIX futures are martingale under the same measure.

7.8.3 A dynamics for S_t

Having specified a dynamics for forward variances, we now define a dynamics for the underlying – how does the former constrain the latter?

As t reaches date T_i the VS volatility for interval $[T_i, T_{i+1}]$, $\widehat{\sigma}_{T_i, T_{i+1}}$ $(t = T_i)$ is known. The dynamics of S_t for $t \in [T_i, T_{i+1}]$ has to comply with this value of $\widehat{\sigma}_{T_i, T_{i+1}}$.

We now specify an SDE for S_t, $t \in [T_i, T_{i+1}]$ that meets the three following requirements:

- The VS, or log-contract, implied volatility at T_i for maturity T_{i+1} is equal to $\widehat{\sigma}_{T_i, T_{i+1}} (T_i)$.

- The probability density of $\frac{S_{T_{i+1}}}{S_{T_i}}$ is independent on S_{T_i}. This is an essential condition for ensuring that future and spot-starting smiles are decoupled. With this provision, prices of cliquets of the form:

$$\sum_i \omega_i f \left(\frac{S_{T_{i+1}}}{S_{T_i}} \right)$$

have no sensitivity to correlations between the Brownian motion driving S_t and those driving forward variances. While these correlations have zero impact on future smiles, they do impact spot-starting vanilla smiles – this is how we decouple spot and future smiles in the model.

- Scenarios of future smiles on $[T_i, T_{i+1}]$, as a function of $\widehat{\sigma}_{T_i, T_{i+1}} (T_i)$, should be set at will.

Consider a path-dependent local volatility dynamics for S_t, $t \in [T_i, T_{i+1}]$, given by:

$$dS_t = (r - q)S_t dt + \sigma^i \left(\frac{S_t}{S_{T_i}} \right) S_t dW_t^S \tag{7.108}$$

where the correlations of W_t^S with W_t^1 and W_t^2 are denoted by ρ_{SX^1} and ρ_{SX^2}. Setting $s_t = \frac{S_t}{S_{T_i}}$, we have:

$$ds_t = (r - q)s_t dt + \sigma^i(s_t) s_t dW_t^S$$
$$s_{T_i} = 1$$

thus the density of $s_{T_{i+1}}$ is indeed independent on S_{T_i}.

We choose the following expression for σ^i:

$$\sigma^i(s) = \sigma_0^i \frac{n^i}{n^i - 1} \frac{(n^i - 1) s^{\beta^i - 1}}{(n^i - 1) + s^{\beta^i - 1}} \tag{7.109}$$

which is parametrized by three numbers: σ_0^i, β^i and n^i.

- σ_0^i is the local volatility for $S = S_{T_i}$.

- β^i controls the skew of maturity T_{i+1}. $\left.\frac{d\sigma^i}{d\ln S}\right|_{S=S_{T_i}} = \sigma_0^i \frac{n^i-1}{n^i}(\beta^i - 1)$. For

  $(\beta^i - 1)$ small, assuming zero interest rate and repo, the ATM skew of maturity T_{i+1} is given, at order one in $(\beta^i - 1)$ by formula (2.50a), page 47:

$$\left.\frac{d\widehat{\sigma}_{K,T_{i+1}}}{d\ln K}\right|_{S_{T_i}} \simeq \left.\frac{1}{2}\frac{d\sigma^i}{d\ln S}\right|_{S_{T_i}} = \frac{\sigma_0^i}{2}\frac{n^i-1}{n^i}(\beta^i - 1) \qquad (7.110)$$

- n^i prevents the divergence of σ for small values of S; the maximum level of volatility is $n^i\sigma_0^i$. In practice, n^i can be used to control other features of the smile, for example the difference between the VS volatility and the ATMF volatility of the smile of maturity T_{i+1}, observed at T_i.

σ_0^i, β^i, n^i can be set at will, as a function of $\widehat{\sigma}_{T_i,T_{i+1}}(T_i)$, provided the log-contract implied volatility at T_i for maturity T_{i+1} is indeed equal to $\widehat{\sigma}_{T_i,T_{i+1}}(T_i)$.

How do we control the future ATMF skew scenarios generated by our model? This is done by expressing how the ATMF skew $\mathcal{S}_{T_{i+1}}(T_i)$ depends on the level of ATMF volatility. Two natural choices are:

- a fixed ATMF skew, irrespective of the level of ATMF volatility: $\mathcal{S}_{T_{i+1}}(T_i) = \mathcal{S}_i$, where we are free to choose the level of the ATMF skew $\mathcal{S}_i$ for each interval $[T_i, T_{i+1}]$.

- a specific dependence of the ATMF skew to the ATMF volatility, for example parametrized by a power-law:

$$\mathcal{S}_{T_{i+1}}(T_i) = \left(\frac{\widehat{\sigma}_{\text{ATMF},T_{i+1}}(T_i)}{\widehat{\sigma}_{\text{ATMF, ref}}^i}\right)^{\gamma^i} \mathcal{S}_{\text{ref}}^i \qquad (7.111)$$

where $\mathcal{S}_{\text{ref}}^i$ and $\widehat{\sigma}_{\text{ATMF, ref}}^i$ are reference levels for ATMF skew and volatility. γ^i, $\mathcal{S}_{\text{ref}}^i$ and $\widehat{\sigma}_{\text{ATMF, ref}}^i$ can be chosen differently for each interval $[T_i, T_{i+1}]$.

Once the type of dependence of $\mathcal{S}_{T_{i+1}}(T_i)$ on $\widehat{\sigma}_{\text{ATMF},T_{i+1}}(T_i)$ is chosen, we need to determine two functions $\sigma_0^i()$, $\beta^i()$ for each interval $[T_i, T_{i+1}]$ such that setting $\sigma_0^i = \sigma_0^i(\widehat{\sigma}_{T_i,T_{i+1}}(T_i))$ and $\beta^i = \beta^i(\widehat{\sigma}_{T_i,T_{i+1}}(T_i))$ produces the desired behavior.

The ability to choose different fixed ATMF skew $\mathcal{S}_i$ or different values of γ^i, $\mathcal{S}_{\text{ref}}^i$ and $\widehat{\sigma}_{\text{ATMF, ref}}^i$ for different intervals $[T_i, T_{i+1}]$ is not superfluous.

It does not make sense to offset the sensistivity to the $[T_i, T_{i+1}]$ future skew with an opposite sensitivity to the future skew for a different interval. Thus, even in the unlikely case that the future skew $\mathcal{S}_i$ implied from market prices of cliquets happens to be constant, we still need to separately calculate and manage the sensitivities to parameters controlling the future skew of each interval $[T_i, T_{i+1}]$ – hence σ_0^i and β^i.

Two examples

Consider an interval $[T_i, T_{i+1}]$. Functions $\sigma_0(\widehat{\sigma}_{T_i,T_{i+1}}(T_i))$ and $\beta(\widehat{\sigma}_{T_i,T_{i+1}}(T_i))$ – we omit the i index to lighten notation – are obtained as follows:

- Select discrete values σ_0^k of σ_0 spanning a sufficiently wide range.

- For each value of σ_0^k find the value β^k of β such that the ATMF skew has the desired value, either constant or specified by (7.111). For each trial value of β, the vanilla smile at T_i for maturity T_{i+1} is obtained by numerically solving the forward equation of the local volatility model (2.7), page 29.

- Numerically solve PDE (5.49), page 173, to generate the VS volatility for maturity T_{i+1}: $\widehat{\sigma}_{VS}^k$.

- Store the couples $(\widehat{\sigma}_{VS}^k, \sigma_0^k)$ and $(\widehat{\sigma}_{VS}^k, \beta^k)$ and proceed to the next value of σ_0.

- Finally, interpolate the discrete couples $(\widehat{\sigma}_{VS}^k, \sigma_0^k)$ to generate the function $\sigma_0(\widehat{\sigma}_{VS})$, and likewise for $\beta(\widehat{\sigma}_{VS})$.

Figure 7.18 shows functions $\sigma_0\,()$ and $\beta\,()$ such that the ATMF skew is fixed, equal to 5%. We use the difference between implied volatilities of the 95% and 105% strikes rather than $\left.\frac{d\widehat{\sigma}_{K,T_{i+1}}}{d\ln K}\right|_{S_{T_i}}$ as a measure of ATMF skew. We have used zero interest rate and repo and a monthly schedule: $T_{i+1} - T_i = 1$ month, and n is set to 3.

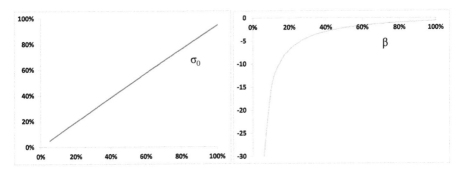

Figure 7.18: σ_0 (left) and β (right) as a function of $\widehat{\sigma}_{T_i,T_{i+1}}(T_i)$, such that the difference of implied volatilities of the 95% and 105% strikes is equal to 5%. We have taken $n_i = 3$ and $T_{i+1} - T_i = 1$ month.

Figure 7.19 shows functions $\sigma_0\,()$ and $\beta\,()$ such that the ATMF skew is of the form in (7.111) with $\gamma = 1$, $\mathcal{S}_{\text{ref}} = 5\%$, $\widehat{\sigma}_{\text{ATMF, ref}} = 20\%$: the ATMF skew is proportional to the ATMF volatility.

The shapes of $\beta(\widehat{\sigma}_{VS})$ are consistent with approximation (7.110) which implies that β is constant for a skew that is proportional to the ATMF volatility while $(\beta - 1)$

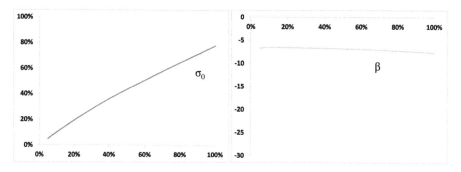

Figure 7.19: σ_0 (left) and β (right) as a function of $\widehat{\sigma}_{T_i,T_{i+1}}(T_i)$, such that the ATMF skew is of the form in (7.111) with $\gamma = 1$, $\mathcal{S}_{\mathrm{ref}} = 5\%$, $\widehat{\sigma}_{\mathrm{ATMF,\ ref}} = 20\%$. We have taken $n_i = 3$ and $T_{i+1} - T_i = 1$ month.

should be inversely proportional to the VS volatility in order to generate a skew that is independent on the level of volatility.[54]

- In the first situation, our discrete forward variance model mimics the two-factor continuous forward variance model, which generates an ATMF skew that is approximately independent on the level of VS volatility. See formula (8.55), page 330, for the ATMF skew at order one in volatility of volatility and Figure 8.4, page 331, for an illustration of the (in)dependence of the ATMF skew on the level of VS volatility.

- In the second situation, with an ATMF skew proportional to the ATMF volatility, we mimic, for short maturities, the behavior of the $\frac{3}{2}$ model. See the corresponding SDE in (8.45), with $\gamma = \frac{3}{2}$, and the short-maturity ATMF skew in (8.49), page 326.

- What if we make the ATMF skew *inversely* proportional to the ATMF volatility, by using (7.111) with $\gamma^i = -1$? We would then be mimicking, for short maturities, the behavior of the Heston model – see expression (6.18a), page 210.

Discrete forward variance models thus afford a great deal of flexibility as to the dependence of the short-maturity future skew to the short future ATMF volatility, while still leaving us the freedom of choosing spot/volatility correlations, which impact spot-starting smiles.

---

[54]For short maturities, approximation (3.28), page 121, shows that the VS implied volatility is equal to σ_0 at order zero in the slope of the local volatility function. For the sake of obtaining the ATMF skew at order one in the slope of the local volatility function, we can replace σ_0 with the VS or ATMF volatility in (7.110).

Two more aspects are worth commenting, before we turn to the vanilla smile.

- What if there are cash-amount dividends? In this case functions $\sigma_0(\xi^i_{T_i})$ and $\beta(\xi^i_{T_i})$ should also depend on S_{T_i}, if they are to ensure (a) that the VS volatility at T_i for maturity T_{i+1} is indeed $\xi^i_{T_i}$, (b) that the ATMF skew at T_i for maturity T_{i+1} still agrees with our specification.

  The case of cash-amount dividends is taken care of economically by (a) making β a function of σ_0, with the dependence obtained in the calibration of the σ_0 and β functions with S_{T_i} set to the forward for maturity T_i, (b) generating the mapping $\sigma_0(\xi^i)$ for a range of values of S_{T_i}, thus obtaining in effect a mapping $\sigma_0(\xi^i, S_{T_i})$. While not exact, step (a) ensures, in practice, that the target forward skew scenarios are obtained with good accuracy.[55]

- While the local volatility function in (7.109) is adequate for generating the desired ATMF skew scenarios, it is not able to generate very large spreads between VS and ATMF volatilities. Also, as t reaches T_i, forward-start options become in effect European options of maturity T_{i+1}; our local volatility function should be such that it is able to match the market smile for maturity T_{i+1} observed at T_i. For these reasons it is a good idea to include an additional quadratic component αs^2.

7.8.4 The vanilla smile

Two mechanisms contribute to the smile of discrete forward models:

- the correlations between the Brownian motions driving S_t and the ξ^i_t

- the local volatility functions $\sigma^i(s)$, which generate the future skews for intervals $[T_i, T_{i+1}]$.

Consider a discrete model with lognormal dynamics (7.99), page 280, for ξ^i and $\sigma^i(s)$ given by (7.109).

Setting $\nu_i = 0$ and $\beta^i = 1, \forall i$ turns our model into a Black-Scholes model with deterministic volatility: on each interval $[T_i, T_{i+1}]$, S_t is lognormal with constant volatility $\sqrt{\xi^i}$.

As $\nu_i \neq 0$ and $(\beta^i - 1) \neq 0$ volatility becomes stochastic. We now derive an expression of the ATMF skew at order one in ν_i and in $(\beta^i - 1)$.

We assume a constant tenor Δ so that our schedule is given by $T_i = i\Delta$. We also use the same values for ν_i, θ_i, α. These numbers then only depend on tenor Δ, through the mapping relationship (7.102), and we thus denote them by $\nu_\Delta, \theta_\Delta, \alpha_{\theta_\Delta}$. Likewise, we omit the i index in σ_0, β.

We use a constant term structure of forward variances: $\xi^i_0 = \hat\sigma^2$, and functions σ_0 and β are chosen so that the ATMF skew of maturity $\mathcal{S}_{T_{i+1}}(T_i)$ does not depend on $\hat\sigma_{T_i,T_{i+1}}(T_i)$ and is equal to $\mathcal{S}_\Delta$.

---

[55]I am indebted to Julien Tijou for developing this enhancement to the original model.

We start from expression (8.29) of vanilla option prices derived in Section 8.4 of Chapter 8 and the resulting expression of the ATMF skew at order one in volatility of volatility (8.32), page 319:

$$\mathcal{S}_T = \frac{1}{2\hat{\sigma}^3 T} \int_0^T \frac{T-t}{T} \frac{\langle d\ln S_t \, d\hat{\sigma}_T^2(t) \rangle_0}{dt} dt \qquad (7.112)$$

where the 0 susbcript means the covariation is evaluated in the unperturbed state, that is with forward variances equal to their values at $t = 0$ and with the instantaneous volatility of S_t read off the initial term structure of VS volatilities. In our context, expanding at order one in volatility of volatility corresponds to expanding at order one both in ν and $(\beta - 1)$.

We calculate $\mathcal{S}_T$ for maturities T that are multiples of tenor Δ: $T = N\Delta$. Using the definition of $\hat{\sigma}_T(t)$, $(T-t)\hat{\sigma}_T^2(t) = \int_t^T \xi_t^\tau d\tau$, where ξ_t^τ is the *instantaneous* forward variance for date τ. (7.112) can be rewritten as:

$$\mathcal{S}_{N\Delta} = \frac{1}{2\hat{\sigma}^3 T^2} \int_0^T \left\langle d\ln S_t \, d\left(\int_t^T \xi_t^\tau d\tau \right) \right\rangle_0$$

$$= \frac{1}{2\hat{\sigma}^3 T^2} \sum_{i=0}^{N-1} \int_{T_i}^{T_{i+1}} \left\langle d\ln S_t \, d\left(\int_t^{T_N} \xi_t^\tau d\tau \right) \right\rangle_0$$

For $\tau \in [T_k, T_{k+1}]$, $\xi_t^\tau = \xi_t^k$. Thus:

$$\int_t^{T_N} \xi_t^\tau d\tau = \int_t^{T_{i+1}} \xi_t^\tau d\tau + \Delta \sum_{j=i+1}^{N-1} \xi_t^j = (T_{i+1} - t)\hat{\sigma}_{T_{i+1}}^2(t) + \Delta \sum_{j=i+1}^{N-1} \xi_t^j$$

$\mathcal{S}_T$ is given by:

$$\mathcal{S}_{N\Delta} = \frac{1}{2\hat{\sigma}^3 T^2} \left(\Delta \sum_{i=0,\, j>i}^{N-1} \int_{T_i}^{T_{i+1}} \langle d\ln S_t \, d\xi_t^j \rangle_0 \right. \qquad (7.113)$$

$$\left. + \sum_{i=0}^{N-1} \int_{T_i}^{T_{i+1}} (T_{i+1} - t)\langle d\ln S_t \, d\hat{\sigma}_{T_{i+1}}^2(t) \rangle_0 \right)$$

The derivation of (7.112) utilizes the assumption that $E\left[\langle d\ln S_t \, d\hat{\sigma}_T^2(t) \rangle \,\middle|\, \ln S \right]$ does not depend on S. We now verify that this holds at order one in ν and $(\beta - 1)$.

- Consider the first line of (7.113). At order one in ν:

$$d\xi_t^j = (2\nu_\Delta)\xi_0^j \, \alpha_{\theta_\Delta} \left((1 - \theta_\Delta) e^{-k_1(T_j - t)} dW_t^1 + \theta_\Delta e^{-k_2(T_j - t)} dW_t^2 \right)$$

For the sake of obtaining the covariation at order one in ν and $(\beta - 1)$, we take $dS_t = (r - q)S_t dt + \hat{\sigma} S_t dW_t^S$ and get:

$$\langle d\ln S_t \, d\xi_t^j \rangle_0 =$$

$$(2\nu_\Delta) \, \hat{\sigma}^3 \alpha_{\theta_\Delta} \left((1 - \theta_\Delta) \, e^{-k_1(T_j - t)} \rho_{SX^1} + \theta_\Delta e^{-k_2(T_j - t)} \rho_{SX^2} \right) dt$$

$$(7.114)$$

where we have used that $\xi_0^j = \hat{\sigma}^2$. $\langle d\ln S_t \, d\xi_t^j \rangle_0$ does not depend on $\ln S_t$.

- Now turn to the second line of (7.113) and consider the contribution of interval $[T_i, T_{i+1}]$. Let us condition the expectation of the covariation with respect to $\xi_{T_i}^i$. At order one in $(\beta - 1)$:

$$\sigma \left(\frac{S_t}{S_{T_i}} \right) = \sigma_0 \left(\xi_{T_i}^i \right) \left(1 + \frac{n-1}{n} \left(\beta \left(\xi_{T_i}^i \right) - 1 \right) \ln \frac{S_t}{S_{T_i}} \right)$$

We can now employ results derived in the perturbative analysis of the local volatility model in Sections 2.4.5 and 2.5.7 of Chapter 2. The local volatility function is of the form (2.44):

$$\sigma(t, S) = \overline{\sigma}(t) + \alpha(t) \ln \frac{S_t}{F_t} \qquad (7.115)$$

with:

$$\overline{\sigma}(t) = \sigma_0 \left(\xi_{T_i}^i \right) \left(1 + \frac{n-1}{n} \left(\beta \left(\xi_{T_i}^i \right) - 1 \right) \ln \frac{F_t}{S_{T_i}} \right) \qquad (7.116a)$$

$$\alpha(t) = \sigma_0 \left(\xi_{T_i}^i \right) \frac{n-1}{n} \left(\beta \left(\xi_{T_i}^i \right) - 1 \right) \qquad (7.116b)$$

We know from Section 2.5.7 that, given a local volatility function of the form in (7.115), $\langle d\ln S_t \, d\hat{\sigma}_{T_{j+1}}^2(t) \rangle_0$ does not depend on S_t, at order one in $\alpha(t)$.

This covariation is already of order one in $(\beta(\xi_{T_i}^i) - 1)$ thus need to be calculated at order *zero* in ν: taking the expectation of $\langle d\ln S_t \, d\hat{\sigma}_{T_{j+1}}^2(t) \rangle_0$ with respect to $\xi_{T_i}^i$ simply amounts to setting $\xi_{T_i}^i = \hat{\sigma}^2$.

Thus the second contribution in (7.113) does not depend on S either, at order one in ν and $(\beta - 1)$.

Since for the sake of calculating the covariation in the second piece of (7.113) $\xi_{T_i}^i$ is frozen, equal to $\hat{\sigma}$, we have, using identity (7.112) for the case of a local volatility function of type (7.115):

$$\int_{T_i}^{T_{i+1}} (T_{i+1} - t) \langle d\ln S_t \, d\hat{\sigma}_{T_{i+1}}^2(t) \rangle_0 = 2\hat{\sigma}^3 \Delta^2 \mathcal{S}_\Delta$$

which allows us to rewrite (7.113) as:

$$\mathcal{S}_{N\Delta} = \frac{1}{N}\mathcal{S}_\Delta + \frac{1}{2\widehat{\sigma}^3 T^2}\Delta \sum_{i=0,\, j>i}^{N-1} \int_{T_i}^{T_{i+1}} \langle d\ln S_t \, d\xi_t^j \rangle_0 \qquad (7.117)$$

Using (7.114) we have:

$$\frac{1}{2\widehat{\sigma}^3 T^2}\Delta \sum_{i=0,\, j>i}^{N-1} \int_{T_i}^{T_{i+1}} \langle d\ln S_t \, d\xi_t^j \rangle_0$$

$$= \frac{1}{2\widehat{\sigma}^3 T^2}\Delta 2\nu_\Delta \widehat{\sigma}^3 \alpha_{\theta\Delta} \sum_{i=0,\, j>i}^{N-1} \int_{T_i}^{T_{i+1}}$$

$$\times \left((1-\theta_\Delta)\, e^{-k_1(T_j-t)} \rho_{SX^1} + \theta_\Delta e^{-k_2(T_j-t)} \rho_{SX^2} \right) dt$$

$$= \nu_\Delta \alpha_{\theta\Delta} \frac{1}{N^2} \sum_{i=0,\, j>i}^{N-1} \frac{1}{\Delta} \int_{T_i}^{T_{i+1}} \left((1-\theta_\Delta)\, e^{-k_1(T_j-t)} \rho_{SX^1} + \theta_\Delta e^{-k_2(T_j-t)} \rho_{SX^2} \right) dt$$

$$= \nu_\Delta \alpha_{\theta\Delta} \left((1-\theta_\Delta)\, \rho_{SX^1}\zeta\,(k_1\Delta, N) + \theta_\Delta\, \rho_{SX^2}\zeta\,(k_2\Delta, N) \right)$$

where we have introduced function $\zeta\,(x, N)$ defined by:

$$\zeta\,(x, N) = \frac{1}{N^2}\sum_{i=0,\, j>i}^{N-1} \int_i^{i+1} \left(e^{-x}\right)^{j-u} du = \frac{e^x - 1}{x}\sum_{n=1}^{N-1}\frac{N-n}{N^2}\left(e^{-x}\right)^n$$

$$(7.118)$$

The final expression of the ATMF skew for maturity $T = N\Delta$, in the discrete two-factor model, at order one in ν and $(\beta - 1)$ is thus:

$$\mathcal{S}_{N\Delta} = \frac{1}{N}\mathcal{S}_\Delta + \frac{1}{2\widehat{\sigma}^3 T^2}\Delta \sum_{i=0,\, j>i}^{N-1} \int_{T_i}^{T_{i+1}} \langle d\ln S_t \, d\xi_t^j \rangle_0 \qquad (7.119a)$$

$$= \frac{1}{N}\mathcal{S}_\Delta + \nu_\Delta \alpha_{\theta\Delta} \left((1-\theta_\Delta)\, \rho_{SX^1}\zeta\,(k_1\Delta, N) + \theta_\Delta\, \rho_{SX^2}\zeta\,(k_2\Delta, N) \right) \qquad (7.119b)$$

- Expression (7.119) is an expansion of the ATMF skew at order one in $(\beta - 1)$ (first piece in (7.119b)) and ν (second piece). Note that $(\beta - 1)$ does not appear explicitly – only the ATMF future skew $\mathcal{S}_\Delta$ for tenor Δ appears in (7.119b). Indeed, from (7.116), the expansion at order one in $(\beta - 1)$ is really an expansion at order one in $\alpha(t)$. Owing to the skew-averaging expression (2.48), page 46, relating $\mathcal{S}_\Delta$ to $\alpha(t)$, ours is equivalently an expansion at order one in $\mathcal{S}_\Delta$.

- The ATMF skew is the sum of two components: the forward-smile contribution and the volatility-of-volatility contribution, which can be separately switched on and off by setting $\mathcal{S}_\Delta$ or ν_Δ equal to 0.

When pricing a cliquet of period Δ, $\mathcal{S}_\Delta$ controls the forward-smile adjustment δP_2 while ν controls the volatility-of-volatility adjustment δP_1 – see the discussion in Section 3.1.6 of Chapter 3.

- (7.119) makes it plain that the ATMF skew of discrete forward variance models is the sum of two contributions: the first piece in (7.119b) is contributed by the foward skew for tenor Δ, $\mathcal{S}_\Delta$, while the covariance of S_t with forward variances ξ_t^i is the second source of skew. Imagine switching off volatility of volatility. We then have:

$$\mathcal{S}_{N\Delta} = \frac{1}{N}\mathcal{S}_\Delta$$

Thus the ATMF skew for maturity T decays like $\frac{1}{T}$.

This is understood by noting that with $\nu = 0$, log-returns $\ln \frac{S_{i\Delta}}{S_{(i-1)\Delta}}$ are independent, thus $\ln \frac{S_{N\Delta}}{S_0}$ is the sum of N independent, identically distributed random variables. The skewness of $\ln \frac{S_{N\Delta}}{S_0}$ then scales like $\frac{1}{\sqrt{N}}$. In Appendix B of Chapter 5, it is shown that, at order one in the skewness s of $\ln S_T$, the ATMF skew $\mathcal{S}_T$ for maturity T is given by expression (5.93), page 194: $\mathcal{S}_T = \frac{s}{6\sqrt{T}}$. We thus get: $\mathcal{S}_{N\Delta} \propto \frac{1}{N}$. Because these results are derived at order one in $\mathcal{S}_\Delta$, they only hold for small values of $\mathcal{S}_\Delta$.

- Imagine taking the limit $\Delta \to 0$. Our model becomes a plain continuous forward variance model. There is only one source of skew: $\mathcal{S}_{N\Delta}$ is generated by the second piece in (7.119b). In expression (7.118) of $\zeta(x, N)$, take the limit $x \to 0$, $N \to \infty$ with Nx fixed, equal to, respectively, $k_1 T$ or $k_2 T$. Converting the sum in (7.118) in an integral, we get:

$$\lim_{\substack{x \to 0,\, N \to \infty \\ Nx = kT}} \frac{e^x - 1}{x} \sum_{n=1}^{N-1} \frac{N - n}{N^2} \left(e^{-x}\right)^n = \int_0^T \frac{T - t}{T^2} e^{-kt}\, dt$$

$$= \frac{kT - \left(1 - e^{-kT}\right)}{(kT)^2}$$

$\mathcal{S}_T$ is then given at order one in ν by:

$$\mathcal{S}_T = \nu \alpha_\theta \left((1 - \theta)\, \rho_{SX^1} \frac{k_1 T - \left(1 - e^{-k_1 T}\right)}{(k_1 T)^2} + \theta\, \rho_{SX^2} \frac{k_2 T - \left(1 - e^{-k_2 T}\right)}{(k_2 T)^2} \right)$$

$$(7.120)$$

where we have simply replaced $\nu_\theta, \theta_\Delta, \alpha_{\theta_\Delta}$ with $\nu, \theta, \alpha_\theta$.

We derive below, in Chapter 8, the expansion of implied volatilities in continuous forward variance models at order two in ν. At order one in ν, we unsurprisingly recover (7.120) – see formula (8.55), page 330.

Numerical examples

The numerical results presented below are obtained with $\Delta = 1$ month, and using:

- parameters in Table 7.8 for the (continuous) two-factor model,[56] which are mapped according to (7.102)

- $\sigma_0()$ and $\beta()$ chosen so that future smiles of maturity Δ exhibit a fixed ATMF skew, with $\mathcal{S}_\Delta = -0.5$ – see Figure 7.18. The latter value of $\mathcal{S}_\Delta$ corresponds to a difference of 5 points of implied volatility for 95% and 105% strikes for maturity Δ. We take $n_i = 3, \forall i$.

- a flat VS volatility $\widehat{\sigma}$ equal to 20%.

ν	θ	k_1	k_2	ρ_{12}	ρ_{SX^1}	ρ_{SX^2}
174%	0.245	5.35	0.28	0%	-75.9%	-48.7%

Table 7.8: Parameters of the (continuous) two-factor model.

These parameter values are realistic – they are such that, in the continuous version of the two-factor model:

- volatilities of VS volatilities approximately decay like $\frac{1}{\sqrt{T}}$ with the volatility of a 3-month VS volatility equal to 100% – see Figure 7.1, page 228.

- the ATMF skew approximately decays like $\frac{1}{\sqrt{T}}$ as well, with the 95%/105% skew equal to 3 points of volatility for maturity 1 year – see Figures 8.3 and 8.5, pages 331 and 332.

In order to gauge the magnitudes of both contributions to the ATMF skew in (7.119b), let us turn off either the forward skew ($\mathcal{S}_\Delta = 0$) or volatility of volatility ($\nu = 0$).

The resulting ATMF skew, expressed as the difference of the implied volatilities for the 95% and 105% strikes, is shown in Figure 7.20, together with the approximate value in (7.119b) and the ATMF skew, when both forward smile and volatility of volatility are turned on.

It is apparent that the decay of the "forward smile" contribution agrees well with the $1/T$ form in (7.119b). The agreement of approximate and actual values of the stochastic volatility components is somewhat less satisfactory, still the increase and subsequent decrease with maturity is well captured by function $\zeta(x, N)$.

While the "approx" curve in the bottom graph of Figure 7.20 is the sum of the components in the top graphs, this is not the case for the "actual" curve; the latter is obtained in a Monte Carlo simulation with both effects turned on. The good agreement of approximate and actual ATMF skews is testament to the fact that order-2 cross terms of the type $(\beta - 1)\nu$ contribute negligibly. $\mathcal{S}_T$ in Figure 7.20 is

---

[56]Parameters in Table 7.8 are those of Set II in Table 7.1, page 229, which we have used in Sections 7.4 and 7.6. We use these parameters again in the following chapter – see Sections 8.7 and 8.8 for smiles generated by the continuous model thus parametrized.

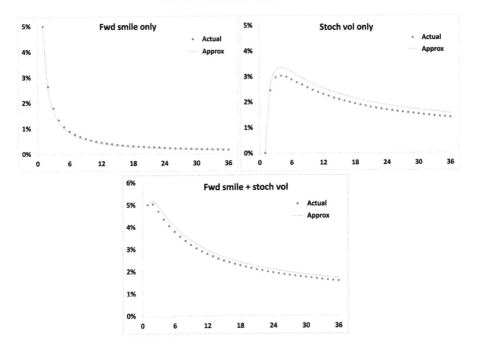

Figure 7.20: Top: $\mathcal{S}_T$, as the 95%/105% skew, obtained either with $\nu = 0$ (left) or with $\mathcal{S}_\Delta = 0$ (right) as a function of T (months), evaluated in a Monte-Carlo simulation of the discrete two-factor model with parameters in Table 7.8 (Actual) and as given by order-one formula (7.119b) (Approx). $\Delta = 1$ month.
Bottom: $\mathcal{S}_T$ when both forward smile ($\mathcal{S}_\Delta = -0.5$) and volatility of volatility ($\nu = 174\%$) are switched on.

non-monotonic as a function of T, however this depends on the relative magnitude of $\mathcal{S}_\Delta$ and ν_Δ - see Figure 7.21 where $\mathcal{S}_T$ is graphed for 3 different values of $\mathcal{S}_\Delta$.

This is illustrated also in Figure 7.22, which shows $\mathcal{S}_T$ for the same parameters as in Figure 7.8, except ν has been halved. For this smaller level of volatility of volatility, agreement with formula (7.119b) is excellent.

7.8.5 Conclusion

- With respect to their continuous counterparts, discrete forward variance models allow separation of the effects of (a) spot/volatility covariance, (b) future smile for a given time scale Δ. In models thus specified, prices of cliquets of the form:

$$\sum_i \omega_i f\left(\frac{S_{T_{i+1}}}{S_{T_i}}\right)$$

where $T_i = i\Delta$, do not depend anymore on spot/volatility correlations.

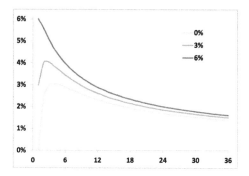

Figure 7.21: $\mathcal{S}_T$, as the 95%/105% skew, as a function of T (months), evaluated in a Monte-Carlo simulation of the discrete two-factor model with parameters in Table 7.8 for 3 different values of $\mathcal{S}_\Delta$, such that the 1-month 95%/105% ATMF skew is equal to 0, 3% and 6%.

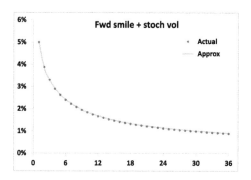

Figure 7.22: $\mathcal{S}_T$, as the 95%/105% skew, as a function of T (months), evaluated in a Monte-Carlo simulation of the discrete two-factor model with the same parameters as in Figure 7.20 except ν has been halved.

One first sets future smiles for maturity Δ, then chooses spot/volatility correlations so as to obtain desired levels of covariances of spot and forward VS/ATMF volatilities, or desired levels for the vanilla ATMF skew. Discrete forward variance models are thus naturally suited to the risk-management of cliquets, such as accumulators.

- For any choice of time scale Δ, simple parameter mappings exist that ensure that instantaneous volatilities of spot or forward-starting VS volatilities in the discrete two-factor model match those of the continuous version of the model.

- Specification of future smiles is very flexible. We give an example of parametrization that allows the user to specify how the ATMF skew for maturity Δ depends on the ATMF volatility for the same maturity. Once functions $\sigma_0()$ and $\beta()$ are tabulated, simulation of the discrete model is as uncomplicated as in the continuous model.

- Discrete forward variance models are also ideally suited to the risk management of payoffs involving both VIX futures and the S&P 500 index. They can be calibrated exactly to VIX smiles, if one so desires, while preserving full flexibility as to forward smile scenarios for the S&P 500 index.

- The vanilla smile of discrete forward variance models is produced by both forward-smile and volatility-of-volatility components. Order-one formula (7.119b) allows for an assessment of the contribution of each effect to the ATMF skew of vanilla options.

Chapter's digest

7.1 Pricing equation

▶ The pricing equation of forward variance models is obtained through a replication argument. It reads:

$$
\frac{dP}{dt} + (r-q) S \frac{dP}{dS} + \frac{\xi^t}{2} S^2 \frac{d^2 P}{dS^2}
$$
$$
+ \frac{1}{2} \int_t^T du \int_t^T du' \nu\,(t,u,u',\xi)\,\frac{d^2 P}{\delta\xi^u \delta\xi^{u'}} + \int_t^T du\,\mu\,(t,u,\xi)\,S\frac{d^2 P}{dS\delta\xi^u} = rP
$$

This SDE admits a probabilistic interpretation. Its solution is given by:

$$
P = E\big[g(S_T) \mid S_t = S,\ \xi_t^u = \xi^u\big]
$$

under a dynamics for S_t, ξ_t^u given by:

$$
\begin{cases}
dS_t &= (r-q)S_t dt + \sqrt{\xi_t^t} S_t dW_t^S \\
d\xi_t^u &= \lambda_t^u dW_t^u
\end{cases}
$$

with λ_t^u and correlations between W_t^S and W_t^u such that:

$$
\lim_{dt \to 0} \frac{1}{dt} E_t\big[d\ln S_t d\xi_t^u\big] = \sqrt{\xi_t^t}\lambda_t^u \frac{1}{dt} E_t\big[dW_t^S dW_t^u\big] = \mu\,(t,u,\xi)
$$
$$
\lim_{dt \to 0} \frac{1}{dt} E_t\big[d\xi_t^u d\xi_t^{u'}\big] = \lambda_t^u \lambda_t^{u'} \frac{1}{dt} E_t\big[dW_t^u dW_t^{u'}\big] = \nu\,(t,u,u',\xi)
$$

🙙 🙙 🙙 🙙 🙙

7.3 N-factor models

▶ Markovian representations of forward variance models are economically obtained by choosing exponential weightings for the driving Brownian motions:

$$
d\xi_t^T = \omega\alpha_w \xi_t^T \sum_i w_i e^{-k_i(T-t)} dW_t^i
$$

▶ N-factor models are simulated by evolving, together with the spot process, N Ornstein-Ühlenbeck processes, which are easily simulated exactly.

▶ The number of driving factors in a model bears no relationship to the number of hedging instruments required; it simply sets the structure and rank of the break-even covariance matrix of the gamma/theta P&L of a hedged position.

🙙 🙙 🙙 🙙 🙙

7.4 A two-factor model

▶ Two factors afford sufficient flexibility as to volatilities and correlations of volatilities. The SDE of ξ_t^T reads:

$$d\xi_t^T = (2\nu)\xi_t^T \, \alpha_\theta \left((1-\theta) \, e^{-k_1(T-t)} dW_t^1 \; + \; \theta e^{-k_2(T-t)} dW_t^2 \right)$$

$$\alpha_\theta = 1/\sqrt{(1-\theta)^2 + \theta^2 + 2\rho_{12}\theta(1-\theta)}$$

where ν is the volatility of a very short volatility. We introduce driftless processes x_t^T:

$$dx_t^T = \alpha_\theta \left[(1-\theta) \, e^{-k_1(T-t)} dW_t^1 \; + \; \theta e^{-k_2(T-t)} dW_t^2 \right]$$

ξ_t^T is given by:

$$\xi_t^T = \xi_0^T f^T \left(t, x_t^T \right)$$

$$f^T(t,x) = e^{\omega x - \frac{\omega^2}{2}\chi(t,T)}$$

with $\omega = 2\nu$ and $\chi(t,T)$ given in (7.35).

▶ The two-factor model can be parametrized so that volatilities of spot-starting volatilities approximately match a power-law decay with maturity. We use the following benchmark:

$$\nu_T^B(t) = \sigma_0 \left(\frac{\tau_0}{T-t} \right)^\alpha$$

This benchmark form, for given values of σ_0, τ_0, α, can be approximately captured in the two-factor model with different sets of parameters.

This additional flexibility is utilized to generate different volatilities of forward-starting volatilities.

▶ While instantaneous forward variances are lognormal, discrete forward variances are not, thus variance swaptions exhibit a slight positive skew.

▶ For set parameters, volatilities of volatilities depend on the shape of the variance curve: they are larger for decreasing term structures of VS volatilities.

▶ The correlation structure of forward volatilities in the two-factor model is poor, as it involves one single time scale: $\frac{1}{k_1-k_2}$. This could motivate the inclusion of additional factors.

❧ ❧ ❧ ❧ ❧

7.6 Options on realized variance

▶ An option on realized variance of maturity T pays a call or put on the realized variance over $[0,T]$, measured with daily log-returns.

▶ The natural hedge instrument is the VS for the residual maturity. A simple model (SM) is built by specifying the dynamics of U_t, defined by:

$$U_t = \frac{Q_t + (T - t)\,\widehat{\sigma}_T^2(t)}{T}$$

Q_t is the quadratic variation at t, and $\widehat{\sigma}_T(t)$ is the VS volatility at t for maturity T. U_t has no drift.

▶ Assuming a lognormal dynamics for U_t given by:

$$\frac{dU_t}{U_t} = 2R_t \frac{T-t}{T}\nu_T(t)\,dW_t$$

yields the price of an option on realized variance, in the form of a simple Black-Scholes formula, where $R_t = \frac{\widehat{\sigma}_T^2(t)}{U_t}$ has been take equal to 1.

$$P(t, U) = P_{\mathrm{BS}}(t, U, \sigma_{\mathrm{eff}}, T)$$

$$\sigma_{\mathrm{eff}}^2 = \frac{1}{T-t} \int_t^T 4\left(\frac{T-\tau}{T}\right)^2 \nu_T^2(\tau)\,d\tau$$

where $\nu_T^2(\tau)$ is the volatility at τ of a VS volatility of maturity T.

▶ Numerical tests whereby the option price computed in the SM is compared to that produced by the two-factor model, parametrized so that the volatilities of volatilities it generates match $\nu_T(\tau)$, show that, for the case of a flat term-structure of VS volatilities, the approximation in the SM is adequate.

▶ When the term structure of VS volatilities is not flat, the approximation $R_t = 1$ is replaced with $R_t = \frac{\widehat{\sigma}_{\tau T}^2(0)}{\widehat{\sigma}_T^2(0)}$, which produces the following amended expression for σ_{eff}, at inception:

$$\sigma_{\mathrm{eff}}^2 = \frac{4}{T} \int_0^T \left(\frac{T-\tau}{T}\right)^2 \left(\frac{\widehat{\sigma}_{\tau T}^2(0)}{\widehat{\sigma}_T^2(0)}\right)^2 \nu_T^2(\tau)\,d\tau$$

▶ The vega-hedge portfolio of an option on realized variance comprises, in addition to a VS of maturity T, a continuum of VSs of intermediate maturities. This vega hedge also functions as a gamma hedge.

▶ This hedge portfolio can be benchmarked against that produced by the two-factor model. The two hedges are slightly different, because of the sensitivity of volatilities of volatilities to the term structure of VS volatilities, in the two-factor model.

▶ The fact that variance does not accrue continuously, but is measured using discrete returns, impacts the value of options on realized variance, especially for short maturities. In case volatilities of volatilities vanish, one is still exposed to the intrinsic variance of the variance estimator itself. This effect is taken care of by using the following expression for σ_{eff} where κ is the (conditional) kurtosis of daily returns.

$$\sigma_{\text{eff}}^2 = \frac{1}{T}\int_0^T 4\left(\frac{T-\tau}{T}\right)^2 \left(\frac{\widehat{\sigma}_{\tau T}^2(0)}{\widehat{\sigma}_T^2(0)}\right)^2 \nu_T^2(\tau)\,d\tau + \frac{2+\kappa}{NT}$$

▶ Upper and lower bounds for prices of vanilla options on realized variance can be derived from the vanilla smile. If breached, a trading strategy consisting of a static position in a realized variance option and a portfolio of vanilla options, together with a dynamic delta position, nets a positive P&L.

▶ Options on forward realized variance cannot be priced in the SM as the latter only takes as ingredient the term structure of volatilities of spot-starting VS volatilities. What is needed, in addition, is the volatility of forward-starting volatilities, which cannot be backed-out of volatilities of spot-starting volatilities, in model-independent fashion. This is illustrated by pricing these options in the two-factor model with parameter sets that generate almost identical prices for spot-starting options. Prices of forward-starting options are different. They are also higher than prices of variance swaptions.

ᘒᗩ ᘒᗩ ᘒᗩ ᘒᗩ ᘒᗩ

7.7 VIX futures and options

▶ The smile of VIX futures can be simply modeled by changing the function that maps processes x_t^T into forward variances ξ_t^T. We introduce the following simple parametrization:

$$f^T(t,x) = (1-\gamma_T)\,e^{\omega_T x - \frac{\omega_T^2}{2}\chi(t,T)} + \gamma_T e^{\beta_T \omega_T x - \frac{(\beta_T \omega_T)^2}{2}\chi(t,T)}$$

Volatility-of-volatility smile parameters $\gamma^T, \beta^T, \zeta^T$, as well as forward variances ξ_0^T, are taken to be piecewise constant, with constant values for all forward variances that underlie a given VIX future. The model is calibrated by choosing $\gamma^T, \beta^T, \zeta^T$ and ξ_0^T so that market values of (a) VIX futures, (b) VIX implied volatilities are matched. This is almost a local volatility model for forward variances.

▶ Different parameter sets of the two-factor model can be employed, resulting in very similar calibration accuracies. What distinguishes these different sets is the different distributions of volatility they generate for VIX futures.

▶ In case VIX futures are needed in the simulation of the two-factor model, they can be efficiently computed through a two-dimensional Gaussian quadrature. In case of very frequent observations, it is preferable to turn to the discrete forward variance models of Section 7.8.2.

▶ Options exist on ETNs whose investment strategies consists in maintaining a long position in the first two nearby VIX futures. These options cannot be priced off VIX smiles in model-independent fashion, as they are very sensitive to the distribution of the volatility of VIX futures.

▶ Forward S&P 500 VSs can be synthesized using VIX instruments: futures and options. It is possible to set up a trading strategy that arbitrages the difference of these VIX-synthesized forward VSs with respect to forward VSs derived from the S&P 500 VS market.

🐦 🐦 🐦 🐦 🐦

7.8 Discrete forward variance models

▶ Discrete variance models arise out of the need to control future smiles independently from the correlation of spot and volatilities, and also to model VIX futures directly.

The specification of these models starts with a schedule of discrete dates T_i, then the model is built in two stages. First, we define a dynamics for discrete forward variances $\xi_t^i = \hat{\sigma}_{T_i,T_{i+1}}^2(t)$, then construct a dynamics for S_t that complies with that of the ξ_t^i.

▶ Since the ξ^i are driftless, as their continuous counterparts ξ^T, the two-factor model can be employed for the ξ^i.

When switching from the continuous to the discrete version of the model, we require that some features remain unchanged, so that volatility-of-volatility risks of different payoffs, calling for different schedules, are still priced at the same levels.

Starting from a parameter set for the continuous model, we generate parameters for the discrete model that ensure that instantaneous volatilities and correlations of spot-starting and forward VS volatilities for maturities T_i match in both models, for forward variances equal to their initial values.

▶ A benefit of discrete forward variance models is that VIX futures can be directly modeled, thus are readily accessible in a simulation of the model. VIX smiles can be calibrated exactly – the resulting model is in fact a local volatility model.

▶ Next we specify a dynamics for S_t that meets the following requirements: (a) the SDE for S_t complies with the dynamics of the ξ^i, (b) the density of $\frac{S_{T_{i+1}}}{S_{T_i}}$ is independent on S_{T_i} so that payoffs $\Sigma_i \omega_i f\left(\frac{S_{T_{i+1}}}{S_{T_i}}\right)$ have zero sensitivity to spot/volatility correlations, (c) the dependence of future skews over intervals $[T_i, T_{i+1}]$ on the corresponding VS volatilities can be set at will.

▶ This is achieved by using a path-dependent local volatility for S_t. We provide a simple parametrization of this local volatility function and give examples of two specifications corresponding to two typical future skew scenarios: future ATMF skews either independent on, or proportional to, ATMF volatilities.

▶ The ATMF skew of discrete forward variance models is generated by two mechanisms: (a) the local volatility functions σ^i, (b) the correlation of S_t and forward variances ξ^i. Working at order one in both the slope of local volatility functions σ^i

and volatility of volatility, we express the ATMF skew for maturities T_i as the sum of two contributions, generated by both effects.

The portion of the ATMF skew generated by local volatility functions σ^i decays as $\frac{1}{T}$. Numerical tests confirm the accuracy of the order-one expansion.

Chapter 8

The smile of stochastic volatility models

What is it that determines the shape of the smile generated by stochastic volatility models? We derive an expansion at order two in volatilities of volatilities, which is easily carried out in the forward variance framework introduced in Chapter 7, and characterize the smile near the money.[1] We derive in particular a general approximate expression for the ATMF skew that is accurate and can be used in practice.

We also present an alternative derivation of the order-one correction based on a representation of European option prices in terms of spot/volatility and volatility/volatility gamma P&Ls.

The characterization of the near-the-money smile in terms of the spot/variance and variance/variance covariance functions is used in Chapter 9 to establish a link between static and dynamic properties of stochastic volatility models.

Finally, efficient techniques for generating vanilla smiles in stochastic volatility models are explored in the appendix.

8.1 Introduction

Any stochastic volatility model – including models based on the dynamics of the instantaneous variance V_t – can be written as a forward variance model. The corresponding pricing equation for a European payoff is given in (7.4), page 219:

$$
\frac{dP}{dt} + (r - q) S \frac{dP}{dS} + \frac{\xi^t}{2} S^2 \frac{d^2 P}{dS^2}
$$
$$
+ \frac{1}{2} \int_t^T du \int_t^T du' \nu\,(t, u, u', \xi) \frac{d^2 P}{\delta \xi^u \delta \xi^{u'}} + \int_t^T du\, \mu\,(t, u, \xi)\, S \frac{d^2 P}{dS \delta \xi^u} = rP
$$

$$(8.1)$$

---

[1]This is based on joint work with Julien Guyon, published in [13].

where the spot/variance and variance/variance covariance functions μ, ν are defined as:

$$\mu\left(t,u,\xi\right) = \lim_{dt\to 0}\frac{1}{dt}E_t\left[d\ln S_t d\xi_t^u\right]$$

$$\nu\left(t,u,u',\xi\right) = \lim_{dt\to 0}\frac{1}{dt}E_t\left[d\xi_t^u d\xi_t^{u'}\right]$$

Remember that μ and ν can depend on the variance curve ξ, but are not allowed to explicitly depend on S: this excludes from our scope mixed local-stochastic volatility models.

Obviously the smile generated by a stochastic volatility model is a product of the joint dynamics of S_t and forward variances ξ_t^T; is it possible to pinpoint precisely which functionals of the covariance functions μ and ν determine the shape of the vanilla smile?

We answer this question by deriving an expansion of vanilla option prices in powers of volatility of volatility. To this end we introduce a dimensionless parameter ε which we use to scale μ and ν according to:

$$\mu \to \varepsilon\mu$$
$$\nu \to \varepsilon^2\nu$$

Once the expansion at the desired order in ε is obtained we set $\varepsilon = 1$.

8.2 Expansion of the price in volatility of volatility

Consider a European option of maturity T whose payoff is $g\left(S_T\right)$. With no loss of generality we take vanishing interest rate and repo. The option's price $P\left(t,S,\xi\right)$ solves the following backward equation:

$$\frac{dP}{dt} + H_t P = 0 \tag{8.2}$$

H_t is given by:
$$H_t = H_t^0 + \varepsilon W_t^1 + \varepsilon^2 W_t^2$$
where operators H_t^0, W_t^1, W_t^2 read:

$$H_t^0 = \frac{\xi^t}{2}\left(\partial_x^2 - \partial_x\right) \tag{8.3a}$$

$$W_t^1 = \int_t^T du\mu\left(t,u,\xi\right)\partial_{x\xi^u}^2 \tag{8.3b}$$

$$W_t^2 = \frac{1}{2}\int_t^T du\int_t^T du'\nu\left(t,u,u',\xi\right)\partial_{\xi^u\xi^{u'}}^2 \tag{8.3c}$$

where $x = \ln S$ and ∂_x, $\partial^2_{x\xi^u}$, $\partial^2_{\xi^u\xi^{u'}}$ stand respectively for $\frac{d}{dx}$, $\frac{d^2}{dx\delta\xi^u}$, $\frac{d^2}{\delta\xi^u\delta\xi^{u'}}$ – note that derivatives with respect to ξ^u are functional derivatives. The terminal condition for P is: $P(T, S, \xi) = g(S)$. Let us write the expansion of P in powers of ε as:

$$P = P_0 + \varepsilon P_1 + \varepsilon^2 P_2 + \cdots \tag{8.4}$$

Rather than inserting expression (8.4) in (8.2), deriving sequential PDEs for the P_i by equating to zero the contribution of each order in ε, and computing the P_i using the Feynman-Kac representation, we use the time-dependent perturbation technique. Integrating PDE (8.2), P is given by:

$$P(t, x, \xi) = U_{tT} g$$

where operator U_{st} with $s \leq t$ is defined by:

$$U_{st} = \lim_{n\to\infty} (1 + \delta t H_{t_0})(1 + \delta t H_{t_1}) \cdots (1 + \delta t H_{t_{n-1}}) \tag{8.5}$$

with $\delta t = \frac{t-s}{n}$ and $t_i = s + i\delta t$. U_{st} can be written as:

$$U_{st} =: \exp\left(\int_s^t H_\tau d\tau\right):$$

where $::$ indicates that the operators inside the colons are time-ordered, as in (8.5). U_{st} satisfies the semi-group property – for $s \leq r \leq t$:

$$U_{st} = U_{sr} U_{rt} \tag{8.6}$$

Let us write $H_t = H_t^0 + \delta H_t$ where H_t^0 is the unperturbed operator corresponding to the Black-Scholes model and δH_t is a perturbation. From (8.5), the expansion of U_{st} in powers of δH reads:

$$U_{st} = U_{st}^0 + \int_s^t d\tau\, U_{s\tau}^0 \delta H_\tau U_{\tau t}^0 + \int_s^t d\tau_1 \int_{\tau_1}^t d\tau_2\, U_{s\tau_1}^0 \delta H_{\tau_1} U_{\tau_1\tau_2}^0 \delta H_{\tau_2} U_{\tau_2 t}^0 + \cdots \tag{8.7}$$

δH_t reads: $\delta H_t = \varepsilon \mathcal{W}_t^1 + \varepsilon^2 \mathcal{W}_t^2$. Inserting this expression in (8.7) and keeping terms up to order two in ε yields:

$$P_0 = U_{tT}^0 g \tag{8.8}$$

$$P_1 = \int_t^T d\tau\, U_{t\tau}^0 \mathcal{W}_\tau^1 U_{\tau T}^0 g \tag{8.9}$$

$$P_2 = \left(\int_t^T d\tau\, U_{t\tau}^0 \mathcal{W}_\tau^2 U_{\tau T}^0 + \int_t^T d\tau_1 \int_{\tau_1}^T d\tau_2\, U_{t\tau_1}^0 \mathcal{W}_{\tau_1}^1 U_{\tau_1\tau_2}^0 \mathcal{W}_{\tau_2}^1 U_{\tau_2 T}^0\right) g \tag{8.10}$$

The expression for U_{st}^0 – the so-called free propagator – is:

$$U_{st}^0 =: \exp\left(\int_s^t H_\tau^0 d\tau\right) := e^{\frac{1}{2}(\int_s^t \xi^\tau d\tau)(\partial_x^2 - \partial_x)} \tag{8.11}$$

where we have removed the time-ordering symbol in the right-hand side as operators H_t^0 for different values of t commute. P_0 in (8.8) is the standard Black-Scholes price: $P(t, x, \xi) = P_{BS}(t, S, \widehat{\sigma})$ with $S = e^x$ and $\widehat{\sigma}^2 = \frac{1}{T-t} \int_t^T \xi^\tau d\tau$. Before computing P_1 and P_2 observe that:

- as is clear from expression (8.11) for U_{st}^0, ∂_x and U_{st}^0 commute. This is equivalent to saying that $\partial_x^n P_0$ is a martingale for all n. Specializing to the cases $n = 1, 2$ this recovers the well-known property that the delta and dollar gammas $S \frac{dP}{dS}$ and $S^2 \frac{d^2 P}{dS^2}$ of a European option in the Black-Scholes model are martingales.

- likewise ∂_{ξ^u} and U_{st}^0 commute unless $u \in [s, t]$:

$$\partial_{\xi^u} U_{st}^0 = U_{st}^0 \partial_{\xi^u} + \mathbb{1}_{u \in [s,t]} \frac{1}{2} \left(\partial_x^2 - \partial_x \right) U_{st}^0 \qquad (8.12)$$

Moreover, $\partial_{\xi^u}^n g = 0$ as g is a function of x only. Applying relationship (8.12) on g with $t = T$ and $s = t$ yields: $\partial_{\xi^u} U_{tT}^0 g = \frac{1}{2} \left(\partial_x^2 - \partial_x \right) U_{tT}^0 g$. This expresses the already-mentioned property that, for a European option in the Black-Scholes model, vega (left-hand side) and gamma (right-hand side) are related: $\frac{dP}{d(\sigma^2(T-t))} = \frac{1}{2} S^2 \frac{d^2 P}{dS^2}$.

Using these two rules, the semi-group property (8.6), and the fact that μ does not depend on x, we get:

$$
\begin{aligned}
P_1 &= \int_t^T d\tau \, U_{t\tau}^0 \mathcal{W}_\tau^1 U_{\tau T}^0 \, g = \int_t^T d\tau \, U_{t\tau}^0 \int_\tau^T du \mu_{\tau,u} \partial_{x\xi^u}^2 U_{\tau T}^0 \, g \\
&= \int_t^T d\tau \, U_{t\tau}^0 \int_\tau^T du \, \mu_{\tau,u} \partial_x \frac{1}{2} \left(\partial_x^2 - \partial_x \right) U_{\tau T}^0 \, g \\
&= \int_t^T d\tau \int_\tau^T du \, \mu_{\tau,u} \partial_x \frac{1}{2} \left(\partial_x^2 - \partial_x \right) U_{t\tau}^0 U_{\tau T}^0 \, g \\
&= \frac{C_t^{x\xi}(\xi)}{2} \partial_x \left(\partial_x^2 - \partial_x \right) P_0
\end{aligned}
$$

where we have used the more compact notation $\mu_{\tau,u} \equiv \mu(\tau, u, \xi)$. The dimensionless quantity $C_t^{x\xi}(\xi)$ is given by:

$$C_t^{x\xi}(\xi) = \int_t^T d\tau \int_\tau^T du \, \mu(\tau, u, \xi) \qquad (8.13a)$$

$$= \int_t^T (T - \tau) \left\langle d\ln S_\tau \, d\widehat{\sigma}_T^2(\tau) \right\rangle \qquad (8.13b)$$

Likewise, the first contribution to P_2 in (8.10) is given by:

$$\int_t^T d\tau\, U_{t\tau}^0 \mathcal{W}_\tau^2 U_{\tau T}^0\, g = \frac{1}{2}\int_t^T d\tau\, U_{t\tau}^0 \int_\tau^T du \int_\tau^T du'\, \nu_{\tau,u,u'} \partial_{\xi^u \xi^{u'}}^2 U_{\tau T}^0\, g$$

$$= \frac{1}{8}\int_t^T d\tau\, U_{t\tau}^0 \int_\tau^T du \int_\tau^T du'\, \nu_{\tau,u,u'} \left(\partial_x^2 - \partial_x\right)^2 U_{\tau T}^0\, g$$

$$= \frac{C_t^{\xi\xi}(\xi)}{8}\left(\partial_x^2 - \partial_x\right)^2 P_0$$

where the dimensionless quantity $C_t^{\xi\xi}(\xi)$ reads:

$$C_t^{\xi\xi}(\xi) = \int_t^T d\tau \int_\tau^T du \int_\tau^T du'\, \nu\left(\tau,u,u',\xi\right) \tag{8.14a}$$

$$= \int_t^T (T-\tau)^2 \left\langle d\widehat{\sigma}_T^2(\tau)\, d\widehat{\sigma}_T^2(\tau)\right\rangle \tag{8.14b}$$

The second contribution to P_2 involves the spot/variance operator $\mathcal{W}^1$ twice:

$$\int_t^T d\tau_1 \int_{\tau_1}^T d\tau_2\, U_{t\tau_1}^0 \mathcal{W}_{\tau_1}^1 U_{\tau_1\tau_2}^0 \mathcal{W}_{\tau_2}^1 U_{\tau_2 T}^0\, g$$

$$= \int_t^T d\tau_1 U_{t\tau_1}^0 \int_{\tau_1}^T d\tau_2 \int_{\tau_1}^T du\, \mu_{\tau_1,u} \partial_{x\xi^u}^2 U_{\tau_1\tau_2}^0 \int_{\tau_2}^T du'\, \mu_{\tau_2,u'} \partial_{x\xi^{u'}}^2 U_{\tau_2 T}^0\, g$$

$$= \frac{1}{2}\int_t^T d\tau_1 U_{t\tau_1}^0 \int_{\tau_1}^T du\, \mu_{\tau_1,u} \int_{\tau_1}^T d\tau_2\, \partial_{x\xi^u}^2 U_{\tau_1\tau_2}^0 \int_{\tau_2}^T du'\, \mu_{\tau_2,u'} \partial_x \left(\partial_x^2 - \partial_x\right) U_{\tau_2 T}^0\, g$$

$$= \frac{1}{2}\int_t^T d\tau_1 U_{t\tau_1}^0 \int_{\tau_1}^T du\, \mu_{\tau_1,u} \partial_{x\xi^u}^2 \int_{\tau_1}^T d\tau_2 \int_{\tau_2}^T du'\, \mu_{\tau_2,u'}\, \partial_x \left(\partial_x^2 - \partial_x\right) U_{\tau_1 T}^0\, g$$

$$= \frac{1}{2}\partial_x^2 \left(\partial_x^2 - \partial_x\right) \int_t^T d\tau_1 U_{t\tau_1}^0 \int_{\tau_1}^T du\, \mu_{\tau_1,u} \partial_{\xi^u} C_{\tau_1}^{x\xi}(\xi) U_{\tau_1 T}^0\, g$$

$$= \frac{1}{4}\partial_x^2 \left(\partial_x^2 - \partial_x\right)^2 \int_t^T d\tau_1 \int_{\tau_1}^T du\, \mu_{\tau_1,u} C_{\tau_1}^{x\xi}(\xi) U_{tT}^0\, g \tag{8.15}$$

$$+ \frac{1}{2}\partial_x^2 \left(\partial_x^2 - \partial_x\right) \left(\int_t^T d\tau_1 \int_{\tau_1}^T du\, \mu_{\tau_1,u} \frac{\delta C_{\tau_1}^{x\xi}(\xi)}{\delta \xi^u}\right) U_{tT}^0\, g$$

$$= \frac{C_t^{x\xi}(\xi)^2}{8}\partial_x^2 \left(\partial_x^2 - \partial_x\right)^2 P_0 + \frac{D_t(\xi)}{2}\partial_x^2 \left(\partial_x^2 - \partial_x\right) P_0 \tag{8.16}$$

The dimensionless quantity $D_t(\xi)$ reads:

$$D_t(\xi) = \int_t^T d\tau \int_\tau^T du\, \mu_{\tau u} \frac{\delta C_\tau^{x\xi}(\xi)}{\delta \xi^u} \tag{8.17a}$$

$$= \int_t^T d\tau \lim_{d\tau \to 0} \frac{1}{d\tau} E_\tau \left[d\ln S_\tau\, dC_\tau^{x\xi} \right] \tag{8.17b}$$

$$= \int_t^T d\tau \int_\tau^T du\, (T-u) \frac{1}{d\tau} \left\langle d\ln S_\tau\, d\left[\frac{\langle d\ln S_u\, d\widehat{\sigma}_T^2(u)\rangle}{du} \right] \right\rangle \tag{8.17c}$$

The alternative expressions in (8.17) for $D_t(\xi)$ follow from the fact that $C_\tau^{x\xi}(\xi)$ is a functional of the variance curve. Indeed, using the definition of $\mu_{\tau u}$:

$$\int_\tau^T du\, \mu_{\tau u} \frac{\delta C_\tau^{x\xi}(\xi)}{\delta \xi^u} = \int_\tau^T du \lim_{d\tau \to 0} \frac{1}{d\tau} E_\tau \left[d\ln S_\tau d\xi^u \right] \frac{\delta C_\tau^{x\xi}(\xi)}{\delta \xi^u}$$

$$= \lim_{d\tau \to 0} \frac{1}{d\tau} E_\tau \left[d\ln S_\tau\, dC_\tau^{x\xi} \right]$$

$$= \int_\tau^T du \int_u^T du' \lim_{d\tau \to 0} \frac{1}{d\tau} E_\tau \left[d\ln S_\tau\, d\mu_{uu'} \right]$$

Note that the definition of $C_{\tau_1}^{x\xi}(\xi)$ in (8.13) has allowed for the following simplification in (8.15):

$$\int_t^T d\tau_1 \int_{\tau_1}^T du\, \mu_{\tau_1,u} C_{\tau_1}^{x\xi}(\xi) = -\int_t^T d\tau_1 \frac{dC_{\tau_1}^{x\xi}(\xi)}{d\tau_1} C_{\tau_1}^{x\xi}(\xi) = \frac{1}{2} C_t^{x\xi}(\xi)^2$$

The final expression of P at order two in ε at $t=0$ is thus:

$$P = \left[1 + \varepsilon \frac{C_0^{x\xi}(\xi_0)}{2} \partial_x(\partial_x^2 - \partial_x) \right.$$

$$+ \varepsilon^2 \left(\frac{C_0^{\xi\xi}(\xi_0)}{8}(\partial_x^2 - \partial_x)^2 + \frac{C_0^{x\xi}(\xi_0)^2}{8} \partial_x^2(\partial_x^2 - \partial_x)^2 + \frac{D_0(\xi_0)}{2} \partial_x^2(\partial_x^2 - \partial_x) \right) \right] P_0 \tag{8.18}$$

The subscript 0 in ξ_0 indicates that $C_0^{x\xi}$, $C_0^{\xi\xi}$, D_0 are evaluated in the unperturbed state, that is using the variance curve observed at time $t=0$.

Discussion

- The corrections to the Black-Scholes price P_0 are obtained in terms of derivatives of P_0 with respect to $x = \ln S$. One can check on (8.18) that the contributions at order 1 and 2 in ε are of the form:

$$\sum_{n=3}^\infty \alpha_n \left(\partial_x^n - \partial_x^2 \right) P_0 \tag{8.19}$$

which is in agreement with expression (5.88), page 193, for general perturbations of the lognormal density of S_T that leave the implied volatility of the log contract unchanged.

In our expansion in powers of ε, forward variances ξ^T are driftless: $E[\xi_T^T] = \xi_0^T$ at each order. Thus VS implied volatilities – equal to log contract implied volatilities since we are in a diffusive setting – stay fixed as ε is varied. This stands in contrast with other types of expansions whose accuracy is marred by the fact that the overall level of implied volatilities in the model shifts as the volatility-of-volatility parameter is increased.

- We have already derived the order-one expansion in the special case of the Heston model – see equation (6.15), page 209 – it is exactly as in (8.18).

- At second order in volatility of volatility, P only depends on 3 dimensionless model-dependent numbers: $C_0^{x\xi}(\xi_0)$, $C_0^{\xi\xi}(\xi_0)$ and $D_0^{\mu}(\xi_0)$ summarize the joint spot/variance dynamics of the model at hand. While $C_0^{x\xi}(\xi_0)$ and $C_0^{\xi\xi}(\xi_0)$ are integrals of the spot/variance and variance/variance covariance functions evaluated on the initial variance curve, $D_0^{\mu}(\xi_0)$ involves an extra degree of model-dependence as it depends on the derivative of $C_t^{x\xi}(\xi)$ with respect to ξ^u: it incorporates the additional information of how $\mu(t, u, \xi)$ changes as the variance curve changes.

- Comparison of expression (8.18) with expression (5.88), page 193, shows that the correction to P_0 at order 1 in ε is of the same form as that generated at order one by the third order cumulant of the distribution of $\ln S_T$. Indeed, the expansion at order one in volatility of volatility can be derived by direct calculation of the third-order cumulant κ_3 of $\ln S_T$, at this order. The interested reader will find the derivation in [11].

- The observant reader will have spotted in (8.18) what looks like the beginning of the expansion of $\exp\left(\varepsilon\frac{C_0^{x\xi}(\xi_0)}{2}\partial_x\left(\partial_x^2 - \partial_x\right)\right)$. This is not fortuitous – more on this in Appendix C, page 347.

8.3 Expansion of implied volatilities

We now convert the expansion for the price (8.18) into an expansion for implied volatilities – the (tedious but straightforward) derivation can be found in [13].

Remarkably, at order ε^2, implied volatilities are exactly quadratic in log-moneyness:[2]

$$\widehat{\sigma}(K,T) = \widehat{\sigma}_{F_T T} + \mathcal{S}_T \ln\left(\frac{K}{F_T}\right) + \frac{\mathcal{C}_T}{2} \ln^2\left(\frac{K}{F_T}\right) + O(\varepsilon^3) \qquad (8.20)$$

The ATMF volatility $\widehat{\sigma}_{F_T T}$, the ATMF skew $\mathcal{S}_T$ and curvature $\mathcal{C}_T$ are given by:

$$\widehat{\sigma}_{F_T T} = \widehat{\sigma}_T \left[1 + \frac{\varepsilon}{4Q}C^{x\xi}\right. \qquad (8.21a)$$
$$\left. + \frac{\varepsilon^2}{32Q^3}\left(12\left(C^{x\xi}\right)^2 - Q\left(Q+4\right)C^{\xi\xi} + 4Q\left(Q-4\right)D\right)\right]$$

$$\mathcal{S}_T = \widehat{\sigma}_T \left[\frac{\varepsilon}{2Q^2}C^{x\xi} + \frac{\varepsilon^2}{8Q^3}\left(4QD - 3\left(C^{x\xi}\right)^2\right)\right] \qquad (8.21b)$$

$$\mathcal{C}_T = \widehat{\sigma}_T \frac{\varepsilon^2}{4Q^4}\left(4QD + QC^{\xi\xi} - 6\left(C^{x\xi}\right)^2\right) \qquad (8.21c)$$

where $\widehat{\sigma}_T$ is the VS volatility for maturity T, $Q = \widehat{\sigma}_T^2 T$ and $C^{x\xi}, C^{\xi\xi}, D$ are compact notations for $C_0^{x\xi}(\xi_0), C_0^{\xi\xi}(\xi_0)$ and $D_0(\xi_0)$ defined in (8.13), (8.14), (8.17). As we use these formulas further below, we will set $\varepsilon = 1$.

- At order one in ε, from (8.21b), the ATMF skew is given by

$$\mathcal{S}_T = \widehat{\sigma}_T \frac{C^{x\xi}}{2(\widehat{\sigma}_T^2 T)^2} \qquad (8.22)$$

  where we have set $\varepsilon = 1$. Whenever spot and variances are uncorrelated $\mathcal{S}_T$ vanishes both at order ε and ε^2, and at all orders – as it should, since it is a well-known result that the smile is symmetric in log-moneyness for uncorrelated spot and variances.[3]

- At order one in ε, $\mathcal{S}_T$ is simply proportional to the doubly-integrated spot/variance covariance function. At this order, one recovers the correction at order one in the cumulant expansion in Appendix B of Chapter 5, contributed by the third-order cumulant, or equivalently the skewness s_T of $\ln S_T$. The interested reader will find in [11] the derivation of s_T at order one in ε:

$$s_T = \frac{3C^{x\xi}}{(\widehat{\sigma}_T^2 T)^{\frac{3}{2}}}$$

  Using then formula (5.93) relating the skewness of $\ln S_T$ to the ATMF skew:

$$\mathcal{S}_T = \frac{s_T}{6\sqrt{T}} \qquad (8.23)$$

  yields (8.22).

---

[2]The cancellation at order ε^2 of higher-order terms in log-moneyness was already noted in [69], for the particular case of a one-factor model. Note that a different convention for the normalization of $\mathcal{C}_T$ is used in [13] .

[3]See footnote 8, page 328.

- Using expression (8.13) for $C^{x\xi}$, (8.22) can be rewritten, more meaningfully, as:

$$\mathcal{S}_T = \frac{1}{2\widehat{\sigma}_T^3 T} \int_0^T \frac{T-\tau}{T} \frac{\langle d\ln S_\tau \, d\widehat{\sigma}_T^2(\tau)\rangle_0}{d\tau} d\tau \qquad (8.24)$$

where the 0 subscript indicates that the instantaneous covariation is evaluated with the initial VS term structure. The ATMF skew is the weighted average of the instantaneous covariance of the spot and the VS volatility for the residual maturity at future dates.

The sagacious reader will remember that this exact same relationship was derived in the context of the local volatility model – see formula (2.89), page 62. We show in the following section that (8.24) follows naturally from a representation of option prices in terms of expectations of spot/volatility and volatility/volatility gamma P&Ls.

Specializing to the case of a flat term structure of VS volatilities equal to $\widehat{\sigma}_T$:

$$\mathcal{S}_T = \frac{1}{\widehat{\sigma}_T^2 T} \int_0^T \frac{T-\tau}{T} \frac{\langle d\ln S_\tau \, d\widehat{\sigma}_T(\tau)\rangle_0}{d\tau} d\tau \qquad (8.25)$$

- At order one the ATMF volatility is given by:

$$\widehat{\sigma}_{F_T T} = \widehat{\sigma}_T + \frac{Q}{2}\mathcal{S}_T \qquad (8.26)$$

In (8.26) we recover, at order one in ε, the relationship between ATMF and VS volatilities that we had already derived in (3.30) at order one in $\mathcal{S}_T$.

- While the exact smile is arbitrage-free by construction, approximate prices (8.18) or implied volatilities (8.20) are not. Expression (8.20), which is quadratic in log-moneyness, is bound to generate arbitrage for very large or small strikes.

Indeed we know from [67] that asymptotically $\widehat{\sigma}^2(K,T)$ is at most an affine function of log-moneyness – see Section 4.3.1. The presence of arbitrage for far-away strikes can also be assessed on the density of $\ln S_T$ directly.

Consider the expansion of P in (8.18). As mentioned in the discussion above, the correction to P_0 is exactly of the form (5.88), obtained in a cumulant expansion of the Gaussian density of $x = \ln \frac{S_T}{F_T}$. This corresponds to a perturbation of the density of x, $\delta\rho(x)$, given by (5.85), which, using (5.83), can also be written as:

$$\frac{\delta\rho(x)}{\rho_0(x)} = \sum_{n=3}^{6} \frac{\delta\kappa_n}{(\widehat{\sigma}_T\sqrt{T})^n\sqrt{n!}} H_n\left(\frac{x + \frac{\widehat{\sigma}_T^2 T}{2}}{\widehat{\sigma}_T\sqrt{T}}\right)$$

where ρ_0 is the unperturbed Gaussian density with volatility $\widehat{\sigma}_T$ and H_n is the Hermite polynomial of degree n, defined in (5.84).

316

Stochastic volatility modeling

Expression (8.18) for P thus corresponds to a density for $\ln \frac{S_T}{F_T}$ which is the unperturbed Gaussian density multiplied by 1 plus a linear combination of Hermite polynomials in $\ln \frac{S_T}{F_T}$ of order up to 6.

For sufficiently large values of S_T this density may – and will – be negative. As we will see shortly, approximation (8.20) is most accurate for strikes near the money. The good agreement of (8.20) with the exact result implies that, in practice, arbitrage in this range of strikes is unlikely.

8.4 A representation of European option prices in diffusive models

We now give an alternative derivation of the order-one expansion that uses a representation of European option prices in general diffusive models in terms of spot/volatility and volatility/volatility gamma P&Ls.

In Section 2.4.1, page 39, we expressed the price of an option in a model with instantaneous volatility σ_2 as the sum of (a) the price in a base model with instantaneous volatility σ_1, (b) the expectation of the integral over the option's lifetime of the option's dollar gamma multiplied by the difference $(\sigma_2^2 - \sigma_1^2)$.

Imagine however that, in addition to delta-hedging, gamma risk is hedged away by dynamically trading vanilla options – or VSs. Our P&L now arises from the spot/volatility cross-gamma and the volatility gamma, and it should be possible to express the price of a European option as the cost of these two gammas. We now make this notion explicit, for general diffusive models, for European options.

Consider the process Q_t defined by:

$$Q_t = Q\left(t, S_t, \omega_t = \frac{1}{T-t} \int_t^T \xi^\tau d\tau\right)$$
$$Q(t, S, \omega) = e^{-rt} P_{\text{BS}}(t, S, \omega)$$

where $P_{\text{BS}}(t, S, \omega)$ is the option's price in a Black-Scholes model where we use variance ω rather than volatility, and ξ^τ are VS forward variances. With respect to the reasoning in Section 2.4.1, this amounts to using the VS of maturity T as a hedge instrument, in addition to S.

$Q(t, S, \omega)$ is the undiscounted price in a Black-Scholes model with constant variance ω, thus obeys the following PDE:

$$\frac{dQ}{dt} + (r - q) S \frac{dQ}{dS} + \frac{\omega}{2} S^2 \frac{d^2Q}{dS^2} = 0 \tag{8.27}$$

Moreover, the relationship between vega and gamma in the Black-Scholes model – see Appendix A of Chapter 5 – reads, in our context:

$$\frac{dQ}{d\omega} = \frac{T-t}{2} S^2 \frac{d^2Q}{dS^2} \tag{8.28}$$

During dt the variation of Q_t is given by:

$$
dQ_t = \frac{dQ}{dt} dt + \frac{dQ}{dS} dS_t + \frac{dQ}{d\omega} \left(\int_t^T \frac{\delta\omega}{\delta\xi^\tau} d\xi_t^\tau \, d\tau + \frac{d\omega}{dt} dt \right) + \frac{1}{2} \frac{d^2Q}{dS^2} \left\langle dS_t^2 \right\rangle
$$
$$
+ \frac{d^2Q}{dSd\omega} \left\langle dS_t d\omega_t \right\rangle + \frac{1}{2} \frac{d^2Q}{d\omega^2} \left\langle d\omega_t^2 \right\rangle
$$

where all derivatives are evaluated at t, S_t, ξ_t.

From the definition of ω, $\frac{d\omega}{dt} = \frac{\omega - \xi_t^t}{T-t} dt$. We then have, using (8.28):

$$
\frac{dQ}{d\omega} \frac{d\omega}{dt} = \frac{dQ}{d\omega} \frac{\omega - \xi_t^t}{T-t} = \frac{\omega - \xi_t^t}{2} S^2 \frac{d^2Q}{dS^2}
$$

Using now (8.27), we have:

$$
\frac{dQ}{dt} dt + \frac{dQ}{dS} dS_t + \frac{dQ}{d\omega} \left(\int_t^T \frac{\delta\omega}{\delta\xi^\tau} d\xi_t^\tau \, d\tau + \frac{d\omega}{dt} \right) + \frac{1}{2} \frac{d^2Q}{dS^2} \left\langle dS_t^2 \right\rangle
$$
$$
= -(r-q)S_t \frac{dQ}{dS} dt + \frac{dQ}{dS} dS_t + \frac{dQ}{d\omega} \int_t^T \frac{\delta\omega}{\delta\xi^\tau} d\xi_t^\tau \, d\tau
$$
$$
+ \frac{1}{2} S^2 \frac{d^2Q}{dS^2} \left((\omega - \xi_t^t) dt - \omega dt + \frac{\left\langle dS_t^2 \right\rangle}{S_t^2} \right)
$$
$$
= \frac{dQ}{dS} \left(dS_t - (r-q)S_t dt \right) + \frac{dQ}{d\omega} \int_t^T \frac{\delta\omega}{\delta\xi^\tau} d\xi_t^\tau \, d\tau
$$

where we have used that $\frac{\left\langle dS_t^2 \right\rangle}{S_t^2} = \xi_t^t$, a property of diffusive models.

The spot/gamma P&L has cancelled out. This is normal; because of the vega/gamma relationship for European options in the Black-Scholes model, the VS used as vega hedge also functions as a gamma hedge. We then have:

$$
dQ_t = \frac{dQ}{dS} \left(dS_t - (r-q)S dt \right) + \frac{dQ}{d\omega} \int_t^T \frac{\delta\omega}{\delta\xi^\tau} d\xi_t^\tau \, d\tau
$$
$$
+ \frac{d^2Q}{dSd\omega} \left\langle dS_t d\omega_t \right\rangle + \frac{1}{2} \frac{d^2Q}{d\omega^2} \left\langle d\omega_t^2 \right\rangle
$$

Let us now take the expectation of dQ_t. The first two contributions vanish as $E[dS_t - (r-q)S_t dt] = 0$ and the ξ_t^τ are martingales:

$$
E[dQ_t | S_t, \xi_t] = \frac{d^2Q}{dSd\omega} \left\langle dS_t d\omega_t \right\rangle + \frac{1}{2} \frac{d^2Q}{d\omega^2} \left\langle d\omega_t^2 \right\rangle
$$

Integrating now this expression on $[0, T]$:

$$E[Q_T] = Q_0 + E\left[\int_0^T e^{-rt}\left(\frac{d^2 P_{\text{BS}}}{dSd\omega}\langle dS_t d\omega_t\rangle + \frac{1}{2}\frac{d^2 P_{\text{BS}}}{d\omega^2}\langle d\omega_t^2\rangle\right)\right]$$

At $t = T$, $Q_T = e^{-rT}P(T, S_T, \omega_T) = e^{-rT}f(S_T)$ where f is the option's payoff. $E[Q_T]$ is thus simply the option's price.

We then have our final representation of the price of a European option in a diffusive model as the sum of the Black-Scholes price with the initial VS volatility, augmented by the expectation of the sum of spot/volatility and volatility/volatility gamma P&Ls, the volatility being the VS volatility for the residual maturity:

$$P = P_{BS}\left(0, S_0, \widehat{\sigma}_T^2(0)\right)$$
$$+ E\left[\int_0^T e^{-rt}\left(\frac{d^2 P_{\text{BS}}}{dSd(\widehat{\sigma}_T^2)}\langle dS_t d\widehat{\sigma}_T^2(t)\rangle + \frac{1}{2}\frac{d^2 P_{\text{BS}}}{(d(\widehat{\sigma}_T^2))^2}\langle d\widehat{\sigma}_T^2(t)d\widehat{\sigma}_T^2(t)\rangle\right)\right]$$
$$(8.29)$$

where $\widehat{\sigma}_T^2(t) = \omega_t$ is the square of the VS volatility of maturity T at time t.

Unlike representation (2.30), page 40, based on spot/spot gamma P&L, (8.29) involves spot/volatility and volatility/volatility gamma P&Ls.

The above derivation is equivalent to considering the P&L of a delta-hedged, vega-hedged (and also gamma-hedged) position, the vega-hedge instruments being VSs.

Can we obtain a similar formula using implied volatilities of other payoffs, for example ATMF options rather than VSs? The answer is no, the reason being that the square of the VS implied volatility is what comes closest to a price – it has zero drift.

8.4.1 Expansion at order one in volatility of volatility

With volatility of volatility switched off, the ξ^τ are frozen, and $\widehat{\sigma}_T^2$ has only a drift: $\langle dS_t d\widehat{\sigma}_T^2\rangle = \langle d\widehat{\sigma}_T^2 d\widehat{\sigma}_T^2\rangle = 0$.

Imagine scaling the volatilities of the ξ^τ by a factor ε. Then $\langle dS_t d\widehat{\sigma}_T^2\rangle$ is of order one in ε, and $\langle d\widehat{\sigma}_T^2 d\widehat{\sigma}_T^2\rangle$ is of order two. At order one in ε, P is thus given by:

$$P = P_0 + E\left[\int_0^T e^{-rt}\frac{d^2 P_{\text{BS}}}{dSd(\widehat{\sigma}_T^2)}\langle dS_t d\widehat{\sigma}_T^2(t)\rangle\right] \qquad (8.30)$$

where the expectation is taken with respect to the density at order zero – that is the lognormal density of a Black-Scholes model with instantaneous volatility $\sqrt{\xi_0^t}$ and $P_0 = P_{BS}(0, S_0, \omega_0 = \widehat{\sigma}_0^2)$.

Using $x = \ln S$ the vega/gamma relationship (8.28) reads:

$$\frac{dP_{\mathrm{BS}}}{d(\widehat{\sigma}_T^2)} = \frac{T-t}{2}\left(\frac{d^2 P_{\mathrm{BS}}}{dx^2} - \frac{dP_{\mathrm{BS}}}{dx}\right)$$

Switching from S_t to $\ln S_t$:

$$P = P_0 + E\left[\int_0^T e^{-rt}\frac{d^2 P_{\mathrm{BS}}}{d\ln Sd(\widehat{\sigma}^2)}\left\langle d\ln S_t d\widehat{\sigma}_T^2\right\rangle\right]$$

$$= P_0 + E\left[\int_0^T e^{-rt}\frac{T-t}{2}\left(\frac{d^3 P_{\mathrm{BS}}}{dx^3} - \frac{d^2 P_{\mathrm{BS}}}{dx^2}\right)\left\langle d\ln S_t d\widehat{\sigma}_T^2\right\rangle\right]$$

Let us assume that $E\left[\left\langle d\ln S_t d\widehat{\sigma}_T^2\right\rangle \mid x\right]$ does not depend on x.

We can then use the property – see Appendix A of Chapter 5 – that, in the Black-Scholes model $E\left[e^{-rt}\frac{d^n P}{dx^n}\right] = \frac{d^n P_0}{dx^n}$ and we get:

$$P = P_0 + \frac{1}{2}E\left[\int_0^T (T-t)\left\langle d\ln S_t d\widehat{\sigma}_T^2\right\rangle\right]\left(\frac{d^3 P_{\mathrm{BS}}}{dx^3} - \frac{d^2 P_{\mathrm{BS}}}{dx^2}\right) \qquad (8.31)$$

This recovers the expansion at order one in (8.18) – we refer the reader to the expression of $C^{x\xi}$ in (8.13). We can then immediately write down the formula for the ATMF skew:

$$\mathcal{S}_T = \frac{1}{2\widehat{\sigma}_T^3(0)T}\int_0^T \frac{T-t}{T}\frac{\left\langle d\ln S_t\, d\widehat{\sigma}_T^2(t)\right\rangle_0}{dt}dt \qquad (8.32)$$

where the subscript 0 signals that the instantaneous covariation, as a function of forward variances, is evaluated using initial values for the latter. This is exactly formula (8.24). Note that we can substitute in (8.32) the VS volatility with the ATMF volatility as they are identical at order one in volatility of volatility.

It is now clear why, in the derivation of the expansion at order two in Section 8.2, we needed the assumption that μ does not depend on S.

Local volatility

We have already encountered formula (8.32) in the context of local volatility in Chapter 2 – see Section 2.5.7, page 61. In that context, $\widehat{\sigma}_T(t)$ is the ATMF volatility, rather than the VS volatility, but that is fine, as at order zero in volatility of volatility they are identical.

The reason (8.32) holds in the local volatility model is we have used a local volatility function of type:

$$\sigma(t,S) = \sigma_0 + \alpha(t)\ln\frac{S}{F_t}$$

From equation (2.88), page 61, at order one in $\alpha(t)$, the covariance of $\widehat{\sigma}^2_{F_T T}$ with $\ln S$ is:

$$\langle d \ln S_t d\widehat{\sigma}^2_T \rangle = 2\sigma_0^2 \left(\frac{1}{T-t} \int_t^T \alpha(\tau)d\tau \right) dt$$

It is independent of S, hence (8.32) applies.

8.4.2 Materializing the spot/volatility cross-gamma P&L

At order one in volatility of volatility, and assuming that the covariance of $\ln S$ and $\widehat{\sigma}^2_T$ does not depend on S, the price of a European payoff is given by (8.31). While $\frac{d^3 P_{BS}}{dx^3} - \frac{d^2 P_{BS}}{dx^2}$ is payoff-dependent, the prefactor involving the weighted average of the spot/volatility covariance is not. Thus, at order one in volatility of volatility, this quantity could be read off any European payoff's market price.

Practically however, backing out of option prices an implied value for the integrated spot/volatility covariance is useful only if the latter can be materialized as a cross-gamma P&L. Does there exist a payoff such that by delta-hedging it and vega-hedging it with VSs we generate as a cross-gamma P&L the integrated spot/volatility covariance in (8.31)? In other words, which is the payoff out of which a measure of the *implied* integrated spot/volatility covariance can be extracted?

This payoff is $\ln^2(S/S_0)$. In the Black-Scholes model the price of payoff $\ln^2(S_T/S_0)$ is given by:

$$P_{BS} = e^{-r\tau} \left[\ln^2 \frac{S}{S_0} + \left(r - q - \frac{\sigma^2}{2} \right)^2 \tau^2 + 2\left(r - q - \frac{\sigma^2}{2} \right)\tau \ln \frac{S}{S_0} + \sigma^2 \tau \right]$$

where $\tau = T - t$. The derivative with respect to σ^2 is given by:

$$\frac{dP_{BS}}{d(\sigma^2)} = -e^{-r(T-t)} (T - t) \left[\ln \frac{S}{S_0} + \left(r - q - \frac{\sigma^2}{2} \right)(T - t) - 1 \right]$$

By choosing $S_0 = S_{t=0} e^{(r-q-\frac{\widehat{\sigma}^2_T(0)}{2})T}$, the VS hedge ratio at $t = 0$ vanishes. The cross-derivative with respect to $\ln S$ and σ^2 is given by:

$$\frac{d^2 P_{BS}}{d(\sigma^2)d\ln S} = -e^{-r(T-t)} (T - t)$$

Consider now delta-hedging and vega-hedging with VSs a short position in this payoff. Summing spot/volatility cross-gamma P&Ls and discounting them to $t = 0$ yields the following expression for our P&L over $[0, T]$:

$$P\&L = e^{-rT} \sum_i (T - t_i)\ \delta \ln S_i\ \delta(\widehat{\sigma}^2_T(t_i)) \tag{8.33}$$

Up to factor e^{-rT}, this is exactly the prefactor in (8.31).

At order one in volatility of volatility the market price of payoff $\ln^2(S_T/S_0)$ minus its Black-Scholes price calculated with the VS volatility at $t = 0$ thus provides a measure of the implied value of the integrated spot/volatility covariance.[4]

As a European payoff, $\ln^2(S/S_0)$ can be replicated with vanilla options. The density $\rho(K)$ of vanilla options of strike K is given by the second derivative of the payoff with respect to S:

$$\rho(K) = \frac{2}{K^2}\left(1 - \ln\frac{K}{S_0}\right)$$

$\rho(K)$ is positive for $K \ll S_0$ and negative for $K \gg S_0$; the price of $\ln^2(S_T/S_0)$ is a global measure of the slope of the smile.

8.5 Short maturities

We resume now our discussion of expansion (8.21) and first consider the limit of vanishing maturities. This special case is worth investigating as the variance curve then collapses to a single object – the instantaneous variance – whose dynamics determines the smile. This allows for a particularly simple characterization of the smile at order two in volatility of volatility.

In addition, while $\hat{\sigma}_{F_T T}, S_T, C_T$ in (8.21) have been derived in an expansion at order two in volatility of volatility, it turns out that the expressions they provide for the ATM volatility, skew and curvature are exact in the limit $T \to 0$.

Let us take the limit $T \to 0$ in equations (8.21). We assume that the covariance functions μ and ν are smooth in $t = 0$. From their expressions in (8.13), (8.14), (8.17), $C^{x\xi}$ is of order T^2, $C^{\xi\xi}$ of order T^3 and D of order T^3.[5] At leading order in T:

$$C^{x\xi} = \frac{T^2}{2}\mu_0$$

$$C^{\xi\xi} = \frac{T^3}{3}\nu_0$$

$$D = \frac{T^3}{6}\mu_0\frac{d\mu_0}{d\xi_0^0}$$

where we have used the compact notation: $\mu_0 = \mu\left(0, 0, \xi_0^0\right)$, $\nu_0 = \nu\left(0, 0, 0, \xi_0^0\right)$.

---

[4]The total P&L at order two in $\delta \ln S$ and $\delta\hat{\sigma}$ also comprises a contribution from the volatility/volatility gamma. We have: $\frac{d^2 P_{\text{BS}}}{(d\sigma^2)^2} = e^{-r\tau}\frac{\tau^2}{2}$. A short position in payoff $\ln^2(S_T/S_0)$, delta-hedged and vega-hedged with VSs thus generates, in addition to (8.33), the volatility/volatility gamma P&L $-e^{-rT}\Sigma_i\frac{(T-t_i)^2}{2}\delta(\hat{\sigma}_T^2(t_i))^2$. This contribution is of second order in volatility of volatility.

[5]Despite what expression (8.17) suggests D is not of order 4. Unlike a standard derivative, a functional derivative has an additional dimension $\frac{1}{T}$: $\frac{\delta C}{\delta\xi^u}$ is such that $dC = \int_0^T \frac{\delta C}{\delta\xi^u}d\xi^u du$.

- In stochastic volatility models – in contrast with jump/Lévy models – the drivers in the dynamics of S_t are Brownian motions: the non-Gaussian character of $\ln S_T$ is only generated by the volatility of volatility. As $T \to 0$ the distribution of $\ln S_T$ does become Gaussian, but exactly how fast it becomes Gaussian determines the behavior of the smile for short maturities. Consider the skewness s_T of $\ln S_T$ and the ATMF skew $\mathcal{S}_T$, which are related at order one in volatility of volatility. Using (8.23), (8.21b) and the above expression for $C^{x\xi}$ we get the following expression for s_T, at order one in volatility of volatility:

$$s_T = \frac{3}{2} \frac{\mu_0}{\hat{\sigma}_T^3} \sqrt{T} \tag{8.34}$$

This expression is instructive: as $T \to 0$, s_T vanishes like $\sqrt{T}$. (8.23) then implies that $\mathcal{S}_T$ tends to a constant. Only if the *skewness* vanishes faster than $\sqrt{T}$ does the short-maturity ATMF *skew* tend to zero.

Keeping only terms at leading order in T and setting $\varepsilon = 1$ we get:

$$\hat{\sigma}_{S,T=0} = \hat{\sigma}_0 \tag{8.35a}$$

$$\mathcal{S}_0 = \hat{\sigma}_0 \frac{1}{4 \left(\xi_0^0\right)^2} \mu_0 \tag{8.35b}$$

$$\mathcal{C}_0 = \hat{\sigma}_0 \frac{1}{4 \left(\xi_0^0\right)^4} \left(\frac{2}{3} \xi_0^0 \mu_0 \frac{d\mu_0}{d\xi_0^0} + \frac{1}{3} \xi_0^0 \nu_0 - \frac{3}{2} \mu_0^2 \right) \tag{8.35c}$$

where $\hat{\sigma}_0 = \sqrt{\xi_0^0}$.

- In the limit $T \to 0$ the ATM volatility is equal to the VS volatility.

- Remembering that $\mu_0 = \frac{\langle d\ln S_0 d\xi_0^0 \rangle}{dt}$ we see from (8.35b) that, as $T \to 0$, the short ATM skew tends to a finite value, which is a direct measure of the instantaneous covariance at $t = 0$ of $\ln S_t$ and the instantaneous variance. Thus, the short spot/variance covariance can be read off the smile in model-independent fashion. Using volatilities rather than variances yields the following expression:

$$\mathcal{S}_0 = \frac{1}{2\hat{\sigma}_0^2} \frac{\langle d\ln S d\hat{\sigma}_0 \rangle}{dt} \tag{8.36}$$

where, from (8.35a), $\hat{\sigma}_0$ is the short ATM volatility. Anticipating on the following chapter and using the general definition of the SSR: $\mathcal{R}_T = \frac{1}{\mathcal{S}_T} \frac{E[d\ln S \, d\hat{\sigma}_{F_T(S)T}]}{E[(d\ln S)^2]}$ – see page 358 – (8.36) is equivalent to the property:

$$\mathcal{R}_0 = 2$$

- The short curvature $\mathcal{C}_0$, on the other hand, depends not only on μ_0 and ν_0, but also involves the quantity $\frac{d\mu_0}{d\xi_0^0}$, which quantifies how the short skew varies as the instantaneous variance changes. $\mu_0 \frac{d\mu_0}{d\xi_0^0}$ can be written in terms of the covariance of $\ln S$ and the short ATMF skew. Using the definition of ν_0 and expressing everything in terms of volatilities, we get:

$$\mathcal{C}_0 = \frac{1}{4\hat{\sigma}_0}\left(\frac{8}{3}\frac{\langle d\ln S d\mathcal{S}_0\rangle}{\hat{\sigma}_0 dt} + \frac{4}{3}\frac{\langle d\hat{\sigma}_0 d\hat{\sigma}_0\rangle}{\hat{\sigma}_0^2 dt} - 8\mathcal{S}_0^2\right) \qquad (8.37)$$

While the short spot/variance covariance can be read off the market smile directly, inverting equation (8.37) does not provide a model-independent value for the short variance of volatility $\langle d\hat{\sigma}_0 d\hat{\sigma}_0\rangle$. In other words it is not possible to read the level of volatility of volatility off the market smile in model-independent fashion. One needs to make an assumption on how the short skew changes as the spot moves; this is quantified in $\langle d\ln S d\mathcal{S}_0\rangle /dt$.

Results (8.36) and (8.37) are practically useful as they characterize near-the-money implied volatilities of general stochastic volatility models in terms of the joint dynamics of financial observables such as $\ln S$, the ATM – or VS – volatility $\hat{\sigma}_0$, and the short ATMF skew $\mathcal{S}_0$.

These relationships are specific to the case of vanishing maturities. They are very general as they also hold for local volatility or mixed local-stochastic volatility models. They can be derived many other ways – see for example [43], [44], [76]. One can in fact get exact expressions for short-dated smiles in general stochastic volatility models at order zero in T – see [7] – and at order one in T – see [56].

Equation (8.36) which relates the spot/short ATMF volatility covariance to the ATMF skew in model-independent fashion is equivalent to the statement that

$$\mathcal{R}_{T=0} = 2$$

in diffusive models. See Section 9.11.1 for a comparison of the SSR in local volatility and stochastic volatility models.

We now consider two examples of volatility dynamics.

8.5.1 Lognormal ATM volatility – SABR model

Assume that the short ATM volatility $\hat{\sigma}_0$ is lognormal; denote by ν its volatility and ρ its correlation with S:

$$dS = \hat{\sigma}_0 S \, dW^S \qquad (8.38a)$$

$$d\hat{\sigma}_0 = \bullet \, dt + \nu\hat{\sigma}_0 \, dW^{\hat{\sigma}_0} \qquad (8.38b)$$

where we leave the drift unspecified as they are immaterial for $T \to 0$. Using expressions (8.36) and (8.37) we get for $T \to 0$:

$$\mathcal{S}_0 = \frac{\rho}{2} \nu \tag{8.39a}$$

$$\mathcal{C}_0 = \frac{1}{6\hat{\sigma}_0} \left(2 - 3\rho^2\right) \nu^2 \tag{8.39b}$$

- In a model where the short ATM volatility $\hat{\sigma}_0$ is lognormal – such as the lognormal model for forward variances of Section 7.4 – the short skew is constant, independent on $\hat{\sigma}_0$, and the curvature is inversely proportional to $\hat{\sigma}_0$.

- This has important implications for the pricing of cliquets. In a model with lognormal instantaneous variances/volatilities, while the short forward ATM volatility at time t, $\hat{\sigma}_0(t)$, is random, the level of short forward skew $\mathcal{S}_0(t)$ is fixed: prices of narrow forward ATM call spreads with short maturities are approximately independent on the level of volatility of volatility.

- The relationship between skew, curvature and volatility of volatility reads:

$$\nu^2 = 3\hat{\sigma}_0 \mathcal{C}_0 + 6\mathcal{S}_0^2 \tag{8.40}$$

The dynamics in (8.38) is the short-maturity limit of the SABR model with $\beta = 1$. We leave it to the reader to check that the ATM skew and curvature obtained from the well-known SABR formula for $T \to 0$ indeed yield (8.39a) and (8.39b) – see [55].

8.5.2 Normal ATM volatility – Heston model

Consider now the case of a normal ATM volatility – to avoid any confusion with the lognormal case denote by σ the normal volatility of volatility:

$$dS = \hat{\sigma}_0 S \, dW^S \tag{8.41a}$$

$$d\hat{\sigma}_0 = \bullet \, dt + \sigma \, dW^{\hat{\sigma}_0} \tag{8.41b}$$

Using (8.36), (8.37) we now get for $T \to 0$:

$$\mathcal{S}_0 = \frac{\rho}{2} \frac{\sigma}{\hat{\sigma}_0} \tag{8.42a}$$

$$\mathcal{C}_0 = \frac{1}{6\hat{\sigma}_0} \left(2 - 5\rho^2\right) \left(\frac{\sigma}{\hat{\sigma}_0}\right)^2 \tag{8.42b}$$

- If the short ATM volatility $\hat{\sigma}_0$ is normal then the short skew is inversely proportional to $\hat{\sigma}_0$ and the curvature is inversely proportional to $\hat{\sigma}_0^3$.

- $\mathcal{S}_0, \mathcal{C}_0, \sigma$ are related through:

$$\left(\frac{\sigma}{\hat{\sigma}_0}\right)^2 = 3\hat{\sigma}_0 \mathcal{C}_0 + 10\mathcal{S}_0^2 \tag{8.43}$$

Compare results in the normal model with those in the lognormal model; imagine that ν and σ are such that the instantaneous volatility of $\hat{\sigma}_0$ at $t = 0$ is identical in both models: $\nu = \frac{\sigma}{\hat{\sigma}_0}$. Comparison of (8.39a, 8.39b) and (8.42a, 8.42b), shows that, while the values of $\mathcal{S}_0$ in both models are identical, values of $\mathcal{C}_0$ are not: the smile curvature depends not only on the spot/variance covariances at $t = 0$ but also on their dependence on the level of volatility. As mentioned in the comment that follows (8.37), extracting the level of volatility of volatility out of market smiles requires a modeling assumption. This is clearly demonstrated in the comparison of (8.40) and (8.43).

- The dynamics in (8.41) is in fact the short-maturity limit of the Heston model – the reader can check that expression (8.42a) for $\mathcal{S}_0$ agrees with (6.18b). Note that σ in Section 6.1 denotes the normal volatility of the instantaneous *variance* – twice the volatility of the instantaneous *volatility*. (8.42a) shows that the Heston model embeds a structural connection between the level of short forward ATM volatility and skew: cliquets of narrow forward ATM call spreads will exhibit a non-vanishing vega. This only occurs because of the hard-wired relationship $\mathcal{S}_0(t) \propto \frac{1}{\hat{\sigma}_0(t)}$.

8.5.3   Vanishing correlation – a measure of volatility of volatility

Imagine there is no correlation between S and the short ATM volatility. Then the ATM skew vanishes – $\mathcal{S}_0 = 0$ – and from (8.37) the smile curvature is given by:

$$\mathcal{C}_0 = \frac{1}{3\hat{\sigma}_0} \frac{\langle d\hat{\sigma}_0 d\hat{\sigma}_0 \rangle}{\hat{\sigma}_0^2 dt} \qquad (8.44)$$

This is correct at second order in volatility of volatility – the first non-trivial contribution when the spot/volatility correlation vanishes. The ATM smile curvature is then a direct measure of the volatility of the short-maturity ATM volatility. Since the spot/volatility correlation vanishes, $\mu(t, u, \xi) = 0$ and the condition that μ not depend on S is not needed anymore.

(8.44) is thus general – it holds in all diffusive models – no assumption has been made about $\nu(t, u, u', \xi)$.

8.6   A family of one-factor models – application to the Heston model

Here, we illustrate how our framework equally applies to first-generation models built on the dynamics of the instantaneous variance V_t – such as those examined in

[69]. Consider a one-factor mean-reverting model of the following type:

$$\begin{cases} dS_t = \sqrt{V_t}S_t dW_t^S \\ dV_t = -k\left(V_t - V^0\right)dt + \sigma V_t^\varphi dW_t^V \end{cases} \tag{8.45}$$

where the correlation between the Brownian motions W^S and W^V is ρ.

The Heston model, covered in Chapter 6, corresponds to $\varphi = \frac{1}{2}$.

Forward variance ξ_t^T is given by: $\xi_t^T = E_t[V_T]$. Taking the conditional expectation of both sides of (8.45) and integrating with respect to t yields:

$$\xi_t^T = E_t\left[V_T\right] = V^0 + \left(V_t - V^0\right)e^{-k(T-t)} \tag{8.46}$$

The (driftless) dynamics of ξ_t^T is:

$$d\xi_t^T = e^{-k(T-t)}\sigma\,\xi_t^{t\,\varphi}\,dW_t^V \tag{8.47}$$

As is typical of first-generation stochastic volatility models, k determines both the term structure of forward variances (8.46) and their volatilities (8.47).

The spot/variance and variance/variance covariance functions are given by:

$$\mu\left(t,u,\xi\right) = \rho\sigma e^{-k(u-t)}\,\xi_t^{t\,\varphi+\frac{1}{2}} \tag{8.48a}$$

$$\nu\left(t,u,u',\xi\right) = \sigma^2 e^{-k(u-t)}e^{-k(u'-t)}\,\xi_t^{t\,2\varphi} \tag{8.48b}$$

From these expressions $C^{x\xi}, C^{\xi\xi}$, D are easily computed and can be inserted in (8.21) to obtain the smile at order two in σ: the reader can check that one recovers the expressions derived in [69]. In particular, using (8.35b), the short ATMF skew is given by:

$$\mathcal{S}_0 = \frac{\rho\sigma}{4}V_{t=0}^{(\varphi-1)} \tag{8.49}$$

Take the particular case of the Heston model – $\varphi = \frac{1}{2}$ – and consider the expansion of the option price at order one. From (8.18), it is given by:

$$P = P_0 + \varepsilon\frac{C_0^{x\xi}(\xi_0)}{2}\left(\partial_x^3 - \partial_x^2\right)P_0$$

Going back to expression (6.15) in Chapter 6, the interpretation of the prefactor is now clear: it is the doubly-integrated spot/variance covariance function $C_0^{x\xi}$.

8.7 The two-factor model

We now consider the model introduced in Section 7.4, defined by the following SDEs:

$$\begin{cases} dS_t = \sqrt{\xi_t^t}S_t dW_t^S \\ d\xi_t^T = 2\nu\xi_t^T\,\alpha_\theta\left((1-\theta)e^{-k_1(T-t)}dW_t^1 + \theta e^{-k_2(T-t)}dW_t^2\right) \end{cases}$$

where ν is the volatility of a VS volatility of vanishing maturity, $\alpha_\theta = 1/\sqrt{(1-\theta)^2 + \theta^2 + 2\rho_{12}\theta(1-\theta)}$ and ρ_{12} is the correlation between W^1 and W^2. We denote by ρ_{SX^1} (ρ_{SX^2}) the correlation between W^S and W^1 (W^2). We would like to answer the following questions:

- Is the accuracy of the order-two approximation (8.21) sufficient for practical purposes?

- Does the two-factor model afford sufficient flexibility as to the type of smiles it is able to generate? Can we obtain a term-structure of ATMF skews that is consistent with typical index smiles?[6]

Among the parameters of the two-factor model, the subset $\nu, \theta, k_1, k_2, \rho_{12}$ determines the dynamics of the VS volatilities in the model. Once these parameters are set, the dynamics of VS volatilities is set; we can then select the additional parameters ρ_{SX^1} and ρ_{SX^2} to generate the desired spot/volatility dynamics and the ensuing vanilla smile. We use in our tests below Set II parameters – see Table 8.1.

ν	θ	k_1	k_2	ρ_{12}
174%	0.245	5.35	0.28	0%

Table 8.1: Numerical values of parameters used in tests of the two-factor model.

Set II generates a term structure of instantaneous volatilities of VS volatilities displayed in Figure 7.1, page 228, which reproduces with good accuracy the power-law dependence (7.40) with $\alpha = 0.4$ and a volatility of a 3-month VS volatility equal to 100%. Volatilities of VS volatilities in the two-factor model are given by (7.39), page 227.

$\mu(t, u, \xi)$ and $\nu(t, u, u', \xi)$ are given by:

$$\mu(t, u, \xi) = 2\nu\xi^u\sqrt{\xi^t}\alpha_\theta\left[\rho_{SX^1}(1-\theta)e^{-k_1(u-t)} + \rho_{SX^2}\theta e^{-k_2(u-t)}\right] \quad (8.50)$$

$$\nu(t, u, u', \xi) = 4\nu^2\xi^u\xi^{u'}\alpha_\theta^2\Big[(1-\theta)^2e^{-k_1(u+u'-2t)} + \theta^2e^{-k_2(u+u'-2t)}$$

$$+ \rho_{12}\theta(1-\theta)\left(e^{-k_1(u-t)}e^{-k_2(u'-t)} + e^{-k_2(u-t)}e^{-k_1(u'-t)}\right)\Big] \quad (8.51)$$

$C^{x\xi}, C^{\xi\xi}, D$ which are multiple integrals of μ and ν can be efficiently calculated numerically by Gaussian quadrature.[7] Because μ and ν are smooth functions of t, u, u', very few points are needed. In the special case of a flat term structure of VS

---

[6]This is an important question for pricing the forward-smile risk of cliquets.

[7]Whenever, as is natural, the variance curve is generated by interpolating $T\hat{\sigma}_T^2$ as a piecewise affine function of T, forward variances ξ^T are piecewise constant. $C^{x\xi}, C^{\xi\xi}, D$ can be computed analytically.

volatilities, ξ_t^u does not depend on u and the integrations can be done analytically. The analytical expressions of $C^{x\xi}$, $C^{\xi\xi}$, D – which we use in our tests below – can be found in [13].

8.7.1 Uncorrelated case

Consider first the case when the correlation of forward variances ξ_t^u with S_t vanishes: $\rho_{SX^1} = \rho_{SX^2} = 0$, thus $\mu\,(\tau, u, \xi) = 0$ and $C^{x\xi} = D = 0$. Expressions (8.21) become:

$$\hat\sigma_{F_TT} = \hat\sigma_T \left(1 - \frac{\varepsilon^2}{32Q^2}\,(Q+4)\,C^{\xi\xi}\right)$$

$$\mathcal{S}_T = 0$$

$$\mathcal{C}_T = \hat\sigma_T \frac{\varepsilon^2}{4Q^3} C^{\xi\xi}$$

As mentioned before, $\mathcal{S}_T$ vanishes both at order one and two in ε – as it should: it is well known that for uncorrelated spot and variances, the smile is symmetric in log-moneyness.[8] Besides, the order-one contributions to $\hat\sigma_{F_TT}$ and $\mathcal{C}_T$ vanish altogether.

Figure 8.1 shows a comparison of exact and approximate smiles for four different maturities, for a flat term structure of VS volatilities at 20%. "Exact" smiles are obtained by Monte Carlo simulation using the gamma/theta technique – see Section 8.10, page 336. As already mentioned, the order-two approximation is bound to differ markedly from the exact result for far-away strikes. Note however how accurate it is for near-the-money strikes.

8.7.2 Correlated case – the ATMF skew and its term structure

We now test the order-two approximation in the correlated case. We still use Set II parameters and take $\rho_{SX^1} = -75.9\%, \rho_{SX^2} = -48.7\%$ – more on the choice of these correlations below. The full list of parameters appears in Table 8.2 below.

This exact same set has been used in the tests of Section 7.8.4 of Chapter 7, devoted to the vanilla smile of discrete forward variance models.

Exact and approximate smiles at order one and two in volatility of volatility are shown in Figure 8.2. As is apparent, while smiles at order one and two in ε have

---

[8]Given a strike K, denote by $K^\star$ the strike with opposite log-moneyness: $K^\star = F_T^2/K$. If forward variances are uncorrelated with S_t, conditional on the path of $\xi_t^t, t \in [0, T]$, S_T is lognormally distributed with a volatility σ given by: $\sigma^2 T = \int_0^T \xi_t^t dt$. The price of a call option of strike K is then equal to $\int C_K^{BS}\,(\sigma)\,\rho\,(\sigma)\,d\sigma$ where $C_K^{BS}\,(\sigma)$ is the Black-Scholes formula with volatility σ and $\rho(\sigma)$ is the density of σ. Denote by $\hat\sigma_K$ the implied volatility for strike K: $C_K^{BS}(\hat\sigma_K) = \int C_K^{BS}\,(\sigma)\,\rho\,(\sigma)\,d\sigma$. Consider now a put option of strike $K^\star$ and use the following relationship between prices of call and put options with opposite log-moneyness in the Black-Scholes model: $P_{K^\star}^{BS} = \frac{F_T}{K}C_K^{BS}$. By definition of $\hat\sigma_{K^\star}, P_{K^\star}^{BS}(\hat\sigma_{K^\star}) = \int P_{K^\star}^{BS}\,(\sigma)\,\rho\,(\sigma)\,d\sigma = \int \frac{F_T}{K}C_K^{BS}\,(\sigma)\,\rho\,(\sigma)\,d\sigma = \frac{F_T}{K}C_K^{BS}(\hat\sigma_K) = P_K^{BS}(\hat\sigma_K)$. Thus $\hat\sigma_{K^\star} = \hat\sigma_K$.

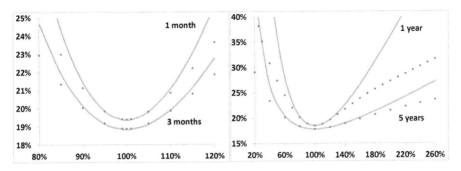

Figure 8.1: Exact (dots) and approximate (continuous line) smiles of the two-factor model with Set II parameters and uncorrelated spot and variances, for four different maturities: 1 month, 3 months, 1 year, 5 years. The term structure of VS volatilities is flat at 20%. The algorithm used is that in Section A.2 of Appendix A.

ν	θ	k_1	k_2	ρ_{12}	ρ_{SX^1}	ρ_{SX^2}
174%	0.245	5.35	0.28	0%	-75.9%	-48.7%

Table 8.2: Numerical values of parameters of the two-factor model.

similar shapes, the overall volatility level – and especially the ATMF volatility – is much better captured at order two in ε.

The accuracy of expression (8.21b) for the ATMF skew $\mathcal{S}_T$, however, is excellent already at order one in ε – as highlighted in [9]. At this order (8.21b) simplifies to:

$$\mathcal{S}_T^{\text{order }1} = \hat{\sigma}_T \frac{\varepsilon}{2Q^2} C^{x\xi} \tag{8.52}$$

In the two-factor model, for a flat term structure of forward variances equal to ξ_0, $\mu(t, u, \xi)$ is given by:

$$\mu(t, u, \xi) = 2\nu \xi_0^{\frac{3}{2}} \alpha_\theta \left[(1 - \theta)\rho_{SX^1} e^{-k_1(u-t)} + \theta \rho_{SX^2} e^{-k_2(u-t)} \right] \tag{8.53}$$

where $\alpha_\theta = \left((1 - \theta)^2 + \theta^2 + 2\rho_{12}\theta(1 - \theta)\right)^{-\frac{1}{2}}$. Plugging this in expression (8.13) for $C^{x\xi}$ yields, after setting $\varepsilon = 1$:

$$\mathcal{S}_T^{\text{order }1} = \frac{\nu \alpha_\theta}{\hat{\sigma}_T^3 T^2} \int_0^T dt \sqrt{\xi_0^t} \int_t^T du \xi_0^u \left[(1 - \theta)\rho_{SX^1} e^{-k_1(u-t)} + \theta \rho_{SX^2} e^{-k_2(u-t)} \right] \tag{8.54}$$

where $\hat{\sigma}_T = \sqrt{\frac{1}{T} \int_0^T \xi_0^t dt}$.

In the case a flat term structure of forward variances/VS volatilities, the double integrals in (8.54) can be done analytically and we get:

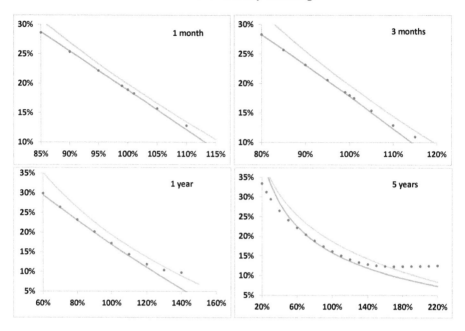

Figure 8.2: Exact (dots) as well as approximate smiles at order one (light line) and order two (dark line) in ε, for parameters in Table 8.2, and a flat term structure of VS volatilities at 20%. The algorithm used is that in Section A.2 of Appendix A.

$$S_T^{\text{order 1}} = \nu\alpha_\theta\left[(1-\theta)\,\rho_{SX^1}\,\frac{k_1 T - (1 - e^{-k_1 T})}{(k_1 T)^2} + \theta\rho_{SX^2}\,\frac{k_2 T - (1 - e^{-k_2 T})}{(k_2 T)^2}\right]$$

$$(8.55)$$

The exact ATMF skew, together with the order-two (8.21b) and order-one expression (8.55) is displayed in Figure 8.3. Figure 8.3 actually shows the difference of the implied volatilities for strikes $0.99 F_T$ and $1.01 F_T$, approximately equal to $-0.02 S_T$. The order-two contribution only marginally improves on the order-one result, which is remarkably accurate.

Observe that in the two-factor model, expression (8.55) for $S_T^{\text{order 1}}$ does not involve the level of VS volatility: at order one in ε, the ATMF skew is unchanged if VS volatilities are rescaled by a common factor.[9] This can be traced to the fact that forward variances in the two-factor model are lognormal. We have already observed this property in the short-maturity limit – see expression (8.39a) for S_0. Because of

---

[9]Expression (8.55) is obtained for a flat term structure of VS volatilities. In the case of a sloping term structure the double integrals in $C^{x\xi}$ cannot be done analytically – the expression for $S_T^{\text{order 1}}$ is more complicated than (8.55) and does depend on the term structure of VS volatilities. However, $S_T^{\text{order 1}}$ is unchanged in a global rescaling of VS volatilities.

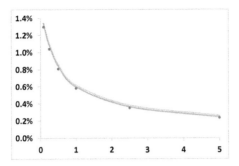

Figure 8.3: Exact (dots) and approximate values of $\widehat{\sigma}_{0.99F_T} - \widehat{\sigma}_{1.01F_T}$ at order one (light line) and two (dark line) in ε, for maturities from 1 month to 5 years. Parameters in Table 8.2 have been used, with a flat term structure of VS volatilities at 20%.

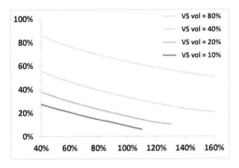

Figure 8.4: Smiles for various levels of VS volatilities, for a one-year maturity, computed in a Monte Carlo simulation of the two-factor model. The term structure of VS volatilities is flat and parameters are those of Table 8.2. As is manifest, ATMF skew levels are practically independent on the level of VS volatility.

the accuracy of $\mathcal{S}_T^{\text{order 1}}$ we expect this behavior to persist in the exact smile: this is illustrated in Figure 8.4, which shows implied volatilities for different levels of VS volatilities, for a one-year maturity.

Figure 8.4 should be contrasted with Figure 6.4, page 211, for the case of the Heston model. As already mentioned, the fact that VS volatilities are approximately normal – rather than lognormal – in the Heston model implies that, at order one in ε, the short ATMF skew is inversely proportional to the short VS volatility.

Term structure of the ATMF skew

How should we choose ρ_{SX^1}, ρ_{SX^2}? Once other parameters are set, ρ_{SX^1}, ρ_{SX^2} will determine both the vanilla smile *and* future smiles. It is necessary that the two-factor model be at least able to generate smiles that are comparable to historically observed smiles, in particular with respect to the term structure of the ATMF skew. This is especially important when pricing cliquets: see the discussion in Section 3.1. Typically equity index smiles display a term structure of the ATMF skew that is well approximated by a power law with an exponent usually around $\frac{1}{2}$ – see examples in Figure 6.5.

The values for ρ_{SX^1}, ρ_{SX^2} in Table 8.2 are such that they generate a term structure for the ATMF skew that is approximately a power law with exponent $\frac{1}{2}$, with $\hat{\sigma}_{0.95F_T} - \hat{\sigma}_{1.05F_T} = 3\%$ for $T = 1$ year. This is illustrated in Figure 8.5.

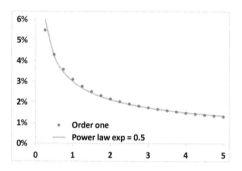

Figure 8.5: The ATMF skew measured as the difference of implied volatilities for strikes $0.95F_T$ and $1.05F_T$ given by expression (8.55) for $\mathcal{S}_T^{\text{order 1}}$ (continuous line) and by a power law benchmark with exponent $\frac{1}{2}$ and $\hat{\sigma}_{0.95F_T} - \hat{\sigma}_{1.05F_T} = 3\%$ for $T = 1$ year (dots), as a function of maturity (years). Maturities run from 3 months to 5 years. Parameters are those of Table 8.2.

We could as well have chosen other values for ρ_{SX^1}, ρ_{SX^2} such that the ATMF skew in the two-factor model generates a power-law-like dependence with a different exponent.

Our freedom is however limited. Indeed, the other parameters in the model – notably k_1, k_2, ρ_{12} – are already set and the triplet $\rho_{12}, \rho_{SX^1}, \rho_{SX^2}$ must make up a valid correlation matrix. This is the case if ρ_{SX^2} is defined as:

$$\rho_{SX^2} = \rho_{12}\rho_{SX^1} + \chi\sqrt{1 - \rho_{12}^2}\sqrt{1 - \rho_{SX^1}^2} \qquad (8.56)$$

where $\chi \in [-1, 1]$.

Still, it is usually possible to cover the range of skew decays that are observed practically. For example, taking $\rho_{SX^1} = -56\%, \rho_{SX^2} = -68\%$ (resp. $\rho_{SX^1} = -95\%$, $\rho_{SX^2} = -31\%$) approximately generates a power law decay for $\mathcal{S}_T$ with an exponent 0.4 (resp. 0.6), with $\hat{\sigma}_{0.95F_T} - \hat{\sigma}_{1.05F_T} = 3\%$ for $T = 1$ year.

8.8 Conclusion

The expansion of implied volatilities at order two is accurate for near-the-money strikes. It can be used for calibrating near-the-money implied volatilities or whenever vanilla implied volatilities are needed as observables, for example in Longstaff-Schwartz algorithms.

Its accuracy deteriorates for longer maturities and larger volatilities of volatilities – note in this respect that we have used realistic levels of volatility of volatility in our tests. The term structure of volatilities of VS volatilities generated by Set II appears in Figure 7.1: a 3-month volatility has a (lognormal) volatility of about 100% while volatilities of 1-year and 5-year VS volatilities are about 50% and 30%, respectively.

With parameters in Table 8.2, 5-year ATMF implied volatilities for a flat VS term structure at 20%, which appear in Figure 8.2, are 16.0% (Monte Carlo simulation) and 15.6% (order two expansion). If we now double ν, the 5-year ATMF volatilities are now 9.8% (Monte Carlo simulation) and 5.1% (order two expansion) – these are however unreasonably large levels of volatility of volatility.[10]

Finally, expression (8.55) for the ATMF skew at order one in ε is remarkably accurate. Values of ρ_{SX^1}, ρ_{SX^2} can be chosen so that the term structure of the ATMF skew is consistent with actual term structures of ATMF skews of equity indexes.

In (8.55) the ATMF skew is given by the product of volatility of volatility and spot/volatility correlations. Rescaling ρ_{SX^1}, ρ_{SX^2} by the same constant and adjusting ν so that products $\rho_{SX^1}\nu$, $\rho_{SX^2}\nu$ are unchanged leaves the ATMF skew unchanged.

One expects the two smiles to differ only by their ATMF volatility and curvature. This is true locally, but not for the global smile.

This is illustrated in Figure 8.6; we have generated smiles for the 3-month and 1-year maturities in the two-factor model with Set II parameters, and with Set II parameters, but with ρ_{SX^1}, ρ_{SX^2} multiplied by 90% and ν divided by 90% (Set II bis). This rescaling of $\rho_{SX^1}, \rho_{SX^2}, \nu$ indeed leaves the ATMF skew unchanged, but affects out-of-the-money volatilities asymmetrically .

8.9 Forward-start options – future smiles

So far we have considered spot-starting smiles. Consider the simple case of an option paying at time T_1 a payout $f\left(\frac{S_{T_2}}{S_{T_1}}\right)$. As explained in Sections 3.1.5 and 3.1.6

---

[10]Unrealistic levels of volatility of volatility may be needed to generate the inordinately large values of forward skew that one implies at times from market prices of cliquets. This drawback is typical of continuous forward variance models – discrete forward variance models, covered in Chapter 7.8, are immune to it.

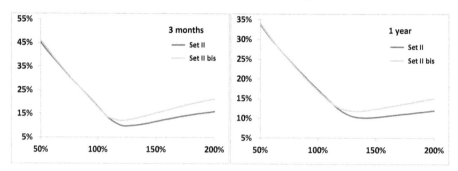

Figure 8.6: 3-month and 1-year smiles in the two-factor model with Set II parameters – see Table 8.2, page 329 – and Set II parameters with ρ_{SX^1}, ρ_{SX^2} multiplied by 0.9 and ν divided by 0.9 (Set II bis). The algorithm used is that in Section A.2 of Appendix A.

the price of such an option incorporates a volatility-of-volatility contribution, δP_1 and forward smile contribution δP_2.

δP_2 quantifies the forward-smile risk. It prices the difference of the market price at $t = T_1$ of the then-vanilla payoff f of maturity T_2 and its price as given in the Black-Scholes model with the log-contract implied volatility $\hat{\sigma}_{T_1 T_2}(T_1)$.

In Section 3.2 we have analyzed how the local volatility models handles forward-smile risk. Calibration of the local volatility model on the market smile at $t = 0$ likely results in a mispricing of δP_2 – and of δP_1 as well. What about stochastic volatility models?

Typically – that is unless we make model parameters explicitly time-dependent – stochastic volatility models are time-homogeneous, thus the instantaneous volatility of $\hat{\sigma}_{T_1 T_2}(t)$ is a function of $T_1 - t$: if our model is properly parametrized, δP_1 is priced correctly. Note in that respect that, already with the simple two-factor model, because a given term-structure of instantaneous volatilities of sport-starting volatilities can be matched with different sets of parameters – see Figure 7.1, page 228– we can use this flexibility to adjust the level of volatilities of forward volatilities somewhat.

Future smiles in stochastic volatility models are similar to spot-starting smiles, but may depend on volatility levels prevailing at future dates. For example, in the Heston model, the short ATMF skew is inversely proportional to the short ATMF volatility. Generally, for the sake of pricing δP_2 – and unless we have strong reasons not to do so – it is preferable not to hard-wire in the model dependencies between risks of a different nature, for example forward volatility and forward-smile risks.

In this respect, the two-factor model is attractive: future ATMF skews do not depend on the level of VS volatilities.[11] For flat future term-structures of VS volatili-

---

[11]They do depend somewhat on the term structure of VS volatilities, but are invariant in a rescaling of VS volatilities.

ties, at order one in volatility of volatility, they are given by (8.55) where T is the residual maturity.

Finally, we refer the reader to the discussion on page 104 for why we do not use the notion of "forward smile".

8.10 Impact of the smile of volatility of volatility on the vanilla smile

Imagine we use the two-factor model in the version of Section 7.7.1 with piecewise-constant volatility-of-volatility parameters $\gamma_T, \beta_T, \zeta_T$ calibrated so that market VIX smiles are matched. How does the smile of volatility of volatility impact the vanilla smile?

The expressions of implied volatilities in the order-two expansion (8.21) involve $C_0^{x\xi}(\xi_0)$, $C_0^{\xi\xi}(\xi_0)$, $D_0(\xi_0)$. $C_0^{x\xi}(\xi_0)$ and $C_0^{\xi\xi}(\xi_0)$, defined in (8.13) and (8.14), only depend on the covariance functions μ and ν evaluated on the initial variance curve ξ_0. In contrast, $D_0(\xi_0)$ – defined in (8.17) – depends on the derivative of μ with respect to ξ, hence is sensitive to the smile of volatility of volatility. $D_0(\xi_0)$ contributes at order ε^2: at order one in volatility of volatility the vanilla smile is unaffected by the smile of volatility of volatility.

VIX smiles are positively sloping: a model calibrated to VIX smiles will generate lower (more negative) values for $\frac{\delta\mu}{\delta\xi T}$, hence a larger (more positive) value for $D_0(\xi_0)$ – remember that μ is negative. From (8.21) we then expect that the ATMF skew will be weaker (less negative) while the ATMF curvature will be larger. This is demonstrated in Figure 8.7.

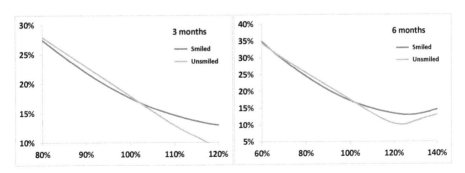

Figure 8.7: Vanilla smiles produced by the two-factor model for different values of the volatility-of-volatility parameters.

We have used Set II parameters along with the following constant values for $\gamma_T, \beta_T, \zeta_T$:

$$\gamma_T = 50\%, \beta_T = 15\%, \zeta_T = 100\% \tag{8.57}$$

These values are typical of VIX smiles – see the values of $\gamma_T, \beta_T, \zeta_T$ calibrated on the VIX smile of June 14, 2011 in Figure 7.9. We have chosen $\zeta_T = 100\%$ so that instantaneous volatilities of forward variances at $t = 0$ match those of the lognormal version of the model ($\gamma^T = \beta^T = 0$, $\zeta_T = 100\%$): the difference in the vanilla smile is then only generated by the smile of volatility of volatility.

Figure 8.7 shows the vanilla smile for 3-month and 6-month maturities both in the smiled (parameters in (8.57)) and unsmiled ($\gamma^T = \beta^T = 0$, $\zeta_T = 100\%$) version of the two-factor model. The predictions from the order-two expansion regarding ATMF skew and curvature are indeed verified.

Appendix A – Monte Carlo algorithms for vanilla smiles

The "exact" results in Figures 8.1 and 8.2 are computed with a Monte Carlo simulation: time is discretized, S_t is simulated with a simple Euler scheme and X_t^1, X_t^2 are simulated exactly – see Section 7.3.1.

The standard technique of averaging vanilla option's payoffs over all Monte Carlo paths produces price estimates that are in practice too noisy. We now present more efficient techniques.

A.1   The mixing solution

Consider an N-factor stochastic volatility model of type (7.11) – the spot/volatility joint dynamics reads:

$$\begin{cases} dS_t = (r - q)\, S_t dt + \sqrt{\xi_t^t} S_t dW_t^S \\ d\xi_t^T = \omega \xi_t^T \sum_i w_i e^{-k_i(T-t)} dW_t^i \end{cases}$$

where we have assumed zero interest rate and repo without loss of generality – otherwise simply replace S_0 with the forward for maturity T in what follows.

Assume first that W^S is uncorrelated with the W^i. We can simulate the W^i first, thus generating ξ_t^t for $t \in [0, T]$, then simulate S_t independently. Conditional on the path of ξ_t^t, S_t is lognormal with deterministic instantaneous volatility $\sqrt{\xi_t^t}$.

Rather than simulating S_t we can thus compute analytically the expectation on W^S:

$$E_{W^S}\left[f(S_T)\,|\,W^i\right] = P_{BS}\left(0, S_0, \widehat{\sigma}^\star\right)$$

where $f(S)$ is the payoff of our vanilla option and $P_{BS}(t, S, \widehat{\sigma})$ is the corresponding Black-Scholes formula for maturity T. The effective volatility $\widehat{\sigma}^\star$ is a function of the path of ξ_t^t – that is of the W^i – and is defined by:

$$\hat{\sigma}^\star = \sqrt{\frac{1}{T}\int_0^T \xi_t^t dt}$$

The price P of the vanilla option in the stochastic volatility model is thus given by:

$$P = E_{W^i, W^S}\left[f(S_T)\right] = E_{W^i}\left[P_{BS}\left(0, S_0, \hat{\sigma}^\star\right)\right] \tag{8.58}$$

Consider now the correlated case. We split W_t^S into two pieces: a portion $\lambda W_t^{\|}$ that is correlated with the W_t^i and an uncorrelated portion $\sqrt{1 - \lambda^2} W_t^{\perp}$:

$$W_t^S = \lambda W_t^{\|} + \sqrt{1 - \lambda^2} W_t^{\perp} \tag{8.59}$$

We have:

$$
\begin{aligned}
d \ln S_t &= -\frac{\xi_t^t}{2} dt + \sqrt{\xi_t^t} dW_t^S \\
&= \left[-\frac{\lambda^2}{2}\xi_t^t dt + \lambda\sqrt{\xi_t^t} dW_t^{\|}\right] + \left[-\frac{1-\lambda^2}{2}\xi_t^t dt + \sqrt{1-\lambda^2}\sqrt{\xi_t^t} dW_t^{\perp}\right]
\end{aligned}
$$

Conditional on the paths of Brownian motions W_t^i, S_T can be rewritten as:

$$S_T = S_0^\star e^{-\frac{(\sigma^\star)^2 T}{2} + \hat{\sigma}^\star \sqrt{T} Z} \tag{8.60}$$

where Z is a standard normal variable and the effective initial spot $S_0^\star$ and effective volatility $\sigma^\star$ are a function of the paths of the W^i:

$$\ln S_0^\star = \ln S_0 - \frac{\lambda^2}{2}\int_0^T \xi_t^t dt + \lambda \int_0^T \sqrt{\xi_t^t} dW_t^{\|} \tag{8.61a}$$

$$\hat{\sigma}^{\star 2} = (1 - \lambda^2)\frac{1}{T}\int_0^T \xi_t^t dt \tag{8.61b}$$

P is computed as:

$$P = E_{W^i}\left[P_{BS}\left(0, S_0^\star, \hat{\sigma}^\star\right)\right]$$

Thus, in the mixing solution we only need to:

- simulate the variance degrees of freedom

- accrue the two integrals $\int_0^T \xi_t^t dt$ and $\lambda \int_0^T \sqrt{\xi_t^t} dW_t^{\|}$

The mixing technique was originally published by G.A. Willard – see [85]. See also [79] for the uncorrelated case. We borrow the "mixing solution" denomination from Alan Lewis – see [69].

The two-factor model

Specializing to our two-factor model, we have:

$$\lambda W_t^{\parallel} = \frac{\rho_{SX^1} - \rho_{12}\rho_{SX^2}}{1 - \rho_{12}^2}W_t^1 + \frac{\rho_{SX^2} - \rho_{12}\rho_{SX^1}}{1 - \rho_{12}^2}W_t^2 \tag{8.62a}$$

$$\lambda = \sqrt{\frac{\rho_{SX^1}^2 + \rho_{SX^2}^2 - 2\rho_{12}\rho_{SX^1}\rho_{SX^2}}{1 - \rho_{12}^2}} \tag{8.62b}$$

The mixing solution consists in analytically integrating on $W_t^{\perp}$. We thus expect maximum efficiency when spot and forward variances have low correlation – see Section A.4 below for practical tests.

A.2 Gamma/theta P&L accrual

We now examine techniques that involve the simulation of S_t. As mentioned above, simply evaluating the vanilla option's payoff produces a noisy estimate of the price.

One can use the final spot price S_T as a control variate. Still a better idea would be to use as control variate the sum of delta P&Ls

$$\sum_i \frac{dP_{BS}}{dS}(t_i, S_i, \hat{\sigma})\left(S_{i+1} - e^{(r-q)(t_{i+1}-t_i)}S_i\right) \tag{8.63}$$

where delta is computed in the Black-Scholes model, with an arbitrary implied volatility $\hat{\sigma}$ – after all, delta-hedging aims at reducing the variance of the final P&L by neutralizing the order-one contribution of the spot variation.[12] While efficient, this solution is costly as computing $\frac{dP_{BS}}{dS}$ entails evaluating the cumulative distribution function of the standard normal distribution at each step of the simulation.

The technique we use is based on representation (2.30), page 40. There, local volatility is used as a base model for computing the price P_2 of a European option in a model whose instantaneous volatility is σ_2. Here we choose as base model the Black-Scholes model with volatility $\hat{\sigma}$; (2.30) becomes:

$$P_2(0, S_0, \bullet) = P_{BS}(0, S_0, \hat{\sigma}) + E_2\left[\int_0^T e^{-rt}\frac{S_t^2}{2}\frac{d^2P_{BS}}{dS^2}(\sigma_{2t}^2 - \hat{\sigma}^2)dt\right] \tag{8.64}$$

where E_2 denotes that the expectation is taken with respect to the dynamics generated by the stochastic volatility model at hand, whose instantaneous volatility process is σ_2.

Equation (8.64) expresses the price of a European option in an arbitrary stochastic volatility model as its price in the Black-Scholes model with implied volatility $\hat{\sigma}$ augmented by the (discounted) expectation of the integrated gamma/theta P&L evaluated with the Black-Scholes gamma – a natural representation from a trading

---

[12](8.63) has vanishing expectation.

point of view. We call $\widehat{\sigma}$ the risk-management volatility. We still need to simulate the spot process, but S_t is only used to compute the Black-Scholes gamma.

Expression (8.64) is practically useful as computing gamma in the Black-Scholes model amounts to evaluating one exponential:

$$S^2 \frac{d^2 P_{BS}}{dS^2} = Se^{-q(T-t)} \frac{1}{\sqrt{2\pi\widehat{\sigma}^2(T-t)}} e^{-\frac{d_1^2}{2}}$$

$$d_1 = \frac{1}{\widehat{\sigma}\sqrt{T-t}} \ln \frac{Se^{(r-q)(T-t)}}{K} + \frac{\widehat{\sigma}\sqrt{T-t}}{2}$$

This cost is more than offset by the increased accuracy brought about by what is, in effect, a perfect delta hedge.

In our simulations we use for $\widehat{\sigma}$ the VS volatility for maturity T. X_t^1, X_t^2, S_t are simulated at discrete times t_i: over each path, the second piece in the right-hand side of (8.64) is evaluated as:

$$\sum_i e^{-rt_i} \frac{S_i^2}{2} \frac{d^2 P_{BS}}{dS^2} (t_i, S_i, \widehat{\sigma}) \left(\xi_{t_i}^{t_i} - \widehat{\sigma}^2\right) \Delta \tag{8.65}$$

where Δ is the time step. Observe that (8.65) involves the instantaneous *implied* quadratic variation $\xi_{t_i}^{t_i} \Delta$ rather than its *realized* value $(S_{i+1}/S_i - 1)^2$ – this also contributes to the accuracy of this technique.

Dynamic adjustment of the implied volatility

Choosing for $\widehat{\sigma}$ the VS volatility of maturity T is somewhat arbitrary. We could choose a value for $\widehat{\sigma}$ so that, on average, the difference $\xi_{t_i}^{t_i} - \widehat{\sigma}^2$ is as small as possible. Still, on each path ξ_t^t may be very different from $\widehat{\sigma}^2$, resulting in large gamma/theta P&Ls and consequently a large variance for (8.65), if S_t happens to be in the vicinity of the option's strike.

Trading intuition suggests that we should dynamically readjust our implied volatility $\widehat{\sigma}$ so that $\widehat{\sigma}^2$ remains close enough to ξ_t^t. Imagine switching at time t from $\widehat{\sigma}_1$ to $\widehat{\sigma}_2$. Write equation (8.64) at time t for volatilities $\widehat{\sigma}_1$ and $\widehat{\sigma}_2$ and subtract one from the other. $P_{\overline{\sigma}}(t)$ cancels out and we get:

$$E_{\widehat{\sigma}_t} \left[\int_t^T e^{-r(u-t)} \frac{S_u^2}{2} \frac{d^2 P_{BS}}{dS^2} (u, S_u, \widehat{\sigma}_1) \left(\overline{\sigma}_u^2 - \widehat{\sigma}_1^2\right) du \right]$$

$$= E_{\widehat{\sigma}_t} \left[\int_t^T e^{-r(u-t)} \frac{S_u^2}{2} \frac{d^2 P_{BS}}{dS^2} (u, S_u, \widehat{\sigma}_2) \left(\overline{\sigma}_u^2 - \widehat{\sigma}_2^2\right) du \right]$$

$$+ \left[P_{BS}(t, S_t, \widehat{\sigma}_2) - P_{BS}(t, S_t, \widehat{\sigma}_1) \right]$$

This equation expresses that we are allowed to switch at time t from $\widehat{\sigma}_1$ to $\widehat{\sigma}_2$, provided we supplement the gamma/theta P&L with the difference $P_{BS}(t, S_t, \widehat{\sigma}_2) - P_{BS}(t, S_t, \widehat{\sigma}_1)$, which, from a trading point of view, is the P&L generated by re-marking our vanilla option to volatility $\widehat{\sigma}_2$.

Denote by τ_k, $k = 1 \ldots n$, the dates at which we switch from volatility $\widehat{\sigma}_{k-1}$ to volatility $\widehat{\sigma}_k$ – these dates can be set path by path dynamically, in the course of the simulation. Set $\tau_{n+1} = T$. The final expression for the option's price in our stochastic volatility model then reads:

$$
P_{\overline{\sigma}}(t = 0) = P_{BS}(0, S_0, \widehat{\sigma}_0)
$$
$$
+ \sum_{k=1}^{n} E_{\overline{\sigma}_t} \left[\int_{\tau_k}^{\tau_{k+1}} e^{-ru} \frac{S_u^2}{2} \frac{d^2 P_{BS}}{dS^2} (u, S_u, \widehat{\sigma}_k) \left(\overline{\sigma}_u^2 - \widehat{\sigma}_k^2 \right) du \right]
$$
$$
+ \sum_{k=1}^{n} e^{-r\tau_k} \left[P_{BS}(\tau_k, S_{\tau_k}, \widehat{\sigma}_k) - P_{BS}(\tau_k, S_{\tau_k}, \widehat{\sigma}_{k-1}) \right]
$$

What is the optimal strategy for choosing times τ_k? As the computational cost of evaluating P_{BS} is appreciable, we should readjust our risk-management volatility only when (a) the instantaneous volatility $\overline{\sigma}_t$ is significantly different from $\widehat{\sigma}_k$, (b) the dollar gamma is large.

For vanilla options, the dollar gamma becomes largest near the option's maturity, for spot values in the neighborhood of the option's strike. As we near the option's maturity it is probably preferable to risk-manage the option at zero implied volatility, so that contributions are only generated by returns that cross the option's strike.

A.3   Timer option-like algorithm

With respect to straight evaluation of the final payoff, accruing the gamma/theta P&L produces a less noisy estimate as it corresponds to delta-hedging our option. Can we take this idea one step further and get rid of the gamma/theta P&L as well? This is achieved by risk-managing our option in timer-wise fashion – we kindly ask the reader to read Appendix A of Chapter 5, page 180, before proceeding further.

Start at time $t_0 = 0$ with a quadratic variation budget Q_0 and zero initial quadratic variation of $\ln S_t$: $Q_{t=0} = 0$. As time advances $Q_0 - Q_t$ decreases. When, at time $t = \tau_1$ $Q_0 - Q_t$ falls below a set threshold we add to the initial budget Q_0 so that at $t = \tau_1^+$ it is equal to Q_1. Remarking our budget from Q_0 to Q_1 generates negative mark-to-market P&L for us.

We proceed likewise, incrementing the quadratic variation budget at (random) times τ_i, $i = 1 \ldots n$, whenever $Q_{i-1} - Q_{\tau_i}$ falls below the threshold, and recording the corresponding mark-to-market P&Ls, until maturity. At T, we pay to the client the option's intrinsic value; any remaining quadratic variation budget generates positive P&L for us.

$P_{\bar{\sigma}}(t=0)$ is thus given by:

$$P_{\bar{\sigma}}(t=0) = e^{-rT}\mathcal{P}_{BS}(S_0 e^{(r-q)(T-t)}, 0; \mathcal{Q}_0)$$
$$+ e^{-rT}\sum_{i=1}^{n} E_{\bar{\sigma}_t}\left[\mathcal{P}_{BS}(F_{\tau_i}^T(S_{\tau_i}), Q_{\tau_i}; \mathcal{Q}_i) - \mathcal{P}_{BS}\left(F_{\tau_i}^T(S_{\tau_i}), Q_{\tau_i}; \mathcal{Q}_{i-1}\right)\right]$$
$$- e^{-rT} E_{\bar{\sigma}_t}\left[\mathcal{P}_{BS}(S_T, Q_T; \mathcal{Q}_n) - f(S_T)\right]$$

(8.66)

where $F_{\tau_i}^T(S_{\tau_i}) = S_{\tau_i} e^{r(T-\tau_i)}$ is the forward for maturity T at time τ_i for spot S_{τ_i}, $\mathcal{Q}_i$ is the quadratic variation budget at $t=\tau_i$ and $f(S)$ is the option's payoff.

$\mathcal{P}_{BS}(S, Q; \mathcal{Q})$, defined in (5.68), page 183, is the Black-Scholes price of our European option, with vanishing interest rate and repo, as a function of the quadratic variation budget $\mathcal{Q}$ and quadratic variation Q:

$$\mathcal{P}_{BS}(S, Q; \mathcal{Q}) = E\left[f\left(Se^{-\frac{\mathcal{Q}-Q}{2}+\sqrt{\mathcal{Q}-Q}Z}\right)\right]$$

where Z is a standard normal variable.

Let us prove that (8.66) is indeed correct. We show on page 186 that $P(t, S_t, Q_t)$, given by:

$$P(t, S_t, Q_t) = e^{-r(T-t)}\mathcal{P}_{BS}(S_t e^{(r-q)(T-t)}, Q_t; \mathcal{Q})$$

is, by construction, a discounted martingale. Indeed:

$$E[dP] = \left(\frac{dP}{dt} + (r-q)S\frac{dP}{dS}\right)dt + \left(\frac{S^2}{2}\frac{d^2P}{dS^2} + \frac{dP}{dQ}\right)\bar{\sigma}_t^2 dt$$ (8.67)

Condition (5.72a) ensures that the second piece in the right-hand side of (8.67) vanishes, thus P is not sensitive to realized volatility. Condition (5.72b) then implies that $E[dP] = rPdt$: $P(t, S_t, Q_t)$ is a discounted martingale. Thus

$$\mathcal{P}_{BS}(F^T(S_t), Q_t; \mathcal{Q}) = e^{r(T-t)}P(t, S_t, Q_t)$$

is a martingale.

In our algorithm $\mathcal{Q}$ is a process that starts from $\mathcal{Q}_0$ at $t=0$ and is piecewise constant, jumping from $\mathcal{Q}_{i-1}$ to $\mathcal{Q}_i$ at times τ_i. Define $\mathcal{P}_t$ as: $\mathcal{P}_t = \mathcal{P}_{BS}(F^T(S_t), Q_t; \mathcal{Q}_t)$: $\mathcal{P}_t$ is a martingale on each interval $]\tau_{i-1}, \tau_i[$ and its discontinuity at times τ_i is :

$$\mathcal{P}_{\tau_i^+} - \mathcal{P}_{\tau_i^-} = \mathcal{P}_{BS}(F^T(S_{\tau_i}), Q_{\tau_i}; \mathcal{Q}_i) - \mathcal{P}_{BS}\left(F^T(S_{\tau_i}), Q_{\tau_i}; \mathcal{Q}_{i-1}\right)$$

Taking expectations, we get the identity:

$$\mathcal{P}_0 = E_{\bar{\sigma}_t}[\mathcal{P}_T] - \sum_{i=1}^{n} E_{\bar{\sigma}_t}\left[\mathcal{P}_{BS}(F^T(S_{\tau_i}), Q_{\tau_i}; \mathcal{Q}_i) - \mathcal{P}_{BS}\left(F^T(S_{\tau_i}), Q_{\tau_i}; \mathcal{Q}_{i-1}\right)\right]$$

This, together with the identity $P_{\bar{\sigma}}(t=0) = e^{-rT}E_{\bar{\sigma}_t}[f(S_T)]$ yields (8.66).

The algorithm that (8.66) expresses is very simple: simulate paths of S, keeping track of the realized quadratic variation, accumulating the Black-Scholes mark-to-market P&Ls in the second line of (8.66) as they occur, as well as the final P&L at maturity.

The strategy for setting the threshold level and the budget increment when it is hit can be optimized. On one hand, incrementing the budget by small quantities will generate small P&Ls at times τ_k and also a small P&L at maturity. On the other hand too many evaluations of these P&Ls slow down the algorithm.

Even without any optimization, this is a very efficient algorithm for computing smiles in stochastic volatility models.

A.4   A comparison

The relative accuracies of the techniques discussed above are illustrated in Figure 8.8, where we have used Set II parameters – see Table 8.2, page 329 – which were used to generate smiles in Figure 8.2. The bottom graphs use Set II parameters, but with $\rho_{SX^1} = \rho_{SX^2} = 0$. The VS volatilities are flat at 20%.

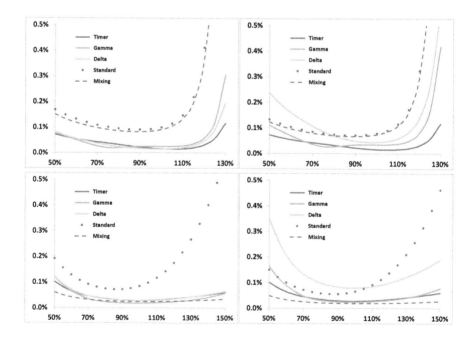

Figure 8.8: Errors in implied volatilities of 1-year vanilla options, in the two-factor model with Set II parameters (top) and Set II parameters with $\rho_{SX^1} = \rho_{SX^2} = 0$ (bottom), for different Monte Carlo algorithms. A VS volatility at 20% has been used.

The left-hand graphs show the errors of one-year implied volatilities, computed as one standard deviation of Monte Carlo prices of vanilla options, divided by the options' vegas to convert the errors in volatility units, with 100000 paths.

"Mixing" denotes the mixing solution, "Standard" denotes the standard technique of evaluating the final payoff, "Delta" refers to the same technique, but where the option is delta-hedged with the Black-Scholes delta computed with $\hat{\sigma} = 20\%$, "Gamma" denotes the estimator in (8.64) where we use $\hat{\sigma} = 20\%$. "Timer" denotes the estimator (8.66); we use as initial budget $\mathcal{Q}_0 = \frac{1}{2}\hat{\sigma}^2 T$; and budgets $\mathcal{Q}_k$ are a function of the VS volatility for the residual maturity: $\mathcal{Q}_k = \frac{1}{2}\hat{\sigma}_T^2(\tau_k)(T - \tau_k)$. $\mathcal{Q}$ is readjusted at time τ_k when the remaining budget falls below $0.03 \bullet \xi_{\tau_k}^{\tau_k}$.

Since the computational costs of these algorithms are different, the same errors appear in the right-hand graphs, rescaled according to $\varepsilon_{\text{rescaled}} = \varepsilon\sqrt{\frac{T}{T_{\text{timer}}}}$, where ε is error of the algorithm considered and T is its computational cost. The rescaled errors then correspond to a fixed computational cost – equal to that of the "Timer" algorithm – rather than a fixed number of Monte Carlo simulations.[13]

As expected, the mixing solution outperforms other techniques in the case of vanishing spot/volatility correlation (bottom graphs). Its effectiveness is greatly reduced in the correlated case (top graphs) to the point where it is barely more accurate than the standard technique, even though spot/volatility correlation levels can hardly be considered extreme. The value of λ in (8.62b) is 90.2%.

As is clear from the left-hand graphs the "Gamma" and "Timer" algorithms outperform the standard technique and have comparable accuracies for near-the-money strikes. The "Delta" technique, though roughly as accurate as the "Gamma" algorithm, is hampered by its computational cost.

It should be mentioned that this discussion is relevant to situations when standard random numbers are used in our Monte Carlo simulation. In this case, the standard deviation of our Monte Carlo estimator is indeed related to the second moment of the random variable whose expectation we are evaluating.

This relationship no longer holds when one uses quasi-random sequences – which is usually the case in practice. It becomes then difficult to assess the accuracy of our Monte Carlo estimate; one typically compares the estimate at hand with a benchmark obtained with a very large number of paths.

The mixing solution produces, by construction, a price estimate for each path that is strictly positive. It is thus possible to imply for all strikes strictly positive and arbitrage-free volatilities – a valuable benefit, even though these volatilities may be inaccurate. This is not the case for other techniques. For example, for far-away strikes, in the "Standard" technique, path contributions vanish, while with the "Timer" and "Gamma" techniques, they may even be negative. Despite this, the "Gamma" and "Timer" algorithms as well as the mixing solution are all good default algorithms for generating vanilla smiles.

---

[13]The Monte Carlo estimate of the option's price is $F = \frac{1}{n}\Sigma_i f_i$ where n is the number of paths used, and f_i is the contribution of path i. When using standard random numbers, the f_i are independent and we have $\text{Stdev}(F) = \frac{1}{\sqrt{n}}\text{Stdev}(f)$, hence the formula for the rescaling.

A.5 Dividends

In the presence of cash-amount dividends, both the "Timer", "Gamma" and mixing solution techniques cannot be used as is, as they make use of the Black-Scholes formula for the vanilla option's price and gamma.

The "Gamma" and "Timer" algorithms can be amended so that they still work with cash dividends. Take as base model the Black-Scholes model with no dividends – better, take as base model a Black-Scholes model with effective proportional dividends such that forwards for all maturities are matched.

Denote by d_k the dividend falling at time t_k, which may generally be a function of $S_{t_k^-}$, and y_k^* the corresponding effective yield such that forwards for all maturities are matched. y_k^* is defined by:

$$y_k^* F^{\tau_k^-}(S_0) = F^{\tau_k^-}(S_0) - F^{\tau_k^+}(S_0)$$

where $F^\tau(S)$ is the forward for maturity τ, for the initial spot value S. We denote by $P_{BS}^\star$ prices computed in the Black-Scholes model with proportional dividends y_k^* – they are still given by the standard Black-Scholes formula.

Going through the derivation on page 40 that led to (2.30) we get an additional contribution to the right-hand side of (8.64), generated by jumps of S_t at dividend dates:

$$E_{\bar\sigma_t}\left[\sum_k e^{-rt_k}\left(P_{BS}^\star(t_k^+, S_{t_k^-} - d_k(S_{t_k^-})) - P_{BS}^\star(t_k^+, (1-y_k^*)S_{t_k^-})\right)\right]$$

$$= E_{\bar\sigma_t}\left[\sum_k e^{-rt_k}\left(P_{BS}^\star\left(t_k^-, \frac{S_{t_k^-} - d_k(S_{t_k^-})}{1-y_k^*}\right) - P_{BS}^\star(t_k^-, S_{t_k^-})\right)\right]$$

where the second line follows from the fact that $P_{BS}^\star(t, S)$ is such that $P_{BS}^\star(t_k^+, (1-y_k^*)S_{t_k^-}) = P_{BS}^\star(t_k^-, S_{t_k^-})$ by construction.

Thus, in the "Gamma" algorithm, (8.65) is supplemented with :

$$\sum_k e^{-rt_k}\left[P_{BS}^\star\left(t_k^-, \frac{S_{t_k^-} - d_k(S_{t_k^-})}{1-y_k^*}, \sigma_0\right) - P_{BS}^\star(t_k^-, S_{t_k^-}, \sigma_0)\right]$$

while in the "Timer" algorithm, (8.66) is supplemented with:

$$\sum_k e^{-rt_k}\left[P_{BS}^\star\left(F_{t_k^-}^T\left(\frac{S_{t_k^-} - d_k(S_{t_k^-})}{1-y_k^*}\right), Q_{t_k}; Q(t_k)\right) - P_{BS}^\star\left(F_{t_k^-}^T(S_{t_k^-}), Q_{t_k}; Q(t_k)\right)\right]$$

where $Q(t_k)$ is the quadratic variation budget at time t_k and $F_t^T(S)$ is the forward at time t, spot S, for maturity T, in the proportional dividend model.

The computational cost of evaluating a Black-Scholes price is equivalent to computing two deltas. This technique can thus be employed when there are few dividends, that is for stocks. What should we do when there are numerous dividends, as in indexes?

A.5.1   An efficient approximation

We now make use of an approximation for vanilla option prices in the Black-Scholes model introduced in Section 2.3.1, page 34, published by Michael Bos and Stephen Vandermark – see [16].

In this approximation dividends are converted into two effective dividends falling at $t = 0$ and at maturity, thus resulting in a negative adjustment of the initial spot value and a positive adjustment of the strike.

For the sake of pricing a vanilla option of strike K, maturity T in the Black-Scholes model, the regular Black-Scholes formula is used, with S, K replaced with $\alpha(T)S - \delta S(T)$ and $K + \delta K(T)$.

We recall here the expressions of $\alpha(T)$, $\delta S(T)$, $\delta K(T)$ already given on page 37. Let y_i and c_i be the yield and cash-amount of the dividend falling at time t_i:$S_{t_i^+} = (1 - y_i)S_{t_i^-} - c_i$. $\alpha(T)$, $\delta S(T)$, $\delta K(T)$ read:

$$
\left\{
\begin{aligned}
\alpha(T) &= \prod_{t_i < T} (1 - y_i) \\
\delta S(T) &= \sum_{t_i < T} \frac{T - t_i}{T} c_i^* \, e^{-(r-q)t_i} \\
\delta K(T) &= \sum_{t_i < T} \frac{t_i}{T} c_i^* \, e^{(r-q)(T-t_i)}
\end{aligned}
\right.
$$

where the effective cash amounts c_i^* are given by: $c_i^* = c_i \displaystyle\prod_{t_i < t_j < T} (1 - y_j)$.

Our (heuristic) recipe is:

- price vanilla options in the stochastic volatility model using effective spot and strike values $\alpha(T)S - \delta S(T)$ and $K + \delta K(T)$ and no dividends,

- imply Black-Scholes volatilities using these effective values as well.

As there are no cash-amount dividends anymore, both the mixing solution, gamma/theta and "Timer" algorithms can be used with no alteration. This approximation for implied volatilities is accurate, both for stock and index smiles – see Figure 8.9 for an example.[14]

Appendix B – local volatility function of stochastic volatility models

Given the smile of a stochastic volatility model, one may need to determine the corresponding local volatility function, for example for the sake of comparing

---

[14]Besides, in case the stochastic volatility model degenerates into the Black-Scholes model – for example with vanishing volatility of volatility – the exact implied volatilities are recovered.

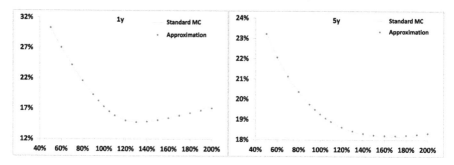

Figure 8.9: 1-year and 5-year smiles as generated by the two-factor model, either in a standard Monte Carlo simulation or using the approximation in Section A.5.1. Market data for the Euro Stoxx 50 index as of October 1, 2014 have been used, with zero repo and interest rate. The two-factor model parameters are: $\nu = 263\%$, $\theta = 11.5\%$, $k_1 = 10.28$, $k_2 = 0.42$, $\rho = 40\%$, $\rho_{SX^1} = -71.8\%$, $\rho_{SX^2} = -21.1\%$.

option prices computed with a stochastic volatility model and a local volatility model calibrated to the same smile. We carry out such a test in Chapter 11.

One may of course generate a vanilla smile using one of the techniques presented above, and then determine the local volatility function with the Dupire formula (2.3) using prices or with (2.19) using implied volatilities.

The local volatility function can however be obtained directly. From (2.6), page 28, the square of the local volatility function is equal to the expectation of the instantaneous variance conditional on the spot value:

$$\sigma(t,S)^2 \;=\; E\big[\xi_t^t | S_t = S\big] \;=\; \frac{E\big[\xi_t^t \delta(S_t - S)\big]}{E\big[\delta(S_t - S)\big]} \tag{8.68}$$

From Section A.1, conditional on the paths of the Brownian motions driving the instantaneous variance ξ_t^t, S_t is lognormally distributed. From equations (8.60) and (8.61):

$$S_t \;=\; S_0^\star e^{-\frac{\widehat{\sigma}_t^{\star 2} t}{2} + \widehat{\sigma}^\star \sqrt{t} Z}$$

where Z is a standard normal variable and the effective spot and volatility $S_0^\star, \sigma^\star$ are given by:

$$\ln S_{0t}^\star \;=\; \ln S_0 - \frac{\lambda^2}{2}\int_0^t \xi_u^u du + \lambda \int_0^t \sqrt{\xi_u^u}\, dW_u^{\|}$$

$$\widehat{\sigma}_t^{\star 2} \;=\; (1-\lambda^2)\frac{1}{t}\int_0^t \xi_u^u du$$

where $\lambda, W_u^{\|}$ are defined in (8.59). We use t subscripts for $S_{0t}^\star$ and $\widehat{\sigma}_t^\star$ as they are processes. Conditional on the paths of Brownian motions driving the instantaneous variance, the density ρ^* of S_t is thus given by:

$$\rho^*(S_t) = \frac{1}{S_t}\frac{1}{\sqrt{2\pi\,\widehat{\sigma}_t^{*2}\,t}}e^{-\frac{1}{2\widehat{\sigma}_t^{*2}t}\left(\ln\frac{S_t}{S_{0t}^*}+\frac{\widehat{\sigma}_t^{*2}t}{2}\right)^2}$$

Calculating the expectations in the right-hand side in (8.68) yields:

$$\sigma(t,S)^2 = \frac{E_\xi\left[\xi_t^t\frac{1}{\sqrt{2\pi\widehat{\sigma}_t^{*2}t}}e^{-\frac{1}{2\widehat{\sigma}_t^{*2}t}\left(\ln\frac{S}{S_{0t}^*}+\frac{\widehat{\sigma}_t^{*2}t}{2}\right)^2}\right]}{E_\xi\left[\frac{1}{\sqrt{2\pi\widehat{\sigma}_t^{*2}t}}e^{-\frac{1}{2\widehat{\sigma}_t^{*2}t}\left(\ln\frac{S}{S_{0t}^*}+\frac{\widehat{\sigma}_t^{*2}t}{2}\right)^2}\right]} \tag{8.69}$$

where the ξ subscripts denote that the expectations are taken with respect to the variance degrees of freedom only.

The recipe for computing $\sigma(t,S)$ in a Monte Carlo simulation is thus:

- Simulate paths for the instantaneous variance, accruing the integrals $\int_0^t \xi_u^u du$ and $\int_0^t \sqrt{\xi_u^u}dW_u^\|$. For each path, store the triplet $(\xi_t^t, S_{0t}^*, \widehat{\sigma}_t^*)$ at times t of interest.

- Compute the expectations in (8.69) as averages over these paths for values of S of interest.

Formula (8.69) was published in [66].

Appendix C – partial resummation of higher orders

Carrying out expansion (8.18) to higher orders in ε is tedious but straightforward. Can we identify, at each order in ε, a subset of terms that could be analytically calculated and resummed? A hint that this may be possible is provided by the contribution of the spot/volatility covariance function evaluated on the initial variance curve, $C_0^{x\xi}(\xi_0)$, to the second-order expansion (8.18):

$$\left(\varepsilon\frac{C_0^{x\xi}(\xi_0)}{2}\partial_x(\partial_x^2-\partial_x)+\varepsilon^2\frac{C_0^{x\xi}(\xi_0)^2}{8}\partial_x^2(\partial_x^2-\partial_x)^2\right)P_0 \tag{8.70}$$

This looks like the beginning of the expansion of $\exp\left(\varepsilon\frac{C_0^{x\xi}(\xi_0)}{2}\partial_x(\partial_x^2-\partial_x)\right)P_0$. The term in ε^2 in (8.70) is generated by the order-two contribution from $\mathcal{W}^1$:

$$\int_t^T d\tau_1\int_{\tau_1}^T d\tau_2\, U_{t\tau_1}^0 \mathcal{W}_{\tau_1}^1 U_{\tau_1\tau_2}^0 \mathcal{W}_{\tau_2}^1 U_{\tau_2 T}^0\, g$$

with $\mathcal{W}_t^1$ given by:

$$\mathcal{W}_t^1 = \int_t^T du \mu\,(t, u, \xi)\, \partial_{x\xi^u}^2$$

where operator $\partial_{x\xi^u}^2$ in $\mathcal{W}_{\tau_1}^1$ is not allowed to act on $\mu\,(\tau_2, u, \xi)$.[15] This amounts to replacing the spot/variance covariance function with its value computed on the initial variance curve: $\mu(t, u, \xi) \equiv \mu(t, u, \xi_0)$.

Let us do the same with the variance/variance covariance function: $\nu(t, u, u', \xi) \equiv \nu(t, u, u', \xi_0)$. Since covariance functions do not depend on forward variances anymore, operator ∂_{ξ^u} can be replaced with $\frac{1}{2}\left(\partial_x^2 - \partial_x\right)$ and the pricing equation (8.1) is replaced with:

$$\partial_t P + \frac{\xi^t}{2}\left(\partial_x^2 - \partial_x\right)P$$
$$+ \frac{\varepsilon}{2}\int_t^T du\,\mu_t^u \partial_x \left(\partial_x^2 - \partial_x\right)P + \frac{\varepsilon^2}{8}\int_t^T\int_t^T du\,du'\,\nu_t^{uu'}\left(\partial_x^2 - \partial_x\right)^2 P = 0$$
$$(8.71)$$

where we assume zero rates and repo, we use the compact notation $\mu_t^u \equiv \mu(t, u, \xi_0)$, $\nu_t^{uu'} = \nu(t, u, u', \xi_0)$, and μ_t^u and $\nu_t^{uu'}$ have been rescaled respectively by ε and ε^2. The solution of (8.71) reads:

$$P = \exp\left(\frac{1}{2}\left[\int_0^T \xi_0^t dt\right]\left(\partial_x^2 - \partial_x\right) + \frac{\varepsilon}{2}\left[\int_0^T dt \int_t^T du\,\mu_t^u\right]\partial_x\left(\partial_x^2 - \partial_x\right)\right.$$
$$\left. + \frac{\varepsilon^2}{8}\left[\int_0^T dt \int_t^T\int_t^T du\,du'\,\nu_t^{uu'}\right]\left(\partial_x^2 - \partial_x\right)^2\right)g$$

$$= e^{\frac{Q}{2}\left(\partial_x^2 - \partial_x\right) + \varepsilon\frac{C_0^{x\xi}(\xi_0)}{2}\partial_x\left(\partial_x^2 - \partial_x\right) + \varepsilon^2\frac{C_0^{\xi\xi}(\xi_0)}{8}\left(\partial_x^2 - \partial_x\right)^2}g \qquad (8.72)$$

$$= e^{\varepsilon\frac{C_0^{x\xi}(\xi_0)}{2}\partial_x\left(\partial_x^2 - \partial_x\right) + \varepsilon^2\frac{C_0^{\xi\xi}(\xi_0)}{8}\left(\partial_x^2 - \partial_x\right)^2}P_0 \qquad (8.73)$$

where $Q = \int_0^T \xi^t dt$. The reader can check by direct substitution that P in (8.72) indeed solves the PDE:

$$\frac{dP}{dt} + \frac{\xi^t}{2}S^2\frac{d^2 P}{dS^2}$$
$$+ \frac{1}{2}\int_t^T du \int_t^T du'\,\nu\,(t, u, u', \xi_0)\frac{d^2 P}{\delta\xi^u\delta\xi^{u'}} + \int_t^T du\,\mu\,(t, u, \xi_0)\,S\frac{d^2 P}{dS\delta\xi^u} = 0$$
$$(8.74)$$

Expanding (8.73) at order two in ε, we recover (8.18) but for the last term, which involves the derivative of μ with respect to forward variances – see the definition of $D_0(\xi_0)$ in (8.17).

---

[15] Action of $\partial_{x\xi^u}^2$ on $\mu\,(\tau_2, u, \xi)$ generates the term $\varepsilon^2 \frac{D_0(\xi_0)}{2}\partial_x^2\left(\partial_x^2 - \partial_x\right)P_0$ in (8.18).

P is easily computed through a Laplace transform. Define

$$x = \ln\frac{S}{K} + (r - q)(T - t)$$

set $P = Se^{-q(T-t)}f(t, x)$ and introduce the Laplace transform $F(t, p)$ of f :

$$F(t, p) = \int_{-\infty}^{+\infty} e^{-px} f(x, t)\, dx$$

The Laplace transform $G(p)$ of g is given, for a call option by:

$$G(p) = \int_{-\infty}^{+\infty} e^{-px} \left(1 - e^{-x}\right)^+ dx = \int_{0}^{\infty} e^{-px}\left(1 - e^{-x}\right) dx = \frac{1}{p\,(p+1)}$$

It is defined for $\mathrm{Re}(p) > 0$. For a put option $G(p)$ is identical, except it is defined for $\mathrm{Re}(p) < -1$. From expression (8.72) $F(t = 0, p)$ is then given by:

$$F(0, p) = \frac{1}{p\,(p+1)}\, e^{\frac{Q}{2}p(1+p)} + \varepsilon\frac{C_0^{x\xi}(\xi_0)}{2}p(1+p)^2 + \varepsilon^2\frac{C_0^{\xi\xi}(\xi_0)}{8}p^2(1+p)^2 \qquad (8.75)$$

Inverting $F(0, p)$ then yields P.

It turns out that, practically, this approximation does not work well and is worse than the order-two expansion in ε. In other words, resumming analytically all terms in the expansion of P in ε that do not involve derivatives of μ and ν with respect to forward variances is not sufficiently accurate. This is probably due to the fact that expression (8.75) for $F(0, p)$ does not correspond to a legitimate density.

Consider a level S_T for the spot at time T and define the log-return

$$z = \ln\frac{S_T}{S} - (r - q)(T - t)$$

The density $\rho\,(S_T)$ is given by $\rho\,(S_T) = e^{r(T-t)}\left.\frac{d^2 P}{dK^2}\right|_{K=S_T}$. Using that $P(t, S) = Se^{-q(T-t)}f(t, x)$ and the definition of z we get:

$$\rho_z(z) = \left(e^x\left(\partial_x + \partial_x^2\right) f\right)_{x=-z}$$

where $\rho_z(z)$ is the density of z. Let us introduce the cumulant-generating function $L(q)$ of ρ_z:

$$
\begin{aligned}
e^{L(q)} &= \int_{-\infty}^{+\infty} e^{-qz}\rho_z(z)dz = \int_{-\infty}^{+\infty} e^{-qz}\left(e^x\left(\partial_x + \partial_x^2\right) f\right)_{x=-z} dz \\
&= q\,(1+q)\,F\left(0, -(1+q)\right)
\end{aligned}
$$

We thus get:

$$L(q) = \frac{Q}{2}q\,(1+q) - \varepsilon\frac{C_0^{x\xi}(\xi_0)}{2}q^2\,(1+q) + \varepsilon^2\frac{C_0^{\xi\xi}(\xi_0)}{8}q^2\,(1+q)^2$$

The conditions that:

- ρ_z integrate to one ($L(0) = 0$),

- the forward for maturity T is matched: $E[S_T] = S_0$ ($L(-1) = 0$)

- the VS volatility for maturity T be matched ($\frac{dL}{dq}\big|_{q=0} = \frac{Q}{2}$)

are obviously satisfied.[16]

The fact that L is a polynomial is a problem, however: a theorem by Marcinkiewicz [72] states that if a cumulant-generating function is a polynomial, its order cannot be greater than 2: the density corresponding to solutions of (8.74) is not positive.[17]

---

[16] See also the discussion in Appendix B of Chapter 5.

[17] This drawback is shared by the density generated by our original expansion at order two in ε; the latter however is a small perturbation of a Gaussian density – see the discussion in Section 8.3.

Chapter's digest

8.2 Expansion of the price in volatility of volatility

▶ We consider forward variance models and derive the following expansion of vanilla option prices at order two in volatility of volatility:

$$
P = \left[1 + \varepsilon \frac{C_0^{x\xi}(\xi_0)}{2} \partial_x \left(\partial_x^2 - \partial_x\right)\right.
$$
$$
\left. + \varepsilon^2 \left(\frac{C_0^{\xi\xi}(\xi_0)}{8}\left(\partial_x^2 - \partial_x\right)^2 + \frac{C_0^{x\xi}(\xi_0)^2}{8}\partial_x^2\left(\partial_x^2 - \partial_x\right)^2 + \frac{D_0(\xi_0)}{2}\partial_x^2\left(\partial_x^2 - \partial_x\right)\right)\right] P_0
$$

in terms of three dimensionless quantities that are readily calculated in any model of interest:

$$
C_t^{x\xi}(\xi) = \int_t^T (T-\tau)\left\langle d\ln S_\tau \, d\widehat{\sigma}_T^2(\tau)\right\rangle
$$
$$
C_t^{\xi\xi}(\xi) = \int_t^T (T-\tau)^2 \left\langle d\widehat{\sigma}_T^2(\tau)\, d\widehat{\sigma}_T^2(\tau)\right\rangle
$$
$$
D_t(\xi) = \int_t^T d\tau \int_\tau^T du \,(T-u)\frac{1}{d\tau}\left\langle d\ln S_\tau \frac{\left\langle d\ln S_u\, d\widehat{\sigma}_T^2(u)\right\rangle}{du}\right\rangle
$$

Our derivation requires that spot/volatility covariances not depend on S.

8.3 Expansion of implied volatilities

▶ This expansion translates into the following expansion of implied volatilities:

$$
\widehat{\sigma}(K,T) = \widehat{\sigma}_{F_T T} + S_T \ln\left(\frac{K}{F_T}\right) + \frac{C_T}{2}\ln^2\left(\frac{K}{F_T}\right)
$$

with:

$$
\widehat{\sigma}_{F_T T} = \widehat{\sigma}_T \left[1 + \frac{\varepsilon}{4Q}C^{x\xi} + \frac{\varepsilon^2}{32Q^3}\left(12\left(C^{x\xi}\right)^2 - Q(Q+4)C^{\xi\xi} + 4Q(Q-4)D\right)\right]
$$
$$
S_T = \widehat{\sigma}_T\left[\frac{\varepsilon}{2Q^2}C^{x\xi} + \frac{\varepsilon^2}{8Q^3}\left(4QD - 3\left(C^{x\xi}\right)^2\right)\right]
$$
$$
C_T = \widehat{\sigma}_T\frac{\varepsilon^2}{4Q^4}\left(4QD + QC^{\xi\xi} - 6\left(C^{x\xi}\right)^2\right)
$$

where $\widehat{\sigma}_T$ is the VS volatility for maturity T, $Q = \widehat{\sigma}_T^2 T$, $C^{x\xi} \equiv C_0^{x\xi}(\xi_0)$, $C^{\xi\xi} \equiv C_0^{\xi\xi}(\xi_0)$, $D = D_0(\xi_0)$.

▶ At order one in volatility of volatility, the ATMF skew is given by:

$$\mathcal{S}_T = \hat{\sigma}_T \frac{C^{x\xi}}{2(\hat{\sigma}_T^2 T)^2}$$

$$= \frac{1}{2\hat{\sigma}_T^3 T} \int_0^T \frac{T-\tau}{T} \frac{\langle d\ln S_\tau \, d\hat{\sigma}_T^2(\tau)\rangle_0}{d\tau} d\tau$$

🐚 🐚 🐚 🐚 🐚

8.4 A representation of European option prices in diffusive models

▶ The expression of the ATMF skew at order one in volatility of volatility can also be obtained from the following general expression of European option prices in diffusive models:

$$P = P_{BS}\left(0, S_0, \hat{\sigma}_T^2(0)\right)$$

$$+ E\left[\int_0^T e^{-rt}\left(\frac{d^2 P_{BS}}{dS d(\hat{\sigma}_T^2)}\langle dS_t d\hat{\sigma}_T^2(t)\rangle + \frac{1}{2}\frac{d^2 P_{BS}}{(d(\hat{\sigma}_T^2))^2}\langle d\hat{\sigma}_T^2(t)d\hat{\sigma}_T^2(t)\rangle\right)\right]$$

▶ The European payoff that materializes the uniformly weighted spot/volatility covariance, at order one in volatility of volatility, is $\ln^2(S/S_0)$.

🐚 🐚 🐚 🐚 🐚

8.5 Short maturities

▶ At short maturities, the ATMF skew and curvature are given, at leading order, by the following general expressions:

$$\mathcal{S}_0 = \frac{1}{2\hat{\sigma}_0^2}\frac{\langle d\ln S d\hat{\sigma}_0\rangle}{dt}$$

$$\mathcal{C}_0 = \frac{1}{4\hat{\sigma}_0}\left(\frac{8}{3}\frac{\langle d\ln S d\mathcal{S}_0\rangle}{\hat{\sigma}_0 dt} + \frac{4}{3}\frac{\langle d\hat{\sigma}_0 d\hat{\sigma}_0\rangle}{\hat{\sigma}_0^2 dt} - 8\mathcal{S}_0^2\right)$$

where $\hat{\sigma}_0$ is the short ATM volatility, $\mathcal{S}_0$ the short ATM skew, and $\mathcal{C}_0$ the short ATM curvature. While the short ATM skew is a model-independent measure of the instantaneous covariance of spot and ATM volatility, the short curvature is not a direct measure of volatility of volatility, as the covariance of spot and ATM skew contributes as well.

▶ Consider a lognormal model for the short ATM volatility, whose lognormal volatility of volatility is ν. We have:

$$\mathcal{S}_0 = \frac{\rho}{2}\nu$$

$$\mathcal{C}_0 = \frac{1}{6\hat{\sigma}_0}\left(2 - 3\rho^2\right)\nu^2$$

The short ATM skew is independent on the level of ATM volatility, while the curvature is inversely proportional to the ATM volatility.

This behavior is typical of the SABR model.

▶ Consider a normal model for the short ATM volatility, whose normal volatility of volatility is σ. We have:

$$\mathcal{S}_0 = \frac{\rho}{2}\frac{\sigma}{\hat{\sigma}_0}$$

$$\mathcal{C}_0 = \frac{1}{6\hat{\sigma}_0}(2 - 5\rho^2)\left(\frac{\sigma}{\hat{\sigma}_0}\right)^2$$

The short ATM skew is inversely proportional to the ATM volatility, while the curvature is inversely proportional to the cube of the ATM volatility.

This behavior is typical of the Heston model.

▶ In case the short ATM volatility is uncorrelated with S, the ATM skew vanishes and the ATM curvature is given by:

$$\mathcal{C}_0 = \frac{1}{3\hat{\sigma}_0}\frac{\langle d\hat{\sigma}_0 d\hat{\sigma}_0 \rangle}{\hat{\sigma}_0^2 dt}$$

❧ ❧ ❧ ❧ ❧

8.7 The two-factor model

▶ At order one in volatility of volatility, the ATMF skew of the two-factor model is given by:

$$\mathcal{S}_T^{\text{order 1}} = \frac{\nu\alpha\theta}{\hat{\sigma}_T^3 T^2}\int_0^T dt\sqrt{\xi_0^t}\int_t^T du\xi_0^u\left[(1-\theta)\rho_{SX^1}e^{-k_1(u-t)} + \theta\rho_{SX^2}e^{-k_2(u-t)}\right]$$

which, for a flat term structure of VS volatilities, translates into:

$$\mathcal{S}_T^{\text{order 1}} = \nu\alpha\theta\left[(1-\theta)\rho_{SX^1}\frac{k_1 T - (1 - e^{-k_1 T})}{(k_1 T)^2} + \theta\rho_{SX^2}\frac{k_2 T - (1 - e^{-k_2 T})}{(k_2 T)^2}\right]$$

$\mathcal{S}_T^{\text{order 1}}$ does not depend on the level of VS volatility. These approximate expressions for the ATMF skew are accurate and can be used for choosing ρ_{SX^1} and ρ_{SX^2} so as to generate the desired decay for the ATMF skew. In particular, it is possible to approximately produce the typical power-law decay of equity index skews.

❧ ❧ ❧ ❧ ❧

8.8 Conclusion

▶ The expansion at order two in volatility of volatility is adequate for near-the-money strikes. Its accuracy deteriorates as one moves to out-of-the-money strikes and longer maturities.

While the order-one expansion is sufficient, in the two-factor model, for generating an accurate estimation of the ATMF skew, for the sake of approximating the ATMF volatility, the second order is needed.

At order one in volatility of volatility, the ATMF skew is determined by the covariance of spot and VS volatilities; changing the spot/volatility correlation while rescaling accordingly volatilities of volatilities so that spot/volatility covariances are unchanged indeed leaves the ATMF skew unchanged.

The effect on out-of-the-money implied volatilities is asymmetrical. Typically, for equity index smiles, implied volatilities for low strikes are roughly unchanged, while those of high strikes are impacted.

🐦 🐦 🐦 🐦 🐦

8.9 Forward-start options – future smiles

▶ Unlike the local volatility model, stochastic volatility models afford more control on the pricing of the risks of forward-start options: volatility-of-volatility and forward-smile risks. In time-homogeneous stochastic volatility models, future smiles are predictable and similar to spot-starting smiles. In the two-factor model, the level of ATMF skew, be it spot-starting of future, is approximately fixed, independent on the level of ATMF volatility.

🐦 🐦 🐦 🐦 🐦

8.10 Impact of the smile of volatility of volatility on the vanilla smile

▶ We use volatility-of-volatility smile parameters derived from calibration on VIX smiles. An upward-sloping volatility-of-volatility smile makes the vanilla smile of the underlying more convex near the money and less steep. This numerical result is supported by the order-two expansion.

🐦 🐦 🐦 🐦 🐦

Appendix A – Monte Carlo algorithms for vanilla smiles

▶ We provide three efficient techniques for generating vanilla smiles in stochastic volatility models.

▶ The Brownian motion driving S_t can be written as $W_t^S = \lambda W_t^{\parallel} + \sqrt{1 - \lambda^2} W_t^{\perp}$, where $W_t^{\perp}$ is uncorrelated to the Brownian motions driving forward variances.

The mixing solution consists in conditioning with respect to the Brownian motions driving forward variances: integration on $W_t^{\perp}$ is analytic.

In the Monte Carlo simulation, forward variances are simulated and two integrals are accrued: $I_1 = \int_0^T \xi_t^t dt$ and $I_2 = \int_0^T \sqrt{\xi_t^t} \lambda dW_t^{\parallel}$. The vanilla option price is obtained as: $P = E_{W^i}\left[P_{BS}\left(0, S_0^{\star}, \widehat{\sigma}^{\star}\right)\right]$ where the expectation is taken on the paths of the Brownian motions driving forward variances, where the effective spot and volatility $S_0^\star$ and $\widehat{\sigma}^\star$ are given by:

$$\ln S_0^\star = \ln S_0 - \frac{\lambda^2}{2}I_1 + I_2$$

$$\widehat{\sigma}^{\star 2} = (1 - \lambda^2)I_1$$

For the two-factor model, we have:

$$\lambda W_t^{\|} = \frac{\rho_{SX^1} - \rho_{12}\rho_{SX^2}}{1 - \rho_{12}^2}W_t^1 + \frac{\rho_{SX^2} - \rho_{12}\rho_{SX^1}}{1 - \rho_{12}^2}W_t^2$$

$$\lambda = \sqrt{\frac{\rho_{SX^1}^2 + \rho_{SX^2}^2 - 2\rho_{12}\rho_{SX^1}\rho_{SX^2}}{1 - \rho_{12}^2}}$$

A set of maturities can be priced at once, by storing the values of I_1, I_2 for maturities of interest.

▶ In the gamma/theta accrual method, the spot and forward variances are simulated, but rather than evaluating the vanilla payoff, one accrues the gamma/theta P&L, calculated with a chosen risk-management volatility $\widehat{\sigma}$. This corresponds in essence to using the delta P&L as a control variate – without calculating delta.

The contribution of a given path to the vanilla price estimator is:

$$P_{BS}(0, S_0, \widehat{\sigma}) + \sum_i e^{-rt_i}\frac{S_i^2}{2}\frac{dP_{BS}}{dS}(t_i, S_i, \widehat{\sigma})\left(\xi_{t_i}^{t_i} - \widehat{\sigma}^2\right)\Delta$$

This algorithm can be optimized by dynamically adjusting the risk-management volatility $\widehat{\sigma}$ to match current levels of realized volatility, in the course of the simulation.

▶ In the timer-like algorithm, we get rid of the gamma/theta P&L altogether. We start at time $t_0 = 0$ with a quadratic variation budget $\mathcal{Q}_0$ which gets eroded by the realized quadratic variation along the simulated path.

Whenever the remaining budget falls below a given threshold, we add to it and the path estimator accrues a mark-to-market P&L given by the difference of two Black-Scholes prices. At maturity, the remaining time value is subtracted from the path estimator.

The resulting path estimator only involves mark-to-market P&Ls.

र र र र र

Appendix B – local volatility function of stochastic volatility models

▶ The local volatility function corresponding to the vanilla smile of a given stochastic volatility model can be very efficiently obtained in a Monte Carlo simulation, without calculating vanilla option prices.

Chapter 9

Linking static and dynamic properties of stochastic volatility models

A stochastic volatility model can be assessed by studying its dynamic properties: the volatilities and correlations of volatilities, be they spot-starting or forward-starting and their joint dynamics with the spot. One can also choose to focus on the (static) smile it produces and examine the strike and maturity dependence of the implied volatilities it generates.

While static and dynamic features of a model are both determined by the joint dynamics of spot and forward variances in the model, how strong is the relationship between both? Which elements of this connection are specific to stochastic volatility models? Would it be possible to parametrize differently a given model so that the vanilla smile is (approximately) unchanged while the dynamics is different?

We establish a link between the rate at which the ATMF skew decays with maturity and the SSR, already introduced in the context of the local volatility model, whose definition we extend to the case of stochastic volatility models. Our analysis will allow us to split stochastic volatility models into two classes.[1]

9.1 The ATMF skew

The ATMF skew vanishes for vanishing volatility of volatility. At order one it is given by (8.52). Setting $\varepsilon = 1$ and using the definition of $C^{x\xi}$ in (8.13) yields:

$$\mathcal{S}_T = \frac{1}{2\sqrt{T}} \frac{1}{\left(\int_0^T \xi_0^\tau d\tau\right)^{3/2}} \int_0^T d\tau \int_\tau^T \mu(\tau, u, \xi_0) du \qquad (9.1)$$

which can be rewritten as in (8.24), page 315:

$$\mathcal{S}_T = \frac{1}{2\hat{\sigma}_T^3 T} \int_0^T \frac{T - \tau}{T} \frac{\langle d\ln S_\tau \, d\hat{\sigma}_T^2(\tau) \rangle_0}{d\tau} d\tau \qquad (9.2)$$

---

[1]These results were first published in [11].

9.2　The Skew Stickiness Ratio (SSR)

The dynamics of a given model is reflected in the covariance of the spot and forward variances. Conditional on a move of the spot, different models generate different scenarios for implied volatilities. This can be quantified by focusing on the ATMF volatility and computing the regression coefficient of $\delta\widehat{\sigma}_{F_T T}$ with respect to $\delta\ln S$.

It seems reasonable and natural to normalize this regression coefficient by the ATMF skew. We thus introduce the Skew Stickiness Ratio (SSR) $\mathcal{R}_T$, defined by:

$$\mathcal{R}_T \;=\; \frac{1}{\mathcal{S}_T}\frac{E[d\ln S\, d\widehat{\sigma}_{F_T(S)T}]}{E[(d\ln S)^2]} \tag{9.3}$$

where the notation $\widehat{\sigma}_{F_T(S)T}$ emphasizes that the strike whose implied volatility we consider is not fixed. Unless necessary we will use the lighter notation $\widehat{\sigma}_{F_T T}$.

$\mathcal{R}_T$ is dimensionless – its value is known for some classes of models:

- In jump-diffusion or Lévy models with independent stationary increments for $\ln S$ – such as the model used in Section 5.3.2 – implied volatilities are a function of moneyness $\frac{K}{S}$ only: as S moves $\widehat{\sigma}_{F_T T}$ is unchanged: $\mathcal{R}_T = 0$, $\forall T$.

- We have already encountered the SSR in the context of the local volatility model. Because implied volatilities are *functions* of (t, S), formula (9.3) simplifies to:

$$\mathcal{R}_T \;=\; \frac{1}{\mathcal{S}_T}\frac{d\widehat{\sigma}_{F_T(S)T}}{d\ln S} \;=\; \frac{1}{\mathcal{S}_T}\left(\left.\frac{d\widehat{\sigma}_{KT}}{d\ln K}\right|_{F_T} + \left.\frac{d\widehat{\sigma}_{KT}}{d\ln S}\right|_{K=F_T}\right)$$

which agrees with the definition used for the SSR in Section 2.5.2. We know from Sections 2.5.3.1 and 2.4.6 that for time-independent local volatility functions for all maturities, or for general local volatility functions in the limit $T \to 0$, $\mathcal{R}_T = 2$. For equity index smiles, $\mathcal{R}_T$ starts from 2 for short maturities and typically reaches values above 2 for long maturities – see for example Figure 2.4, page 59, and Figure 9.9, page 380.

We now compute $\mathcal{R}_T$ in a general stochastic volatility model at lowest order in volatility of volatility. $\mathcal{S}_T$ vanishes for vanishing volatility of volatility and starts with a term of order one. To get $\mathcal{R}_T$ at lowest order we thus need to compute the covariance $E[dS\, d\widehat{\sigma}_{F_T T}]$ at order one. From (8.21a), the difference between ATMF and VS volatilities is of order one. For the purpose of computing $E[dS\, d\widehat{\sigma}_{F_T T}]$ at order one, we can thus conveniently replace the ATMF volatility with the VS volatility:

$$E[dS\, d\widehat{\sigma}_{F_T T}] \;\simeq\; E[dS\, d\widehat{\sigma}_T]$$

From the definition of the VS volatility $\hat{\sigma}_T^2(t) = \frac{1}{T-t}\int_t^T \xi_t^u du$ we have:

$$E[d\ln S_t d\hat{\sigma}_T(t)] = \frac{1}{2(T-t)\hat{\sigma}_T(t)} \int_t^T E[d\ln S_t \, d\xi_t^u] du$$

$$= \frac{1}{2(T-t)\hat{\sigma}_T(t)} \int_t^T \mu(t,u,\xi_0) du \, dt$$

Setting $t = 0$ and dividing by $E[(d\ln S)^2]$ which is equal to $\xi_0^0 dt$, and by S_T, we get the following expression for $\mathcal{R}_T$ at lowest non-trivial order in volatility of volatility:

$$\mathcal{R}_T = \frac{\int_0^T \xi_0^\tau d\tau}{T\xi_0^0} \frac{T\int_0^T \mu(0,u,\xi_0)du}{\int_0^T d\tau \int_\tau^T \mu(\tau,u,\xi_0)\, du} \tag{9.4}$$

9.3   Short-maturity limit of the ATMF skew and the SSR

Let us take the limit $T \to 0$ in expression (9.1). We get:

$$\mathcal{S}_0 = \lim_{T\to 0} \frac{1}{2\sqrt{T}} \frac{1}{\left(\int_0^T \xi_0^\tau d\tau\right)^{3/2}} \int_0^T d\tau \int_\tau^T \mu(\tau,u,\xi_0)\, du = \frac{\mu(0,0,\xi_0)}{4(\xi_0^0)^{3/2}}$$

This recovers (8.36): at order one in volatility of volatility the short ATMF skew is a direct measure of the instantaneous spot/volatility covariance. Turning now to the SSR:

$$\mathcal{R}_0 = \lim_{T\to 0} \frac{\int_0^T \xi_0^\tau d\tau}{T\xi_0^0} \frac{T\int_0^T \mu(0,u,\xi_0)du}{\int_0^T d\tau \int_\tau^T \mu(\tau,u,\xi_0)\, du} = \lim_{T\to 0} \frac{T\int_0^T du}{\int_0^T d\tau \int_\tau^T du}$$

$$= 2$$

Thus, in stochastic volatility models, the short limit of the SSR is 2, as in the local volatility model. This "2" is the 2 in the denominator of (8.36).

9.4   Model-independent range of the SSR

Let us assume that the term structure of VS volatilities is flat – $\xi_0^\tau \equiv \xi_0$ – and that the model at hand is time-homogeneous, that is the spot/variance covariance function $\mu(\tau,u,\xi)$ only depends on $u - \tau$. We now use the economical notation $\mu(u - \tau)$:

$$\mu(\tau,u,\xi_0) = \lim_{d\tau\to 0} \frac{1}{d\tau} E\left[d\ln S_\tau d\xi_\tau^u\right] \equiv \mu(u-\tau)$$

One of the integrals in (9.1) can be done analytically and we get the following simple expressions for $\mathcal{S}_T$ and $\mathcal{R}_T$:

$$\mathcal{S}_T = \frac{1}{2\xi_0^{3/2}T^2} \int_0^T (T-t)\mu(t)dt \tag{9.5}$$

$$\mathcal{R}_T = \frac{\int_0^T \mu(t)dt}{\int_0^T (1-\frac{t}{T})\mu(t)dt} \tag{9.6}$$

These expressions for $\mathcal{S}_T$ and $\mathcal{R}_T$ only involve the spot/variance covariance function μ – can we derive general properties without assuming a specific form for $\mu(t)$?

Let us make the natural assumption that $\mu(t)$ decays monotonically towards zero as $t \to \infty$. Define $g(\tau)$ as:

$$g(\tau) = \int_0^\tau \mu(t)dt \tag{9.7}$$

$\mathcal{S}_T$ and $\mathcal{R}_T$ can be rewritten as:

$$\mathcal{S}_T = \frac{1}{2\xi_0^{3/2}T^2} \int_0^T g(\tau)d\tau \qquad \mathcal{R}_T = \frac{g(T)}{\frac{1}{T}\int_0^T g(\tau)d\tau} \tag{9.8}$$

$g(\tau) = 0$ for $\tau = 0$ and is either increasing concave if $\mu(t) \geq 0$, or decreasing convex if $\mu(t) \leq 0$.

- $\mathcal{R}_T$ is the ratio of $g(T)$ to its average value over $[0, T]$, thus $\mathcal{R}_T \geq 1$.

- $g(\tau)$ is either positive and concave, or negative and convex. Thus, $\frac{g(\tau)}{g(T)} \geq \frac{\tau}{T}$. This yields: $\mathcal{R}_T = \frac{1}{\frac{1}{T}\int_0^T \frac{g(\tau)}{g(T)})d\tau} \leq \frac{1}{\frac{1}{T}\int_0^T \frac{\tau}{T}d\tau} = 2.$

Thus, for a time-homogeneous model such that its spot/variance covariance function decays monotonically, and for a flat term structure of VS volatilities, we get, at order one in volatility of volatility, the following model-independent range for $\mathcal{R}_T$:

$$\mathcal{R}_T \in [1, 2] \tag{9.9}$$

In diffusive stochastic volatility models – with the assumptions we have made – the SSR cannot go below 1. To get lower values for $\mathcal{R}_T$ one presumably needs to incorporate a jump or Lévy component in the process for $\ln \mathcal{S}_t$.

(9.9) strictly holds for a flat term structure of VS volatilities. Glancing again at the definition of the SSR in (9.3) we can see that the numerator is proportional to the instantaneous volatility of S_t while the denominator is proportional to its square, thus the SSR is inversely proportional to the short VS volatility.

It can be made artificially small or large by shifting the short end of the variance curve – one should bear this in mind when assessing the SSR of a given market smile.

An example of the impact of the term structure of VS volatilities is discussed on page 366.

For $T \to 0$, assuming that $\mu(t)$ is smooth as $t \to 0$, $g(\tau) = \tau\mu(0)$: expression (9.8) for $\mathcal{R}_T$ again yields:

$$\mathcal{R}_0 = 2$$

9.5   Scaling of ATMF skew and SSR – a classification of models

To investigate further the connection between $\mathcal{S}_T$ and $\mathcal{R}_T$ we need a characterization of the rate of decay of $\mu(t)$. Let us assume that for $t \to \infty$ $\mu(t)$ decays with a characteristic exponent γ:

$$\mu(t) \propto \frac{1}{t^\gamma} \tag{9.10}$$

Consider $g(\tau) = \int_0^\tau \mu(t)dt$. For large τ, it is equal to $C + \alpha\tau^{1-\gamma}$, where C, α are constants. If $\gamma > 1$ it tends towards C while for $\gamma < 1$ it is equivalent to $\alpha\tau^{1-\gamma}$.

Now turn to $\int_0^T g(\tau)d\tau$: for large T it is equal to $B+CT+\frac{\alpha}{2-\gamma}T^{2-\gamma}$. For $T \to \infty$ this quantity is equivalent to CT if $\gamma > 1$ while it is equivalent to $\frac{\alpha}{2-\gamma}T^{2-\gamma}$ if $\gamma < 1$. As a result we get, using formulas (9.8), two types of behavior for $\mathcal{S}_T$ and $\mathcal{R}_T$, which leads to a division of stochastic volatility models into two classes.

For long maturities:

- (Type I) If $\gamma > 1$:

$$\mathcal{S}_T \propto \frac{1}{T} \quad \text{and} \quad \lim_{T\to\infty} \mathcal{R}_T = 1 \tag{9.11}$$

- (Type II) If $\gamma < 1$:

$$\mathcal{S}_T \propto \frac{1}{T^\gamma} \quad \text{and} \quad \lim_{T\to\infty} \mathcal{R}_T = 2 - \gamma \tag{9.12}$$

We leave it to the reader to check that exponential decay of $\mu(t)$ produces Type I behavior.

Both Type I and Type II scalings are compactly summarized in the following relationship. For $T \to \infty$:

$$\mathcal{S}_T \propto \frac{1}{T^{2-\mathcal{R}_\infty}} \tag{9.13}$$

9.6 Type I models – the Heston model

The fact that for a fast-decaying spot/variance covariance function $\mathcal{S}_T$ decays like $\frac{1}{T}$ is not unexpected. Indeed in this case, in the limit of long time intervals, increments of $\ln S_t$ become independent and identically distributed. Cumulants of $\ln S_t$ then scale linearly with T, thus the skewness s_T of $\ln S_T$ scales like $\frac{1}{\sqrt{T}}$. Consequently, at order one in volatility of volatility, or equivalently at order one in s_T, the skew/skewness relationship (8.23) implies that the ATMF skew scales like $\frac{1}{T}$.[2]

The Heston model provides an example of Type I behavior. $\mu(t)$ is exponentially decaying – see expressions (8.48) – hence we expect Type I scaling (9.35). Let us verify this by using expressions (6.19) for $\widehat{\sigma}_{F_T T}$ and $\mathcal{S}_T$ at order one in volatility of volatility for long maturities.

The ATMF skew is given by (6.19a):

$$\mathcal{S}_T = \frac{\rho\sigma}{2kT}\frac{1}{\sqrt{V^0}} \tag{9.14}$$

$\mathcal{S}_T$ indeed has Type I scaling. Consider now the SSR. $\widehat{\sigma}_{F_T T}$ in (6.19a) is a function of V – its covariance with $d\ln S$ is thus given by:

$$\frac{E\left[d\widehat{\sigma}_{F_T T}d\ln S_t\right]}{E\left[(d\ln S_t)^2\right]} = \frac{\partial\widehat{\sigma}_{F_T T}}{\partial V}\frac{E\left[dVd\ln S\right]}{Vdt}$$
$$= \frac{1}{2kT\sqrt{V^0}}\frac{E\left[dVd\ln S\right]}{Vdt}$$
$$= \frac{1}{2kT\sqrt{V^0}}\rho\sigma \tag{9.15}$$

where in $\frac{\partial\widehat{\sigma}_{F_T T}}{\partial V}$ we have kept the contribution at zeroth order in volatility of volatility to get the covariance of $d\widehat{\sigma}_{F_T T}$ and $d\ln S_t$ at order one. Dividing (9.14) by (9.15) yields $\mathcal{R}_T = 1$: we have confirmed by hand that the Heston model is indeed of Type I – this was already observed in [8].

---

[2]This $\frac{1}{T}$ scaling of the ATMF skew is also shared by jump-diffusion and Lévy models, at order one in the skewness of $\ln S_T$, as in these models, increments of $\ln S_t$ are indeed independent and identically distributed. The skewness of $\ln S_T$ exactly scales like $\frac{1}{\sqrt{T}}$, and at order one in this skewness, the skew scales like $\frac{1}{T}$.

However, while $\lim_{T\to\infty}\mathcal{R}_T = 1$ in Type I models, the behavior of $\mathcal{R}_T$ in jump/Lévy models is different. Because implied volatilities are a function of K/S only, $\mathcal{R}_T = 0$, $\forall T$.

9.7 Type II models

In Type I models, the long-maturity scaling of the ATMF skew and the limit of the SSR are fixed and do not depend on γ. Depending on the size of γ, the long-maturity regime sets in for shorter or longer maturities, but the limiting behavior of the ATMF skew and of the SSR is universal and bears no trace of the precise underlying dynamics.

Type II models are richer as both the scaling of $\mathcal{S}_T$ and the limit of $\mathcal{R}_T$ are non-trivial and reflect the characteristic exponent of the decay of the spot/variance covariance function $\mu(t)$.

Moreover, the ATMF skew of market smiles typically decays algebraically, with an exponent around $-\frac{1}{2}$. This calls for a Type II model; hence the following natural questions:

- Is it practically possible to build a Type II model?

- Is the *dynamics* of market smiles consistent with Type II?

Type II scaling in the two-factor model
Consider an N-factor model of the type studied in Section 7.3:

$$\begin{cases} dS_t = (r-q)S_t dt + \sqrt{\xi_t^t} S_t dW_t^S \\ d\xi_t^T = \omega\xi_t^T \sum_i w_i e^{-k_i(T-t)} dW_t^i \end{cases}$$

Assuming a flat term structure of VS volatilities the model is time-homogeneous and the spot/variance covariance function reads:

$$\mu(\tau) = \omega\xi_0^{\frac{3}{2}} \sum_i w_i \rho_{iS} e^{-k_i\tau}$$

where ρ_{iS} is the correlation between W^S and W^i. Inserting this expression into formulas (9.5) and (9.6) yields the following expressions for the ATMF skew and the SSR:

$$\mathcal{S}_T = \frac{\omega}{2} \sum_i w_i \rho_{iS} \frac{k_i T - (1 - e^{-k_i T})}{(k_i T)^2} \tag{9.16a}$$

$$\mathcal{R}_T = \frac{\sum_i w_i \rho_{iS} \frac{1-e^{-k_i T}}{k_i T}}{\sum_i w_i \rho_{iS} \frac{k_i T - (1-e^{-k_i T})}{(k_i T)^2}} \tag{9.16b}$$

$\mu(\tau)$ is a linear combination of exponentials. As $\tau \to \infty$, $\mu(\tau) \propto e^{-\min_i k_i T}$. Thus for $T \to \infty$ this model is of type I: $\mathcal{S}_T \propto \frac{1}{T}$ and $\lim_{T\to\infty} \mathcal{R}_T = 1$; this can be checked explicitly on expressions (9.16). The fact that $\mu(\tau)$ is a linear combination of exponentials is dictated by the property that this form allows for a Markov representation of the model with a number of state variables – besides S_t – equal to the number of exponentials – see Section 7.1.

We could try and use a model whose dynamics for forward variances reads:

$$d\xi_t^T = \omega\xi_t^T \frac{1}{(T-t+\theta)^\gamma}dW \qquad (9.17)$$

where θ is a (small) offset parameter. In this model $\mu(\tau)$ has the desired power law scaling for large τ: $\mu(\tau) \propto \frac{1}{\tau^\gamma}$. Unfortunately we lose the Markov representation: it is no longer possible to express the set of forward variances ξ^T as a function of a finite number of state variables. Such a model is not usable in practice.

Luckily, by suitably choosing model parameters it is possible to get Type II scaling in an N-factor model – in fact a two-factor model – over a range of maturities that is sufficient for practical purposes, even though Type I scaling eventually kicks in for (very) long maturities.

Let us use the two-factor model of Section 8.7 with the parameters listed in Table 8.2, page 329. Remember that (a) $\nu, \theta, k_1, k_2, \rho$ have been selected so that the volatilities of VS volatilities decay approximately as a power law with exponent 0.4, (b) ρ_{SX^1} and ρ_{SX^2} have been chosen so that the ATMF skew decays approximately as a power law with exponent $\frac{1}{2}$. This is illustrated in Figure 8.5.

At order one in volatility of volatility the ATMF skew is given by expression (8.54), page 329. The SSR is calculated using expression (9.4) and we have:

$$\mathcal{S}_T = \frac{\nu\alpha_\theta}{\hat\sigma_T^3 T^2} \int_0^T dt \sqrt{\xi_0^t} \int_t^T \xi_0^u \left[(1-\theta)\rho_{SX^1}e^{-k_1(u-t)} + \theta\rho_{SX^2}e^{-k_2(u-t)}\right] du \qquad (9.18)$$

$$\mathcal{R}_T = \frac{1}{\sqrt{\xi_0^0}} \frac{\hat\sigma_T^2 T \int_0^T \xi_0^t \left[(1-\theta)\rho_{SX^1}e^{-k_1t} + \theta\rho_{SX^2}e^{-k_2t}\right] dt}{\int_0^T dt \sqrt{\xi_0^t} \left(\int_t^T \xi_0^u \left[(1-\theta)\rho_{SX^1}e^{-k_1(u-t)} + \theta\rho_{SX^2}e^{-k_2(u-t)}\right] du\right)} \qquad (9.19)$$

For a flat term structure of VS volatilities:

$$\mathcal{S}_T = \nu\alpha_\theta \left[(1-\theta)\rho_{SX^1}\frac{k_1 T - (1-e^{-k_1 T})}{(k_1 T)^2} + \theta\rho_{SX^2}\frac{k_2 T - (1-e^{-k_2 T})}{(k_2 T)^2}\right] \qquad (9.20)$$

$$\mathcal{R}_T = \frac{(1-\theta)\rho_{SX^1}\frac{1-e^{-k_1 T}}{k_1 T} + \theta\rho_{SX^2}\frac{1-e^{-k_2 T}}{k_2 T}}{(1-\theta)\rho_{SX^1}\frac{k_1 T - (1-e^{-k_1 T})}{(k_1 T)^2} + \theta\rho_{SX^2}\frac{k_2 T - (1-e^{-k_2 T})}{(k_2 T)^2}} \qquad (9.21)$$

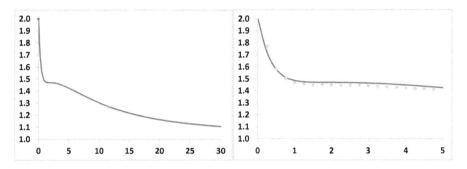

Figure 9.1: The SSR of the two-factor model, computed with formula (9.20) and parameters in Table 8.2, page 329. Left: maturities up to 30 years – right: zoom on maturities less than 5 years. The dots are the result of a Monte Carlo evaluation of the SSR – see Section 9.8 below.

$\mathcal{R}_T$, computed using (9.21), appears in Figure 9.1, as a function of T.

Consider first the left-hand graph. The short-maturity limit of the SSR is 2, as shown in Section 9.3. As $T \to \infty$, the SSR tends to 1 – as it should. Notice however the shoulder for a value of the SSR around 1.5, which appears in more detail in the right-hand graph. The SSR for maturities from 1 to 5 years is stable around 1.5. For this range of maturities, the model obeys the Type II scaling rules in (9.36): the value of the SSR is 2 minus the characteristic exponent of the decay of the ATMF skew – in our case $\frac{1}{2}$.

Note in the right-hand graph how the approximate expression (9.16b), obtained at order one in volatility of volatility, agrees well with the actual value of the SSR.

Another example is shown in Figure 9.2: here we have chosen ρ_{SX^1}, ρ_{SX^2} so that the ATMF skew decays approximately with a characteristic exponent equal to 0.3 for maturities up to 5 years (left-hand graph). As the right-hand side graph shows, we get an SSR around 1.7, i.e. $2 - 0.3$ over this range.[3]

Formula (9.16b) for the SSR has been derived for a flat term structure of VS volatilities. When this is not the case, the prefactor $\xi^u \sqrt{\xi^t}$ in expression (8.50), page 327, for $\mu(t, u, \xi)$ in the two-factor model is not constant and, at order one in volatility of volatility, both the numerator and denominator of $\mathcal{R}_T$ have to be calculated by numerical integration.

The SSR thus depends on the term structure of VS volatilities. For example, for a VS term structure that increases (decreases) from 20% to 25% for $T = 1$ year, the SSR for this maturity is 1.46 (1.50). Let us take $T = 5$ years; for a VS term structure that increases (decreases) from 20% to 30%, the SSR is 1.43 (1.49). These values

---

[3]With parameters $\nu, \theta, k_1, k_2, \rho$ already set so as to generate the desired term structure of volatilities of VS volatilities, ρ_{SX^1}, ρ_{SX^2} are the only handles left to control the term structure of (a) the ATMF skew, (b) the SSR. Introducing more factors allows for more flexibility, but in the author's experience two factors are sufficient for capturing typical market ATMF skews and SSR.

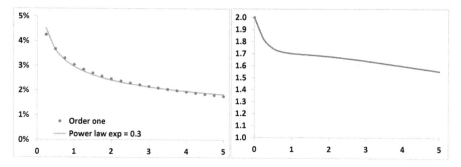

Figure 9.2: Left: The ATMF skew measured as the difference of implied volatilities for strikes $0.95F_T$ and $1.05F_T$ at order one in volatility of volatility (dots, formula (9.16a)) and a power law benchmark with exponent 0.3 (line). Right: the SSR at order one for maturities up to 5 years computed with expression (9.16b). The parameters are those of Table 8.1, with $\rho_{SX^1} = -35\%$, $\rho_{SX^2} = -83\%$.

are obtained in a Monte Carlo simulation of the two-factor model – see Section 9.8 below – with the same parameters as in Figure 9.1. For all practical purposes, the accuracy of formula (9.16b) for $\mathcal{R}_T$ is thus adequate as long as VS term structures are not too steep.

The definition of the SSR in (9.3) involves in the denominator the instantaneous variance of S at $t = 0$. In case short VS volatilities are, say, substantially larger than longer-dated ones, we get a smaller value for the SSR than that given by expression (9.16b). For the purpose of evaluating the SSR at $t = 0$, approximation (9.16b) is thus incorrect.

However, it is the average level of spot/volatility covariance generated by the model up to the maturity of our exotic option, rather than its particular value at $t = 0$, that is practically relevant. It is then advisable to replace the short-dated variance in the denominator of (9.3) with the square of the VS volatility for maturity T. Approximation (9.16b) is then almost exact.

In conclusion, while the two-factor model is strictly speaking a Type I model, by suitably choosing parameters we obtain Type II scaling for a range of maturities that is practically relevant. For these maturities we are able to generate a decay of the ATMF skew with an exponent less than 1 and the relationship between $\mathcal{S}_T$ and $\mathcal{R}_T$ in (9.13) is approximately obeyed.

Type II scaling in reality
We are able to obtain Type II scaling in a two-factor model, but are actual market smiles consistent with Type II behavior? This is characterized in (9.36) by two features: one static, one dynamic:

- The ATMF skew decays with a non-trivial exponent: $\mathcal{S}_T \propto \frac{1}{T^\gamma}$ with $\gamma < 1$

- The SSR is different than 1 and is related to γ: for long maturities $\mathcal{R}_T \to 2 - \gamma$

The decay of the ATMF skew of market smiles is consistent with Type II, but what about the SSR? From the expression of the SSR in (9.3) we define the realized SSR as:

$$\mathcal{R}_T^r = \frac{\sum_i \ln \frac{S_{i+1}}{S_i} (\hat{\sigma}_{T,i+1} - \hat{\sigma}_{T,i})}{\sum_i \mathcal{S}_{T,i} (\ln \frac{S_{i+1}}{S_i})^2} \tag{9.22}$$

where $\hat{\sigma}_{T,i}$ (resp. $\mathcal{S}_{T,i}$) is the ATMF implied volatility (resp. ATMF skew) at time i for residual maturity T.

$\mathcal{R}_T^r$ for $T = 2$ years is shown in Figure 9.3 for the Euro Stoxx 50 and S&P 500 indexes. We have used for simplicity ATM rather than ATMF volatilities.

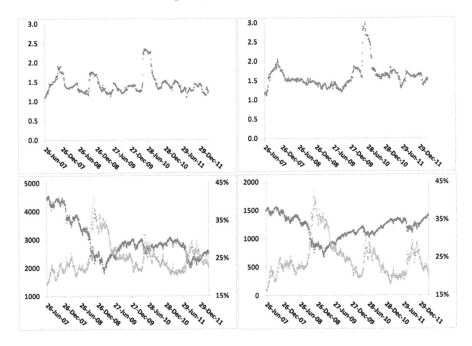

Figure 9.3: Top: 3-months realized SSR ($\mathcal{R}_T^r$) for the Euro Stoxx 50 (left) and S&P 500 (right) indexes, for $T = 2$ years. Bottom: the underlying (dark dots, left axis) and 2-year ATM implied volatility (lighter dots, right axis).

For both indexes, the realized SSR hovers around 1.5, with occasional spikes generated by simultaneous and opposite moves of S and $\hat{\sigma}_{F_T(S)T}$. The magnitude of the spikes and of the fluctuations around 1.5 is not relevant; it would be larger (resp. smaller) had we estimated $\mathcal{R}_T^r$ on a shorter (resp. longer) window than 3 months – the duration of the spikes is indeed about 3 months. Notice also how similar ATM implied volatilities are for both indexes. Realized values for the SSR for longer-dated maturities – say $T = 5$ years – are comparable to the 2-year case.

In conclusion, the dynamics of equity market smiles is consistent with Type II behavior: $\mathcal{S}_T \propto \frac{1}{T^\gamma}$ with $\gamma < 1$ and for large T, $\gamma + \mathcal{R}_T$ is approximately equal to 2.

9.8 Numerical evaluation of the SSR

At order one in volatility of volatility, for the sake of computing the SSR the ATMF volatility can be substituted with the VS volatility and this leads to the analytic expression (9.16a) for a flat term structure of VS volatilities.

Generally, however, the SSR can be easily computed numerically. In the two-factor model, the ATMF volatility is a function of X^1, X^2. At $t = 0$:

$$\widehat{\sigma}_{F_T T} \equiv \widehat{\sigma}_{F_T T}\left(X_0^1, X_0^2\right)$$

Expanding at first order in dX^1, dX^2:

$$d\widehat{\sigma}_{F_T T} = \frac{d\widehat{\sigma}_{F_T T}}{dX^1}dX^1 + \frac{d\widehat{\sigma}_{F_T T}}{dX^2}dX^2$$

From the definition of the SSR in (9.3) and using that $E[(d\ln S_t)^2] = \xi_0^0 dt$, $E[d\ln S_t dX_t^1] = \rho_{SX^1}\sqrt{\xi_0^0}dt$ and likewise for X^2, we get:

$$\mathcal{R}_T = \frac{1}{\mathcal{S}_T}\frac{E[d\ln S \, d\widehat{\sigma}_{F_T T}]}{E[(d\ln S)^2]} = \frac{1}{\mathcal{S}_T}\frac{1}{\sqrt{\xi_0^0}}\left(\frac{d\widehat{\sigma}_{F_T T}}{dX^1}\rho_{SX^1} + \frac{d\widehat{\sigma}_{F_T T}}{dX^2}\rho_{SX^2}\right)$$

Thus $\mathcal{R}_T$ can be simply evaluated numerically by computing $\widehat{\sigma}_{F_T T}$ with two different initial values for (X, Y):

$$\mathcal{R}_T \simeq \frac{1}{\mathcal{S}_T}\frac{1}{\sqrt{\xi_0^0}}\frac{\widehat{\sigma}_{F_T T}\left(X_0^1 + \varepsilon\rho_{SX^1}, X_0^2 + \varepsilon\rho_{SX^2}\right) - \widehat{\sigma}_{F_T T}\left(X_0^1, X_0^2\right)}{\varepsilon}$$

with ε a small offset. Typically we take $X_0^2 = X_0^1 = 0$.

In stochastic volatility models defined by the dynamics of the instantaneous variance V_t, the SSR is simply computed by shifting V as volatilities for a fixed moneyness are a function of V. For example, in the Heston model, using the notations of Chapter 6, $\mathcal{R}_T$ is simply given by:

$$\mathcal{R}_T \simeq \frac{1}{\mathcal{S}_T}\frac{\widehat{\sigma}_{F_T T}\left(V + \varepsilon\rho\sigma\right) - \widehat{\sigma}_{F_T T}\left(V\right)}{\varepsilon}$$

9.9 The SSR for short maturities

In both Type I and Type II models the short-maturity limit of the SSR is 2.

In a stochastic volatility model, in the limit $T \to 0$, the VS and ATM volatilities are identical. We will thus use the notation $\hat{\sigma}_0$ for both.

From the expression of the SSR in (9.3), in the special case of vanishing maturities, we can derive two definitions of the *realized* SSR according to whether we choose:

- the realized value of the instantaneous variance $(\ln \frac{S_{i+1}}{S_i})^2$

- or the implied value of the realized variance $\hat{\sigma}_{T,i}^2 \Delta t$

in the denominator of (9.3).

For reasons that will be clearer shortly, we choose the second convention. For small T we define $\mathcal{R}_T^{r,\text{short}}$ as:

$$\mathcal{R}_T^{r,\text{short}} = \frac{\sum_i \ln \frac{S_{i+1}}{S_i} (\hat{\sigma}_{T,i+1} - \hat{\sigma}_{T,i})}{\Delta t \sum_i S_{T,i} \hat{\sigma}_{T,i}^2} \tag{9.23}$$

where Δt is the duration of one (trading) day.

$\mathcal{R}_T^r$ for $T = 1$ month is shown in Figure 9.4 for the Euro Stoxx 50 and S&P 500 indexes.

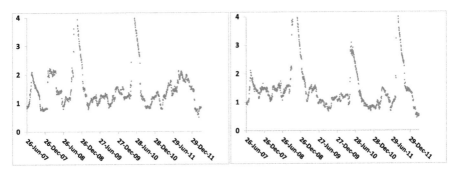

Figure 9.4: 3-months realized SSR ($\mathcal{R}_T^r$) for the Euro Stoxx 50 (left) and S&P 500 (right) indexes, for $T = 1$ month.

The average value of $\mathcal{R}_T^r$ is substantially lower than the model-independent "implied" value of 2. Does this point to a discrepancy between the dynamics of stochastic models and the dynamics of market smiles?

Consider the example of realized and implied volatilities – typically, for equity indexes, VS implied volatilities are systematically higher on average than realized volatilities. This in itself does not signal a modeling inconsistency. What is important is that the difference between both volatilities can be materialized as the P&L of an option position – here a VS.

The natural question is thus: is it possible to materialize the difference $2 - \mathcal{R}_T^r$ as the P&L of an option position?

9.10   Arbitraging the realized short SSR

The SSR expresses the implied spot/volatility covariance in units of the ATMF skew. Arbitraging $\mathcal{R}_T^{r,\text{short}}$ entails being able to materialize this covariance as a P&L. In what follows we set $T = 1$ month and assume zero interest rate and repo for simplicity.

To single out the realized spot/volatility covariance as a carry P&L we risk-manage our option position using the lognormal model of Section 8.5.1. In the limit $T \to 0$ the variance curve collapses to a single volatility $\hat{\sigma}_0$.

Since any stochastic volatility model calibrated to the market smile yields the same implied value for the instantaneous covariance of S and $\hat{\sigma}_0$ we could use any dynamics for $\hat{\sigma}_0$. However, in the lognormal model the short ATM skew is independent on $\hat{\sigma}_0$, a property that is also approximately shared by market smiles.

Choosing the lognormal model for risk-managing our option position then reduces chances that the mark-to-market P&L generated by remarking the ATM skew to market is large.

9.10.1   Risk-managing with the lognormal model

Consider the limit $T \to 0$ in the pricing equation (8.1): P becomes a function of $t, S, \hat{\sigma}_0$: $P(t, S, \xi) \equiv P(t, S, \hat{\sigma}_0)$. In the lognormal model for the instantaneous volatility the pricing equation is given by:

$$\frac{dP}{dt} + \frac{\hat{\sigma}_0^2}{2} S^2 \frac{d^2 P}{dS^2} + \frac{\nu^2}{2} \hat{\sigma}_0^2 \frac{d^2 P}{d\hat{\sigma}_0^2} + \rho \nu \hat{\sigma}_0^2 S \frac{d^2 P}{dS d\hat{\sigma}_0} = 0 \qquad (9.24)$$

where ν is the (lognormal) volatility of $\hat{\sigma}_0$ and ρ the correlation of $\hat{\sigma}_0$ with S.

We have derived in Section 8.5.1 the smile at order two in volatility of volatility, in the limit $T \to 0$. At this order, expressions (8.39a) and (8.39b) for the ATM skew and curvature yield the following expression for the price of a vanilla option of strike K, maturity T:

$$P = P_{BS}\left(t, S, \hat{\sigma}(x), K, T\right) \qquad (9.25a)$$

$$\hat{\sigma}(x) = \hat{\sigma}_0 + \mathcal{S}x + \frac{\mathcal{C}}{2}x^2 \qquad (9.25b)$$

where P_{BS} is the Black-Scholes formula, $x = \ln\left(\frac{K}{S}\right)$ and $\mathcal{S}, \mathcal{C}$ are given by:

$$\mathcal{S} = \frac{\rho \nu}{2} \qquad \mathcal{C} = \frac{2 - 3\rho^2}{6\hat{\sigma}_0}\nu^2$$

For a market smile of the form (9.25b) ρ, ν are given by:

$$\rho = \frac{2\mathcal{S}}{\sqrt{3\hat{\sigma}_0\mathcal{C} + 6\mathcal{S}^2}} \qquad \nu = \sqrt{3\hat{\sigma}_0\mathcal{C} + 6\mathcal{S}^2} \qquad (9.26)$$

In the limit $T \to 0$, $\mathcal{S}$ and $\mathcal{C}$, derived at order two in volatility of volatility, do not depend on T and the ATM implied volatility is identical to the VS volatility. This identity is exact; indeed consider a diffusive model whose instantaneous volatility at $t = 0$ is σ. Using expression (8.64) with $\hat{\sigma} = 0$, the price of a European option whose payoff is $f(S)$ is given, at order one in T by: $P = f(S_0) + \frac{\sigma^2 T}{2} S_0^2 \frac{d^2 f}{dS_0^2}$ where S_0 is the spot value at $t = 0$. Denote now by $\hat{\sigma}$ the implied volatility of this option. At order one in T, $\hat{\sigma}$ is such that $P = f(S_0) + \frac{\hat{\sigma}^2 T}{2} S_0^2 \frac{d^2 f}{dS_0^2}$. This yields $\hat{\sigma} = \sigma$: the implied volatilities of all European payoffs f such that $\frac{d^2 f}{dS_0^2} \neq 0$ are identical, and equal to the instantaneous volatility.[4]

Consider now a long position in a delta-hedged European option, risk-managed with the lognormal model. We will calibrate $\mathcal{S}$, $\mathcal{C}$ – or equivalently ρ, ν – to the market smile near the money on a daily basis. Denote by $\Pi(t, S, \hat{\sigma}_0, \mathcal{S}, \mathcal{C})$ the value of our European option position. We do not consider for now the mark-to-market P&L generated by a change of $\mathcal{S}$, $\mathcal{C}$ – or equivalently of ρ, ν – and instead focus on the carry P&L. This P&L during δt, at order one in δt and two in δS and $\delta \hat{\sigma}_0$, is given by:

$$P\&L = \frac{d\Pi}{dt}\delta t + \frac{d\Pi}{d\hat{\sigma}_0}\delta\hat{\sigma}_0 + \frac{1}{2}\frac{d^2\Pi}{dS^2}\delta S^2 + \frac{1}{2}\frac{d^2\Pi}{d\hat{\sigma}_0^2}(\delta\hat{\sigma}_0)^2 + \frac{d^2\Pi}{dSd\hat{\sigma}_0}\delta S\delta\hat{\sigma}_0 \quad (9.27)$$

While the delta hedge removes the term in δS, there remains a contribution in $\delta\hat{\sigma}_0$ as we are not vega-hedged.

Using (9.24), at order one in δt and two in δS and $\hat{\sigma}_0$ our P&L is given by:

$$P\&L = \frac{d\Pi}{d\hat{\sigma}_0}\delta\hat{\sigma}_0 \quad (9.28a)$$

$$+ \frac{1}{2}S^2\frac{d^2\Pi}{dS^2}\left(\left(\frac{\delta S}{S}\right)^2 - \hat{\sigma}_0^2\delta t\right) + \frac{1}{2}\hat{\sigma}_0^2\frac{d^2\Pi}{d\hat{\sigma}_0^2}\left(\left(\frac{\delta\hat{\sigma}_0}{\hat{\sigma}_0}\right)^2 - \nu^2\delta t\right) \quad (9.28b)$$

$$+ S\hat{\sigma}_0\frac{d^2\Pi}{dSd\hat{\sigma}_0}\left(\frac{\delta S}{S}\frac{\delta\hat{\sigma}_0}{\hat{\sigma}_0} - \rho\nu\hat{\sigma}_0\delta t\right) \quad (9.28c)$$

From (9.25b) the ATMF skew $\mathcal{S}$ is $\frac{\rho\nu}{2}$ – this is equivalent to the statement $\mathcal{R}_{T=0} = 2$. The last piece (9.28c) can thus be rewritten as:

$$S\hat{\sigma}_0\frac{d^2\Pi}{dSd\hat{\sigma}_0}\left(\frac{\delta S}{S}\frac{\delta\hat{\sigma}_0}{\hat{\sigma}_0} - \rho\nu\hat{\sigma}_0\delta t\right) = S\hat{\sigma}_0\frac{d^2\Pi}{dSd\hat{\sigma}_0}\left(\frac{\delta S}{S}\frac{\delta\hat{\sigma}_0}{\hat{\sigma}_0} - 2\mathcal{S}\hat{\sigma}_0\delta t\right)$$

$$= S\hat{\sigma}_0^2\mathcal{S}\frac{d^2\Pi}{dSd\hat{\sigma}_0}\left(\frac{\frac{\delta S}{S}\frac{\delta\hat{\sigma}_0}{\hat{\sigma}_0}}{S\hat{\sigma}_0^2\delta t} - 2\right)\delta t \quad (9.29a)$$

$$= S\hat{\sigma}_0^2\mathcal{S}\frac{d^2\Pi}{dSd\hat{\sigma}_0}\left(\mathcal{R}_{T=0}^{r,\text{short}} - 2\right)\delta t \quad (9.29b)$$

---

[4]This is not the case for vanilla options with strikes different than S_0 as for these options $\frac{d^2 f}{dS_0^2} = 0$ – the contribution of order one in T vanishes.

From (9.29a) it is apparent that, in defining the realized SSR, the short implied – rather than the realized – variance should be used in the denominator, in the definition of $\mathcal{R}_T^{r,\text{short}}$, hence expression (9.23) for $\mathcal{R}_T^{r,\text{short}}$.

9.10.2   The realized skew

The P&L above can be equivalently expressed as a difference between the implied skew $\mathcal{S}$ and the instantaneous *realized* skew $\mathcal{S}^r$ which we define as:

$$\mathcal{S}^r = \frac{1}{2\hat{\sigma}_0 \delta t} \frac{\delta S}{S} \frac{\delta \hat{\sigma}_0}{\hat{\sigma}_0} = \left(\frac{\mathcal{R}_{T=0}^{r,\text{short}}}{2}\right) \mathcal{S} \qquad (9.30)$$

Expression (9.30) shows that the relative mismatch of realized to implied skew is equal to that of realized to implied SSR. The cross gamma/theta P&L 9.29b reads:

$$2S\hat{\sigma}_0^2 \frac{d^2\Pi}{dS d\hat{\sigma}_0} \left(\mathcal{S}^r - \mathcal{S}\right) \delta t \qquad (9.31)$$

Note that the P&L in (9.28) is not the total P&L incurred on our option position. Additional P&L is generated by daily recalibration of $\mathcal{S}$ and $\mathcal{C}$. Only if this P&L is small and if the contributions in (9.28a) and (9.28b) are vanishing or negligible are we able to isolate the P&L of interest (9.29b) or equivalently (9.31).

9.10.3   Splitting the theta into three pieces

Expression (9.25a) for P is correct at order two in volatility of volatility: P does not exactly solve (9.24). Can we still use (9.25a) for P&L accounting? (9.24) expresses that the theta $\frac{dP}{dt}$ can be broken up in three pieces which match each of the second-order gamma contributions in (9.28).

In our model theta is equal to the Black-Scholes theta since $\hat{\sigma}(x)$ in (9.25b) does not depend on T. In the Black-Scholes model all of the theta is ascribed to the spot gamma – the second piece in the right-hand side of (9.28) – with a break-even level, the implied volatility, which is strike-dependent. In contrast, in the stochastic volatility model we are using, this theta is distributed across three gammas with break-even levels that are *not* strike-dependent. Checking whether (9.24) holds amounts to checking how well the following equality holds

$$\frac{\hat{\sigma}(x)^2}{2} S^2 \frac{d^2 P_{BS}}{dS^2} = \frac{\hat{\sigma}_0^2}{2} S^2 \frac{d^2 P}{dS^2} + \frac{\nu^2}{2} \hat{\sigma}_0^2 \frac{d^2 P}{d\hat{\sigma}_0^2} + \rho \nu \hat{\sigma}_0^2 S \frac{d^2 P}{dS d\hat{\sigma}_0} \qquad (9.32)$$

where the three contributions in the right-hand side are called, respectively: spot theta, vol theta, cross theta.

Let us take the typical example of a one-month maturity smile with $\hat{\sigma}_0 = 20\%$, $\mathcal{S} = -0.7$, $\mathcal{C} = 0.4$. With these parameters, the implied volatilities of the 90%, 100%, 110% strikes are, respectively, 27.6%, 20%, 13.5%. From (9.26) we have: $\rho = -78.5\%$, $\nu = 178.3\%$.

Figure 9.5 shows on the left the three thetas as a function of K, for $S = 100$. The sum of these three thetas, along with the Black-Scholes theta – the left-hand side of (9.32) where a different implied volatility is used for each strike – is shown on the right.

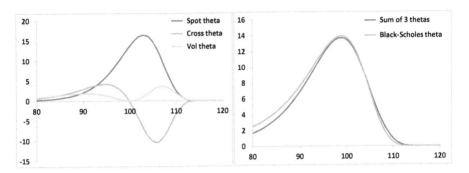

Figure 9.5: Left: the three terms in the right-hand side of (9.32) as a function of K, for $S = 100$. Right: the sum of the three thetas compared to the Black-Scholes theta $\frac{\hat{\sigma}(x)^2}{2} S^2 \frac{d^2 P_{BS}}{dS^2}$.

As we can see our approximation applied to the lognormal model does a decent job at splitting the Black-Scholes theta in three pieces, whose break-even levels are independent on the option's strike. The three thetas do not exactly add up to the Black-Scholes theta for strikes far out of the money, but the agreement is satisfactory for strikes between 95% and 105%, which is the range we use in our tests.

As is clear from Figure 9.5, the spot theta dominates by far – as our objective is to isolate the cross theta, we need to ensure that the spot theta of our position vanishes. The vol theta, on the other hand, is much smaller and almost cancels for a spread position.

9.10.4   Backtesting on the Euro Stoxx 50 index

We now use historical implied volatilities of the Euro Stoxx 50 index from April 2007 to March 2012 to backtest the following dynamical option trading strategy. We sell one-month options of strike 95% and buy the appropriate number of one-month options of strike 105% so that the spot gamma vanishes. We delta-hedge this position until the next (trading) day, when it is unwound and a new position is started. In order to approximately maintain a constant level of cross-gamma, we trade a constant notional – 100€ – of the 95%-strike option. For typical levels of 1-month market skew, for one 95% option sold, we need to buy about 0.5 options of strike 105%.

We use daily one-month implied volatilities for strikes 95%, 100%, 105% to determine the skew $\mathcal{S}$ and curvature $\mathcal{C}$ and back out ρ and ν using (9.26). These, together with the ATM volatility $\hat{\sigma}_0$, are fed to the lognormal model which we use

(a) to determine the ratio of the 105% options to 95% options, (b) to compute the delta.

Calibrated values of $\mathcal{S}$ (multiplied by -10) and $\mathcal{C}$ appear in the left-hand graph of Figure 9.6. $\mathcal{C}$ is very noisy – indeed over the strike range $[95\%, 105\%]$, the one-month smile is almost a straight line: in our tests we thus set the product $\mathcal{C}$ equal to $\frac{2}{\widehat{\sigma}_0}$ – the precise value of $\mathcal{C}$ used hardly affects our results since we only use 95% and 105% strikes.

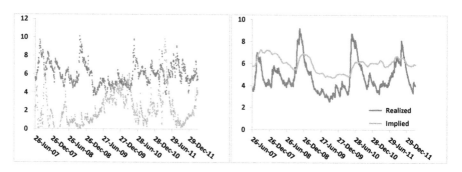

Figure 9.6: Left: daily values of skew $\mathcal{S}$ (darker dots) and curvature $\mathcal{C}$ (lighter dots) of one-month smiles of the Euro Stoxx 50 index. Right: 3-months exponentially weighted moving averages of the implied ATM skew $\mathcal{S}$ and its realized counterpart $\mathcal{S}^{\text{real}}$ for a one-month maturity – see (9.30). Values of $\mathcal{S}$ in both graphs, and of $\mathcal{S}^{\text{real}}$ have been multiplied by -10 to correspond approximately to the 95%/105% skew in volatility points.

The right-hand graph shows the implied and realized 1-month skew for the Euro Stoxx 50 index. In our historical sample, the average value of the *implied* 95%/105% skew is 6 points of volatility, while the average value of its *realized* counterpart – defined in (9.30) – is 4.8 points; this is 20% less than its implied counterpart.

Consequently, from (9.30), the average value of the realized SSR is about 20% lower than its implied value of 2; this agrees with the average value of realized SSR in Figure 9.4.

Our daily P&L between day i and day $i+1$ is very plainly given by:

$$P\&L_{\text{Total}} = \left[\Pi_i(t_{i+1}, S_{i+1}, \widehat{\sigma}_{0i+1}, \mathcal{S}_{i+1}, \mathcal{C}_{i+1}) - \Pi_i(t_i, S_i, \widehat{\sigma}_{0i}, \mathcal{S}_i, \mathcal{C}_i)\right]$$
$$- \frac{d\Pi_i}{dS_i}(S_{i+1} - S_i)$$

where Π_i is the market value of the option portfolio purchased at time t_i. This P&L can be broken down in three contributions:

$$P\&L_{\text{Total}} = P\&L_{\text{Carry}}^{\text{vega-hedged}} + P\&L_{\text{Vega}} + P\&L_{\text{MtM}}^{\mathcal{S},\mathcal{C}}$$

where:

$$P\&L_{\text{Carry}}^{\text{vega-hedged}} = \left[\Pi_i(t_{i+1}, S_{i+1}, \widehat{\sigma}_{0i+1}, \ \mathcal{S}_i, \mathcal{C}_i) - \Pi_i(t_i, S_i, \widehat{\sigma}_{0i}, \ \mathcal{S}_i, \mathcal{C}_i)\right]$$
$$- \frac{d\Pi_i}{dS_i}(S_{i+1} - S_i) - \frac{d\Pi_i}{d\widehat{\sigma}_{0i}}(\widehat{\sigma}_{0i+1} - \widehat{\sigma}_{0i})$$

$$P\&L_{\text{Vega}} = \frac{d\Pi_i}{d\widehat{\sigma}_{0i}}(\widehat{\sigma}_{0i+1} - \widehat{\sigma}_{0i})$$

$$P\&L_{\text{MtM}}^{\mathcal{S},\mathcal{C}} = \Pi_i(t_{i+1}, S_{i+1}, \widehat{\sigma}_{0i+1}, S_{i+1}, \mathcal{C}_{i+1}) - \Pi_i(t_{i+1}, S_{i+1}, \widehat{\sigma}_{0i+1}, \ \mathcal{S}_i, \mathcal{C}_i)$$

$P\&L_{\text{MtM}}^{\mathcal{S},\mathcal{C}}$ is the mark-to-market P&L generated by recalibrating $\mathcal{S}$ and $\mathcal{C}$ to the market smile at time t_{i+1}.

At order one in $(t_{i+1} - t_i)$ and order two in $(S_{i+1} - S_i)$ and $(\widehat{\sigma}_{0i+1} - \widehat{\sigma}_{0i})$, $P\&L_{\text{Carry}}^{\text{Vega-hedged}}$ is the sum of the three gamma/theta contributions in (9.28b) and (9.28c).

Our option position has vanishing spot gamma by construction. Moreover, inspection of the left-hand graph of Figure 9.5 suggests that the volatility gamma/theta P&L is small: ideally P&L (9.28b) will be negligible so that $P\&L_{\text{Carry}}^{\text{Vega-hedged}}$ closely tracks the quantity of interest, that is the cross-gamma/theta P&L (9.28c) which we aim to single out. Our P&L is however polluted by the contributions of $P\&L_{\text{MtM}}^{\mathcal{S},\mathcal{C}}$ and $P\&L_{\text{Vega}}$.

How well $(P\&L_{\text{Carry}}^{\text{vega-hedged}} + P\&L_{\text{MtM}}^{\mathcal{S},\mathcal{C}})$, that is $(P\&L_{\text{Total}} - P\&L_{\text{Vega}})$ correlates to the cross-gamma/theta P&L is assessed in the left-hand graph of Figure 9.7.

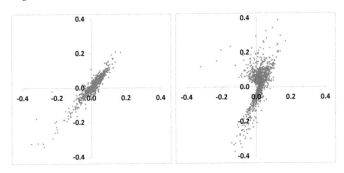

Figure 9.7: Scatter plots of daily values of $P\&L_{\text{Total}} - P\&L_{\text{Vega}}$ (left-hand graph) and $P\&L_{\text{Total}}$ (right-hand graph) as a function of P&L (9.28c).

$P\&L_{\text{MtM}}^{\mathcal{S},\mathcal{C}}$ contributes in fact little noise; the reason we do not get a straight line is mostly due to higher-order terms in $P\&L_{\text{Carry}}^{\text{vega-hedged}}$. Still, the total daily P&L corrected for the vega contribution captures the daily cross gamma/theta P&L with acceptable accuracy.

Inclusion of the vega contribution – see right-hand graph – reduces the correlation of both P&Ls but, as we will now see, the cumulative vega P&L is small enough that our strategy's P&L is still mostly attributable to the cross gamma/theta P&L.

Figure 9.8 shows the cumulative P&Ls of our arbitrage strategy. That the bulk of the P&L is contributed by the cross gamma/theta P&L is manifest.

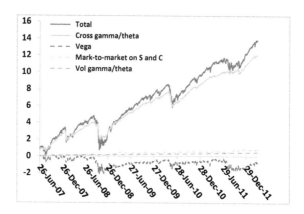

Figure 9.8: Cumulative P&Ls of the realized skew arbitrage strategy.

The volatility gamma/theta P&L, as well $P\&L_{\text{MtM}}^{S,C}$ and $P\&L_{\text{Vega}}$, are both small, even though the latter is very noisy. As a rule, sharp market moves generate simultaneous downward moves of the spot and upward moves of implied volatilities. Our gamma-neutral position has negative vega. On these days, our option position loses money both because of our positive cross-gamma and negative vega positions: this is confirmed by the simultaneous drops in the cumulative cross gamma/theta and vega P&Ls in Figure 9.8. These drops indeed coincide with peaks of the realized skew seen in Figure 9.6.

Remember that in our historical sample the *realized* ATM skew is on average 20% lower than the *implied* ATM skew – or equivalently that the realized SSR is 20% lower than its model-independent value of 2. Our position has vanishing spot gamma; moreover Figure 9.8 shows that the volatility gamma/theta P&L is small. Thus, almost all of the theta is generated by the cross theta. The cumulative theta P&L of our strategy is 45.7€. 20% of this, that is about 9€ is a number that is indeed consistent with the cumulative cross gamma/theta P&L.

9.10.5   The "fair" ATMF skew

In conclusion, just as the difference between realized and implied volatility can be arbitraged – that is materialized as the P&L of an option strategy – for short maturities, the difference between the realized value of the SSR and its model-independent value of 2 – or equivalently the difference between the realized ATM skew and the implied ATM skew – can be materialized, to a good approximation, as the P&L of a dynamical option strategy.[5]

---

[5]In practice, we would presumably buy a little more 105% strike options in order to reduce our vega position. This would generate a positive spot gamma position – which helps in case of large market

There is then no inconsistency in the fact that $\mathcal{R}^{\text{real}} < 2$. The short ATM skew quantifies the *implied* value of the instantaneous spot/volatility covariance, which may differ from its *realized* value. In our sample, it is on average 20% lower than its *implied* counterpart. Our trading strategy approximately materializes this difference as a P&L.

We have shown that this is equivalent to materializing as a P&L the difference between implied skew and realized skew:

$$\mathcal{S}^r \;=\; \frac{1}{2\widehat{\sigma}_0 \delta t} \left\langle \frac{\delta S}{S} \frac{\delta \widehat{\sigma}_0}{\widehat{\sigma}_0} \right\rangle$$

Note that the fair level of the ATM skew is not determined by the covariance of the spot and *realized* volatility; rather it is given by the covariance of spot and ATM *implied* volatility, a circumstance that may surprise at first.

Imagine there is no options' market – our only hedge instrument is the underlying itself – and we are asked to quote "fair" vanilla option prices, "fair" meaning that we do not make or lose money on average. We then need to model the process of the instantaneous historical realized variance – or equivalently the process of expected future historical variances, that is forward variances. We use pricing equation (7.4), which expresses that, on average no money is made or lost as we delta-hedge our option. The only difference now is that ξ^τ is no longer a market VS forward variance, but instead the expected future realized instantaneous variance.[6]

The results of Chapter 8 apply: at order one in volatility of volatility the "fair" ATM skew is given by the weighted integral of the *historical* covariance of spot and future (expected) *realized* variance.[7]

Now assume that there exists instead a market of ATM options – or variance swaps. Just as in the above backtest, we use these options to cancel the spot gamma, thus removing the sensitivity to *realized* volatility.

Since the vanilla hedge has to be dynamically readjusted, our position becomes sensitive to the joint dynamics of the spot and *implied* volatilities. Practically, this is

---

moves but costs some theta otherwise. Using the 100% strike along with the 95% and 105% strikes in order to cancel both gamma and vega does not work, as the resulting position is long the 95% and 105% strikes and short the 100% strike: this is mostly a volatility gamma/theta position. Also, unwinding and restarting a new position on a daily basis is not practical: in our backtest, factoring in a bid/offer spread of 0.2 points of volatility on each leg of our spread position wipes out the strategy's P&L.

[6]Interestingly the ξ^τ are still driftless. The ξ^τ in (7.4) are driftless because they can be delta-hedged by taking a position in market instruments (VSs) that require no financing. In the present context, the ξ^τ are driftless in the historical probability measure, just because they are expectations (of future realized variances).

[7]One can also avoid modeling variances altogether by using the hedged Monte Carlo technique of Bouchaud, Potters and Sestovic – see [17], or equivalently, a method proposed by Bruno Dupire. It consists in (a) using consecutive sequences of historical returns as Monte Carlo paths, (b) simulating the daily delta-hedging – at a given implied volatility – of a vanilla option, (c) finding the implied volatility such that the average of the final payoff minus the P&L from the delta hedge equals the Black-Scholes price. Underlying this technique is the unstated – and strong – assumption of stationarity: we are averaging over different volatility regimes and the conditionality on spot level or past return history is lost.

materialized, at order one in volatility of volatility, as the cross gamma of spot and *implied* volatility.

9.10.6   Relevance of model-independent properties

It is not clear that other model-independent properties can be established, without resorting to more or less reasonable additional assumptions. For example, Peter Carr and Roger Lee show in [27] that if implied volatilities of power payoffs[8] are uncorrelated with the spot process, the density of the realized quadratic variation up to T can be extracted from the vanilla smile of maturity T, hence payoffs on realized variance can be replicated by dynamical trading in vanilla options.

Are these model-independent rules practically relevant? This is assessed by studying whether a violation of these rules can be arbitraged, i.e. materialized as a P&L. It is not clear in particular that a violation of (9.13) could be practically arbitraged.

This issue is connected to the general question of the practical relevance of calibration. Our study of the arbitrage of the short SSR is a sobering illustration of how difficult it can be to lock the value of a model parameter – in our case the covariance of spot and ATM implied volatility – by dynamically trading vanilla options.

Only when one is able to do so does it make sense to entrust a model with the task of backing the value of a dynamical parameter out of the vanilla smile.

9.11   Conclusion

Because the skew in stochastic volatility models is generated by the covariance of spot and forward variances, some features of the underlying spot/variance dynamics can be recovered from the resulting smile. Provided some reasonable assumptions hold – in particular time homogeneity – for a flat term structure of VS volatilities, at order one in the volatility of volatility, the Skew Stickiness ratio is bounded above and below:

$$\mathcal{R}_T \in [1, 2]$$

We also show that $\mathcal{R}_T$ – a quantity that characterizes the model's dynamics – is related to the scaling of the ATMF skew $\mathcal{S}_T$ with maturity a static property of the smile. For short maturities, $\mathcal{R}_T$ tends to the universal value of 2. For long maturities its scaling depends on the characteristic exponent γ of the underlying

---

[8]See Section 4.3 for the definition of power payoffs. In the derivation of [27] the assumption of no correlation between S and the instantaneous volatility is made. What matters however from a trading point of view is that S and implied volatilities of power payoffs be uncorrelated. Only then does the spot/implied volatility cross-gamma P&L vanish.

spot/variance covariance function. Stochastic volatility models fall into one of two classes depending on the value of γ.

- Type I models are such that $\gamma > 1$: for large T, $\mathcal{S}_T \propto \frac{1}{T}$ and $\mathcal{R}_T \to 1$

- Type II models are such that $\gamma < 1$: for large T $\mathcal{S}_T \propto \frac{1}{T^\gamma}$ and $\mathcal{R}_T \to 2 - \gamma$

Thus, Type II models allow for a slower decay of the ATMF skew – which is consistent with market skews, for which $\gamma \simeq \frac{1}{2}$ – and a non-trivial value of the long-maturity limit of the SSR, which, again, is compatible with the observed dynamics of market smiles. Moreover, the characteristic exponent of the spot/variance covariance function can be backed out of the scaling of $\mathcal{S}_T$ and the long-maturity limit of $\mathcal{R}_T$.

This connection between $\mathcal{S}_T$ and $\mathcal{R}_T$ is a distinguishing feature of time-homogeneous stochastic volatility models – it is summarized by the following formula: for long maturities,

$$\mathcal{S}_T \propto \frac{1}{T^{2-\mathcal{R}_\infty}}$$

In time-homogeneous Jump-Lévy models, by contrast, while for long maturities $\mathcal{S}_T \propto \frac{1}{T}$, $\mathcal{R}_T$ vanishes for all T.

While putting together a genuine Type II model is difficult – because one needs to reconcile a non-trivial scaling of the spot/variance covariance function with the requirement of a low-dimensional Markov representation – Type II behavior can be achieved for a decent range of maturities in a model driven by simple Ornstein-Uhlenbeck processes.

We have provided evidence that suitable parametrization of a simple two-factor model generates Type II scaling for a range of maturities that is practically relevant – while also enforcing the desired scaling of volatilities of volatilities.

Having analyzed how dynamical properties of spot and time-homogeneous stochastic volatility models are related to the smile they produce, the following questions naturally arise:

- How do they compare with dynamical properties of a local volatility model calibrated to the same smile?

- How do they compare with the *realized* behavior of spot and implied volatilities?

9.11.1 SSR in local and stochastic volatility models – and in reality

The behavior of $\mathcal{R}_T$ as a function of T is structurally different in stochastic volatility models than in the local volatility model. In either Type I or Type II stochastic volatility models, the SSR starts from 2, then *decreases* either towards 1 or towards a non-trivial value in $[1, 2]$. In the local volatility model, instead, the SSR

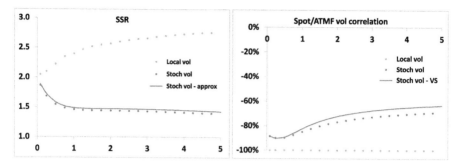

Figure 9.9: Left: the SSR as a function of maturity (years) in the two-factor stochastic volatility model (dark dots) and in the local volatility model (light dots) calibrated on the smile generated by the stochastic volatility model. Parameters in Table 8.2, page 329, have been used. The solid line corresponds to the SSR of the stochastic volatility model calculated using expression (9.21), at order one in volatility of volatility. Right: correlation of spot and ATMF volatilities. The solid line corresponds to the correlation of spot and VS – rather than ATMF – volatilities in the two-factor model, which is easily evaluated exactly.

starts from 2 for short maturities, then *increases* for longer maturities, for typical equity index smiles – see Figure 2.4, page 59.

This is illustrated in Figure 9.9. We have used the parameters in Table 8.2 and a flat term structure of VS volatilities at 20% (same parameters as those used in Figure 9.1). The curves in Figure 9.9 are thus obtained with the same vanilla smile – shown in Figure 8.2, page 330, for select maturities.

In both models the SSR starts from the model-independent value of 2 for short maturities. Equivalently, the implied regression coefficient of the short ATMF volatility on the spot is model-independent and given by the ATMF skew.

Parameters in Table 8.2 are such that the ATMF skew approximately decays with the characteristic exponent $\gamma = \frac{1}{2}$. For the longer maturities in our graph, the SSR tends to 1.5 in the stochastic volatility model (approximately $2 - \gamma$), while for the local volatility model it tends to a value close to 3.

For a smile such that the ATMF skew decays as a power law, the approximate expression (2.81), page 56, of $\mathcal{R}_\infty$ in the local volatility model:

$$\mathcal{R}_\infty = \frac{2 - \gamma}{1 - \gamma}$$

indeed yields $\mathcal{R}_\infty = 3$ for $\gamma = \frac{1}{2}$.

The spot/ATMF volatility covariance

The SSR is a useful indicator as it measures the implied covariance of $\ln S$ and the ATMF volatility $\hat{\sigma}_{F_T T}$ – in units of the ATMF skew – generated by the pricing model. Spot/volatility cross-gammas are one of the main risks of exotic equity payoffs.[9] Rather than directly comparing the implied spot/volatility covariance with its realized counterpart, we can convert the *realized* covariance of $\ln S$ and $\hat{\sigma}_{F_T T}$ into a *realized* SSR.

Model-generated and realized SSR can then be compared to assess whether the pricing model is sufficiently conservative.

The realized SSR for the Euro Stoxx 50 and S&P 500 indexes, for $T = 2$ years appear in Figure 9.3 – the estimator for the realized SSR is given by expression (9.22). For both indexes, the average realized SSR is around 1.5. For other indexes, the realized SSR can be very different. Figure 9.10 shows historical values of the realized 6-month value of $\mathcal{R}_T$ for $T = 1$ year, for the S&P 500 and Nikkei indexes. While the SSR for the S&P 500 is fairly stable around 1.5, for the Nikkei it reaches at times very negative values.

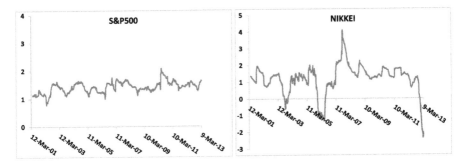

Figure 9.10: The sliding 6-month realized SSR for $T = 1$ year, for the S&P 500 (left) and Nikkei (right) indexes.

The large negative dip of the SSR of the Nikkei at the end of 2012 can be traced to the impact of the vega hedging of autocalls on the liquidity of the Nikkei vanilla option market.

Autocalls[10] provide the buyer a negative volatility exposure which vanishes when the spot price goes above an upper threshold, whereupon the option expires. Upon selling an autocall the dealer hedges his/her vega position by selling vanilla

---

[9]What matters is the cross-gamma of the hedged position – see the discussion in Section 1.3.

[10]Consider an autocall of notional N on an underlying S. At inception the autocall buyer pays N. Periodically – say every 3 months – the buyer receives a fixed coupon proportional to N provided the spot value is below a given threshold, typically 105% of the initial spot value. If, on a coupon date, the spot value is above this threshold, the buyer receives his coupon, gets N back, and the autocall expires. At maturity T the buyer gets back N unless at some point during the autocall's life the spot crosses a lower barrier, typically 60% of the initial spot value, in which case the payoff at maturity is $N \min(\frac{S_T}{S_0}, 1)$.

Autocalls are typically written on an index or on the worst of 3/4 stocks: $S_t \equiv \min_i (S_t^i)$. The (large) coupon of an autocall compensates the buyer for the risk that he may not recover the invested notional, if the underlying drops significantly.

options. The presence of the upper threshold generates a large spot/volatility cross-gamma: as the spot moves up, the dealer needs to buy back vanilla options. In normal circumstances, as the spot moves up, implied volatilities decrease, and the unwinding of the vega hedge should generate on average positive cross-gamma P&L. This cross-gamma P&L is offset by a theta contribution which is quantified by the SSR of the pricing model:

$$
\begin{aligned}
P\&L &= -S\widehat{\sigma}_T \frac{d^2\Pi}{dSd\widehat{\sigma}_T} \left(\frac{\delta S}{S} \frac{\delta\widehat{\sigma}_T}{\widehat{\sigma}_T} - \mathcal{R}_T \mathcal{S}_T \frac{\sigma^2}{\widehat{\sigma}_T} \delta t \right) \\
&= -S\mathcal{S}_T \frac{d^2\Pi}{dSd\widehat{\sigma}_T} \left(\mathcal{R}_T^r \sigma_r^2 - \mathcal{R}_T \sigma^2 \right) \delta t
\end{aligned} \tag{9.33}
$$

Figure 9.9 illustrates how different the SSR of the local volatility model and the SSR of a stochastic volatility model are. When carrying a long spot/volatility cross-gamma position, it thus seems preferable to price with the local volatility model, while with a short spot/volatility cross-gamma position, a stochastic volatility model should be used.

We refer the reader to a similar discussion in Section 12.6 of Chapter 12, page 482, in the more general context of mixed local-stochastic volatility models.

9.11.2   Volatilities of volatilities

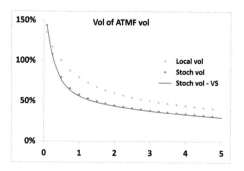

Figure 9.11: Volatility of ATMF volatilities as a function of maturity (years) in the two-factor stochastic volatility model (dark dots) and in the local volatility model (light dots) calibrated on the smile generated by the stochastic volatility two-factor model. Parameters in Table 8.2 have been used. The solid line corresponds to volatilities of VS – rather than ATMF – volatilities in the two-factor model calculated exactly using expression (7.39), page 227.

What about volatilities of volatilities? Volatilities of ATMF volatilities for both models appear in Figure 9.11. The term structure of volatilities of volatilities of the two-factor stochastic volatility model has been examined in Section 7.4. As for the local volatility model, because $\widehat{\sigma}_{F_T T}$ is a function of S, the volatility of $\widehat{\sigma}_{F_T T}$ can be expressed using the SSR. The instantaneous lognormal volatility of $\widehat{\sigma}_{F_T T}$ is given by (see Section 2.5.5):

$$\text{vol}(\widehat{\sigma}_{F_T T}) \; = \; \mathcal{R}_T \mathcal{S}_T \frac{\widehat{\sigma}_{F_0 0}}{\widehat{\sigma}_{F_T T}}$$

where the short ATMF volatility $\widehat{\sigma}_{F_0 0}$ is also equal to the instantaneous volatility of S.

For $T \to 0$ $\mathcal{R}_T$ tends to 2 and the volatility of the short ATMF volatility is twice the ATMF skew – we recover result (2.85) of Section 2.5.5.

For longer maturities, one can use the approximate expression (2.83):

$$\text{vol}(\widehat{\sigma}_{F_T T}) \; \simeq \; \left(\mathcal{S}_T + \frac{1}{T} \int_0^T \mathcal{S}_\tau d\tau \right) \frac{\widehat{\sigma}_{F_0 0}}{\widehat{\sigma}_{F_T T}}$$

which only involves the term structure of the ATMF skew.

Again, in contrast with the two-factor model, the local volatility model is not time-homogeneous; volatilities of volatilities depend not only on $T - t$, but also on t. Typically, volatilities of volatilities in the local volatility model are larger for small t/smaller for large t, compared with those of a time-homogeneous stochastic volatility model calibrated on the same smile.

9.11.3 Carry P&L of a partially vega-hedged position

Consider a delta- and vega-hedged option position. Assume that there is only one implied volatility $\widehat{\sigma}$, in addition to S. The gamma/theta P&L of the hedged position reads:[11]

$$P\&L \; = \; -\frac{1}{2} \frac{d^2 P}{dS^2} \left(\delta S^2 - \bullet \delta t \right) - \frac{1}{2} \frac{d^2 P}{d\widehat{\sigma}^2} \left(\delta \widehat{\sigma}^2 - \bullet \delta t \right) - \frac{d^2 P}{dS d\widehat{\sigma}} \left(\delta S \delta \widehat{\sigma} - \bullet \delta t \right)$$

The suitability of the model parametrization is assessed by comparing the model-generated values of the spot/volatility and volatility/volatility covariances – denoted by $\bullet \delta t$ – with their realized counterparts, the latter evaluated using historical data for S and $\widehat{\sigma}$.

In practice delta-hedging is performed daily – *and delta is calculated using current market implied volatilities* – but vega-hedging may not, because of larger bid/offer costs. Vega-hedging and delta-hedging occur asynchronously. Which second order spot gamma, spot/volatility cross-gamma and volatility/volatility cross-gamma P&Ls does one then materialize?[12]

---

[11] Typically, either a term structure of implied volatilities $\widehat{\sigma}_T$ – in the case of the two-factor model – or all vanilla implied volatilities $\widehat{\sigma}_{KT}$ – when using the local volatility model or the local-stochastic volatility models of Chapter 12 – are taken as underliers in addition to S, and contribute to the P&L.

[12] This issue calls to mind that of spot/spot cross-gammas for underlyings trading in different time zones – see [12].

Imagine we vega-hedge our option position periodically every n days – say one week ($n = 5$). We make the reasonable assumption that gammas and cross-gammas are constant during this period.

Rather than plodding through the calculation of this P&L, we choose a more enlightening route: start from the P&L generated by hedging once per period, then remove terms that can be offset by daily delta-hedging. We denote by $\delta\widehat{\sigma}_i$ and δS_i the respective daily increments over $[t_{i-1}, t_i]$.

Spot gamma

Assuming we delta-hedge only once per period, the spot gamma P&L reads:

$$
\begin{aligned}
\frac{1}{2}\frac{d^2 P}{dS^2}\left(\Sigma_i \delta S_i\right)^2 &= \frac{1}{2}\frac{d^2 P}{dS^2}\left(\Sigma_i \delta S_i\right)\left(\Sigma_j \delta S_j\right)\\
&= \frac{1}{2}\frac{d^2 P}{dS^2}\left(\Sigma_i \delta S_i^2\right) + \frac{d^2 P}{dS^2}\Sigma_i\left(\Sigma_{j<i}\delta S_j\right)\delta S_i
\end{aligned}
$$

The prefactor in front of δS_i in the second piece involves spot increments that precede δS_i, thus are known at time t_{i-1}. This P&L linear in δS_i can then be cancelled by a delta strategy $\Delta_{t_{i-1}}\delta S_i$ with $\Delta_{t_{i-1}} = \frac{1}{2}\frac{d^2 P}{dS^2}\left(\Sigma_{j<i}\delta S_j\right)$. Only the first piece remains and we get the usual result for the spot gamma P&L over one period:

$$
\frac{1}{2}\frac{d^2 P}{dS^2}\left(\Sigma_i \delta S_i^2\right)
$$

It involves the spot variance, measured using daily increments – as expected.

Volatility gamma

Since no vega-hedging happens during the n-day period, the volatility gamma P&L is simply:

$$
\frac{1}{2}\frac{d^2 P}{d\widehat{\sigma}^2}\left(\Sigma_i \delta\widehat{\sigma}_i\right)^2
$$

that is, the variance of volatility is sampled according to our weekly schedule – as expected.

Spot/volatility cross-gamma

The spot/volatility cross-gamma P&L reads:

$$
\frac{d^2 P}{dSd\widehat{\sigma}}\left(\Sigma_i \delta S_i\right)\left(\Sigma_j \delta\widehat{\sigma}_j\right) = \frac{d^2 P}{dSd\widehat{\sigma}}\Sigma_i\left(\Sigma_{j<i}\delta\widehat{\sigma}_j\right)\delta S_i + \frac{d^2 P}{dSd\widehat{\sigma}}\Sigma_i\left(\Sigma_{j\geq i}\delta\widehat{\sigma}_j\right)\delta S_i
$$

The first portion of the right-hand side can be offset by a delta position; our final P&L over one period reads:

$$
\frac{d^2 P}{dSd\widehat{\sigma}}\Sigma_i\left(\Sigma_{j\geq i}\delta\widehat{\sigma}_j\right)\delta S_i \tag{9.34}
$$

which involves the product of spot increments and all subsequent volatility increments.

The spot/volatility estimator in (9.34) is what should be used to measure realized spot/volatility covariance, for the sake of comparing realized and model-implied levels. How does it differ from the usual estimator that applies to the situation of synchronous delta and vega-hedging?

Imagine that during the vega-hedging period S and $\hat{\sigma}$ vary by δS, $\delta \hat{\sigma}$ and that both experience a trend: $\delta S_i = \frac{\delta S}{n}$, $\delta \hat{\sigma}_i = \frac{\delta \hat{\sigma}}{n}$.

Then, using the above expressions, the spot gamma P&L is equal to $\frac{1}{n}\left(\frac{1}{2}\frac{d^2 P}{dS^2}\delta S^2\right)$ – that is delta-hedging has reduced it by a factor $\frac{1}{n}$ – and the volatility gamma P&L is $\frac{1}{2}\frac{d^2 P}{d\hat{\sigma}^2}\delta\hat{\sigma}^2$.

As for the cross-gamma P&L, it is equal to $\frac{n+1}{2n}\left(\frac{d^2 P}{dSd\hat{\sigma}}\delta S\delta\hat{\sigma}\right) \simeq \frac{1}{2}\frac{d^2 P}{dSd\hat{\sigma}}\delta S\delta\hat{\sigma}$, for n large.

Thus, using current implied volatilities in the calculation of the delta – even without trading any intermediate vega hedges – has reduced the realized cross-gamma P&L by a factor $\simeq \frac{1}{2}$.

Chapter's digest

9.1 The ATMF skew

▶ At order one in volatility of volatility, the ATMF skew is given, as a function of the spot/variance covariance function by:

$$\mathcal{S}_T = \frac{1}{2\sqrt{T}} \frac{1}{\left(\int_0^T \xi_0^\tau d\tau\right)^{3/2}} \int_0^T d\tau \int_\tau^T \mu\left(\tau, u, \xi_0\right) du$$

❧ ❧ ❧ ❧ ❧

9.2 The Skew Stickiness Ratio (SSR)

▶ The SSR is defined as the instantaneous regression coefficient of the ATMF volatility on $\ln S$, in units of the ATMF skew:

$$\mathcal{R}_T = \frac{1}{\mathcal{S}_T} \frac{E[d\ln S \, d\widehat{\sigma}_{F_T(S)T}]}{E[(d\ln S)^2]}$$

In jump-diffusion models $\mathcal{R}_T = 0, \forall T$. In the local volatility model, since the ATMF volatility is a function of S: $\mathcal{R}_T = \frac{1}{\mathcal{S}_T} \frac{d\widehat{\sigma}_{F_T(S)T}}{d\ln S}$.

▶ At order one in volatility of volatility, $\mathcal{R}_T$ is given, as a function of the spot/variance covariance function, by:

$$\mathcal{R}_T = \frac{\int_0^T \xi_0^\tau d\tau}{T\xi_0^0} \frac{T \int_0^T \mu(0, u, \xi_0) du}{\int_0^T d\tau \int_\tau^T \mu\left(\tau, u, \xi_0\right) du}$$

❧ ❧ ❧ ❧ ❧

9.3 Short-maturity limit of the ATMF skew and the SSR

▶ For $T \to 0$, $\mathcal{S}_0 = \frac{\mu(0,0,\xi_0)}{4(\xi_0^0)^{3/2}}$ and $\mathcal{R}_0 = 2$. The short-maturity limit of the SSR is 2, as in the local volatility model.

❧ ❧ ❧ ❧ ❧

9.4 Model-independent range of the SSR

▶ Assuming the VS term structure is flat and the spot/variance covariance function is time homogeneous – $\mu\left(\tau, u, \xi_0\right) \equiv \mu\left(u - \tau\right)$ – one derives the following expressions for ATMF skew and SSR:

$$\mathcal{S}_T = \frac{1}{2\xi_0^{3/2}T^2}\int_0^T (T-t)\mu(t)dt$$

$$\mathcal{R}_T = \frac{\int_0^T \mu(t)dt}{\int_0^T(1-\frac{t}{T})\mu(t)dt}$$

Making the asumption that $\mu(t)$ decays monotonically towards zero as $t \to \infty$, we get model-independent lower and upper bounds on $\mathcal{R}_T$:

$$\mathcal{R}_T \in [1,2]$$

Because of its definition the SSR is inversely proportional to the short VS volatility. One should bear in mind this dependence when assessing SSRs of market smiles.

ৰৰ ৰৰ ৰৰ ৰৰ ৰৰ

9.5 Scaling of ATMF skew and SSR – a classification of models

▶ Depending on the rate of decay of $\mu(t)$ for $t \to \infty$, two long-maturity regimes for the ATMF skew and SSR can be defined, which lead to the division of stochastic volatility models into two classes. Assuming that $\mu(t) \propto \frac{1}{t^\gamma}$ for $t \to \infty$, for long maturities:

- (Type I) If $\gamma > 1$:

$$\mathcal{S}_T \propto \frac{1}{T} \quad \text{and} \quad \lim_{T\to\infty} \mathcal{R}_T = 1 \tag{9.35}$$

- (Type II) If $\gamma < 1$:

$$\mathcal{S}_T \propto \frac{1}{T^\gamma} \quad \text{and} \quad \lim_{T\to\infty} \mathcal{R}_T = 2-\gamma \tag{9.36}$$

Exponential decay of $\mu(t)$ produces Type I behavior. In Type I models, in the long-maturity regime, $\mathcal{S}_T$ and $\mathcal{R}_T$ bear no signature of the rate of decay of μ.

For both types of models, the long-maturity ATMF skew and SSR are related through:

$$\mathcal{S}_T \propto \frac{1}{T^{2-\mathcal{R}_\infty}}.$$

ৰৰ ৰৰ ৰৰ ৰৰ ৰৰ

9.6 Type I models – the Heston model

▶ The Heston model produces Type I behavior. Its long-maturity SSR is 1, which can be checked using the order-one expansion of the ATMF skew in Chapter 6.

๘ ๘ ๘ ๘ ๘

9.7 Type II models

▶ Even though the two-factor model is strictly of Type 1, since $\mu(t)$ decays exponentially for large t, we can still achieve Type II scaling of the ATMF skew – and the corresponding value of the SSR – on a range of maturities that is practically relevant.

For a flat term-structure of VS volatilities, at order one in volatility of volatility, both quantities are given by:

$$
\mathcal{S}_T = \nu \alpha_\theta \left[(1 - \theta) \rho_{SX^1} \frac{k_1 T - (1 - e^{-k_1 T})}{(k_1 T)^2} + \theta \rho_{SX^2} \frac{k_2 T - (1 - e^{-k_2 T})}{(k_2 T)^2} \right]
$$

$$
\mathcal{R}_T = \frac{(1 - \theta) \rho_{SX^1} \frac{1 - e^{-k_1 T}}{k_1 T} + \theta \rho_{SX^2} \frac{1 - e^{-k_2 T}}{k_2 T}}{(1 - \theta) \rho_{SX^1} \frac{k_1 T - (1 - e^{-k_1 T})}{(k_1 T)^2} + \theta \rho_{SX^2} \frac{k_2 T - (1 - e^{-k_2 T})}{(k_2 T)^2}}
$$

▶ The realized behavior of equity index smiles is consistent with Type II.

๘ ๘ ๘ ๘ ๘

9.8 Numerical evaluation of the SSR

▶ The SSR of the two-factor model is easily evaluated numerically in a Monte Carlo simulation by simply shifting the initial values of processes X_t and Y_t.

๘ ๘ ๘ ๘ ๘

9.9 The SSR for short maturities

▶ The value of the realized SSR of short-maturity equity index smiles is usually substantially lower than the model-independent value of stochastic volatility models. Can this difference be materialized as the P&L of a trading strategy?

๘ ๘ ๘ ๘ ๘

9.10 Arbitraging the realized short SSR

▶ We use a lognormal model for the short ATM volatility. In our two asset-model – spot and short ATM volatility – the difference between the realized SSR and its model-independent value of 2 is materialized as a cross-gamma/theta P&L.

▶ We backtest a dynamical delta-hedged option strategy on the Euro Stoxx 50 index that consists in maintaining a short-skew position around the money. The resulting P&L approximately captures the cross-gamma/theta P&L corresponding to the difference between implied and realized skew, even though the residual vega of our position impacts the cumulative P&L negatively on large downward moves of the spot.

ᒲ ᒲ ᒲ ᒲ ᒲ

9.11 Conclusion

▶ The behavior of $\mathcal{R}_T$ as a function of T is structurally different in stochastic volatility models than in the local volatility model. In the local volatility model, for typical equity index smiles, the SSR starts from 2 for short maturities, then *increases* for longer maturities.

For an ATMF skew that decays algebraically with exponent γ, in the local volatility model $\mathcal{R}_\infty = \frac{2-\gamma}{1-\gamma}$, while in a stochastic volatility model, $\mathcal{R}_\infty = 2 - \gamma$. For the typical value $\gamma = \frac{1}{2}$, the SSR of the local volatility model for long maturities is $\mathcal{R}_\infty = 3$, compared to $\mathcal{R}_\infty = 1.5$ for a stochastic volatility model.

▶ Instantaneous volatilities of volatilities of long-dated vanilla options are also larger in the local volatility model than in a stochastic volatility model.

▶ In practice, vega hedging occurs less frequently than delta hedging, thus the spot/volatility cross/gamma-theta P&L is not materialized exactly. Assuming that deta-hedging is performed daily, using actual implied volatilities, while vega-hedging is performed less frequently, we obtain the approximate result that, in case of a trend in both spot and implied volatility in between two vega rehedges, half of the cross-gamma P&L is materialized.

Chapter 10

What causes equity smiles?

In Chapter 8 we have characterized stochastic volatility smiles and how they are related to the model's specification: where do skew and curvature come from and how we can compute them approximately?

Then in Chapter 9 we have focused on dynamical aspects of smiles in stochastic volatility models: how do ATMF volatilities move in these models? From the bare knowledge of the smile generated by a stochastic volatility model can we say anything about the joint dynamics of spot and implied volatilities in the model?

"But what about actual equity smiles?" is the restive reader bound to ask. What is it that is responsible for their skew and curvature? Large historical drawdowns in equity indexes are often purported to be responsible for the strong negative skew of implied index smiles. Is this correct? Are vanilla smiles in any way related to statistical properties of historical returns? Which payoffs do the latter impact? These are the questions we address.

The (un)related topic of jump-diffusion and Lévy models is dealt with in Appendix A.

10.1 The distribution of equity returns

Figure 10.1 shows the cumulative distribution of normalized negative and positive returns of the Dow Jones index. We have used daily closing quotes from January 1, 1900 to July 20, 2014.[1]

We compute daily returns $r_i = \frac{S_{i+1}}{S_i} - 1$. The square root of the second moment of non-centered negative returns is 1.15% and that of non-centered positive returns is 1.08%. We separately normalize negative and positive returns by the square roots of second-order moments so that the non-centered second moment of normalized positive returns $\bar{r}_i$ equals 1, as does that of negative returns. The lowest (normalized) negative return in our sample is -19.6 and the highest positive return is 14.3.[2]

---

[1]We gratefully acknowledge the website *http://stooq.com* for making these data available.

[2]With the second moments of non-centered positive and negative $\bar{r}_i$ equal to 1, the standard deviation of the $\bar{r}_i$ is almost exactly equal to 1.

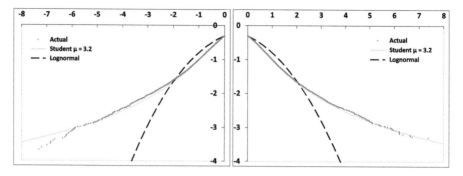

Figure 10.1: Logarithm ($\log_{10}$) of the empirical distribution function of normalized negative (left-hand graph) and positive (right-hand graph) returns of the Dow Jones index, together with the lognormal and Student distribution functions.

We then rank negative returns and define the empirical distribution function of normalized negative returns $\bar{r}$ as:

$$P\left[\bar{r} \leq \bar{r}_i\right] = \frac{1}{2}\frac{i}{N^-}$$

where N^- is the number of negative returns in our sample ($N^- = 14103$) and the normalization factor $\frac{1}{2}$ ensures that $P\left[\bar{r} \leq 0\right] = \frac{1}{2}$. We do the same for positive returns ($N^+ = 15657$). In Figure 10.1, $\log_{10}\left(P\left[\bar{r} \leq r\right]\right)$ (left-hand graph) and $\log_{10}\left(P\left[\bar{r} \geq r\right]\right)$ (right-hand graph) are graphed as a function of r.

Together with the empirical distribution, we show two distributions: (a) the lognormal distribution function, (b) the Student distribution function with $\mu = 3.2$.

The Student distribution is well-suited to the modeling of fat-tailed random variables; its density is defined by:

$$\rho_\mu\left(x\right) = \frac{\Gamma\left(\frac{1+\mu}{2}\right)}{\sqrt{\mu\pi}\,\Gamma\left(\frac{\mu}{2}\right)}\frac{1}{\left(1 + \frac{x^2}{\mu}\right)^{\frac{1+\mu}{2}}} \tag{10.1}$$

where μ is customarily called "number of degrees of freedom".

The smaller μ the thicker the tails. For large values of x the density scales like $\frac{1}{x^{1+\mu}}$, thus the one-sided cumulative distribution function scales like $\frac{1}{x^\mu}$. Only moments of order smaller than μ exist.

The variance of a Student random variable is $\frac{\mu}{\mu-2}$ and its kurtosis is $\frac{6}{\mu-4}$. For $\mu \leq 4$ the fourth moment diverges and so does the second moment for $\mu \leq 2$; for $\mu \to \infty$ the Student density converges to the Gaussian density.

The value $\mu = 3.2$ is obtained by least-square minimization of the difference of the logarithms of the empirical and Student distributions. When comparing empirical and Student distributions in Figure 10.1, one should observe that while the empirical curve for negative returns consists of more than 14000 points, only a hundred of them correspond to values of $\bar{r} \leq -4$.

Student distribution functions for different values of μ are shown in Figure 10.2. Any value in the interval $[3, 4]$ provides an acceptable fit to empirical data.[3]

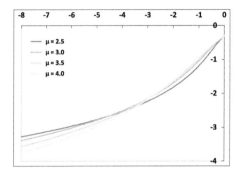

Figure 10.2: Logarithm ($\log_{10}$) of the left tail of the distribution function of a Student random variable, normalized so that its variance is 1, for different values of μ.

Thus our preliminary conclusions are:

- The empirical distribution of realized equity index returns is well approximated by a Student distribution with μ typically in the interval $[3, 4]$, with tail probabilities much larger than those of a lognormal distribution. Figure 10.1 shows that the lognormal density underestimates the probability of a move larger than 4 standard deviations by a factor of 100.

- Negative and positive returns have very similar second-order moments and tail parameters. In contrast with what is frequently heard, there is no evidence that negative returns have thicker tails than positive ones.

These conclusions are by no means specific to the Dow Jones index, they apply broadly to all equity indexes. Figure 10.3 shows another example; a Hong Kong-based index – the Hang Seng China Enterprises Index (HSCEI), which is roughly twice as volatile as the Dow Jones index, using data from 1993 to 2014.

---

[3]For $\mu < 4$, the kurtosis κ of Student-distributed returns diverges. Obviously, the kurtosis estimator applied to any sample yields a finite number. However, the divergence of κ manifests itself in the fact that occasional large returns generate large jumps in the estimator for κ. Tails of the Student distribution for $\mu < 4$ are thick enough that these events occur sufficiently frequently that defining an average finite value for κ reliably becomes practically impossible. With respect to our Dow Jones data, even though our best fit yields values of $\mu < 4$, this does not mean that $\kappa = \infty$, as the very rare and large events that cause the divergence of κ for $\mu < 4$ may not be present in our sample.

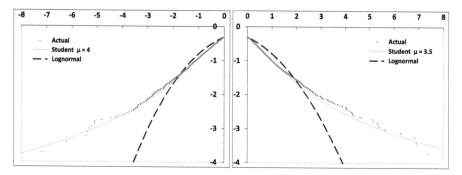

Figure 10.3: Logarithm ($\log_{10}$) of the empirical distribution function of normalized negative (left-hand graph) and positive (right-hand graph) returns of the HSCEI index, together with the lognormal and Student distribution functions.

10.1.1 The conditional distribution

Even though empirical curves in Figure 10.1 are remarkably smooth, one could question the pertinence of fitting a single distribution to one century worth of data, during which very dissimilar volatility regimes have existed.

In the analysis of the P&L of a delta-hedged option position in Chapter 1, daily returns r_i are modeled as:

$$r_i = \sigma_i \sqrt{\delta t} z_i \qquad (10.2)$$

where σ_i is the instantaneous volatility and z_i are iid random variables with unit variance.

The rationale for this ansatz is that the variability of the probability distribution of r_i is condensed in that of the scale factor σ_i – the z_i are identically distributed.

When σ_i is stochastic the resulting distribution of the r_i is non-Gaussian even though the z_i may be. How much of the thickness of the tails of daily returns is generated by (a) the randomness of σ_i, (b) the distribution of z_i?

(10.2) is a natural ansatz but accessing σ_i is difficult in practice. Figure 10.4 shows the same graphs as in Figure 10.1, except each return is normalized by the historical volatility calculated using the 200 previous returns, rather than that evaluated over the whole historical sample.

As expected, normalizing returns by an (heuristic) estimate of σ_i reduces the tail thickness, which is manifested in larger values for μ. A least-squares fit produces $\mu = 3.8$ for negative returns – a modest increase with respect to $\mu = 3.2$ – while we get $\mu = 6$ for positive returns.[4]

---

[4]The fact that the conditional value of μ for negative returns is little changed with respect to its unconditional value may be due to the fact that large negative returns are more likely to occur irrespective of current volatility levels: normalization by the latter does not shrink their tails substantially.

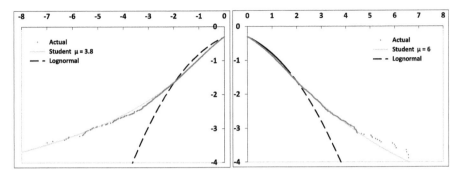

Figure 10.4: Logarithm ($\log_{10}$) of the empirical distribution function of conditional normalized negative (left-hand graph) and positive (right-hand graph) returns of the Dow Jones index, together with the lognormal and Student distribution functions.

10.2 Impact of the distribution of daily returns on derivative prices

Empirical distributions of returns – even conditional ones – have substantially fatter tails than the lognormal distribution. This has prompted some to argue that using lognormal models – or more generally diffusive stochastic volatility models – for pricing derivatives is inappropriate. Others have claimed that it is precisely the fat-tailed/non-lognormal nature of returns that accounts for the volatility smile. Is either of these statements correct?

Remember we have argued in Chapters 8 and 9 – see in particular the discussion in Section 9.10.5, page 376 – that the ATMF skew of the volatility smile is generated by the covariance of spot and *implied* volatilities. Do fat-tailed returns alter this picture?

We now investigate the following questions:

- Is the fat-tailed nature of returns manifested in any way in equity smiles? Stated differently, is there any trace of the one-day smile in smiles of standard maturities?

- Which payoffs are specifically sensitive to tails of daily returns/the one-day smile?

We answer these questions by using a model that enables us to separate the effects of (a) the distribution of daily returns, (b) the effect of stochastic (implied) volatilities.

10.2.1   A stochastic volatility model with fat-tailed returns

We use the two-factor model introduced in Section 7.4, page 226. While the model for forward variances ξ_t^τ is continuous, the spot returns are discrete and are simulated over one-day time intervals.

We cannot use a Student distribution for $\ln\left(\frac{S_{t+\Delta}}{S_t}\right)$, as $E_t\left[S_{t+\Delta}\right]$ would be infinite, thus we use the Student distribution for $\frac{S_{t+\Delta}}{S_t} - 1$ and set:

$$S_{t+\Delta} = S_t\left[1 + (r - q)\,\Delta + \sigma_t\delta Z\right] \tag{10.3}$$

where σ_t is given by:

$$\sigma_t = \sqrt{\frac{1}{\Delta}\int_t^{t+\Delta}\xi_t^\tau\,d\tau}$$

and the dynamics of forward variances ξ^τ in the two-factor model is given by equation (7.28), page 226:[5]

$$d\xi_t^\tau = 2\nu\xi_t^\tau\,\alpha_\theta\left((1-\theta)\,e^{-k_1(\tau-t)}dW_t^1 + \theta e^{-k_2(\tau-t)}dW_t^2\right)$$

whose solution is given in equation (7.33), page 227, in terms of two Ornstein-Ühlenbeck processes X^1, X^2, whose increments over $[t, t+\Delta]$ are given by:

$$\delta X^i = \int_t^{t+\Delta}e^{-k_i(t+\Delta-u)}dW_u^i$$

In (10.3) δZ is a random variable with mean 0 and variance Δ. In the standard version of the two-factor model δZ is simply replaced by: $\delta Z \equiv \delta W^S = \int_t^{t+\Delta}dW_u^S$ and the covariances of the Gaussian random variables δW^S and δX^i are given by:

$$E\left[\delta X^i\delta X^j\right] = \rho_{ij}\frac{1 - e^{-(k_i+k_j)\Delta}}{k_i + k_j} \qquad E\left[\delta X^i\delta W^S\right] = \rho_{iS}\frac{1 - e^{-k_i\Delta}}{k_i}$$

Here we need to draw δZ from a two-sided Student distribution with different values of μ for the left and right tails.

While in empirical distributions, probabilities for negative and positive returns are almost exactly equal to $\frac{1}{2}$, we would like to be able to set these probabilities, or in other words the price of the one-day at-the-money digital option: $E\left[1_{\delta Z\geq 0}\right]$, that is, have a handle on the one-day ATM skew – see the discussion in Section 10.2.2 below.

Let us call p_+, p_- the probabilities of positive and negative returns, and μ_+, μ_- the parameters of the corresponding Student distributions. We define δZ as:

$$\begin{cases}\delta Z = \sigma_+\sqrt{\Delta}\left|X_{\mu_+}\right| & \text{with probability } p_+\\\delta Z = -\sigma_-\sqrt{\Delta}\left|X_{\mu_-}\right| & \text{with probability } p_-\end{cases}$$

---

[5]No need for volatility-of-volatility smile here.

where X_μ denotes a Student random variable with μ degrees of freedom. The density of δZ is given by:

$$
\begin{cases}
\rho(\delta Z) = \dfrac{2p_+}{\sigma_+\sqrt{\Delta}}\,\rho_{\mu_+}\!\left(\dfrac{\delta Z}{\sigma_+\sqrt{\Delta}}\right) & \text{with probability } p_+ \\[2ex]
\rho(\delta Z) = \dfrac{2p_-}{\sigma_-\sqrt{\Delta}}\,\rho_{\mu_-}\!\left(\dfrac{\delta Z}{\sigma_-\sqrt{\Delta}}\right) & \text{with probability } p_-
\end{cases}
$$

The values of σ_+, σ_- must be such that:

$$
E[\delta Z] = 0 \qquad E[(\delta Z)^2] = \Delta \tag{10.4}
$$

Since the standard deviation of a Student random variable is $\sqrt{\frac{\mu}{\mu-2}}$ it is more natural to rewrite σ_+, σ_- as:

$$
\sigma_+ = \sqrt{\frac{\mu_+ - 2}{\mu_+}}\,\zeta_+ \qquad \sigma_- = \sqrt{\frac{\mu_- - 2}{\mu_-}}\,\zeta_-
$$

The conditions in (10.4) read:

$$
p_+\zeta_+\,\alpha_+ - p_-\zeta_-\,\alpha_- = 0
$$
$$
p_+\zeta_+^2 + p_-\zeta_-^2 = 1
$$

where $\alpha_+ = \dfrac{2}{\sqrt{\pi}}\dfrac{\sqrt{\mu_+ - 2}}{\mu_+ - 1}\dfrac{\Gamma\left(\frac{1+\mu_+}{2}\right)}{\Gamma\left(\frac{\mu_+}{2}\right)}$ and likewise for α_-. The solution is:

$$
\zeta_+ = \frac{p_-\alpha_-}{\sqrt{p_+\left(p_-\alpha_-\right)^2 + p_-\left(p_+\alpha_+\right)^2}} \qquad \zeta_- = \frac{p_+\alpha_+}{\sqrt{p_+\left(p_-\alpha_-\right)^2 + p_-\left(p_+\alpha_+\right)^2}}
$$

and δZ is given by:

$$
\begin{cases}
\delta Z = \zeta_+\sqrt{\frac{\mu_+ - 2}{\mu_+}}\sqrt{\Delta}\,|X_{\mu_+}| & \text{with probability } p_+ \\[2ex]
\delta Z = -\zeta_-\sqrt{\frac{\mu_- - 2}{\mu_-}}\sqrt{\Delta}\,|X_{\mu_-}| & \text{with probability } p_-
\end{cases}
$$

Student random variables can be generated in a number of ways. In our setting, we need (a) to correlate δZ with $\delta X^1, \delta X^2$ and (b) to be able to degenerate δZ into a Gaussian random variable to recover the standard form of the two-factor model.

It is then simpler to start with the Brownian increments δW^S supplied by the Monte Carlo engine of our standard two-factor model and map them into δZ according to:

$$
\delta Z = \sqrt{\Delta}\,f\left(\frac{\delta W^S}{\sqrt{\Delta}}\right) \tag{10.5}
$$

with f given by:

$$
\begin{cases}
x \leq \mathcal{N}_G^{-1}(p_-) & f(x) = \zeta_- \sqrt{\frac{\mu_- - 2}{\mu_-}}\, \mathcal{N}_{\mu_-}^{-1}\left(\frac{\mathcal{N}_G(x)}{2p_-}\right) \\
x \geq \mathcal{N}_G^{-1}(p_-) & f(x) = \zeta_+ \sqrt{\frac{\mu_+ - 2}{\mu_+}}\, \mathcal{N}_{\mu_+}^{-1}\left(\frac{1}{2} + \frac{\mathcal{N}_G(x) - p_-}{2p_+}\right)
\end{cases}
\tag{10.6}
$$

where $\mathcal{N}_G$ is the cumulative distribution function of the standard normal variable, $\mathcal{N}_G^{-1}$ its inverse, and $\mathcal{N}_\mu^{-1}$ is the inverse cumulative distribution function of a Student random variable with μ degrees of freedom.[6]

Note that if $p_+ = p_- = \frac{1}{2}$, $\lim_{\mu_+,\mu_- \to \infty} \delta Z = \delta W$, as expected.

Rescaling spot/volatility correlations
We can now simulate the joint dynamics of S_t and forward variances using the Monte Carlo engine of our two-factor model and are one step away from vanilla smiles.

As we vary μ_+, μ_- to assess the influence of the one-day smile we must ensure that other features of the model are unchanged. As we replace δW^S with δZ in (10.3) the instantaneous volatilities of VS volatilities are unaffected, as is the instantaneous volatility of S, but spot/volatility covariances are altered.

We thus need to define new rescaled correlations ρ_{1S}^*, ρ_{2S}^* so that the instantaneous correlations – or covariances – of δZ and δX^1 and of δZ and δX^2 using ρ_{1S}^*, ρ_{2S}^* are identical to the covariances in the standard two-factor model parametrized with ρ_{1S}, ρ_{2S}, that is $E\left[\delta W^S \delta X^1\right]$ and $E\left[\delta W^S \delta X^2\right]$:

$$
E_*\left[\delta Z \delta X^i\right] = E\left[\delta W^S \delta X^i\right]
$$

where E_* denotes the expectation evaluated using the new correlations. The formula for rescaling correlations is easily obtained. Spot/volatility correlations are rescaled uniformly, according to:

$$
\frac{\rho_{iS}^*}{\rho_{iS}} = \frac{\Delta}{E\left[\delta Z \delta W^S\right]} = \frac{1}{\int_{-\infty}^{+\infty} \phi(x)\, x f(x)\, dx}
\tag{10.7}
$$

where f is the mapping function in (10.6) and ϕ is the probability density of the standard normal variable.

The integral in the denominator of (10.7) is evaluated numerically once. We have $\rho_{iS}^* \geq \rho_{iS}$.[7]

---

[6] $\mathcal{N}_G$, $\mathcal{N}_G^{-1}$, $\mathcal{N}_\mu^{-1}$ are readily available in standard numerical libraries. $f(x)$ should be cached – simulating δZ is then only marginally more expensive than simulating δW^S.

[7] We have $\int \phi(x)x^2 dx = 1$ and $\int \phi(x)f(x)^2 dx = 1$, where f is the mapping function defined in (10.6). From Cauchy-Schwarz we get: $\int \phi(x)\, x f(x)\, dx \leq 1$. Unless we go to very low values of μ, ρ_{iS}^*/ρ_{iS} is not too large. For example, taking $\mu_+ = \mu_- = \mu$ and $p_+ = p_- = 1/2$ we have $\rho_{iS}^*/\rho_{iS} = 1.01$ for $\mu = 6$, 1.03 for $\mu = 4$, 1.09 for $\mu = 3$, 1.2 for $\mu = 2.5$.

10.2.2 Vanilla smiles

We now use the model described above to assess the impact of the one-day smile on the vanilla smile.

Smiles are obtained by straight pricing of vanilla payoffs. None of the techniques mentioned in Appendix A of Chapter 8 can be used here – be they the mixing solution, gamma/theta or timer techniques – as they rely on the assumption of a diffusion for S_t that is on the fact that the expansion of the P&L at order 2 in δS is exact for short time intervals – see the discussion below.

We use throughout a flat variance curve: $\xi_0^\tau = \xi_0 = 20\%^2$ and parameters in Table 10.1.[8]

ν	θ	k_1	k_2	ρ	ρ_{SX^1}	ρ_{SX^2}
257%	0.151	8.96	0.46	40%	-74.6%	-13.7%

Table 10.1: Parameters of the two-factor model

We first start with identical probabilities for positive and negative returns ($p_+ = p_- = \frac{1}{2}$) and use identical values for tail parameters: $\mu_+ = \mu_- = \mu$.

Figure 10.5 displays three-month and one-year smiles in the standard and fat-tailed version of the two-factor model.

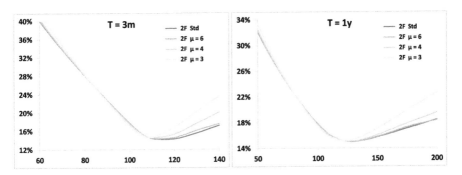

Figure 10.5: Smiles in the two-factor model with Student-distributed returns for different values of μ, $p_+ = p_- = 1/2$ and parameters in Table 10.1 along with the smile in the standard version of the model ($\mu = \infty$).

The near-ATMF smile is hardly affected by tails of daily returns, which mostly impact far out-of-the-money calls. The effect of μ is better appreciated by turning off stochastic volatility: the smile for maturity T is then generated by the 1-day smile only. The latter, together with the 1-year smile, is shown in Figure 10.6.

---

[8]These were typical of the Euro Stoxx 50 index as of July 2014.

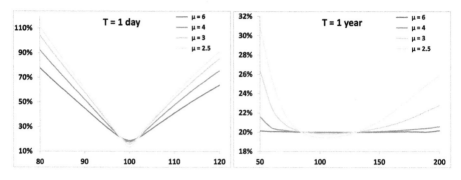

Figure 10.6: 1-day (left) and 1-year (right) smile for different values of $\mu = \mu_+ = \mu_-$ with $p_+ = p_- = 1/2$ and $\sigma = 20\%$ – no stochastic volatility ($\nu = 0$).

We now keep identical values for μ_+ and μ_- but use different values for p_+, p_- so that the ATM skew of the 1-day smile is non-vanishing. The resulting 1-year smile appears in Figure 10.7 for $p_+ = 30\%, 40\%, 50\%, 60\%, 70\%$.

Again let us turn off stochastic volatility; the 1-day and 1-year smiles are shown in Figure 10.8.

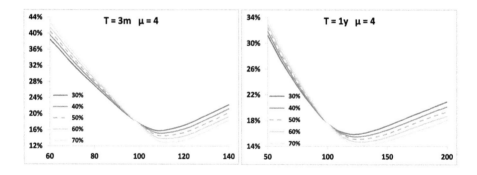

Figure 10.7: Smile in the two-factor model with Student-distributed returns for different values of p_+, with $\mu_+ = \mu_- = 4$ and parameters in Table 10.1.

It is apparent that the difference $p_+ - p_-$ drives the ATM skew of the 1-day smile, which almost vanishes for $p_+ = p_- = \frac{1}{2}$. The fact that the one-day ATM skew steepens – and implied volatilities for strikes larger than 100 become lower – for larger values of p_+ is understood by noting that p_+ is the (undiscounted) price of a one-day digital that pays 1 if tomorrow's spot value is larger than today's.

Denoting by $\mathcal{C}_K$ the undiscounted price of a call option of strike K, we have $\mathcal{C}_K = \mathcal{C}_K^{BS}(\widehat{\sigma}_K)$ where $\mathcal{C}_K^{BS}$ is the Black-Scholes price and $\widehat{\sigma}_K$ the implied volatility

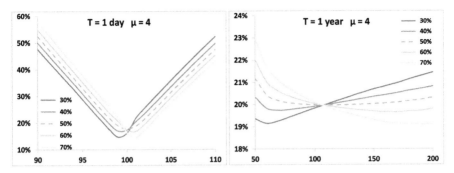

Figure 10.8: 1-day (left) and 1-year (right) smile for different values of p_+, with $\mu_+ = \mu_- = 4$ and $\sigma = 20\%$ – no stochastic volatility ($\nu = 0$).

of strike K. Thus:

$$
\begin{aligned}
p_+ &= -\frac{dC_K}{dK} \\
&= -\left.\frac{dC_K^{BS}}{dK}\right|_{\widehat{\sigma}_K} - \left.\frac{dC_K^{BS}}{d\widehat{\sigma}}\right|_{\widehat{\sigma}_K} \frac{d\widehat{\sigma}_K}{dK}
\end{aligned}
$$

According to this formula, increasing p_+ with $\xi_0^\tau = \xi_0$ unchanged – thus leaving the ATM volatility almost unchanged – has the effect that $\frac{d\widehat{\sigma}_K}{dK}$ decreases ($\frac{dC_K^{BS}}{d\widehat{\sigma}}$ is positive), which is indeed what occurs in the left-hand graph of Figure 10.8. Note we have deliberately used an unrealistically wide range for p_+.[9]

How does the contribution of the 1-day ATMF skew to the ATMF skew of maturity T scale with T? Figure 10.9 shows the 95/105 skew as a function of T, together with a $\frac{1}{T}$ fit, with $\nu = 0$: the only source of smile is the 1-day smile.

The observant reader will remember coming across this scaling before, in the context of discrete forward variance models, in Section 7.8 of Chapter 7 – consider expression (7.119b), page 294, of the ATMF skew for maturity $N\Delta$, where Δ is the time scale of the discrete model.

These models generate the ATMF skew through two mechanisms: (a) the ATMF skew for time scale Δ, (b) the covariance of spot and forward variances – not unlike our fat-tailed stochastic volatility model.

See also Appendix A for more on this scaling.

10.2.3 Discussion

Let us assume no additional degrees of freedom besides S, and vanishing interest rate and repo without loss of generality. The P&L of a short delta-hedged option

---

[9]See the discussion of the impact of the skew on digital options in Section 1.3.1, page 19.

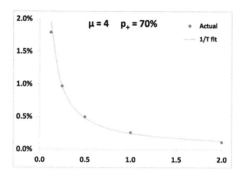

Figure 10.9: Difference of the 95% and 105% implied volatilities, as a function of maturity – in years – (dots) along with a $1/T$ fit (line), in the two-factor model with $\nu = 0$ (no stochastic volatility) with $\mu_+ = \mu_- = 4$ and $p_+ = 70\%$.

position during $\delta t = \Delta$ is:

$$P\&L \;=\; -\left(P\left(t+\Delta, S + \delta S\right) - P\left(t, S\right)\right) + \frac{dP}{dS}\delta S \qquad (10.8)$$

where $P\left(t, S\right)$ is our pricing function. In the models used so far P can be expressed as an expectation, thus :

$$P\left(t, S\right) \;=\; E\left[P\left(t+\Delta, S_{t+\Delta}\right) | S_t = S\right]$$

Δ is assumed sufficiently small that the normal/lognormal distinction is not practically relevant. We then have:

$$P\left(t, S\right) \;=\; E\left[P\left(t+\Delta, S\left(1 + \sigma\delta Z\right)\right)\right] \qquad (10.9)$$

where δZ is a random variable with vanishing mean and variance Δ, and σ is the instantaneous volatility. Inserting expression (10.9) into (10.8) and expanding in powers of δZ yields:

$$
\begin{aligned}
P\&L \;=\; & -\left(P\left(t+\Delta, S + \delta S\right) - E_{t,S}\left[P\left(t+\Delta, S\left(1 + \sigma\delta Z\right)\right)\right]\right) + \frac{dP}{dS}\delta S \\
\;=\; & -\frac{S^2}{2}\frac{d^2 P}{dS^2}\left(\frac{\delta S^2}{S^2} - \sigma^2\Delta\right) - \sum_{k>2}\frac{S^k}{k!}\frac{d^k P}{dS^k}\left(\left(\frac{\delta S}{S}\right)^k - \sigma^k E\left[\delta Z^k\right]\right)
\end{aligned}
$$

$$(10.10)$$

In (10.10) we single out the order-two contribution in δS, as when we take the limit of short time intervals between successive delta rehedges, only this term survives in the cumulative P&L. Indeed consider an option of maturity T; the cumulative P&L over the option's lifetime is the sum of $\frac{T}{\Delta}$ P&Ls of the form (10.10). In the continuous-time diffusive models considered so far $\frac{\delta S}{S} \propto \sqrt{\Delta}$, thus the order-two contribution in (10.10) is of order Δ, hence generates a cumulative P&L that is finite. The kth contribution in (10.10), however, is of order $\Delta^{\frac{k}{2}}$, generating a cumulative P&L that scales like $\frac{T}{\Delta}\Delta^{\frac{k}{2}}$, hence tends to zero as $\Delta \to 0$.

- In previous chapters we have implicitly assumed that Δ was sufficiently small that the $\Delta \to 0$ limit was relevant, hence typical P&L expressions so far only included the first term in (10.10).[10]

- When $\Delta > 0$ this is no longer the case, as terms of order $k > 2$ contribute. These involve the quantity $E\left[\delta Z^k\right]$ which depends on the particular distribution of δZ. The dependence of vanilla option's prices to the latter is illustrated in figures 10.6 and 10.8 without stochastic volatility, that is with a constant σ, and in figures 10.5 and 10.7 with stochastic volatility.[11]

The conclusion from figures 10.5 and 10.7 is that, except for exaggerate values of tail parameters, and possibly very short maturities, the impact of the conditional one-day smile on smiles for standard maturities is small. In particular the ATMF skew is predominantly generated by the covariance of spot and forward variances. Only a minute portion of it can be traced to the one-day smile or, equivalently, the conditional distribution of daily returns.

Moreover, high-strike – rather than low-strike – implied volatilities seem to be impacted the most.

The steep ATMF skews observed for equity indexes for typical maturities have thus nothing to do with the fact that historical distributions of daily equity returns may or may not exhibit large drawdowns. Rather, they supply an estimate of the implied level of spot/volatility covariances, just like VS implied volatilities supply an estimate of the implied level of spot volatility.

Recall expression (8.24), page 315, relating the ATMF skew to the integrated covariance of spot and VS volatility for the residual maturity, at order one in volatility of volatility:

$$\mathcal{S}_T = \frac{1}{2\hat{\sigma}_T^3 T} \int_0^T \frac{T-\tau}{T} \frac{\langle d\ln S_\tau \, d\hat{\sigma}_T^2(\tau)\rangle}{d\tau} d\tau$$

What about very short maturities? Smiles in figures 10.5 and 10.7 imply that vanilla smiles for maturities of the order of a few days will exhibit a large sensitivity

---

[10]Note that in jump-diffusion models $\frac{dS}{S}$ does not scale like $\sqrt{\Delta}$ anymore and all orders of δS^k contribute to the P&L, even for arbitrarily small values of Δ – see Section 10.3 below.

[11]When using the fat-tailed stochastic volatility model of Section 10.2.1, the total P&L comprises contributions from $\delta \xi^\tau$ as well. The order-two contributions read $-\frac{S^2}{2}\frac{d^2P}{dS^2}\left(\frac{\delta S^2}{S^2} - \sigma^2\Delta\right) - \int_t^T S\frac{d^2P}{dSd\xi^u}\left(\frac{\delta S}{S}\delta\xi^u - \mu(t,u,\xi)\Delta\right)du - \frac{1}{2}\iint_t^T \frac{d^2P}{\delta\xi^u\delta\xi^{u'}}\left(\delta\xi^u\delta\xi^{u'} - \nu(t,u,u',\xi)\Delta\right)dudu'$ where μ and ν are defined in equation (7.2), page 218. The rescaling of spot/volatility correlations in (10.7) guarantees that, just as in the constant volatility case, the values of all instantaneous covariances do not depend on the distribution of δZ. Only higher-order contributions to the P&L will depend on the distribution of δZ.

to the one-day smile. Beside very short-dated vanilla options, are there other, longer-dated, options that are sensitive to the one-day smile?

Path-dependent options that involve daily returns are the natural candidates. One such payoff is the variance swap – see Chapter 5 for an introduction.

10.2.4   Variance swaps

In Section 5.3.4, page 162, we have compared the P&L of a delta-hedged log contract with the payoff of the VS and have shown that the difference is generated by terms δS^k with $k > 2$.

In the limit of short returns, in diffusive models, the implied volatilities $\hat{\sigma}_{\mathrm{VS},T}$ of the VS and $\hat{\sigma}_T$ of the log contract coincide. For daily returns, in a diffusive model, the difference $\hat{\sigma}_{\mathrm{VS},T} - \hat{\sigma}_T$ should still be negligible, while it is expected to be sizeable in fat-tailed models. Formula (5.38), page 162, for example, expresses the difference between $\hat{\sigma}_{\mathrm{VS},T}$ and $\hat{\sigma}_T$ generated by terms of order 3 in δS as a function of the (unconditional) skewness of daily returns.

Table 10.2 lists the values of $\hat{\sigma}_{\mathrm{VS},T} - \hat{\sigma}_T$ in our fat-tailed two-factor model, for different parameter configurations, for a 1-year VS.

μ	∞	6	4	3	
$\nu = 0$	0%	0%	0.02%	0.16%	
$\nu = 257\%$	0.02%	0.04%	0.10%	0.29%	

p_+	30%	40%	50%	60%	70%
$\mu = 4, \nu = 257\%$	−0.11%	0%	0.10%	0.23%	0.40%

Table 10.2: Top: $(\hat{\sigma}_{\mathrm{VS},T} - \hat{\sigma}_T)$ for $T = 1$ year as a function of μ for $p_+ = p_- = 1/2$ with $\nu = 0$ (no stochastic volatility) and $\nu = 257\%$ – corresponds to smiles in Figure 10.5. Bottom: $(\hat{\sigma}_{\mathrm{VS},T} - \hat{\sigma}_T)$ as a function of p_+ for $\mu = 4, \nu = 257\%$ – these parameters correspond to smiles in Figure 10.7.

We can see in the top section of Table 10.2 that for $\mu = \infty$ – which corresponds to the standard version of the two-factor model – even for a large volatility of volatility, VS and log-contract volatilities are essentially identical. They become appreciably different as we widen the tails of returns (we decrease μ), and as we vary p^+, thus increasing the ATMF skew (bottom section of table).[12]

The *relative* difference between log-contract and VS volatilities produced by our model for values of μ and p^+ that are consistent with historical distributions of daily returns ($\mu \simeq 4, p^+ = 50\%$) is $0.1\%/20\% = 0.5\%$.

---

[12]Because of expression (10.3) for the dynamics of S_t in the fat-tailed version of the two-factor model, the implied VS volatilities we input in the model are those of payoff $\Sigma_i (S_{i+1}/S_i - 1)^2$ rather than of payoff $\Sigma_i \ln^2 (S_{i+1}/S_i)$. This is not a problem; forward variances for this alternative definition of the VS are still driftless.

This is roughly the order of magnitude of the realized value of $\hat{\sigma}_{\text{VS},T} - \hat{\sigma}_T$ in the backtest of Chapter 5 – see Figure 5.1, page 166.

10.2.5 Daily cliquets

Daily cliquets are cliquets written on daily returns. An example is a one-year option on the Euro Stoxx 50 index that pays a daily coupon equal to the put payoff $(k - \frac{S_{i+1}}{S_i})^+$ where S_i, S_{i+1} are two consecutive daily closing values of the index.

Typical strikes for these daily puts are in the range of 75% to 90%. Daily cliquets are also typically of the knock-out type – the option expires once one coupon has paid off and the premium is paid on a quarterly basis – making them very similar to CDS contracts.

Some popular variants involve put spreads – daily cliquets whose coupons are capped at some specified level.

Table 10.3 shows prices for a 1-year cliquet of daily puts struck at 80% (no knock-out feature) for various values of μ_- with no stochastic volatility ($\nu = 0$). We have taken $\mu_+ = 4$ and $p_+ = p_- = \frac{1}{2}$ as well as $\sigma = 20\%$. The resulting one-day smiles appear in Figure 10.10.

μ_-	∞	6	4	3	2.5	2.2
	0.00%	0.00%	0.02%	0.15%	0.43%	0.62%

Table 10.3: Prices for a 1-year daily cliquet struck at 80% as a function of μ_- with $\mu_+ = 4, p_+ = p_- = \frac{1}{2}, \sigma = 20\%$ and no volatility of volatility.

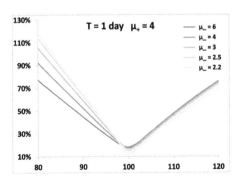

Figure 10.10: 1-day smile for different values of μ_- for $\mu_+ = 4, p_+ = p_- = 1/2, \sigma = 20\%$.

Obviously, daily cliquets are worthless unless very large daily returns can be generated in the model. We have included in Table 10.3 the value $\mu_- = 2.2$ as values of μ_- this low are typically needed to match market prices.

This implied value of μ_- is thus much lower than its historical counterpart, which lies in the range $[3, 4]$. However, there is good reason for pricing daily cliquets very conservatively. Daily cliquets can be vega-hedged but cannot be replicated by delta and gamma-hedging as each coupon's maturity is one day. This leaves the seller saddled with the risk of a potentially very large negative unhedgeable daily P&L. Daily cliquets are in fact insurance-type products that enable trading desks to exchange and mitigate stress-test risk.

Prices in Table 10.3 are calculated with $\nu = 0$. Once stochastic volatility is switched on prices of daily cliquets increase, but are still mostly dependent on μ_-. For example, setting $\nu = 257\%$ increases the rightmost price in Table 10.3 by about 5 bps.[13]

10.3 Conclusion

- While historical (unconditional) distribution densities of equity returns exhibit fat tails, much thicker than can be achieved with typical diffusive models, the skew of vanilla smiles is overwhelmingly generated by the covariance of the spot with the implied volatility of the residual maturity, rather than the one-day smile.

- Unlike vanilla options, path-dependent options that involve daily returns do exhibit some sensitivity to the one-day smile, modest in the case of VSs, strong for daily cliquets.

- These results have been obtained using a fat-tailed version of the two-factor stochastic volatility model. This is the two-factor forward variance model of Chapter 7, minimally adjusted so that:

  - daily returns are drawn with a two-sided Student distribution, rather than a Gaussian distribution. μ_+, μ_- along with p_+ afford good control of the one-day conditional smile.

  - spot/volatility covariances remain unchanged. This is achieved by uniformly rescaling the native spot/volatility correlations.

---

[13]Stochastic volatility has the effect that the scales of successive returns are correlated, thus making non-knock-out daily cliquets worth more than their knock-out counterparts. In the absence of stochastic volatility, because the probability of a knock-out event is so low, prices of both versions are practically identical.

Appendix A – jump-diffusion/Lévy models

This section is devoted to jump-diffusion/Lévy models – when are they called for and what do they do that stochastic volatility doesn't?

A.1   A stress-test reserve/remuneration policy

In the P&L of a delta-hedged option position, the order-one contribution in δS is cancelled by the delta, so the P&L starts with a δS^2 term. The first step in model building focuses on terms of order δS^2 – the gamma P&L. The essence of the Black-Scholes pricing equation lies in the provision of a deterministic theta term to offset the gamma P&L, given a break-even volatility σ:

$$P\&L = -\frac{1}{2}S^2\frac{d^2P}{dS^2}\left(\frac{\delta S^2}{S^2} - \sigma^2\delta t\right) \tag{10.11}$$

Mathematically, the solution P of the (parabolic) pricing equation has a probabilistic interpretation as the expectation of the payoff under a dynamics for S_t which is a diffusion. When other assets beside S_t – for example variances – are modeled as diffusive processes, other contributions to the P&L arise, of the same form as (10.11).[14]

As discussed in Section 1.2 of Chapter 1, the risk associated with gamma P&L is still sizeable and needs to be hedged away. This is done by dynamically trading vanilla options. With regard to δS, the residual risk is thus contributed by terms of order 3 and above.

Section 10.2.1 above has been devoted to the construction of a model that allows us to price these terms by separating the effect of the scale of δS from that of its distribution. We first choose a time scale – in our case one day – and write the return as:

$$\frac{\delta S}{S} = \sigma\delta Z$$

The instantaneous volatility σ – the scale of the return – is modeled with a two-factor forward variance model, and the distribution of δZ – or equivalently the conditional distribution of δS – is taken care of by the one-day smile parameters of the model.

Imagine we do not require this much sophistication and are looking for a way of pricing the effect of large returns using very simple assumptions. For example, we wish to assess the overall impact of typical adverse scenarios, or stress tests, on our hedged option position – that is the exotic option together with its vanilla hedge – and adjust the exotic option's price accordingly or set aside a reserve.

---

[14]The P&L of a delta-hedged, vega-hedged, position reads as in (7.3a), page 218, in the case of a forward variance model, or as in (2.105), page 69, in the case of the local volatility model.

This reserve is either (a) intended to offset on average the impact of the adverse scenarios and is released gradually to the trading desk or (b) charged by the bank to the trading desk on an ongoing basis to pay for the cost of allocating capital against adverse scenarios.

We already considered such a simple adjustment for variance swaps in Chapter 5 – see equation (5.42), page 166, for a VS together with its vanilla hedge. Consider then a shock of fixed relative magnitude J on our underlying:

$$S \rightarrow S(1+J)$$

The P&L it generates on a short delta-hedged option position is:

$$P\&L = -\big(P(t, S(1+J)) - P(t, S)\big) + JS\frac{dP}{dS} \tag{10.12}$$

where $P(t, S)$ is our pricing function, and the second piece in (10.12) the contribution of our delta hedge. Expanding in powers of J:

$$P\&L = -\frac{1}{2}S^2\frac{d^2P}{dS^2}J^2 - \frac{1}{6}S^3\frac{d^3P}{dS^3}J^3 + \cdots$$

Note that this P&L includes a δS^2 portion, which acts as an extra contribution to the volatility of S. Let us now assign an annualized frequency λ to our stress test. There occur on average $\lambda\delta t$ shocks during the time interval $[t, t+\delta t]$, whose P&L impact will be offset, on average, by a reserve given by:

$$\lambda\delta t\left[\big(P(t, S(1+J)) - P(t, S)\big) - JS\frac{dP}{dS}\right] \tag{10.13}$$

(10.13) can be interpreted in three ways:

- either as a theta contribution to offset on average the P&L generated by the stress-test scenario, which is assumed to occur with intensity λ.

- or as a tax levied on the trading desk by the bank to pay for the cost of capital allocated to stress-test risk. Assuming this capital is proportional to the stress-test P&L, with a proportionality coefficient β, and that the rate of return required by the bank on its capital is μ, the amount charged to the desk is then of the form (10.13) with $\lambda = \beta\mu$; λ is no longer interpreted as an intensity. Depending on the sign of its contribution to the overall stress-test P&L, a trading desk would thus be either taxed or rewarded.

- or as a minimal return we require on our consumption of stress-test limit. Trading desks are usually not charged directly for stress-test P&L, but are assigned stress-test P&L limits. This limit can be managed at the desk level by requiring that, for a given consumption of stress-test budget, commensurate revenue be generated. This amounts to demanding that our carry P&L comprises, on top of the usual gamma/theta P&L, a piece given by (10.13) with λ the (annualized) rate of remuneration of stress-test-budget usage.

Consider an option of maturity T. To calculate a reserve policy we use the initial estimate at time t, spot value S, of the P&L impact of a shock to compute an adjustment ΔP to the option price. ΔP is given by:

$$\Delta P(t,S) = \lambda(T-t)\left[(P(t,S(1+J))-P(t,S))-JS\frac{dP}{dS}\right] \quad (10.14)$$

The price we quote at time t for our derivative is:

$$P+\Delta P$$

ΔP is proportional to $T-t$: at time passes the reserve is released and converges to 0 at $t=T$.

$\Delta P(t,S)$ is a reserve policy evaluated at t with the initial spot value S. Consider a move δS. While we started out with an amount $\Delta P(t,S)$, the new reserve we should now be holding is $\Delta P(t,S+\delta S)$. To generate this extra cash, we need to delta-hedge ΔP.

ΔP however includes no provision for the financing cost of its own delta hedge, and no theta to offset its own gamma.

These limitations are typical when one uses an ad-hoc reserve policy rather than a full-blown model. We now derive a pricing equation that takes care of these issues.

A.2 Pricing equation

We would like our P&L during two delta rehedges to read as in (10.11), but with an additional theta contribution given by (10.13):

$$P\&L = -\frac{1}{2}S^2\frac{d^2P}{dS^2}\left(\frac{\delta S^2}{S^2}-\sigma^2\delta t\right)+\lambda\left[(P(t,S(1+J))-P(t,S))-JS\frac{dP}{dS}\right]\delta t \quad (10.15)$$

Proceeding as in Section 1.1, page 2, we can write down the corresponding pricing equation at once:

$$\frac{dP}{dt}+(r-q)S\frac{dP}{dS}+\frac{\sigma^2}{2}S^2\frac{d^2P}{dS^2}+\lambda\left[(P(t,S(1+J))-P(t,S))-JS\frac{dP}{dS}\right]=rP \quad (10.16)$$

With respect to the Black-Scholes equation, the last piece in the left-hand side provides for an additional theta to offset, on average, the P&L impact of our stress-test scenario.

Mathematically, the solution of (10.16) can be expressed as the expectation of the option's payoff under a dynamics for S_t that consists of a diffusion together with Poisson jumps:

$$dS_t = (r-q)S_tdt+\sigma S_tdW_t+JS_t(dN_t-\lambda dt) \quad (10.17)$$

where N_t is a counting process. Unlike the pricing equations we have encountered so far, (10.16) involves non-local terms: $\frac{dP}{dt}$ is a function not only of derivatives of P

with respect to S, which characterize P in the vicinity of S, but also on the value of P for $S\left(1+J\right)$.

The amplitude J of the stress-test scenario can be made a random variable. Let us call $\rho(J)$ its distribution. Our pricing equation becomes:

$$\frac{dP}{dt} + (r-q)\,S\frac{dP}{dS} + \frac{\sigma^2}{2}S^2\frac{d^2P}{dS^2}$$

$$+ \lambda \int_{-1}^{\infty} \rho(J)\left[\left(P(t,S(1+J)) - P(t,S)\right) - JS\frac{dP}{dS}\right]dJ \; = \; rP \quad (10.18)$$

Solving it

For a constant volatility σ and vanilla option payoffs (10.18) can be solved easily since it is homogeneous in $\ln S$. The procedure is similar to that in Appendix A of Chapter 8, page 347.

Define $\tau = T - t$ and:

$$x \; = \; \ln\frac{S}{K} + (r-q)\tau$$

where K is the strike of the vanilla option, set $P\left(t,S\right) = Se^{-q\tau}f(\tau,x)$ and introduce the Laplace transform $F(\tau,p)$ of f:

$$F\left(\tau,p\right) \; = \; \int_{-\infty}^{+\infty}e^{-px}f\left(\tau,x\right)dx$$

Replacing P by $Se^{-q\tau}f(\tau,x)$ in (10.18) yields:

$$-\frac{df}{d\tau} + \frac{\sigma^2}{2}\left(\frac{df}{dx} + \frac{d^2f}{dx^2}\right)$$

$$+ \lambda \int_{-\infty}^{+\infty}\rho^*(u)\left(e^u f(x+u,\tau) - f - (e^u - 1)\left(f + \frac{df}{dx}\right)\right)du \; = \; 0$$

where $\rho^*(u)$ is the density of

$$u = \ln(1+J)$$

Taking now the Laplace transform of both sides leads to the following ODE for $F\left(\tau,p\right)$:

$$-\frac{dF}{d\tau} + \left[\frac{\sigma^2}{2}p(1+p) + \lambda\left(\psi(p) - (1+p)\psi(0)\right)\right]F \; = \; 0 \qquad (10.19)$$

where $\phi(p)$ is defined as:

$$\psi(p) \; = \; \int_{-\infty}^{+\infty}\rho^*(u)\left(e^{(1+p)u} - (1+p)u - 1\right)du$$

The initial condition for F for $\tau = 0$, that is $t = T$, is provided by the option's payoff. For a call option $f(0, x) = (1 - e^{-x})^+$ thus:

$$F(0, p) = \int_{-\infty}^{+\infty} e^{-px} \left(1 - e^{-x}\right)^+ dx = \frac{1}{p(1 + p)}$$

$F(0, p)$ is defined for $\text{Re}(p) > 0$. For a put option $F(0, p)$ is identical except the condition is $\text{Re}(p) < -1$.

Integrating (10.19) we then get:

$$F(\tau, p) = \frac{1}{p(1 + p)} e^{\tau H(p)} \tag{10.20}$$

$$H(p) = \frac{\sigma^2}{2} p(1 + p) + \lambda\big(\psi(p) - (1 + p)\psi(0)\big)$$

Inverting $F(\tau, p)$ yields option prices.

Using as pricing function the solution P of (10.18) rather than $P^0 + \Delta P$ ensures that:

- the additional theta the model pays us in between two delta rehedges is exactly (10.13)

- $\frac{dP}{dS}$ incorporates the delta hedge of the reserve policy

- the P&L generated by jumps on the reserve policy itself is accounted for

One can back Black-Scholes implied volatilities out of $P(t, S)$ – what does the resulting smile look like? We now briefly analyze its ATMF skew.

A.3 ATMF skew

In order to use the perturbative result of Appendix B of Chapter 5 we need the density ρ_z of z_T, defined by:

$$z_T = \ln \frac{S_T}{S} - (r - q)(T - t)$$

The density of S_T is given by: $\rho(S_T) = e^{r(T-t)} \left.\frac{d^2 P}{dK^2}\right|_{K=S_T}$. Using that $P(t, S) = S e^{-q(T-t)} f(t, x)$ and the definition of z we get:

$$\rho_z(z) = \left(e^x \left(\partial_x + \partial_x^2\right) f\right)_{x=-z}$$

The cumulant-generating function of ρ_Z, $L\left(\tau,q\right)$, which is the logarithm of the characteristic function of ρ_z, is given by:

$$e^{L(\tau,q)} = \int_{-\infty}^{+\infty} e^{-qz}\rho_z(z)dz \qquad (10.21)$$

$$= \int_{-\infty}^{+\infty} e^{-qz}\left(e^x\left(\partial_x + \partial_x^2\right)f(\tau,x)\right)_{x=-z}dz$$

$$= q\left(1+q\right)F\left(\tau,-(1+q)\right)$$

In what follows we will be sitting at $t = 0$; we thus set $\tau \equiv T$. From expression (10.20) we thus get $L(T,q) = TH\left(p = -(1+q)\right)$.

$$L(T,q) = T\left(\frac{\sigma^2}{2}q(1+q) + \lambda\left(\psi(-(1+q)) + q\psi(0)\right)\right)$$

$$= T\left(\phi(q) + q\phi(-1)\right) \qquad (10.22)$$

where $\phi(q)$ is given by:

$$\phi(q) = \frac{\sigma^2}{2}q^2 + \psi(-(1+q))$$

$$= \frac{\sigma^2}{2}q^2 + \int_{-\infty}^{+\infty}\lambda\rho^*(u)\left(e^{-qu} + qu - 1\right)du \qquad (10.23)$$

Using J rather than u, $L\left(T,q\right)$ equivalently reads:

$$L(T,q) = T\left(\frac{\sigma^2}{2}q(1+q) + \int_{-1}^{\infty}\lambda\rho(J)\left((1+J)^{-q} - 1 + qJ\right)dJ\right) \qquad (10.24)$$

Note that $L\left(T,q\right)$ scales linearly with T: this is true of all processes for $\ln S$ with independent stationary increments, which is obviously the case for process (10.17).

As already mentioned in Appendix C of Chapter 8, and using equation (5.82), page 191, $L(T,q)$ has the following properties:

- the condition that the density of z_T integrate to 1 is $L\left(T,0\right) = 0$
- the condition that the forward of S_T be $Se^{(r-q)T}$ is $L(T,-1) = 0$
- the log-contract implied volatility is given by: $\widehat{\sigma}_T^2 = \frac{2}{T}\left.\frac{dL}{dq}\right|_{q=0}$

By construction, as can be checked using expression (10.22) for L, the two conditions are satisfied. When using an approximate form for L we need to make sure they still hold. Using (10.22) we get:

$$\widehat{\sigma}_T^2 = \sigma^2 + 2\lambda\overline{e^u - u - 1} = \sigma^2 + 2\lambda\overline{J - \ln\left(1+J\right)}$$

where $\overline{X}$ stands for the mean of random variable X.

The VS implied volatility, $\hat{\sigma}_{\text{VS},T}$, on the other hand, is simply given by the quadratic variation of $\ln S$:[15]

$$\hat{\sigma}^2_{\text{VS},T} = \sigma^2 + \lambda \overline{\ln(1+J)^2}$$

For $J = 0$, the jump-diffusion model reduces to the Black-Scholes model and $L(q)$ is a polynomial of order 2. To analyze the ATMF skew of jump-diffusion models, let us assume that J is small and let us expand L in powers of J, stopping at the lowest non-trivial order. We get from (10.24):

$$L(T,q) = T\left(\frac{\sigma^2 + \lambda \overline{J^2}}{2}q(1+q) - \frac{\lambda \overline{J^3}}{6}q(1+q)(2+q)\right) \qquad (10.25)$$

The contribution of order 2 in J merely shifts the volatility level: $\sigma^2 \to \sigma^2 + \lambda \overline{J^2}$, but the model remains of the Black-Scholes type – we need to go to order 3 in J. $L(T,q)$ in (10.25) complies with the two conditions above. The log-contract implied volatility, at this order, is:

$$\hat{\sigma}^2_T = \frac{2}{T}\frac{dL}{dq}\bigg|_{q=0} = \sigma^2 + \lambda \overline{J^2}$$

For small values of J the ATMF skew $\mathcal{S}_T$, at order one in s, is given by (5.93), page 194:

$$\mathcal{S}_T = \frac{s}{6\sqrt{T}}$$

s is the skewness of $\ln S_T$, that is $s = \frac{\kappa_3}{(\hat{\sigma}^2_T T)^{3/2}}$ where κ_3 is the cumulant of order 3. Cumulants are defined through the expansion of $L(q)$ in powers of q:

$$L(T,q) = \sum_{n=1}^{\infty} \frac{(-1)^n}{n!}\kappa_n(T)q^n$$

From (10.25), $\kappa_3 = \lambda \overline{J^3}T$, thus $s = \frac{\lambda \overline{J^3}T}{(\sigma^2 + \lambda \overline{J^2})^{3/2}T^{3/2}}$. We thus get the following approximate formula for $\mathcal{S}_T$:

$$\mathcal{S}_T = \frac{\lambda \overline{J^3}}{6\left(\sigma^2 + \lambda \overline{J^2}\right)^{3/2}T} = \frac{\lambda \overline{J^3}}{6\hat{\sigma}^3_T T} \qquad (10.26)$$

Thus, for small jump sizes, the ATMF skew of jump-diffusion models decays like $\frac{1}{T}$.

This scaling is illustrated in Figure 10.9, page 402. While data in Figure 10.9 are obtained with the fat-tailed two-factor model of Section 10.2.1, which moreover is a discrete model, because stochastic volatility is turned off, it generates independent stationary (discrete) increments for $\ln S$, hence the $\frac{1}{T}$ scaling of the ATMF skew.[16]

---

[15]We recover the expressions for $\hat{\sigma}_T$ and $\hat{\sigma}_{\text{VS},T}$ in Section 5.3.2.

[16]For $T \to 0$, approximation (10.26) for $\mathcal{S}_T$ diverges, which is not the case of the actual ATMF skew. Why does the approximation break down in the short-T limit? Expanding $L(T,q)$ at order one in κ_3

A.4 Jump scenarios in calibrated models

By using equation (10.16) to risk-manage the exotic option at hand, we obtain a price and a delta-hedging strategy so that the P&L of our short option position together with its delta hedge in between two delta rehedges reads as in (10.15). In addition to the usual second-order gamma/theta P&L, we have an extra theta contribution, which corresponds to a regular release of the stress-test reserve policy where the stress test is specified by the frequency and distribution of the jumps in our model.

This would be fine if we were just delta-hedging our option: we quote as initial price $P(t = 0, S_0)$, the solution of (10.16), run our delta hedge and we are done.

In reality we use other derivatives – typically vanilla options – as hedge instruments, thus the price charged to the client needs to incorporate a stress-test reserve policy that offsets on average the stress-test P&L

$$P\&L = -\big(P(t, S(1 + J)) - P(t, S)\big) + JS\frac{dP}{dS}$$

of the *global* position, rather than that of the *naked* exotic option: P is the value of the *hedged* position. Consequently the hedging vanilla options need to be risk-managed using the same model.

Assume we are using a market model, for example the local volatility model, or one of the admissible local-stochastic volatility models of Chapter 12. We start with the pricing equation of one such model and insert in it the contribution from jumps – the last term in the left-hand sides of equation (10.16) or (10.18).

Is the resulting carry P&L still of the standard gamma/theta form typical of diffusive market models, with an additonal theta corresponding to our stress-test/jump scenario? Let us assume here that we are using the local volatility model.

The pricing equation for $P^{LV}(t, S, \sigma)$ is (10.16) with σ replaced with the local volatility $\sigma(t, S)$, chosen so that market prices of vanilla options are recovered.[17]

With respect to the original pricing equation (2.102), page 68, of the local volatility model, we have an extra contribution to $\frac{dP^{LV}}{dt}$ coming from jumps.

---

amounts to perturbing ρ_z around a Gaussian density with Hermite polynomials – see expression (5.83), page 191. The coefficient of H_n is $\frac{\delta\kappa_n}{\Sigma^n\sqrt{n!}}$, which in our case is proportional to $T/T^{\frac{n}{2}} = T^{1-\frac{n}{2}}$, i.e. diverges for $n = 3$: this is no longer a small perturbation.

[17]Space prevents us from discussing here the calibration of such a model. In the presence of jumps, we can carry out the same derivation as in Section 2.2.1, page 27, to generate the following forward equation for vanilla option prices, which replaces (2.7):

$$\frac{dC}{dT} + (r - q - \lambda J) K \frac{dC}{dK} - \frac{\sigma^2(T, K)}{2} K^2 \frac{d^2C}{dK^2} - \lambda \left[(1 + J) C\left(\frac{K}{1 + J}, T\right) - C(K, T)\right]$$
$$= -(q + \lambda J) C$$

From this we get the expression for $\sigma^2(T, K)$:

$$\sigma^2(T, K) = \sigma_{\text{loc}}^2(K, T) - 2\lambda \frac{\left[(1 + J) C_{K/(1+J), T} - C_{K, T}\right] - JC + \lambda JK \frac{dC}{dK}}{K^2 \frac{d^2C}{dK^2}} \tag{10.27}$$

In Section 2.7 of Chapter 2 we have analyzed the carry P&L of a delta-hedged/vega-hedged position and have shown that it has the usual gamma/theta form.

In a local volatility model with jumps, implied volatilities are still a function of t, S and $\sigma(t, S)$: $\widehat{\sigma}_{KT} = \Sigma_{KT}^{LV}(t, S, \sigma)$, except Σ_{KT}^{LV} is a different function than in Section 2.7 as jumps are taken into account.

Going back to page 67, the reader can check that the derivation in the jump case is identical, except we have the following additional term in the right-hand side of equation (2.105), page 69:

$$+\lambda \left[\left(P^{LV}(t, S(1+J), \sigma) - P^{LV}(t, S, \sigma) \right) - JS \frac{P^{LV}}{dS} \right] \delta t$$

Expressing everything in terms of $P(t, S, \widehat{\sigma}_{KT}) = P^{LV}(t, S, \sigma\,[t, S, \widehat{\sigma}_{KT}])$, using (2.99), this P&L can be rewritten as:

$$+\lambda \left[\left(P\big(t, S(1+J), \widehat{\sigma}_{KT} + \Delta\Sigma_{KT}^{LV}(t, S)\big) - P\big(t, S, \widehat{\sigma}_{KT}\big) \right) \right.$$
$$\left. - JS \left(\frac{dP}{dS} + \frac{dP}{d\widehat{\sigma}_{KT}} \bullet \frac{d\Sigma_{KT}^{LV}}{dS} \right) \right] \delta t \qquad (10.28)$$

where $\Sigma_{KT}^{LV}(t, S)$ is the (additive) jump of implied volatility $\widehat{\sigma}_{KT}$ generated at time t, spot S, by a relative jump of S of magnitude J, keeping the local volatility function – equal to $\sigma\,[t, s, \widehat{\sigma}_{KT}]$ – unchanged:

$$\Delta\widehat{\sigma}_{KT}(t, S) = \Sigma_{KT}^{LV}(t, S(1+J), \sigma\,[t, s, \widehat{\sigma}_{KT}]) - \widehat{\sigma}_{KT}$$

The additional theta (10.28) is thus proportional to the P&L generated by *joint jumps* in spot and implied volatilities, minus the contribution from the delta-hedge.

The conclusion of this section is that, depending on the model we start with, adding jumps on S in the pricing equation results in general in a stress-test scenario that involves jumps not only on S but also on implied volatilities. In the local volatility model, the jump in implied volatilities is dictated by the jump scenario for S and the smile used for calibration.[18]

---

where $\sigma_{\text{loc}}(K, T)$ is the local volatility in the case with no jumps, given by the Dupire formula (2.3), page 26. We can check that, expanding the right-hand side of (10.27) at order two in J, we get:

$$\sigma^2(T, K) = \sigma_{\text{loc}}^2(K, T) - \lambda J^2$$

For numerical aspects related to (a) the calibration of the local volatility model with jumps, and (b) pricing in jump/diffusion models, we refer the reader to [2] and [82].

[18]How do we gain leverage on the jump scenario for implied volatilities? This is a hard question. We would need to define a set of local volatilities $\sigma_n(t, S)$ where n is the number of jumps before t. Denote by $P_n(t, S)$ the option price. P_n solves equation (10.16) or (10.18) with $\sigma(t, S)$ replaced by $\sigma_n(t, S)$

A.5 Lévy processes

Imagine we add up different independent Poisson processes, each with its own intensity λ_i and jump density ρ_i^*. In expression (10.23) for $\phi(q)$, $\lambda\rho^*(u)$ is replaced with:

$$\lambda\rho^*(u) \;\to\; \sum_i \lambda_i \rho_i^*(u)$$

There are two situations:

- uninteresting: if $\Sigma_i\lambda_i$ is finite, this boils down to an effective Poisson process with intensity $\lambda = \Sigma_i\lambda_i$ and jump density $\rho^* = \frac{\Sigma_i\lambda_i\rho_i^*}{\Sigma_i\lambda_i}$.

- interesting: $\Sigma_i\lambda_i$ is infinite – jumps occur infinitely frequently. The expression of $\phi(q)$ is:

$$\phi(q) \;=\; \frac{\sigma^2}{2}q^2 + \int_{-\infty}^{+\infty} \left(e^{-qu} + qu - 1\right)k(u)du \qquad (10.29)$$

The process for z_t is a non-trivial Lévy process and (10.29) is a particular form of the Lévy-Khinchine representation, with $k(u) = \Sigma_i\lambda_i\rho_i^*(u)$.

Mathematically, a Lévy process is a process with stationary independent increments. As such, the class of Lévy processes trivially contains Brownian motion and Poisson processes, but includes many other processes. We refer the interested reader to the many textbooks on this topic – see [33] for applications to derivatives pricing.

A.6 Conclusion

Jump-diffusion – and more generally Lévy – processes are useful tools for embedding in derivatives' prices a stress-test reserve policy corresponding to the scenarios that these processes express. As tools for modeling the actual dynamics of securities or for pricing derivatives, they are, in the author's view, not adequate, for mostly two reasons:

- the assumption of independent increments is violated in historical returns, and in models as well, in the latter because instantaneous volatility – which sets the scale of returns – is stochastic and needs to be so. As a representation of actual returns, jump/Lévy processes are a fairly unrealistic construct.

---

and the contribution from jumps is replaced with:

$$\lambda\left[(P_{n+1}(t, S(1+J)) - P_n(t, S)) - JS\frac{dP_n}{dS}\right]$$

We have a set of nested equations for the P_n – we obviously need to make additional assumptions in order to be able to calibrate $\sigma_n(t, S)$ to the market.

The case with two local-volatility functions, for the pre-default and post-default states, is relevant for stocks.

- jump processes are much more difficult to correlate than diffusions, except for those Lévy processes that have a representation in terms of time-changed Brownian motion.

Jump-diffusion models should then not be considered a bona fide characterization of the actual or implied dynamics of real securities but viewed as a means of calculating – and releasing – a reserve policy based on ad-hoc stress-test scenarios.

Using a diffusive process for pricing does not mean we actually assume that securities behave as diffusions. A diffusive pricing equation is merely a technical device for embedding in a derivative's price the time value needed to offset second order gamma P&Ls with given break-even levels.

Practically, properly offsetting a trading book's second-order sensitivities is already quite a challenge. Hedging P&Ls generated by higher-order moments is in practice hardly possible. The best we can hope for is either:

- (a) price them in a model that realistically models the tail behavior of actual returns

- or (b) estimate them using ad-hoc stress-test scenarios.

Case (b) is taken care of by jump/Lévy models.

As for case (a), we have explained in Section 10.2.1 above how to economically adjust an existing stochastic volatility model, with the benefit that the richness of the dynamics of volatilities is preserved – while achieving a realistic modeling of tail behavior.

Chapter's digest

10.1 The distribution of equity returns

▶ Unconditional distributions of daily returns of equity indexes exhibit fat tails that are well-approximated with a Student distribution, with similar parameters for left and right tail distributions in the range $[3, 4]$.

▶ Conditional distributions, whereby returns are rescaled by an estimate of realized volatility, still exhibit fat-tailed distributions, with somewhat thinner tails – still much thicker than in the lognormal distribution.

🦭 🦭 🦭 🦭 🦭

10.2 Impact of the distribution of daily returns on derivative prices

▶ To study the effect of fat tails of daily returns – or equivalently the effect of the one-day smile – we develop a fat-tailed version of the two-factor model. The conditional distribution of daily returns is a two-sided Student distribution. The one-day smile is parametrized by (a) the parameters of the left and right tails, (b) the probability that returns are positive, which sets the one-day ATM skew.

▶ Forward variances are simulated in the two-factor model, as usual, and so are increments δW^S for the Brownian motion that drives S_t. S_t is simulated on a daily schedule according to:

$$S_{t+\Delta} = S_t \left[1 + (r - q)\Delta + \sigma_t \delta Z\right]$$

where $\delta Z = \sqrt{\Delta} f\left(\frac{\delta W^S}{\sqrt{\Delta}}\right)$. The mapping function is given by:

$$\begin{cases} x \leq \mathcal{N}_G^{-1}(p_-) & f(x) = \zeta_- \sqrt{\frac{\mu_- - 2}{\mu_-}} \mathcal{N}_{\mu_-}^{-1}\left(\frac{\mathcal{N}_G(x)}{2p_-}\right) \\ x \geq \mathcal{N}_G^{-1}(p_-) & f(x) = \zeta_+ \sqrt{\frac{\mu_+ - 2}{\mu_+}} \mathcal{N}_{\mu_+}^{-1}\left(\frac{1}{2} + \frac{\mathcal{N}_G(x) - p_-}{2p_+}\right) \end{cases}$$

▶ Spot/volatility correlations need to be rescaled so that covariances of spot and implied volatilities remain identical in both standard and fat-tailed versions of the two-factor model. This rescaling is uniform with the ratio of correlations in the fat-tailed model to those in the standard model given by:

$$\frac{\rho_{iS}^*}{\rho_{iS}} = \frac{1}{\int_{-\infty}^{+\infty} \phi(x) \, x f(x) \, dx}$$

where $\phi(x)$ is the probability density of the standard normal variable.

▶ Vanilla smiles exhibit little sensitivity to the fat tails of daily returns, except for short maturities and out-of-the-money volatilities – mostly for high strikes. ATMF skews are hardly affected.

▶ The one-day ATM skew does impact vanilla smiles, but the size of its contribution decays as $\frac{1}{T}$.

▶ The conclusion is that, for standard maturities, the fat-tailed nature of the conditional distribution of daily returns hardly impacts the ATMF skew, which is predominantly the product of the covariance of spot and implied volatilities.

▶ Derivatives whose payoffs involve daily returns are sensitive to the effect of the one-day smile – typical examples include variance swaps and daily cliquets.

❦ ❦ ❦ ❦ ❦

Appendix A – jump-diffusion/Lévy models

▶ Pricing equations based on jump-diffusion processes naturally arise when one needs to factor in the price of a derivative (a) the average impact of specific stress-test scenarios on the portfolio of the derivative together with its delta (and possibly its vega) hedge, (b) the cost of capital set aside to cover for the P&L generated by stress-test scenarios, (c) a minimal rate of return on the consumption of stress-test limits.

▶ Jumps increase the volatility in the model but also generate a smile. At the lowest non-trivial order in jump size – order 3 – the ATMF skew is given by:

$$\mathcal{S}_T = \frac{\lambda \overline{J^3}}{6\hat{\sigma}_T^3 T}$$

thus decays as $\frac{1}{T}$.

▶ Jump-diffusion and Lévy models are natural tools for embedding in the price of a derivative – and releasing it as time elapses – a reserve policy reflecting the average cost of specific stress-test scenarios.

▶ They are ill-suited to the modeling of the one-day smile as (a) their scale is fixed, (b) they cannot be easily correlated with other processes.

Chapter 11

Multi-asset stochastic volatility

Multi-asset stochastic volatility models are not mere juxtapositions of single-asset models. Obviously we need to specify an asset/asset correlation matrix, just as in a multi-asset Black-Scholes model.

In addition to the individual spot/volatility and volatility/volatility covariance functions whose role we have examined in Chapter 8, multi-asset stochastic volatility models involve cross spot/volatility and volatility/volatility correlations. What is the effect of these new parameters?

We propose a parametrization of the global correlation matrix based on observable physical quantities, study the ATMF skew of a basket and the correlation swap.

In the course of our discussion we use the example of the multi-asset local volatility model, the most popular multi-asset stochastic volatility model.

11.1 The short ATMF basket skew

This section focuses on the short-maturity ATMF basket skew. Consider a basket B consisting of n assets, with weights α_i:

$$B = \sum_i \alpha_i S_i$$

and let us define relative weights $w_i = \frac{\alpha_i S_i}{B}$: $\Sigma_i w_i = 1$. The variation of B is:

$$\frac{dB}{B} = \sum_i w_i \frac{dS_i}{S_i} \tag{11.1}$$

While the α_i are fixed, the w_i are not; as we will be studying the short-maturity case we will make the approximation that the w_i are constant. From the results of Section 8.5, the short ATMF implied volatility $\hat{\sigma}_B$ and skew $\mathcal{S}_B$ are given by – see formula (8.36), page 322:

$$\hat{\sigma}_B = \sigma_B \tag{11.2a}$$

$$\mathcal{S}_B = \frac{1}{2\hat{\sigma}_B^2} \frac{\langle d\ln B \, d\hat{\sigma}_B \rangle}{dt} \tag{11.2b}$$

where σ_B is the instantaneous volatility of the basket which, from (11.1) is given by:

$$\sigma_B^2 = \sum_{ij} w_i w_j \rho_{ij} \sigma_i \sigma_j$$

σ_i is the instantaneous volatility of asset i – equal to the short ATMF volatility $\hat{\sigma}_i$ – and ρ_{ij} is the correlation of S_i and S_j. The relationship between the short ATMF basket volatility and the short ATMF volatilities of the constituents reads:

$$\hat{\sigma}_B = \sqrt{\sum_{ij} w_i w_j \rho_{ij} \sigma_i \sigma_j} = \sqrt{\sum_{ij} w_i w_j \rho_{ij} \hat{\sigma}_i \hat{\sigma}_j} \tag{11.3}$$

$d\hat{\sigma}_B$ is given by:

$$d\hat{\sigma}_B = \frac{1}{\hat{\sigma}_B} \sum_{ij} w_i w_j \rho_{ij} \hat{\sigma}_i d\hat{\sigma}_j$$

and we get:

$$\frac{1}{2\hat{\sigma}_B^2} \frac{\langle d \ln B \, d\hat{\sigma}_B \rangle}{dt} = \frac{1}{2\hat{\sigma}_B^3} \sum_{ijk} w_i w_j w_k \rho_{ij} \hat{\sigma}_i \frac{\langle d \ln S_k \, d\hat{\sigma}_j \rangle}{dt}$$

$$= \frac{1}{2\hat{\sigma}_B^3} \sum_{ij} w_i w_j^2 \rho_{ij} \hat{\sigma}_i \frac{\langle d \ln S_j \, d\hat{\sigma}_j \rangle}{dt} + \frac{1}{2\hat{\sigma}_B^3} \sum_{i \, j \neq k} w_i w_j w_k \rho_{ij} \hat{\sigma}_i \frac{\langle d \ln S_k \, d\hat{\sigma}_j \rangle}{dt}$$

The basket skew is thus given by:

$$\mathcal{S}_B = \frac{1}{\hat{\sigma}_B^3} \sum_{ij} w_i w_j^2 \rho_{ij} \hat{\sigma}_i \hat{\sigma}_j^2 \, \mathcal{S}_j + \frac{1}{2\hat{\sigma}_B^3} \sum_{i \, j \neq k} w_i w_j w_k \rho_{ij} \hat{\sigma}_i \frac{\langle d \ln S_k \, d\hat{\sigma}_j \rangle}{dt} \tag{11.4}$$

where we have separated terms involving the covariance on an asset with its own volatility and have used (11.2b) to relate $\langle d \ln S_j \, d\hat{\sigma}_j \rangle$ to $\mathcal{S}_j$, the ATMF skew of basket component S_j. Expression (11.4) shows that the basket ATMF skew is generated partly by the ATMF skew of the basket components and partly by the covariance of each component with other components' ATMF volatilities.

11.1.1 The case of a large homogeneous basket

Assume that all volatilities and correlations are equal and that the n components of the basket are equally weighted: $\hat{\sigma}_i \equiv \hat{\sigma}$, $\rho_{ij} \equiv \rho_{SS}$, $S_i = S$, $w_i = \frac{1}{n}$. Expressions (11.3) and (11.4) for the basket ATMF volatility and skew simplify to:

$$\hat{\sigma}_B = \sqrt{\frac{1 + (n-1)\rho_{SS}}{n}} \, \hat{\sigma} \tag{11.5a}$$

$$\mathcal{S}_B = \frac{1 + (n-1)\rho_{SS}}{n} \frac{\hat{\sigma}^3}{\hat{\sigma}_B^3} \left[\frac{\mathcal{S}}{n} + \frac{n-1}{n} \frac{1}{2\hat{\sigma}^2} \frac{\langle d \ln S \, d\hat{\sigma} \rangle_{\text{cross}}}{dt} \right] \tag{11.5b}$$

where $\langle d\ln S\, d\widehat{\sigma}\rangle_{\text{cross}}$ denotes the instantaneous covariation of a basket component with another component's ATMF volatility. Observe that the portion contributed to $\mathcal{S}_B$ by each component's ATMF skew scales like $\frac{1}{n}$, hence goes to zero for a large basket.

For a large basket ($n \gg 1$) – which is the case of most equity indexes – we get the following simpler formulas:

$$\widehat{\sigma}_B \simeq \sqrt{\rho_{SS}}\,\widehat{\sigma} \tag{11.6a}$$

$$\mathcal{S}_B \simeq \frac{1}{\sqrt{\rho_{SS}}}\frac{1}{2\widehat{\sigma}^2}\frac{\langle d\ln S\, d\widehat{\sigma}\rangle_{\text{cross}}}{dt} \tag{11.6b}$$

Equation (11.6b) makes it clear that the short ATMF skew of a large basket only depends on cross-asset spot-volatility covariances. Consider the following opposite cases:

- $\langle d\ln S\, d\widehat{\sigma}\rangle_{\text{cross}} = 0$: a component's volatility is uncorrelated with other components. In this case $\mathcal{S}_B = 0$.

- $\langle d\ln S\, d\widehat{\sigma}\rangle_{\text{cross}} = \langle d\ln S\, d\widehat{\sigma}\rangle_{\text{diag}}$ where $\langle d\ln S\, d\widehat{\sigma}\rangle_{\text{diag}}$ is the instantaneous covariation of a basket component with its own implied volatility. This situation is obtained for example by assuming that volatilities of all assets are driven by a single Brownian motion: the $\widehat{\sigma}_i$ are 100% correlated. Using:

$$\mathcal{S} = \frac{1}{2\widehat{\sigma}^2}\frac{\langle d\ln S\, d\widehat{\sigma}\rangle_{\text{diag}}}{dt} \tag{11.7}$$

we get:

$$\mathcal{S}_B \simeq \frac{1}{\sqrt{\rho_{SS}}}\mathcal{S} \tag{11.8}$$

The basket skew is larger than the component's skew.

Before turning to the real case, we examine the prediction of the local volatility model.

11.1.2 The local volatility model

A multi-asset local volatility model is a peculiar stochastic volatility model in that it only takes as inputs asset-asset correlations. Implied volatilities are *functions* of spot and time: $\widehat{\sigma} = \widehat{\sigma}(t, S)$, thus are driven by the same Brownian motion as the spot process. For a homogeneous basket, such that the local volatility functions of all components are identical, this entails that:

$$\langle d\ln S\, d\widehat{\sigma}\rangle_{\text{cross}} = \rho_{SS}\langle d\ln S\, d\widehat{\sigma}\rangle_{\text{diag}}$$

Using again (11.7), expressions (11.5) become:

$$\widehat{\sigma}_B = \sqrt{\frac{1 + (n-1)\rho_{SS}}{n}}\,\widehat{\sigma} \simeq \sqrt{\rho_{SS}}\,\widehat{\sigma} \tag{11.9a}$$

$$\mathcal{S}_B = \sqrt{\frac{1 + (n-1)\rho_{SS}}{n}}\,\mathcal{S} \simeq \sqrt{\rho_{SS}}\,\mathcal{S} \tag{11.9b}$$

One consequence of (11.9) is that, in the local volatility model, for short maturities, the ratios of ATMF basket skew to ATMF component skew and of ATMF basket volatility to ATMF component volatility are identical and equal to $\sqrt{\rho_{SS}}$.

$$\frac{\mathcal{S}_B}{\mathcal{S}} \simeq \frac{\widehat{\sigma}_B}{\widehat{\sigma}} \simeq \sqrt{\rho_{SS}} \qquad (11.10)$$

Figure 11.1 provides an illustration of how well relationship (11.10) is actually obeyed, for a range of maturities. We have taken $n = 10$, $\rho_{SS} = 50\%$, and have used for the 10 components the same smile.[1]

Observe how $\frac{\widehat{\sigma}_B}{\widehat{\sigma}}$ stays remarkably close to its theoretical value $\sqrt{\frac{1+(n-1)\rho_{SS}}{n}} = 74.2\%$, even for long maturities. $\frac{\mathcal{S}_B}{\mathcal{S}}$, on the other hand, is in fact lower than its short-maturity value (11.9b).

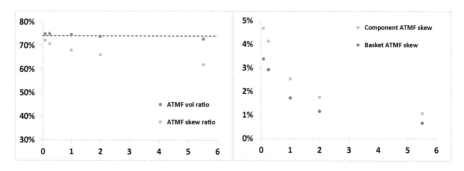

Figure 11.1: Left: ratios $\frac{\widehat{\sigma}_B}{\widehat{\sigma}}$ and $\frac{\mathcal{S}_B}{\mathcal{S}}$ as a function of maturity (years) in the local volatility model for an equally weighted basket of 10 assets with the same smile. The dashed line indicates the theoretical short-maturity value $\sqrt{\frac{1+(n-1)\rho_{SS}}{n}}$. Right: the ATMF skew expressed as the difference of the implied volatilities of the 95% and 105% strikes.

11.1.3   The basket skew in reality

Typically, implied correlations levels for equity indexes – derived from (11.6a) – are $\rho_{SS} \simeq 60\%$: ATMF index implied volatilities are about 25% lower than stock implied volatilities. Market ATMF skews of indexes are stronger than stock skews. They are typically about 25% larger – this is in fact roughly what formula (11.8) for the case of 100% correlated volatilities yields.

This stands in stark contrast with the prediction of the local volatility model: the short-maturity approximate formula (11.10) implies that the index skew should be about 20% *lower* than the stock skew; values of $\frac{\mathcal{S}_B}{\mathcal{S}}$ in Figure 11.1 – obtained with $\rho_{SS} = 50\%$ – show that the actual ratio in the local volatility model is lower still.

---

[1]We have used for the components the smile of the Euro Stoxx 50 index of June 12th, 2012, which is steeper than typical stock skews – using weaker skews wouldn't alter our conclusions.

11.1.4 Digression – how many stocks are there in an index?

Formulas (11.5a) and (11.5b) for $\hat{\sigma}_B$ and S_B apply to the case of an equally weighted basket; the large-basket regime is reached when $n \gg 1$. Equity indexes are not equally weighted baskets. Given a particular index, which value of n should be used in (11.5a) and (11.5b) for assessing whether the asymptotic regime is applicable?

Let us assume that the volatilities of the index components are all equal to σ.[2] σ_B^2 is given by:

$$\sigma_B^2 = \sigma^2 \sum_{ij} w_i w_j \rho_{ij} = \sigma^2 \Big(\sum_i w_i^2 + \rho_{SS} \sum_{i \neq j} w_i w_j \Big) \qquad (11.11)$$

Comparing this formula with (11.5a), we see that the effective number of components of an index, $n^\star$, should be defined as:[3]

$$n^\star = \frac{1}{\Sigma_i w_i^2} \qquad (11.12)$$

Table 11.1 lists the value of n and $n^\star$ for a few equity indexes; $n^\star$ can be sizeably smaller than n.

	S&P 500	Euro Stoxx 50	NIKKEI	KOSPI	FTSE	SMI	CAC	Russell 2000
n	500	50	225	200	101	20	40	2000
$n^\star$	143	37	47	17	36	8	20	1056

Table 11.1: Actual (n) and effective ($n^\star$) number of components of some equity indexes. In (11.12) values of w_i as of August 6, 2013 have been used.

11.2 Parametrizing multi-asset stochastic volatility models

Given a set of single-asset models, we need to define (a) correlations for the spot processes, (b) a parametrization of cross spot/volatility and volatility/volatility correlations.

Rather than directly parametrizing cross factor/factor and spot/factor correlations, whose significance depends on the specifics of the factor structure, we aim for a parsimonious parametrization of "physical" quantities.

Such parametrization can be used regardless of the particular factor structure of the model at hand.

---

[2]Or less crudely that there is no systematic correlation between w_i and σ_i.
[3]The same value of $n^\star$ can be extracted from the second component of σ_B^2 since $\Sigma_i w_i = 1$. Interestingly, $n^\star$ is used by economists as a measure of market concentration under the name of Herfindahl-Hirschman Index.

We know from Chapter 8 that, at order two in volatilities of volatilities, the smile is determined by the spot/variance and variance/variance covariance functions $\mu(t,u,\xi) = \frac{\langle d\ln S_t\, d\xi_t^u\rangle}{dt}$ and $\nu(t,u,u',\xi) = \frac{\langle d\xi_t^u\, d\xi_t^{u'}\rangle}{dt}$. We then focus on these objects and define cross spot/factor and factor/factor correlations so that cross-covariance functions are related simply to their diagonal counterparts.

11.2.1   A homogeneous basket

Consider the case a basket of n assets, each driven by the two-factor model of Section 7.4 with identical parameters $\nu, \theta, k_1, k_2, \rho_{X_1X_2}$ (previously noted ρ_{12}), ρ_{SX_1}, ρ_{SX_2}. For the sake of readability, we decide to carry the factor indices as subscripts rather than superscripts, in the present section.

The instantaneous spot/variance and variance/variance covariance functions of an asset S_t and its forward variances ξ_t^u are given by expressions (8.50) and (8.51), page 327:

$$\mu^{\text{diag}}(t,u,\xi) = 2\nu\alpha_\theta\sqrt{\xi^t}\xi^u\left((1-\theta)\rho_{SX_1}^{\text{diag}}e^{-k_1(u-t)} + \theta\rho_{SX_2}^{\text{diag}}e^{-k_2(u-t)}\right)$$
(11.13)

$$\nu^{\text{diag}}(t,u,u',\xi) = 4\nu^2\xi^u\xi^{u'}\alpha_\theta^2\left[(1-\theta)^2\rho_{X_1X_1}^{\text{diag}}e^{-k_1(u+u'-2t)}\right.$$
(11.14)
$$+ \theta^2\rho_{X_2X_2}^{\text{diag}}e^{-k_2(u+u'-2t)}$$
$$\left.+ \rho_{X_1X_2}^{\text{diag}}\theta(1-\theta)\left(e^{-k_1(u-t)}e^{-k_2(u'-t)} + e^{-k_2(u-t)}e^{-k_1(u'-t)}\right)\right]$$

where the *diag* subscript indicates that these covariance functions apply to spot and variances of the same asset. For the diagonal case, we have:

$$\rho_{SX_1}^{\text{diag}} = \rho_{SX_1}, \quad \rho_{SX_2}^{\text{diag}} = \rho_{SX_2}, \quad \rho_{X_1X_1}^{\text{diag}} = 1, \quad \rho_{X_2X_2}^{\text{diag}} = 1, \quad \rho_{X_1X_2}^{\text{diag}} = \rho_{X_1X_2}$$

Upon replacing *diag* with *cross* superscripts, $\mu_{\text{cross}}, \nu_{\text{cross}}$ are given by expressions (11.13) and (11.14) as well, except the *diag* superscript is replaced with *cross*.

Correlations $\rho_{X_1X_2}, \rho_{SX_1}, \rho_{SX_2}$ are set. We need to define $\rho_{SX_1}^{\text{cross}}, \rho_{SX_2}^{\text{cross}}, \rho_{X_1X_1}^{\text{cross}}, \rho_{X_2X_2}^{\text{cross}}, \rho_{X_1X_2}^{\text{cross}}$ as well as the asset-asset correlation ρ_{SS}.

Our aim is to parametrize the multi-asset model so that μ^{cross} and ν^{cross} are defined in terms of their diagonal counterparts. Specifically, we require that the term structure of μ^{cross} and ν^{cross} – that is their dependency on $u-t$ and $u'-t$ – match that of μ^{diag} and ν^{diag}, once corrected for the effect of the term structure of forward variances. Note that prefactors $\sqrt{\xi^t}\xi^u$ and $\xi^u\xi^{u'}$ in (11.13) and (11.14) would no longer appear if we worked with instantaneous spot/variance and variance/variance *correlations* rather than *covariances*.

We introduce parameter $\chi_{S\sigma}$ and define cross-correlations $\rho_{SX_1}^{\text{cross}}, \rho_{SX_2}^{\text{cross}}$ through:

$$\begin{cases} \rho_{SX_1}^{\text{cross}} = \chi_{S\sigma}\,\rho_{SX_1}^{\text{diag}} \\ \rho_{SX_2}^{\text{cross}} = \chi_{S\sigma}\,\rho_{SX_2}^{\text{diag}} \end{cases}$$

$\chi_{S\sigma}$ simply measures how much larger or smaller cross spot/volatility correlations are with respect to their diagonal counterparts. Using (11.13) we can check that, for a flat term structure of VS volatilities: $\mu^{\text{cross}}(t,u,\xi) = \chi_{S\sigma}\,\mu^{\text{diag}}(t,u,\xi)$ or, equivalently:

$$\rho^{\text{cross}}(S,\widehat{\sigma}_T) = \chi_{S\sigma}\,\rho^{\text{diag}}(S,\widehat{\sigma}_T)$$

where $\rho^{\text{diag}}(S,\widehat{\sigma}_T)$ is the instantaneous correlation between S and its implied VS volatility of maturity T and $\rho^{\text{cross}}(S,\widehat{\sigma}_T)$ is its cross-counterpart.

Turning now to ν, we similarly introduce coefficient $\chi_{\sigma\sigma}$ and define $\rho^{\text{cross}}_{X_1X_1}, \rho^{\text{cross}}_{X_2X_2}, \rho^{\text{cross}}_{X_1X_2}$ as:

$$\begin{cases} \rho^{\text{cross}}_{X_1X_1} &= \chi_{\sigma\sigma}\,\rho^{\text{diag}}_{X_1X_1} \\ \rho^{\text{cross}}_{X_2X_2} &= \chi_{\sigma\sigma}\,\rho^{\text{diag}}_{X_2X_2} \\ \rho^{\text{cross}}_{X_1X_2} &= \chi_{\sigma\sigma}\,\rho^{\text{diag}}_{X_1X_2} \end{cases}$$

Thus,

$$\rho^{\text{cross}}_{X_1X_1} = \chi_{\sigma\sigma}, \quad \rho^{\text{cross}}_{X_2X_2} = \chi_{\sigma\sigma}, \quad \rho^{\text{cross}}_{X_1X_2} = \chi_{\sigma\sigma}\,\rho_{X_1X_2}$$

One can check on (11.14) that with this parametrization, the instantaneous correlation between ξ_i^u and $\xi_j^{u'}$ where i,j denote different assets is equal to $\chi_{\sigma\sigma}$ times the correlation of ξ_i^u and $\xi_i^{u'}$. For a flat term structure of VS volatilities $\nu^{\text{cross}}(t,u,u',\xi) = \chi_{\sigma\sigma}\nu^{\text{diag}}(t,u,u',\xi)$ or, equivalently:

$$\rho^{\text{cross}}(\widehat{\sigma}_T,\widehat{\sigma}_{T'}) = \chi_{\sigma\sigma}\,\rho^{\text{diag}}(\widehat{\sigma}_T,\widehat{\sigma}_{T'})$$

where $\rho^{\text{diag}}(\widehat{\sigma}_T,\widehat{\sigma}_{T'})$ is the instantaneous correlation between VS volatilities of maturities T,T' of the same underlying and $\rho^{\text{cross}}(\widehat{\sigma}_T,\widehat{\sigma}_{T'})$ is a similarly-defined quantity for correlations of VS volatilities of different underlyings.

Finally we introduce the (uniform) correlation between spot processes, ρ_{SS}. Each asset is associated to 3 Brownian motions: W_i^S, W_i^1, W_i^2. The global correlation matrix is thus of dimension $3n \times 3n$. What are the conditions on $\rho_{SS}, \chi_{\sigma\sigma}, \chi_{S\sigma}$ so that it is positive?

Conditions on $\rho_{SS}, \chi_{\sigma\sigma}, \chi_{S\sigma}$

Let us compute the eigenvalues of matrix Ω, defined by: $\Omega = \frac{1}{dt}\langle dU\,dU^{\intercal}\rangle$ where $U^{\intercal} = (W_1^S \cdots W_n^S,\ W_1^1 \cdots W_n^1,\ W_1^2 \cdots W_n^2)$.

Let λ be an eigenvalue of Ω, associated to the eigenvector T, whose components we write $T^{\intercal} = (s_1 \cdots s_n,\ x_1 \cdots x_n,\ y_1 \cdots y_n)$. Expressing that $\Omega T = \lambda T$ we get, for $1 \le i \le n$:

$$\begin{cases} \left(s_i + \rho_{SS}\overline{\Sigma}_j s_j\right) + \rho_{SX_1}\left(x_i + \chi_{S\sigma}\overline{\Sigma}_j x_j\right) + \rho_{SX_2}\left(y_i + \chi_{S\sigma}\overline{\Sigma}_j y_j\right) &= \lambda s_i \\ \rho_{SX_1}\left(s_i + \chi_{S\sigma}\overline{\Sigma}_j s_j\right) + \left(x_i + \chi_{\sigma\sigma}\overline{\Sigma}_j x_j\right) + \rho_{X_1X_2}\left(y_i + \chi_{\sigma\sigma}\overline{\Sigma}_j y_j\right) &= \lambda x_i \\ \rho_{SX_2}\left(s_i + \chi_{S\sigma}\overline{\Sigma}_j s_j\right) + \rho_{X_1X_2}\left(x_i + \chi_{\sigma\sigma}\overline{\Sigma}_j x_j\right) + \left(y_i + \chi_{\sigma\sigma}\overline{\Sigma}_j y_j\right) &= \lambda y_i \end{cases}$$

where $\overline{\Sigma}_j$ is a shorthand notation for $\Sigma_{j\neq i}$ and we have removed the *diag* superscripts in $\rho_{X_1X_2}^{\text{diag}}$, $\rho_{SX_1}^{\text{diag}}$, $\rho_{SX_2}^{\text{diag}}$ for notational economy. The correlation structure is invariant under permutation of the i,j indices and in particular under translation: $\rho(W_i^\circ, W_j^\bullet) = \rho(W_{i+k\,\text{mod}\,n}^\circ, W_{j+k\,\text{mod}\,n}^\bullet)\ \forall k$ where $\circ, \bullet$ stand for $S, 1, 2$.

We thus diagonalize Ω in the basis of eigenvectors of the discrete translation operator. T is parametrized with the 4 numbers θ_k, s, x, y:

$$T^{\mathsf{T}} = (se^{i\theta_k}\cdots se^{in\theta_k}, xe^{i\theta_k}\cdots xe^{in\theta_k}, ye^{i\theta_k}\cdots ye^{in\theta_k})$$

with $\theta_k = \frac{2k\pi}{n}$ where $k = 0\cdots n-1$ and $i = \sqrt{-1}$. We have:

$$\sum_{j=0,j\neq i}^{n-1} e^{ij\theta_k} = \sum_{j=i+1}^{n-1+i} e^{ij\theta_k} = e^{ii\theta_k}\sum_{j=1}^{n-1} e^{ij\theta_k} = \begin{cases} (n-1)e^{ii\theta_k} & k=0 \\ -e^{ii\theta_k} & k\neq 0 \end{cases}$$

We thus define $A_{SS}(k)$ by:

$$A_{SS}(k) = \left(1 + \rho_{SS}\Sigma_{j=1}^{j=n-1}e^{ij\theta_k}\right) = \begin{cases} 1+(n-1)\rho_{SS} & k=0 \\ 1-\rho_{SS} & k\neq 0 \end{cases} \tag{11.15}$$

$A_{\sigma\sigma}(k)$, $A_{S\sigma}(k)$ are defined similarly in terms of $\chi_{\sigma\sigma}, \chi_{S\sigma}$. We have:

$$\begin{cases} A_{\sigma\sigma}(k=0) = 1+(n-1)\chi_{\sigma\sigma} \\ A_{\sigma\sigma}(k\neq 0) = 1-\chi_{\sigma\sigma} \end{cases} \text{ and } \begin{cases} A_{S\sigma}(k=0) = 1+(n-1)\chi_{S\sigma} \\ A_{S\sigma}(k\neq 0) = 1-\chi_{S\sigma} \end{cases}$$

The resulting system for s, x, y is:

$$\begin{cases} A_{SS}s + \rho_{SX_1}A_{S\sigma}x + \rho_{SX_2}A_{S\sigma}y = \lambda s \\ \rho_{SX_1}A_{S\sigma}s + A_{\sigma\sigma}x + \rho_{X_1X_2}A_{\sigma\sigma}y = \lambda x \\ \rho_{SX_2}A_{S\sigma}s + \rho_{X_1X_2}A_{\sigma\sigma}x + A_{\sigma\sigma}y = \lambda y \end{cases}$$

The condition is thus that the following symmetric matrix

$$\omega(k) = \begin{pmatrix} A_{SS} & \rho_{SX_1}A_{S\sigma} & \rho_{SX_2}A_{S\sigma} \\ \rho_{SX_1}A_{S\sigma} & A_{\sigma\sigma} & \rho_{X_1X_2}A_{\sigma\sigma} \\ \rho_{SX_2}A_{S\sigma} & \rho_{X_1X_2}A_{\sigma\sigma} & A_{\sigma\sigma} \end{pmatrix}$$

be positive. This implies in particular that $A_{SS}(k) \geq 0$, $A_{\sigma\sigma}(k) \geq 0\ \forall k$, which places the following bounds on $\rho_{SS}, \chi_{\sigma\sigma}$:

$$-\frac{1}{n-1} \leq \rho_{SS}, \chi_{\sigma\sigma} \leq 1$$

These conditions, in the equity context, are not very restrictive.

We now define a "correlation matrix" $\rho(k)$ obtained by rescaling the symmetric matrix $\omega(k)$: $\rho_{ij}(k) = \frac{\omega_{ij}(k)}{\sqrt{\omega_{ii}(k)\omega_{jj}(k)}}$:

$$\rho(k) = \begin{pmatrix} 1 & \zeta\rho_{SX_1} & \zeta\rho_{SX_2} \\ \zeta\rho_{SX_1} & 1 & \rho_{X_1X_2} \\ \zeta\rho_{SX_2} & \rho_{X_1X_2} & 1 \end{pmatrix} \tag{11.16}$$

where $\zeta(k)$ is given by:

$$\zeta(k) = \frac{A_{S\sigma}(k)}{\sqrt{A_{SS}(k)A_{\sigma\sigma}(k)}} = \begin{cases} \dfrac{1+(n-1)\chi_{S\sigma}}{\sqrt{(1+(n-1)\rho_{SS})(1+(n-1)\chi_{\sigma\sigma})}} & k=0 \\[3mm] \dfrac{1-\chi_{S\sigma}}{\sqrt{(1-\rho_{SS})(1-\chi_{\sigma\sigma})}} & k\neq 0 \end{cases}$$

$\rho(k)$ in (11.16) is in fact the correlation matrix of the single-asset case, but with spot/volatility correlations rescaled by the factor $\zeta(k)$, which may be smaller or larger than 1 depending on the values of $\rho_{SS}, \chi_{\sigma\sigma}, \chi_{S\sigma}$.

$\chi_{S\sigma}$ is allowed to be larger than 1, unlike ρ_{SS} and $\chi_{\sigma\sigma}$. This will be needed in situations when the index skew is much steeper than the component skew – see the discussion in Section 11.1.1.

The case $\zeta(k) \equiv 1$ is obtained by taking $\chi_{S\sigma} = \chi_{\sigma\sigma} = \rho_{SS}$. $\rho(k)$ is then equal to the single-asset correlation matrix thus the positivity of the $3n \times 3n$ global covariance matrix is ensured. In this model, cross-correlations (spot/spot, spot/volatility, volatility/volatility) are all equal to ρ_{SS} times their diagonal counterparts. This is the case of the multi-asset local volatility model.[4]

Conclusion

Given correlations $\rho_{SX_1}, \rho_{SX_1}, \rho_{X_1X_2}$ for the single-asset case, we parametrize the $3n \times 3n$ global correlation matrix of the multi-asset case by introducing three additional parameters: $\rho_{SS}, \chi_{\sigma\sigma}, \chi_{S\sigma}$.

ρ_{SS} is the spot/spot correlation. $\chi_{\sigma\sigma}, \chi_{S\sigma}$ define cross spot/volatility and volatility/volatility covariance functions in terms of their diagonal counterparts.

ρ_{SS} and $\chi_{\sigma\sigma}$ are restricted to the interval $[-\frac{1}{n-1}, 1]$. The necessary and sufficient condition on $\rho_{SS}, \chi_{\sigma\sigma}, \chi_{S\sigma}$ for the global correlation matrix to be positive is that the following effective correlation matrix for the single-asset case:

$$\begin{pmatrix} 1 & \zeta\rho_{SX_1} & \zeta\rho_{SX_2} \\ \zeta\rho_{SX_1} & 1 & \rho_{X_1X_2} \\ \zeta\rho_{SX_2} & \rho_{X_1X_2} & 1 \end{pmatrix}$$

be positive for the two following values of ζ:

$$\zeta = \frac{1+(n-1)\chi_{S\sigma}}{\sqrt{(1+(n-1)\rho_{SS})(1+(n-1)\chi_{\sigma\sigma})}} \tag{11.17a}$$

$$\zeta = \frac{1-\chi_{S\sigma}}{\sqrt{(1-\rho_{SS})(1-\chi_{\sigma\sigma})}} \tag{11.17b}$$

in addition to the $\zeta = 1$ case. This is the case if $|\zeta\rho_{SX_1}| \leq 1$, $|\zeta\rho_{SX_2}| \leq 1$ and:

$$-1 \leq \frac{\rho_{X_1X_2} - \zeta^2\rho_{SX_1}\rho_{SX_2}}{\sqrt{1-\zeta^2\rho_{SX_1}^2}\sqrt{1-\zeta^2\rho_{SX_2}^2}} \leq 1$$

---

[4]The local volatility model's assumption of setting volatility/volatility correlations identical to spot/spot correlations is not very realistic. Typically, correlations of implied volatilities of underlyings belonging to the same index are usually larger than the correlations of the underlyings themselves.

It is easy to check, going backwards through the derivation in Section 11.2.1, that these necessary conditions are also sufficient: we have effectively built the $3n$ eigenvectors of Ω.[5]

While in what follows we will be using a homogeneous model, in practice, parameters $\nu, \theta, k_1, k_2, \rho_{X_1X_2}, \rho_{SX_1}, \rho_{SX_2}$ are different for different components. In particular, even though we typically use identical values for $\nu, \theta, k_1, k_2, \rho_{X_1X_2}$ so that volatilities of volatilities are identical, we may use different values for $\rho_{SX_1}^{\text{diag}}, \rho_{SX_2}^{\text{diag}}$. Consider two components S_i, S_j. Should we set $\rho_{S^iX_1^j}^{\text{cross}} = \chi_{S\sigma}\rho_{S^jX_1^j}^{\text{diag}}$ or $\rho_{S^iX_1^j}^{\text{cross}} = \chi_{S\sigma}\rho_{S^iX_1^i}^{\text{diag}}$?

When setting the cross spot/volatility correlations of S_i and $\widehat{\sigma}_j^T$, it is more reasonable to derive the T-dependence of $\rho(S_i, \widehat{\sigma}_j^T)$ from the term-structure of $\rho(S_j, \widehat{\sigma}_j^T)$ rather than that of $\rho(S_i, \widehat{\sigma}_i^T)$. We would thus set:

$$\rho_{S^iX_1^j}^{\text{cross}} = \chi_{S\sigma}\,\rho_{S^jX_1^j}^{\text{diag}}$$

and likewise for $\rho_{S^iX_2^j}^{\text{cross}}$. Going back to implied volatilities, this is equivalent to setting:

$$\rho(S^i, \widehat{\sigma}_T^j) = \chi_{S\sigma}\,\rho(S^j, \widehat{\sigma}_T^j)$$

Likewise, with respect to volatility/volatility correlations, we could set

$$\text{either } \rho(\widehat{\sigma}_T^i, \widehat{\sigma}_{T'}^j) = \chi_{\sigma\sigma}\,\rho(\widehat{\sigma}_T^i, \widehat{\sigma}_{T'}^i)$$
$$\text{or } \rho(\widehat{\sigma}_T^i, \widehat{\sigma}_{T'}^j) = \chi_{\sigma\sigma}\,\rho(\widehat{\sigma}_T^j, \widehat{\sigma}_{T'}^j)$$

In practice, volatility/volatility correlations are sufficiently similar across underlyings that either choice results in very similar values of $\chi_{\sigma\sigma}$ – see the examples in Table 11.2 below.

The global correlation matrix we thus build for a non-homogeneous basket is not guaranteed to be positive – but is not expected to be very negative either; in case it is we use the algorithm proposed in [61] to generate the closest positive correlation matrix.[6]

11.2.2 Realized values of $\chi_{S\sigma}$ and $\chi_{\sigma\sigma}$

Typical realized values of $\chi_{S\sigma}$ and $\chi_{\sigma\sigma}$ for indexes appear in Table 11.2. For each pair of underlyings (S^1, S^2), two values of $\chi_{S\sigma}$ and $\chi_{\sigma\sigma}$ have been computed, obtained by averaging respectively either $\rho(\widehat{\sigma}_T^1, S^2)/\rho(\widehat{\sigma}_T^1, S^1)$ or

---

[5]Ω has 3 non-degenerate eigenvalues (obtained with $k = 0$) and 3 degenerate eigenvalues (obtained with $k \neq 0$, thus each with $(n-1)$ degeneracy).

[6]The algorithm in [61] generates the correlation matrix that is closest to a candidate input matrix. In the formula for the matrix distance, appropriate weights can be used so as to enforce, for example, that diagonal correlations $\rho_{S^iX_1^i}^{\text{diag}}, \rho_{S^iX_2^i}^{\text{diag}}, \rho_{X_1^iX_2^i}^{\text{diag}}$, as well as spot/spot correlations are least altered.

$\rho(\widehat{\sigma}_T^2, S^1)/\rho(\widehat{\sigma}_T^2, S^2)$ over several maturities. Likewise, for the determination of $\chi_{\sigma\sigma}$, two values are obtained by averaging either $\rho(\widehat{\sigma}_T^1, \widehat{\sigma}_{T'}^2)/\rho(\widehat{\sigma}_T^1, \widehat{\sigma}_{T'}^1)$ or $\rho(\widehat{\sigma}_T^1, \widehat{\sigma}_{T'}^2)/\rho(\widehat{\sigma}_T^2, \widehat{\sigma}_{T'}^2)$, for all (T, T') couples.

Note that it is important to use asynchronous estimators – lest we underestimate correlations; see [12].

	June 2008 - June 2013					June 2003 - June 2008				
	S&P500 Stoxx50	Stoxx50 FTSE	S&P500 NIKKEI	Stoxx50 NIKKEI	NIKKEI KOSPI	S&P500 Stoxx50	Stoxx50 FTSE	S&P500 NIKKEI	Stoxx50 NIKKEI	NIKKEI KOSPI
ρ	83%	89%	55%	60%	67%	71%	86%	45%	59%	66%
$\chi_{S\sigma}$	87%	95%	59%	63%	85%	78%	89%	56%	61%	108%
$\chi_{S\sigma}$	86%	93%	71%	78%	78%	75%	91%	53%	77%	65%
$\chi_{\sigma\sigma}$	84%	92%	54%	60%	72%	72%	85%	31%	40%	50%
$\chi_{\sigma\sigma}$	84%	92%	55%	61%	73%	71%	85%	31%	41%	50%

Table 11.2: Historical spot/spot correlations and values of $\chi_{S\sigma}$ and $\chi_{\sigma\sigma}$ for different pairs of indexes, measured on two 5-year samples. $\chi_{S\sigma}$ and $\chi_{\sigma\sigma}$ are evaluated by averaging pairwise ratios of spot/volatility and volatility/volatility correlations using implied ATMF volatilities with maturities 3 months, 6 months, 1 year, 2 years. The asynchronous estimator in [12] has been used.

Empirically, diagonal volatility/volatility correlations across indexes are very similar, thus the two ways of estimating $\chi_{\sigma\sigma}$ yield very similar value; we can thus use an average of the two estimates.

This is less the case for $\chi_{S\sigma}$. Indeed, historical regimes of spot/volatility correlations for the Nikkei index, for example, can be quite different from those of the S&P 500 and Euro Stoxx 50 indexes. This is also reflected in the different behavior of their SSRs – see Figure 9.10, page 381.

11.3 The ATMF basket skew

After focusing in Section 11.1 on short maturities, we derive now an approximation of the basket ATMF skew at order one in volatility of volatility by using expression (8.22) which involves the spot/variance covariance function evaluated at order one in volatility of volatility. We use the notation $\widehat{\sigma}_T^B$ and $\mathcal{S}_T^B$ for the basket VS volatility and ATMF skew and denote *basket* forward variances by ζ_t^u.

We will assume that the underlying model is time-homogeneous and that the VS term structures of all components are flat and identical.

From equation (8.22), page 314, $\mathcal{S}_T^B$ is given by:

$$\mathcal{S}_T^B = \hat{\sigma}_T^B \frac{C_B^{x\xi}(T)}{2\left((\hat{\sigma}_T^B)^2 T\right)^2} \tag{11.18}$$

where $C_B^{x\xi}(T)$ is the doubly-integrated spot/variance covariance function:

$$C_B^{x\xi}(T) = \int_0^T dt \int_t^T du\, \mu_B\,(t,u) \tag{11.19}$$

$$\mu_B\,(t,u) = \lim_{dt \to 0} \frac{1}{dt} E_t\left[d\ln B_t d\zeta_t^u\right] \tag{11.20}$$

where $B_t = \Sigma_i w_i S_{it}$. $\mu_B\,(t,u)$ is evaluated at order one in volatility of volatility on the initial basket variance curve.

Let us assume that the basket is homogeneous and equally weighted; spot correlations $\rho_{ij}, i \neq j$ are all identical, equal to ρ and we use the correlation parametrization of Section 11.2.1.

We also make the approximation that weights are frozen, equal to their initial values, which we denote by w_{i0}. Basket forward variances ζ_t^T are given by:

$$\zeta_t^u = E_t[\sum_{ij} w_i w_j \rho_{ij} \sqrt{\xi_{iu}^u}\sqrt{\xi_{ju}^u}] \tag{11.21}$$

$$= \sum_{ij} \rho_{ij} w_{i0} w_{j0} E_t\left[\sqrt{\xi_{iu}^u}\sqrt{\xi_{ju}^u}\right] \tag{11.22}$$

Thus

$$d\zeta_t^u = \sum_{ij} w_{i0} w_{j0} \rho_{ij} E_t\left[\frac{\sqrt{\xi_{ju}^u}}{\sqrt{\xi_{iu}^u}} d\xi_{iu}^u\right]$$

At order one in volatility of volatility, it is sufficient to replace the prefactor inside the expectation with its value in the unperturbed state, that is using forward variances values at $t = 0$. At this order:

$$d\zeta_t^u = \sum_{ij} w_{i0} w_{j0} \rho_{ij} \frac{\sqrt{\xi_{j0}^u}}{\sqrt{\xi_{i0}^u}} E_t\left[d\xi_{iu}^u\right] = \sum_{ij} w_{i0} w_{j0} \rho_{ij} \frac{\sqrt{\xi_{j0}^u}}{\sqrt{\xi_{i0}^u}} d\xi_{it}^u$$

Let us now compute $\mu_B\,(t,u)$. We have:

$$\frac{dB_t}{B_t} = \sum_i w_{k0} \frac{dS_{k,t}}{S_{k,t}}$$

We then get the expression of $\mu_B\,(t,u)$ as a function of the diagonal spot/variance covariance function, which we denote compactly by $\mu(u-t) = \frac{1}{dt}\langle d\ln S_{it} d\xi_{it}^u\rangle$, as we have assumed a time-homogeneous model:

$$
\mu_B(t, u) = \frac{1}{dt} \langle d \ln B_t d\zeta_t^u \rangle = \sum_{ijk} w_{i0} w_{j0} w_{k0} \rho_{ij} \frac{\sqrt{\xi_{j0}^u}}{\sqrt{\xi_{i0}^u}} \frac{\langle d \ln S_{kt} d\xi_{it}^u \rangle}{dt}
$$

$$
= \sum_{ij} w_{i0}^2 w_{j0} \rho_{ij} \frac{\sqrt{\xi_{j0}^u}}{\sqrt{\xi_{i0}^u}} \mu(u - t) + \chi_{S\sigma} \sum_{i \neq k, j} w_{i0} w_{j0} w_{k0} \rho_{ij} \frac{\sqrt{\xi_{j0}^u}}{\sqrt{\xi_{i0}^u}} \mu(u - t)
$$

For a homogeneous basket with flat and identical term structures of VS volatilities, this simplifies to:

$$
\mu_B(t, u) = \frac{1 + (n - 1)\chi_{S\sigma}}{n} \frac{1 + (n - 1)\rho_{SS}}{n} \mu(u - t)
$$

Using expression (11.19) for $C_B^{x\xi}(T)$:

$$
C_B^{x\xi}(T) = \left[\frac{1 + (n - 1)\chi_{S\sigma}}{n} \right] \left[\frac{1 + (n - 1)\rho_{SS}}{n} \right] \int_0^T dt \int_0^T du\, \mu(u - t)
$$

Finally, the ATMF skew of a homogeneous basket is given, at order one in volatility of volatility, by:

$$
\mathcal{S}_T^B = \frac{\hat{\sigma}_{0T}^B}{2\left((\hat{\sigma}_{0T}^B)^2 T\right)^2} \left[\frac{1 + (n - 1)\chi_{S\sigma}}{n} \right] \left[\frac{1 + (n - 1)\rho_{SS}}{n} \right] \int_0^T dt \int_t^T du\, \mu(u - t)
$$

$$(11.23)$$

where $\hat{\sigma}_{0T}^B$ is the basket VS volatility at order zero in volatility of volatility, that is with zero volatility of volatility. Assuming that the VS term structures of all components are flat, and equal to $\hat{\sigma}^2$, and that the basket is equally weighted, $\hat{\sigma}_{0T}^B$ is given by:

$$
(\hat{\sigma}_{0T}^B)^2 = \frac{1 + (n - 1)\rho_{SS}}{n} \hat{\sigma}^2
$$

Using this expression of $\hat{\sigma}_{0T}^B$ in (11.23) and remembering that the single-asset skew is given by expression (8.22), page 314, with $C^{x\xi}$ given by (8.13):

$$
\mathcal{S}_T = \frac{\hat{\sigma}}{2(\hat{\sigma}^2 T)^2} \int_0^T dt \int_t^T du\, \mu(u - t)
$$

we get our final expression for the basket skew, as a function of the components' skew:

$$
\mathcal{S}_T^B = \mathcal{S}_T \sqrt{\frac{n}{1 + (n - 1)\rho_{SS}}} \left[\frac{1 + (n - 1)\chi_{S\sigma}}{n} \right] \qquad (11.24)
$$

Thus an implied value for $\chi_{S\sigma}$ can be backed out of the ratio of basket skew to component skew.

Let us check that in the limit $T \to 0$, (11.24) yields back the short-maturity result (11.5b) for $\mathcal{S}_T^B$. Using the fact that $\int_0^T dt \int_t^T du = \frac{T^2}{2}$ and that $\mathcal{S}_{T=0} = \frac{\mu(0)}{4\hat{\sigma}_{T=0}^3}$, expression (11.23) for $\mathcal{S}_T^B$ yields:

$$
\begin{aligned}
\mathcal{S}_{T=0}^B &= \frac{1}{4\left(\hat{\sigma}_{T=0}^B\right)^3} \left[\frac{1+(n-1)\chi_{S\sigma}}{n}\right]\left[\frac{1+(n-1)\rho_{SS}}{n}\right]\mu(0) \\
&= \frac{1}{4\left(\hat{\sigma}_{T=0}^B\right)^3} \left[\frac{1+(n-1)\chi_{S\sigma}}{n}\right]\left[\frac{1+(n-1)\rho_{SS}}{n}\right] 4\hat{\sigma}_{T=0}^3 \mathcal{S}_{T=0} \\
&= \mathcal{S}_{T=0}\left(\frac{\hat{\sigma}_{T=0}}{\hat{\sigma}_{T=0}^B}\right)^3 \left[\frac{1+(n-1)\chi_{S\sigma}}{n}\right]\left[\frac{1+(n-1)\rho_{SS}}{n}\right] \qquad (11.25)
\end{aligned}
$$

where we have omitted the 0 subscript in $\hat{\sigma}_{0T}^B$, as, for $T=0$, $\hat{\sigma}_{0T}^B = \hat{\sigma}_T^B$. (11.25) is identical to (11.5b) for $T=0$.

Again, expression (11.24) for $\mathcal{S}_T^B$ involves the factor $\frac{1+(n-1)\chi_{S\sigma}}{n}$: the portion contributed to the basket skew by the components' skew scales like $\frac{1}{n}$ and the bulk of the basket skew is generated by cross spot/volatility correlations.

11.3.1 Application to the two-factor model

In the two-factor model the dynamics of each component reads:

$$
\begin{cases}
dS_t &= \sqrt{\xi_t^t} S_t dW_t^S \\
d\xi_t^T &= 2\nu\xi_t^T \alpha_\theta \left((1-\theta)e^{-k_1(T-t)}dW_t^1 + \theta e^{-k_2(T-t)}dW_t^2\right)
\end{cases}
$$

where ν is the lognormal volatility of a very short-dated implied VS volatility, and $\alpha_\theta = \left((1-\theta)^2 + \theta^2 + 2\rho_{X_1 X_2}\theta(1-\theta)\right)^{-1/2}$. The cross spot/variance and variance/variance correlations are parametrized using $\chi_{S\sigma}$ and $\chi_{\sigma\sigma}$ defined in Section 11.2.1 and the correlations between the S_i are equal to ρ.

For a flat term structure of VS volatilities equal to $\hat{\sigma}$, the spot/variance covariance function $\mu(\tau)$ is given by:

$$
\mu(\tau) = (2\nu)\hat{\sigma}^3\alpha_\theta \left((1-\theta)\rho_{SX_1}e^{-k_1\tau} + \theta\rho_{SX_2}e^{-k_2\tau}\right) \qquad (11.26)
$$

and the component's ATMF skew at order one in volatility of volatility is given by expression (8.55):

$$
\mathcal{S}_T = \nu\alpha_\theta \left[(1-\theta)\,\rho_{SX^1}\frac{k_1 T - (1-e^{-k_1 T})}{(k_1 T)^2} + \theta\rho_{SX^2}\frac{k_2 T - (1-e^{-k_2 T})}{(k_2 T)^2}\right] \qquad (11.27)
$$

Using (11.24) the basket skew – at order one in volatility of volatility – is given by:

$$
\begin{aligned}
\mathcal{S}_T^B = \nu\alpha_\theta\sqrt{\frac{n}{1+(n-1)\rho_{SS}}}&\left[\frac{1+(n-1)\chi_{S\sigma}}{n}\right] \\
\times &\left[(1-\theta)\,\rho_{SX^1}\frac{k_1 T - (1-e^{-k_1 T})}{(k_1 T)^2} + \theta\rho_{SX^2}\frac{k_2 T - (1-e^{-k_2 T})}{(k_2 T)^2}\right] \qquad (11.28)
\end{aligned}
$$

While, for the sake of calculating $\mathcal{S}_T^B$ we have needed $\hat{\sigma}_T^B$ at order zero in volatility of volatility only, it can be calculated exactly – with the assumption of frozen weights:

$$
\left(\hat{\sigma}_T^B\right)^2 = \frac{1}{T}\int_0^T \sum_{ij} w_{i0}w_{j0}\rho_{ij} E\left[\sqrt{\xi_{it}^t}\sqrt{\xi_{jt}^t}\right] dt
$$

$$
= \frac{1}{T}\int_0^T \sum_{ij} w_{i0}w_{j0}\rho_{ij}\sqrt{\xi_{i0}^t}\sqrt{\xi_{j0}^t}\, h_{ij}(t)dt
$$

with $h_{ij}(\tau)$ given by: $h_{ii}(\tau) = 1$, $h_{i\neq j}(\tau) = h(\tau)$, where $h(\tau)$ reads:

$$
h(\tau) = e^{-\nu^2\alpha_\theta^2(1-\chi_{\sigma\sigma})\left((1-\theta)^2\frac{1-e^{-2k_1\tau}}{2k_1}+\theta^2\frac{1-e^{-2k_2\tau}}{2k_2}+2\theta(1-\theta)\rho_{X_1X_2}\frac{1-e^{-(k_1+k_2)\tau}}{k_1+k_2}\right)}
$$

(11.29)

For flat and identical term-structures of VS volatilities:

$$
\left(\hat{\sigma}_T^B\right)^2 = \hat{\sigma}^2 \frac{1+(n-1)\rho_{SS}\frac{1}{T}\int_0^T h(\tau)\,d\tau}{n}
$$

(11.30)

11.3.2 Numerical examples

We now test the accuracy of formulas (11.24), (11.30) for $\mathcal{S}_T^B$, $\hat{\sigma}_T^B$ with the two-factor model.

To generate stock-like parameters, we start from the index-like parameters in Table 8.2, page 329, and reduce ρ_{SX_1}, ρ_{SX_2} by about 30% so as to reduce the ATMF skew by the same relative amount. The other parameters are left unchanged. The basket component's parameters we obtain appear in Table 11.3.

ν	θ	k_1	k_2	$\rho_{X_1X_2}$	ρ_{SX_1}	ρ_{SX_2}
174%	0.245	5.35	0.28	0%	−53.0%	−33.9%

Table 11.3: Basket component's parameters in the two-factor model.

The component's term-structure of volatilities of volatilities is that corresponding to Set II in Figure 7.2, page 229. The VS term structure of volatilities is flat at 30%. We take $n = 10$ components and will use $\rho_{SS} = 60\%$ for the component's correlations. We use vanishing rate and repo.

Let us first set $\chi_{S\sigma} = 0$. The basket smile – with $\chi_{\sigma\sigma} = 0$ – is shown in Figure 11.2, along with the component's smile for comparison. The basket ATMF skew almost vanishes – a reflection of the fact that the basket ATMF skew is mostly generated by cross spot/volatility correlations.

We now use more reasonable values for $\chi_{S\sigma}$. Figure 11.3 shows basket smiles generated with $(\chi_{S\sigma} = 80\%,\ \chi_{\sigma\sigma} = 80\%)$ and $(\chi_{S\sigma} = 115\%,\ \chi_{\sigma\sigma} = 80\%)$.

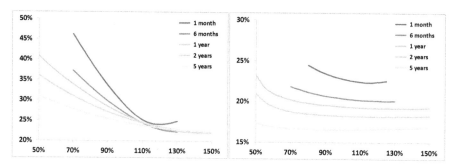

Figure 11.2: Left: component's smile with parameters of Table 11.3. Right: basket smile with $\rho_{SS} = 60\%$, $\chi_{S\sigma} = 0$ and $\chi_{\sigma\sigma} = 0$.

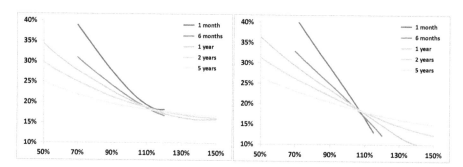

Figure 11.3: Basket smile with: $\rho_{SS} = 60\%$, $\chi_{\sigma\sigma} = 80\%$. Left: $\chi_{S\sigma} = 80\%$. Right: $\chi_{S\sigma} = 115\%$.

Comparing with the left-hand graph in Figure 11.2 it is apparent that the basket ATMF skew with ($\chi_{S\sigma} = 115\%$, $\chi_{\sigma\sigma} = 80\%$) is now steeper than the component's ATMF skew.

How accurate are formulas for (11.24),(11.30) for $\mathcal{S}_T^B$, $\hat{\sigma}_T^B$?

Basket ATMF skew

Figure (11.4) shows the 95/105 ATMF basket skew ($\hat{\sigma}_{K=0.95S_0,T}^B - \hat{\sigma}_{K=1.05S_0,T}^B$), either calculated in a Monte Carlo simulation, or evaluated using (11.24), with $\mathcal{S}_T$ given either by the actual component's ATMF skew, or given by (11.27), for two different values of $\chi_{S\sigma}$. The left-hand graph confirms that expression (11.24) is very accurate. The small overestimation of $\mathcal{S}_T^B$ in the right-hand graph is due to the slight overestimation of the component's skew in (11.27), evidenced in Figure 8.3, page 331.

Figure (11.5) provides another confirmation that the basket ATMF skew is controlled by $\chi_{S\sigma}$. Here we have varied $\chi_{\sigma\sigma}$ while keeping $\rho_{SS} = 60\%$ and $\chi_{S\sigma} = 80\%$ fixed. The ATMF skew hardly changes when $\chi_{\sigma\sigma}$ is varied. For the three values

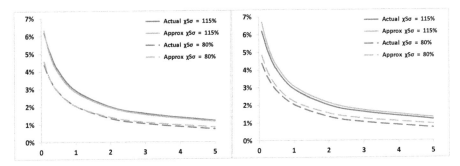

Figure 11.4: Left: basket 95/105 skew in volatility points (Actual) compared to formula (11.24) (Approx) where the actual component's 95/105 skew has been used. Right: basket 95/105 skew (Actual) compared to formula (11.28) (Approx). Maturities are in years. The two values of $\chi_{S\sigma} = 80\%, 115\%$ have been used. All other parameters are kept constant, including $\chi_{\sigma\sigma} = 80\%$.

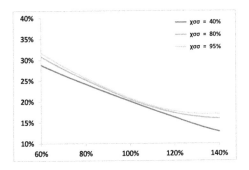

Figure 11.5: One-year basket smiles for different values of $\chi_{\sigma\sigma}$. All other parameters are kept constant: $\rho_{SS} = 60\%$ and $\chi_{S\sigma} = 80\%$.

of $\chi_{\sigma\sigma}$ used: $40\%, 80\%, 95\%$, the 95/105 one-year skew values are respectively: $2.01\%, 2.05\%, 2.06\%$.

Using identical and flat term structures of VS volatilities for the basket components, as well as identical sets of parameters, leads to the particularly simple formulas (11.24) and (11.28) but the derivation of $\mathcal{S}_T^B$ in the general case presents no particular difficulty. One only needs to express $\mu_B(t, u)$ as a function of the component's spot/volatility covariance functions $\mu(t, u, \xi)$, which are given by expression (8.50) in the two-factor model.

Basket VS volatility

$\hat{\sigma}_T^B$, either evaluated in a Monte Carlo simulation or given by (11.30), is graphed in Figure 11.6 for different values of $\chi_{S\sigma}$. In expression (11.30), with the approximation

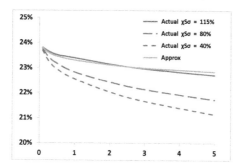

Figure 11.6: $\widehat{\sigma}_T^B$, either evaluated in a Monte Carlo simulation of the two-factor model, for three values of $\chi_{S\sigma}$: 40%, 80%, 115%, or given by expression (11.30). Maturities are in years. The component's parameters are listed in Table 11.3 and $\rho_{SS} = 60\%$, $\chi_{\sigma\sigma} = 80\%$.

of frozen weights, only spot/spot (ρ_{SS}) and volatility/volatility ($\chi_{\sigma\sigma}$) correlations appear.

Figure 11.6 makes it clear that this assumption is not adequate: $\widehat{\sigma}_T^B$ does depend on cross spot/volatility correlations.

11.3.3   Mimicking the local volatility model

Can our two-factor stochastic volatility model mimic a multi-asset local volatility model? In the local volatility model, $\rho^{\text{cross}}(S, \widehat{\sigma}_T) = -\rho_{SS}$ and $\rho^{\text{cross}}(\widehat{\sigma}_T, \widehat{\sigma}_{T'}) = \rho_{SS}$. We have: $\rho^{\text{diag}}(S, \widehat{\sigma}_T) = -1$ and $\rho^{\text{diag}}(\widehat{\sigma}_T, \widehat{\sigma}_{T'}) = 1.$[7] Thus, in the local volatility model:

$$\rho^{\text{cross}}(S, \widehat{\sigma}_T) = \rho_{SS}\, \rho^{\text{diag}}(S, \widehat{\sigma}_T)$$
$$\rho^{\text{cross}}(\widehat{\sigma}_T, \widehat{\sigma}_{T'}) = \rho_{SS}\, \rho^{\text{diag}}(\widehat{\sigma}_T, \widehat{\sigma}_{T'})$$

Thus, with regard to spot/volatility and volatility/volatility correlations, the local volatility model can be viewed as a particular breed of multi-asset stochastic volatility with $\chi_{S\sigma}$, $\chi_{\sigma\sigma}$ given by:

$$\chi_{S\sigma} = \rho_{SS} \qquad\qquad\qquad (11.31a)$$
$$\chi_{\sigma\sigma} = \rho_{SS} \qquad\qquad\qquad (11.31b)$$

With this choice of cross-parametrization the values of ζ in (11.17) are both equal to 1, thus positivity conditions are satisfied.

---

[7]We use here $\rho^{\text{diag}}(S, \widehat{\sigma}_T) = -1$ as we are considering the typical case of negatively sloping equity smiles. We could equivalently have written $\rho^{\text{diag}}(S, \widehat{\sigma}_T) = 1$.

It is easy to check that in case spot/spot correlations are not all equal, the correlation structure of the multi-asset local volatility model is still given by (11.31) with ρ_{SS}, $\chi_{S\sigma}$, $\chi_{\sigma\sigma}$ adorned with ij superscripts, as $\rho^{\text{cross}}(S^i, \widehat{\sigma}_T^j) = -\rho_{SS}^{ij}$ and $\rho^{\text{cross}}(\widehat{\sigma}_T^i, \widehat{\sigma}_{T'}^j) = \rho_{SS}^{ij}$.

We now test this mapping with the same parameter values used in Section 11.3.2; $\rho_{SS} = 60\%$. We use two-factor-model parameters in Table 11.3 to generate a vanilla smile. Using this as the component's vanilla smile, we calibrate the component's local volatility function and price the basket smile using $\rho_{SS} = 60\%$.

We then compare the resulting basket smile with that produced by our multi-asset two-factor model with $\chi_{\sigma\sigma} = \chi_{S\sigma} = \rho_{SS} = 60\%$. Results appear in Figure 11.7.

It is apparent that the mapping (11.31) is accurate for short maturities; for longer maturities implied volatilities in the stochastic volatility model are lower than in the local volatility model. This is not surprising: because volatilities – especially forward volatilities – are more volatile in the stochastic volatility model than in the local volatility model: the level of "effective" spot/spot correlation is lower than ρ_{SS}. This affects all multi-asset products, including the correlation swap which we now study.

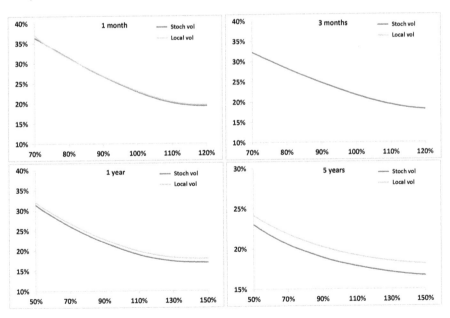

Figure 11.7: Comparison of basket smiles in local volatility and stochastic volatility models, for several maturities. The mapping of local to stochastic volatility is obtained with $\chi_{\sigma\sigma} = \chi_{S\sigma} = \rho_{SS} = 60\%$. The basket consists of $n = 10$ components, with identical smiles given by parameters in Table 11.3.

11.4   The correlation swap

Consider a basket of stocks or indexes. A correlation swap of maturity T pays at T the average pairwise realized correlation of the basket components minus a fixed strike $\widehat{\rho}$

$$\frac{1}{n(n-1)}\sum_{i\neq j}\rho_{ij} - \widehat{\rho} \tag{11.32}$$

where $\widehat{\rho}$ is set so that the swap's initial value vanishes. Correlation ρ_{ij} of S_i,S_j is defined with the standard realized correlation estimator, using daily log-returns:

$$\rho_{ij} = \frac{\sum r_k^i r_k^j}{\sqrt{\sum r_k^{i\,2}}\sqrt{\sum r_k^{j\,2}}} \tag{11.33}$$

where $r_k^i = \ln(S_k^i/S_{k-1}^i)$ and the sums runs from $k=1$ to $k=N$, where N is the number of returns used for estimating covariances and variances in (11.33).

n is the number of securities: 2 or 3 when the components are indexes and up to 50 for a correlation swap on the constituents of the Euro Stoxx 50 index. We have used in (11.32) the typical equal weighting.

Strike $\widehat{\rho}$ is also called the implied correlation of the swap. Indeed, in a constant volatility model, for daily returns, with all spot/spot correlations ρ_{SS} equal and a large number of returns – thus a long maturity – $\widehat{\rho}=\rho_{SS}$.

For shorter maturities, (11.33) is biased. Let us make the assumption of centered log-returns: $r_k^i = \sigma^i\sqrt{\Delta t}Z_k^i$, where σ^i is the (constant) volatility of asset i, Δt the interval between two spot observations – here one day, and the Z_k^i are iid standard normal random variables.

The bias and standard deviation of the correlation estimator are derived in Appendix A, at lowest order in $\frac{1}{N}$:

$$E[\rho_{ij}] = \rho_{SS}\left(1 - \frac{1-\rho_{SS}^2}{2N}\right) \tag{11.34}$$

$$\text{Stdev}(\rho_{ij}) = \frac{1}{\sqrt{N}}\left(1 - \rho_{SS}^2\right) \tag{11.35}$$

where N is the number of returns in the historical sample. Typically the bias in $E[\rho_{ij}]$ is about one point of correlation for $\rho_{SS}=60\%$ and $T=1$ month.

Correlation swaps were introduced as a means of trading correlation and making implied correlation an observable parameter.

As a measure of correlation, however, averaging pairwise correlations results in a poorly defined estimator.

One would expect of an adequately defined estimator that its standard deviation vanishes either in the limit of a large sample size ($N\to\infty$) or as the number of basket constituents is increased ($n \to \infty$). This is not the case for the correlation swap estimator.

Making the same assumption of constant volatility centered log-returns as above, the bias and standard deviation of the correlation swap estimator in the limit $n \to \infty$ are given, at lowest order in $\frac{1}{N}$, by:

$$\lim_{n \to \infty} E\left[\frac{1}{n(n-1)}\sum_{i \neq j}\rho_{ij}\right] = \rho_{SS}\left(1 - \frac{1 - \rho_{SS}^2}{2N}\right)$$

$$\lim_{n \to \infty} \text{Stdev}\left(\frac{1}{n(n-1)}\sum_{i \neq j}\rho_{ij}\right) = \frac{1}{\sqrt{N}}\left(1 - \rho_{SS}^2\right)\frac{\sqrt{2}\,|\rho_{SS}|}{1 + \rho_{SS}} \qquad (11.36)$$

Observe that the right-hand side of (11.36) ($n = \infty$) is only marginally smaller than that of (11.35) ($n = 2$). For the sake of creating an estimator of average realized correlation, it would have been more judicious to average covariances and divide them by the average of variances – the standard deviation of the resulting correlation estimator tends to zero as the number of constituents is increased.

One would hope that the correlation swap is the perfect instrument for calibrating spot/spot correlation levels, but it is not so. Let us assume that our basket is homogeneous and let σ_t^i be the instantaneous (lognormal) volatility of S_t^i. In the limit of short returns, $\widehat{\rho}$ is given by:

$$\widehat{\rho}(T) = E[\rho_{ij}] = \rho_{SS}\, E\left[\frac{\int_0^T \sigma_t^i \sigma_t^j\, dt}{\sqrt{\int_0^T \sigma_t^{i^2} dt}\,\sqrt{\int_0^T \sigma_t^{j^2} dt}}\right] \qquad (11.37)$$

where we explicitly keep track of the maturity dependence of $\widehat{\rho}$.

In case instantaneous volatilities σ_t^i and σ_t^j are collinear – $\sigma_t^j = \lambda \sigma_t^i$ – then $\widehat{\rho}$ is indeed equal to ρ_{SS}. In all other cases, $\widehat{\rho} \leq \rho_{SS}$, and the amount by which $\widehat{\rho}$ differs from ρ_{SS} depends on volatilities and correlations of instantaneous volatilities. Instantaneous volatilities are unobservable quantities – how does this manifest itself practically?

Risk-managing the correlation swap in the Black-Scholes model

Consider risk-managing a correlation swap on two underlyings S^1, S^2 with the Black-Scholes model. Since the dollar gammas and vegas of the correlation swap do not depend on S^1, S^2, it is natural to use variance swaps as hedge instruments.

Denote by $\sigma_{1t}, \sigma_{2t}, \rho_t$ the realized volatilities and correlation of log returns over $[0, t]$ and $\widehat{\sigma}_{1t}, \widehat{\sigma}_{2t}$ implied VS volatilities at time t for maturity T, which we are using as implied volatilities in our Black-Scholes model. Denote by $\widehat{\rho}_0$ the (implied) correlation level at which we are risk-managing the correlation swap, which we keep constant.

The value at t of the correlation swap in the Black-Scholes model is given by:[8]

$$P = \frac{t\rho_t\sigma_{1t}\sigma_{2t} + (T-t)\widehat{\rho}_0\widehat{\sigma}_{1t}\widehat{\sigma}_{2t}}{\sqrt{t\sigma_{1t}^2 + (T-t)\widehat{\sigma}_{1t}^2}\sqrt{t\sigma_{2t}^2 + (T-t)\widehat{\sigma}_{2t}^2}}$$

At $t = 0$, $P = \widehat{\rho}_0$ and $\frac{dP}{d(\widehat{\sigma}_{1t}^2)} = \frac{dP}{d(\widehat{\sigma}_{2t})} = 0$: the correlation swap has no sensitivity to implied volatilities at inception. We use variances rather than volatilities as state variables, since variance swaps provide perfect delta hedges with respect to variances. Let us now examine the correlation swap's sensitivities at later times.

Consider a short position in a correlation swap vega-hedged with variance swaps of maturity T. Let us calculate the P&L over $[t, t + \delta t]$ generated by variations $\delta(\widehat{\sigma}_{1t}^2)$, $\delta(\widehat{\sigma}_{2t}^2)$ of implied VS variances, at order two. Computing second derivatives of P with respect to $\widehat{\sigma}_{1t}^2$, $\widehat{\sigma}_{2t}^2$ is straightforward.

Consider the particular situation when realized volatilities and correlation over $[0, t]$ match the implied values at time t: $\sigma_{1t} = \widehat{\sigma}_{1t}$, $\sigma_{2t} = \widehat{\sigma}_{2t}$, $\rho_t = \widehat{\rho}_0$ – this allows for a more compact expression or the P&L. The P&L at order two in $\widehat{\sigma}_{1t}^2$, $\widehat{\sigma}_{2t}^2$ – for a short position – reads:

$$\begin{aligned}
P\&L &= \frac{\widehat{\rho}_0}{8}\frac{t(T-t)}{T^2}\left(\frac{\left(\delta(\widehat{\sigma}_{1t}^2)\right)^2}{(\widehat{\sigma}_{1t}^2)^2} + \frac{\left(\delta(\widehat{\sigma}_{2t}^2)\right)^2}{(\widehat{\sigma}_{2t}^2)^2} - 2\frac{\delta(\widehat{\sigma}_{1t}^2)\delta(\widehat{\sigma}_{2t}^2)}{\widehat{\sigma}_{1t}^2\widehat{\sigma}_{2t}^2}\right) \\
&= \frac{\widehat{\rho}_0}{8}\frac{t(T-t)}{T^2}\left(\frac{\delta(\widehat{\sigma}_{1t}^2)}{\widehat{\sigma}_{1t}^2} - \frac{\delta(\widehat{\sigma}_{2t}^2)}{\widehat{\sigma}_{2t}^2}\right)^2
\end{aligned} \qquad (11.38)$$

Had we used different values for σ_{1t}, σ_{2t}, ρ_t, (11.38) would have been replaced with a more complicated quadratic form of $\delta(\widehat{\sigma}_{1t}^2)$, $\delta(\widehat{\sigma}_{2t}^2)$, but the prefactor $\frac{t(T-t)}{T^2}$ would have remained.

(11.38) confirms – for the special case of realized volatilities and correlations at time t matching their implied values – that if $\frac{\delta\widehat{\sigma}_{1t}}{\widehat{\sigma}_{1t}} = \frac{\delta\widehat{\sigma}_{2t}}{\widehat{\sigma}_{2t}}$ no P&L is generated by the variation of VS volatilities.[9]

Correlation swaps are thus exotic volatility instruments that depend on the dynamics of forward variances. Practically, this manifests itself through volatility/volatility cross-gamma P&Ls.

11.4.1 Approximate formula in the two-factor model

We now compute $\widehat{\rho}(T)$ in (11.37) at order two in volatility of volatility in the two-factor model, for two underlyings with the same parameters, with a flat VS term

---

[8]We have made the assumption of very short returns, so that the contribution from the risk-neutral drift of $\ln S_t$ is negligible – this is adequate for daily returns of equities.

[9]In contrast with options on realized variance, there is no way that the vega hedge of the correlation swap may also function as gamma hedge since VSs on S^1 and S^2 generate gamma P&Ls proportional to $(\delta S^1)^2$ and $(\delta S^2)^2$ while the gamma P&L of the correlation swap also includes a $\delta S^1 \delta S^2$ term. In addition to the P&L in (11.38), our P&L thus comprises a spreaded position of diagonal spot gammas against spot cross-gammas, accompanied by their respective thetas.

structure, in the limit of short returns. The instantaneous variance ξ_t^t is given by:

$$\xi_t^t = \xi_0^t e^{2\nu x_t^t - 2\nu^2 E\left[(x_t^t)^2\right]}$$

where x_t^T has been defined in (7.30), page 226. For $T = t$:

$$x_t^t = \alpha_\theta \left[(1 - \theta) X_t^1 + \theta X_t^2\right] \qquad (11.39)$$

and $\alpha_\theta = \left((1 - \theta)^2 + \theta^2 + 2\rho_{X_1 X_2}\theta(1 - \theta)\right)^{-1/2}$. Remember ν is the volatility of a very short volatility. Expanding at order two in ν:

$$\xi_t^t = \xi_0^t \left(1 + 2\nu x_t^t + 2\nu^2 \left[(x_t^t)^2 - E[(x_t^t)^2]\right]\right) \qquad (11.40)$$

In an expansion of $\widehat{\rho}(T)$ at order two in ν, the ν^2 term in (11.40) is multiplied by a constant: when evaluating its expectation its contribution vanishes. For the sake of calculating $\widehat{\rho}(T)$ at order two in ν, the expansion of ξ_t^t at order one is thus sufficient:

$$\xi_t^t = \xi_0^t \left(1 + 2\nu x_t^t\right)$$

We consider two underlyings each with a flat VS term structure and denote by x_t, y_t the x_t^t processes for, respectively, the first and second underlying. At second order in ν, $\widehat{\rho}(T)$ reads:

$$\widehat{\rho}(T) = \rho_{SS} E \left[\frac{\int_0^T \sqrt{1 + 2\nu x_t}\sqrt{1 + 2\nu y_t}\, dt}{\sqrt{\int_0^T (1 + 2\nu x_t)\, dt}\sqrt{\int_0^T (1 + 2\nu y_t)\, dt}} \right]$$

Working out the expansion at order two in ν we get:

$$\begin{aligned}
\widehat{\rho}(T) &= \rho_{SS} E \left[\frac{1 + \nu\left(\bar{x} + \bar{y}\right) - \frac{\nu^2}{2}\overline{(x - y)^2}}{1 + \nu\left(\bar{x} + \bar{y}\right) - \frac{\nu^2}{2}\left(\bar{x} - \bar{y}\right)^2} \right] \\
&= \rho_{SS} E \left[1 - \frac{\nu^2}{2}\left(\overline{(x - y)^2} - \overline{x - y}^2\right) \right]
\end{aligned}$$

which at order two in ν can be rewritten as:

$$\widehat{\rho}(T) = \rho_{SS}\, e^{-\frac{\nu^2}{2} E\left[\overline{(x-y)^2} - \overline{x-y}^2\right]} \qquad (11.41)$$

where:

$$\bar{x} = \frac{1}{T} \int_0^T x_t\, dt$$

$$\overline{(x - y)^2} = \frac{1}{T} \int_0^T (x_t - y_t)^2\, dt$$

$$\overline{x - y}^2 = \left(\frac{1}{T} \int_0^T (x_t - y_t)\, dt\right)^2$$

Using the exponential in (11.41) ensures that for large values of ν, $\widehat{\rho}$ at most vanishes, but does not become negative. From the definitions above we have the pathwise inequality $\overline{(x-y)^2} \geq \overline{x-y}^2$ thus $E\left[\overline{(x-y)^2}\right] \geq E\left[\overline{x-y}^2\right]$.

(11.41) thus also ensures that the property $\widehat{\rho} \leq \rho_{SS}$ that expression (11.37) for $\widehat{\rho}(T)$ implies, is obeyed in our expansion at second order in ν. If forward variances for the two underlyings are collinear processes – $x_t = y_t$ – we recover that $\widehat{\rho} = \rho_{SS}$.

Carrying out the computation for the two-factor model with x_t^i given by (11.39) yields:

$$\widehat{\rho}(T) = \rho_{SS}\, e^{-(1-\chi_{\sigma\sigma})\left(\nu^2 T\right)\frac{1}{T}\int_0^T f(t)dt} \tag{11.42}$$

with the dimensionless $f(t)$ given by:

$$
\begin{aligned}
f(t) = {}& \alpha_\theta^2\Bigg[(1-\theta)^2\frac{1-e^{-2k_1 t}}{2k_1 T}\left(1-2\frac{1-e^{-k_1(T-t)}}{k_1 T}\right) \\
& + \theta^2\frac{1-e^{-2k_2 t}}{2k_2 T}\left(1-2\frac{1-e^{-k_2(T-t)}}{k_2 T}\right) \\
& + 2\rho_{X_1 X_2}\theta\,(1-\theta)\frac{1-e^{-(k_1+k_2)t}}{(k_1+k_2)\,T}\left(1-\frac{1-e^{-k_1(T-t)}}{k_1 T}-\frac{1-e^{-k_2(T-t)}}{k_2 T}\right)\Bigg]
\end{aligned}
$$

11.4.2 Examples

We use parameter values in Table 11.3, page 435, for each of our two underlyings, which correspond to the smiles in Figure 11.2. We take $\rho_{SS} = 60\%$ and use three different values for $\chi_{\sigma\sigma}$: $40\%, 60\%, 80\%$.

$\widehat{\rho}(T)$ appears in Figure 11.8, both evaluated numerically in a Monte Carlo simulation, and as given by formula (11.42). The right-hand side of Figure 11.8 shows the same curves for $\chi_{\sigma\sigma} = 60\%$, together with the values generated by a local volatility model parametrized as in the previous section, that is calibrated on the smile generated by the two-factor model, and with $\rho_{SS} = 60\%$.

Observe first that the accuracy of the second-order expansion in volatility of volatility is satisfactory, even though it is systematically biased low, except for short maturities. For short maturities $\widehat{\rho}(T)$ in (11.42) surges above the exact value: this is due to the fact that we have carried out our expansion in the continuous-time version of the correlation swap.

Thus $\widehat{\rho}(T)$ in (11.42) is not subject to the bias in (11.34), which instead affects the actual Monte Carlo estimation – as it uses discrete daily returns – and is material for small values of N, that is short maturities.

The right-hand graph is a compelling illustration of how different a local volatility and a stochastic volatility model can be, even though they are calibrated on the same smile.

In the local volatility model, $\widehat{\rho}(T)$ hardly depends on T: this is likely due to the fact that forward variances – especially long-dated ones – have very little volatility.

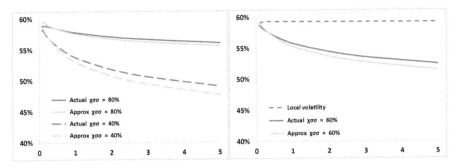

Figure 11.8: Left: $\widehat{\rho}(T)$ for maturities $T = 1$ month to 5 years, evaluated in a Monte Carlo simulation (Actual) and with formula (11.42) at second order in volatility of volatility (Approx), for $\chi_{\sigma\sigma} = 80\%$ and $\chi_{\sigma\sigma} = 40\%$. Right: same curves for $\chi_{\sigma\sigma} = 60\%$, along with $\widehat{\rho}(T)$ given by the local volatility model, calibrated on the smile generated by the two-factor model with parameters in Table 11.3.

Contrast this with the graphs in Figure 11.7. While for the basket smile the mapping $\chi_{\sigma\sigma} = \chi_{S\sigma} = \rho_{SS} = 60\%$ makes both models equivalent, they are not with respect to the correlation swap.

Figure 11.8 shows that the longer the maturity, the larger the impact of volatility of volatility, thus the lower $\widehat{\rho}(T)$. This deviates from market practice: typically the term structure of implied correlation swap correlations rises, rather than decreasing.

Finally, we have taken $\chi_{S\sigma} = 80\%$ in our numerical tests; using different values results, as expected, in no change of the actual and approximate values of $\widehat{\rho}(T)$.

11.5   Conclusion

- Life in the multi-asset local volatility framework used to be simple. Say we chose $\rho_{SS} = 60\%$ as correlation for the driving Brownian motions, priced multi-asset options and backed out the implied correlation in a multi-asset Black-Scholes model. Depending on the payoff at hand – say, an ATM basket call option, a VS on the basket, a correlation swap, a forward on the worst-performing asset – we would get different numbers, but clustered within a small range below 60%.

- This is not so with multi-asset stochastic volatility: exotic payoffs are sensitive to spot/volatility and volatility/volatility correlation levels, which, unlike in the local volatility model, can now be set separately. The richness of correlation risks manifests itself in the fact that implied spot/spot correlations backed out in a multi-asset Black-Scholes model for different payoffs will now vary widely.

- In our simple parametrization, in addition to ρ_{SS}, we introduce two quantities: $\chi_{S\sigma}$ and $\chi_{\sigma\sigma}$. By choosing $\chi_{\sigma\sigma} = \chi_{S\sigma} = \rho_{SS}$ we are able to mimic the local volatility model, for basket European payoffs, for short maturities.

- Do $\chi_{S\sigma}$ and $\chi_{\sigma\sigma}$ have implied counterparts? We have shown that only a small fraction of the basket skew is contributed by the components' skew; the bulk of it is determined by $\chi_{S\sigma}$. Thus, the ratio of basket-to-component skew is a measure of the implied value of $\chi_{S\sigma}$, once ρ_{SS} has been set. Similarly $\chi_{\sigma\sigma}$ determines the strike of the correlation swap, again once ρ_{SS} has been set.

It would be useful to have good approximations for the basket VS and ATMF volatilities. It doesn't help that baskets are typically of the arithmetic – rather than geometric – type: the weights of each asset's return in the basket's return are not constant. Thus even simple objects such as basket VS and ATMF volatilities depend in a complicated way on $\rho_{SS}, \chi_{S\sigma}, \chi_{\sigma\sigma}$, let alone payoffs on worst-ofs.

Appendix A – bias/standard deviation of the correlation estimator

We present here the derivation of formulas (11.34) and (11.35), for the case of Gaussian-distributed returns with constant volatility. Since volatility is constant, we can normalize returns by their standard deviation and evaluate correlations with normalized returns. Consider the i-th and j-th underlyings and denote by x_i^τ, x_j^τ their τ-th daily return. x_i^τ, x_j^τ are assumed to be iid standard normal random variables with $\langle x_i^\tau x_j^\tau \rangle = \rho$ if $i \neq j$ and $\langle x_i^\tau x_i^\tau \rangle = 1$ where ρ is the correlation between the two underlyings.

The correlation estimator for the (i, j) couple is:

$$\widehat{\rho}_{ij} = \frac{C_{ij}}{\sqrt{C_{ii}C_{jj}}}$$

with

$$C_{ij} = x_i . x_j \equiv \frac{1}{N}\Sigma_\tau x_i^\tau x_j^\tau$$

where N is the number of returns in our historical sample. C_{ij} is the average of independent random variables $x_i^\tau x_j^\tau$. As $N \to \infty$, fluctuations around its mean $\langle C_{ij} \rangle = \rho$ tend to zero, so let us write

$$C_{i\neq j} = \rho + \varepsilon_{ij} \quad \varepsilon_{ij} = x_i . x_j - \rho$$
$$C_{ii} = 1 + \varepsilon_{ii} \quad \varepsilon_{ij} = x_i . x_i - 1$$

We have:

$$\widehat{\rho}_{ij} = \frac{\rho + \varepsilon_{ij}}{\sqrt{1+\varepsilon_{ii}}\sqrt{1+\varepsilon_{jj}}}$$

We now expand $\widehat{\rho}_{ij}$ in powers of fluctuations ε_{ij}. By construction the ε_{ij} are centered: $\langle\varepsilon_{ij}\rangle = 0$. The first non-trivial contribution to the bias of $\widehat{\rho}_{ij}$ is thus generated by second-order terms. At second order in the ε_{ij} we have:

$$\widehat{\rho}_{ij} = \rho - \left(\rho\frac{\varepsilon_{ii}+\varepsilon_{jj}}{2} - \varepsilon_{ij}\right) - \left(\varepsilon_{ij}\frac{\varepsilon_{ii}+\varepsilon_{jj}}{2} - \rho\left(\frac{3}{8}(\varepsilon_{ii}^2+\varepsilon_{jj}^2)+\frac{1}{4}\varepsilon_{ii}\varepsilon_{jj}\right)\right)$$
(11.43)

At this order, the expectation of the correlation estimator is given by:

$$\langle\widehat{\rho}_{ij}\rangle = \rho - \left\langle\varepsilon_{ij}\frac{\varepsilon_{ii}+\varepsilon_{jj}}{2} - \rho\left(\frac{3}{8}(\varepsilon_{ii}^2+\varepsilon_{jj}^2)+\frac{1}{4}\varepsilon_{ii}\varepsilon_{jj}\right)\right\rangle$$

This formula involves moments of Gaussian variables x_i^τ, x_j^τ. The Wick theorem yields the following expressions for 4th order moments:

$$\langle\varepsilon_{ii}^2\rangle = \frac{2}{N} \quad \langle\varepsilon_{ii}\varepsilon_{jj}\rangle = \frac{2\rho^2}{N} \quad \langle\varepsilon_{ii}\varepsilon_{ij}\rangle = \frac{2\rho}{N} \quad \langle\varepsilon_{ij}^2\rangle = \frac{1+\rho^2}{N}$$

$$\langle\varepsilon_{ij}\varepsilon_{kk}\rangle = \frac{2\rho^2}{N} \quad \langle\varepsilon_{ij}\varepsilon_{kl}\rangle = \frac{2\rho^2}{N} \quad \langle\varepsilon_{ij}\varepsilon_{il}\rangle = \frac{\rho+\rho^2}{N}$$

where indices i, j, k, l are all different.

We then get, from (11.43):

$$\langle\widehat{\rho}_{ij}\rangle = \rho - \left(\frac{2\rho}{N} - \rho\left(\frac{3}{8}\frac{4}{N}+\frac{1}{4}\frac{2\rho^2}{N}\right)\right)$$

$$= \rho\left(1 - \frac{1-\rho^2}{2N}\right)$$

which is the result in (11.34).

We now turn to the variance of $\widehat{\rho}_{ij}$. The first non-trivial contribution is generated by the square of the order-one correction to ρ in (11.43), and is of order $\frac{1}{N}$. For the sake of calculating the variance of $\widehat{\rho}_{ij}$ we can then ignore the bias just derived, which generates a contribution of order $\frac{1}{N^2}$. Using the expressions of 4th order moments above, we have:

$$\left\langle\left(\rho\frac{\varepsilon_{ii}+\varepsilon_{jj}}{2} - \varepsilon_{ij}\right)^2\right\rangle = \frac{1}{N}(1-\rho^2)^2$$

from which (11.35) follows.

Consider now the correlation estimator used in the definition of the correlation swap, namely the average of all pairwise correlation estimators:

$$\widehat{\rho} = \frac{1}{n(n-1)}\sum_{i\neq j}\widehat{\rho}_{ij}$$

Assuming all pairwise correlations are equal to ρ, at order one in the fluctuations ε_{ij}:

$$\hat{\rho} = \rho - \frac{1}{n(n-1)} \sum_{i \neq j} \left(\rho \frac{\varepsilon_{ii} + \varepsilon_{jj}}{2} - \varepsilon_{ij} \right) \tag{11.44}$$

Calculating now the expectation of the square of the right-hand side of (11.44) and taking the limit $n \to \infty$ yields expression (11.36) for the standard deviation of $\hat{\rho}$.

Chapter's digest

11.1 The short ATMF basket skew

▶ We consider the case of an equally weighted basket of n components. Assuming identical correlations ρ_{SS} among components, the ATM implied volatility and skew of the basket are given, for short maturities, by:

$$\widehat{\sigma}_B = \sqrt{\frac{1 + (n-1)\rho_{SS}}{n}}\,\widehat{\sigma}$$

$$\mathcal{S}_B = \frac{1 + (n-1)\rho_{SS}}{n}\frac{\widehat{\sigma}^3}{\widehat{\sigma}_B^3}\left[\frac{\mathcal{S}}{n} + \frac{n-1}{n}\frac{1}{2\widehat{\sigma}^2}\frac{\langle d\ln S\, d\widehat{\sigma}\rangle_{\mathrm{cross}}}{dt}\right]$$

where $\widehat{\sigma}$ and $\mathcal{S}$ are the components' ATMF volatility and skew, assumed to be identical for all components, and $\langle d\ln S\, d\widehat{\sigma}\rangle_{\mathrm{cross}}$ is the cross spot/volatility covariance.

Only a fraction $\frac{1}{n}$ of the basket ATMF skew is contributed by the ATMF skew of the components. The bulk of the basket skew is generated by cross spot/volatility correlations.

▶ Specializing to the case of a large homogeneous basket:

$$\widehat{\sigma}_B \simeq \sqrt{\rho_{SS}}\,\widehat{\sigma}$$

$$\mathcal{S}_B \simeq \frac{1}{\sqrt{\rho_{SS}}}\frac{1}{2\widehat{\sigma}^2}\frac{\langle d\ln S\, d\widehat{\sigma}\rangle_{\mathrm{cross}}}{dt}$$

▶ For vanishing cross spot/volatility correlations, the basket skew vanishes.

▶ If cross spot/volatility covariances are identical to their diagonal counterparts, $\mathcal{S}_B \simeq \frac{1}{\sqrt{\rho_{SS}}}\mathcal{S}$: the basket skew is larger than the component's.

▶ In the local volatility model:

$$\widehat{\sigma}_B \simeq \sqrt{\rho_{SS}}\,\widehat{\sigma}$$

$$\mathcal{S}_B \simeq \sqrt{\rho_{SS}}\mathcal{S}$$

The basket skew is smaller than the component's skew.

❧ ❧ ❧ ❧ ❧

11.2 Parametrizing multi-asset stochastic volatility models

▶ We wish to parametrize a multi-asset stochastic volatility model (a) with few additional parameters, (b) in a manner that does not depend on the model's factor structure. We introduce dimensionless numbers $\chi_{S\sigma}$ and $\chi_{\sigma\sigma}$ to specify the cross spot/volatility and volatility/volatility covariance functions, in terms of their diagonal counterparts:

$$\rho^{\text{cross}}(S, \widehat{\sigma}_T) = \chi_{S\sigma}\, \rho^{\text{diag}}(S, \widehat{\sigma}_T)$$

$$\rho^{\text{cross}}(\widehat{\sigma}_T, \widehat{\sigma}_{T'}) = \chi_{\sigma\sigma}\, \rho^{\text{diag}}(\widehat{\sigma}_T, \widehat{\sigma}_{T'})$$

For a homogeneous basket, the conditions on $\chi_{S\sigma}, \chi_{\sigma\sigma}$ so that the global correlation matrix is positive are easily expressed. While $\chi_{\sigma\sigma} \leq 1$, $\chi_{S\sigma}$ can go above 1.

▶ Realized values of $\chi_{S\sigma}$ and $\chi_{\sigma\sigma}$ can be measured using historical data for spot and implied volatilities.

☙ ☙ ☙ ☙ ☙

11.3 The ATMF basket skew

▶ At order one in volatility of volatility, for a homogeneous basket, the ATMF skew is given by:

$$\mathcal{S}_T^B = \mathcal{S}_T \sqrt{\frac{n}{1 + (n-1)\rho_{SS}}} \left[\frac{1 + (n-1)\chi_{S\sigma}}{n} \right]$$

where ρ_{SS} is the correlation among basket components and $\mathcal{S}_T$ their ATMF skew.

▶ Numerical experiment show that this formula for the basket skew is accurate, and that the latter hardly depends on $\chi_{\sigma\sigma}$.

▶ The approximate expression for the basket VS volatility derived, as is the case for $\mathcal{S}_T^B$, with the assumption of frozen weights, is, on the other hand, not accurate. In reality it does depend on $\chi_{S\sigma}$.

▶ The multi-asset local volatility model is mimicked by setting $\chi_{S\sigma}$ and $\chi_{\sigma\sigma}$ equal to the spot/spot correlation.

In practice, while the shape of the resulting smile indeed matches that of the multi-asset local volatility model, the level of basket ATMF volatility is shifted downwards, especially for long maturities.

☙ ☙ ☙ ☙ ☙

11.4 The correlation swap

▶ The correlation swap is an exotic volatility instrument. It is very sensitive to the correlation of the volatility processes of the basket's components, thus to parameter $\chi_{\sigma\sigma}$.

At lowest order in volatility of volatility, the fair strike of the correlation swap, for the case of a homogeneous pair of underlyings, is given by:

$$\widehat{\rho}(T) = \rho_{SS}\, e^{-(1-\chi_{\sigma\sigma})(\nu^2 T)\frac{1}{T}\int_0^T f(t)dt}$$

where $f(t)$ is a simple function that involves model parameters.

▶ Numerical experiments show that the accuracy of our approximate formula is acceptable. For long maturities, the correlation swap's fair strike lies much lower than the correlation of spot processes. This contrasts with the behavior in the local volatility model, which yields a fair strike almost equal to the correlation of spot processes.

Chapter 12

Local-stochastic volatility models

This chapter should be considered as a natural sequel to Chapter 2 on the local volatility model – which we urge the reader to read if she or he has not done so – and Chapter 7 on forward variance models.

Local-stochastic volatility models are market models that possess a Markovian representation in terms of t, S plus a few additional state variables.

We begin by motivating their study, then cover their calibration to market smiles before we get to practical modeling issues: what does the carry P&L look like in these models? Which models can be used in trading applications? How can we adjust spot/volatility and volatility/volatility break-even levels?

12.1   Introduction

The local volatility model is a market model for the spot and vanilla options, or, equivalently, implied volatilities $\widehat{\sigma}_{KT}$ – it is covered in detail in Chapter 2. A market model takes as inputs any non-arbitrageable configuration of all hedging instruments – spot and vanilla options – generates a delta on each of them, and is characterized by the covariance structure it generates for the assets it models. The latter translates into a genuine gamma/theta analysis of the carry P&L.

The local volatility model is the simplest of all market models for vanilla options, as all instruments have a one-dimensional Markovian representation in terms of t, S. With this frugality comes, however, a total lack of control on the dynamics of implied volatilities generated by the model.

Forward variance models are surveyed in Chapter 7. They are market models for S together with a (one-dimensional) term structure of implied volatilities, for example VS or ATMF implied volatilities. Unlike local volatility, forward variance models afford a great deal of flexibility as to the dynamics of implied volatilities they are able to generate. However, while their parameters – volatilities of volatilities and spot/volatility correlations – can be chosen so as to best match a given market smile, they typically cannot be calibrated exactly to the full set of implied volatilities $\widehat{\sigma}_{KT}$.

What we are really aiming for is a market model that lets us specify – at least partially and posssibly indirectly – the joint dynamics of the spot and implied volatilities. Local-stochastic volatility models are a modest step in that direction.

There is a natural reason for considering them. In practice, only models having a low-dimensional Markovian representation can be employed. Starting with the simplest, the local volatility model, which is Markovian in t, S, which model is next in the hierarchy of market models? The answer is local-stochastic volatility models. They can be defined as market models that possess a Markovian representation in terms of t, S plus a few additional state variables, for example the factors X_t^1, X_t^2 of the two-factor model of Chapter 7.

We only consider diffusive models, which, practically, means that their carry P&L is characterized by their break-even levels for gammas and cross-gammas.

12.2   Pricing equation and calibration

12.2.1   Pricing

In local-stochastic volatility models – which we also call mixed models – we choose the following ansatz for the instantaneous volatility of S_t:

$$\sigma_t = \sqrt{\zeta_t^t}\, \sigma(t, S_t) \tag{12.1}$$

ζ_t^t is a positive process that has a Markovian representation in terms of a small number of factors.[1]

Let us assume that ζ_t^t is a process driven by the two-factor model of Section 7.4. From equation (7.28), page 226, the SDEs for S_t and ζ_t^T read:

$$\begin{cases} dS_t = (r - q)S_t dt + \sigma(t, S_t)\sqrt{\zeta_t^t} S_t dW_t^S \\ d\zeta_t^T = 2\nu \zeta_t^T \alpha_\theta \left((1 - \theta)\, e^{-k_1(T-t)} dW_t^1 + \theta e^{-k_2(T-t)} dW_t^2 \right) \end{cases} \tag{12.2}$$

where $\alpha_\theta = 1/\sqrt{(1 - \theta)^2 + \theta^2 + 2\rho_{12}\theta\,(1 - \theta)}$. We use the notation ζ_t^T rather than ξ_t^T, as ζ_t^T is no longer a forward variance. In the mixed model forward variances are given by:

$$\xi_t^T = E_t\left[\sigma_T^2\right] = E_t\left[\zeta_T^T \sigma(T, S_T)^2\right]$$

They are not known analytically.

---

[1]Though typical, there is nothing mandatory about making the instantaneous volatility the *product* of stochastic and local volatility components.

The pricing equation in the mixed model is almost identical to that of the underlying stochastic volatility model, but for the local volatility component. Starting from equation (7.4), page 219, for the n-factor model, we get the pricing equation for the mixed model:

$$\frac{dP}{dt} + (r-q)\, S \frac{dP}{dS} + \frac{\zeta^t \sigma(t,S)^2}{2} S^2 \frac{d^2 P}{dS^2}$$

$$+ \frac{1}{2}\int_t^T du \int_t^T du'\, \nu\,(t,u,u',\zeta)\, \frac{d^2 P}{\delta\zeta^u \delta\zeta^{u'}} + S\sigma(t,S)\int_t^T du\, \mu\,(t,u,\zeta)\, \frac{d^2 P}{dS\delta\zeta^u} = rP$$

$$(12.3)$$

with $\nu\,(t,u,u',\zeta)$ and $\mu\,(t,u,\zeta)$ given by expressions (8.50) and (8.51), page 327, for the special case of the two-factor model:

$$\mu(t,u,\xi) = 2\nu\xi^u \sqrt{\xi^t}\alpha_\theta \left[\rho_{SX^1}(1-\theta)e^{-k_1(u-t)} + \rho_{SX^2}\theta e^{-k_2(u-t)} \right] \qquad (12.4)$$

$$\nu(t,u,u',\xi) = 4\nu^2 \xi^u \xi^{u'} \alpha_\theta^2 \Big[(1-\theta)^2 e^{-k_1(u+u'-2t)} + \theta^2 e^{-k_2(u+u'-2t)}$$

$$+ \rho_{12}\theta(1-\theta)\left(e^{-k_1(u-t)}e^{-k_2(u'-t)} + e^{-k_2(u-t)}e^{-k_1(u'-t)} \right) \Big]$$

ν – which we term "volatility of volatility" – is the volatility of a VS volatility with vanishing maturity. Practically, it functions as a scale factor of volatilities of volatilities.

12.2.2   Is it a price?

Equation (12.3) is derived from the corresponding pricing equation (7.4) of forward variance models through the ansatz:

$$\sqrt{\zeta_t^t} \rightarrow \sqrt{\zeta_t^t}\, \sigma(t,S_t) \qquad (12.5)$$

The pricing equation of forward variance models arises from a replication analysis – see Section 7.1, page 217 – hence the P&L of a delta-hedged position has the usual gamma/theta expression, with well-defined break-even levels.

In local-stochastic volatility models, we use the seemingly innocuous ansatz (12.5). However, equation (12.3) does not arise from a replication-based argument, thus there is no reason that solving it produces a price – that is that the P&L of a hedged position is of the usual gamma/theta form.

If it is not, what will it be, may the reader ask? The carry P&L could include, in addition to gamma/theta terms, additional contributions that cause P&L leakage. This happens for example if we use an arbitrary pricing function $P\,(t,S,\hat\sigma_{KT})$; in this case the model is not usable.

This will need to be assessed a posteriori – in Section 12.3. As it turns out, most local-stochastic volatility models are not usable models.

Simulation

In the original version of the underlying stochastic volatility model, initial values of forward variances are analytically calibrated to the term structure of log-contract volatilities: $\xi_{t=0}^T = \frac{d}{dT}(T\hat{\sigma}_T^2)$, or numerically to the term structure of ATMF volatilities, or the term structure of implied volatilities for a given moneyness.

Because the underlying stochastic volatility model has a Markovian representation in terms of two factors, solving SDE (12.2) or PDE (12.3) boils down to the simulation of 3 processes: S_t, X_t^1, X_t^2.

The SDE for S_t, X_t^1, X_t^2 are:

$$
\begin{cases}
dS_t = (r - q)S_t dt + \sqrt{\zeta_t^t}\sigma(t, S_t)S_t dW_t^S \\
dX_t^1 = -k_1 X_t^1 dt + dW_t^1 \\
dX_t^2 = -k_2 X_t^2 dt + dW_t^2
\end{cases}
\tag{12.6}
$$

with $X_{t=0}^1 = X_{t=0}^2 = 0$. X_t^1, X_t^2 are Ornstein-Ühlenbeck processes that are easily simulated exactly – see Section 7.3.1, page 222.

ζ_t^t in (12.6) is given by:

$$
\begin{cases}
\zeta_t^t = \zeta_0^t f\left(t, X_t^1, X_t^2\right) \\
f\left(t, x_1, x_2\right) = e^{2\nu\alpha_\theta[(1-\theta)x_1+\theta x_2]-\frac{(2\nu)^2}{2}\chi(t,t)}
\end{cases}
\tag{12.7}
$$

where $\chi(t, T \geq t)$ is given by expression (7.35), page 227.

The only remaining task left is calibration of the local volatility function $\sigma(t, S)$.

12.2.3   Calibration to the vanilla smile

Consider a diffusive model and σ_t the instantaneous volatility in that model; the condition that vanilla option prices be matched at $t = 0$ is given in (2.6), page 28:

$$
E\left[\sigma_t^2|S_t = S\right] = 2 \left. \frac{\frac{dC}{dT} + qC + (r - q)K\frac{dC}{dK}}{K^2 \frac{d^2C}{dK^2}} \right|_{\substack{K=S \\ T=t}} = \sigma_{\text{Mkt}}(t, S)^2
\tag{12.8}
$$

where $C(K, T)$ is the market price for a call option of strike K, maturity T, and $\sigma_{\text{MKT}}(t, S)$ is the local volatility function associated to the market smile.

The simplest way of complying with (12.8) is to choose $\sigma_t \equiv \sigma_{\text{MKT}}(t, S)$ – this is the local volatility model.

In mixed models, (12.8) translates into:

$$
E\left[\zeta_t^t\sigma(t, S)^2|S_t = S\right] = \sigma_{\text{Mkt}}(t, S)^2
$$

thus $\sigma(t, S)$ is given by:

$$\sigma(t, S)^2 = \frac{\sigma_{\text{Mkt}}(t, S)^2}{E\left[\zeta_t^t | S_t = S\right]} \tag{12.9}$$

This is a self-consistent equation for $\sigma(t, S)$: the unknown local volatility function appears both explicitly in the left-hand side, and implicitly in the right-hand side, in the expectation in the denominator. It is not clear how the solution (12.9) should be approached.

Algorithms for solving (12.9) start with the discretization of time and proceed forward, starting from $t = 0$. Imagine the density $\varphi(t, S, X)$ is known at time t, where S is the underlying and X represents the state variables of the underlying stochastic volatility model. We use $\varphi(t, S, X)$ to calculate the conditional expectation in the denominator of (12.9). (12.9) then yields the time-t slice of the local volatility function, which we use to build the density $\varphi(t + \delta t, S, X)$, and so on.

How should φ be calculated practically? It can be done in two ways, depending on the dimensionality of the underlying stochastic volatility model.

- For a (very) small number of factors one can use the PDE technique described below – it is really usable for a one-factor model and consists in solving the (two-dimensional) forward equation for the joint density $\varphi(t, S, X)$.

- For a larger number of factors, the method of choice is the particle method, which we outline next, first introduced by Pierre Henry-Labordère and Julien Guyon in [52]. This algorithm is much more straightforward than the PDE technique and is immune to the curse of dimensionality.

12.2.4 PDE method

Let us assume that ζ_t is driven by a one-factor model – typically, through a Markovian representation as a function of an Ornstein-Ühlenbeck (OU) process X_t:

$$\begin{cases} \zeta_t^t = \zeta_0^t f(t, X_t) \\ dX_t = -kX_t dt + dZ_t, \ X_0 = 0 \end{cases}$$

The SDE for S_t is:

$$dS_t = (r - q) S_t dt + \sqrt{\zeta_0^t} f(t, X_t) \, \sigma(t, S_t) S_t dW_t \tag{12.10}$$

and we denote by ρ the correlation between Z_t and W_t.

Consider the density $\varphi(t, S, X) = E\left[\delta(S - S_t)\delta(X - X_t)\right]$. Equation (12.9) for $\sigma(t, S)$ can be rewritten as:

$$\sigma(t, S)^2 = \frac{\sigma_{\text{Mkt}}(t, S)^2}{\zeta_0^t} \frac{\int_{-\infty}^{+\infty} \varphi(t, S, X) \, dX}{\int_{-\infty}^{+\infty} \varphi(t, S, X) f(t, X) \, dX} \tag{12.11}$$

$\varphi\,(t \geq 0, S, X)$ is obtained by solving the forward Kolmogorov equation:

$$\frac{d\varphi}{dt} = \mathcal{L}\varphi \qquad\qquad (12.12a)$$

$$\mathcal{L} = \mathcal{L}_S + \mathcal{L}_X + \mathcal{L}_{SX} \qquad\qquad (12.12b)$$

with the initial condition:

$$\varphi\,(t = 0, S, X) = \delta\,(S - S_0)\,\delta(X - X_0)$$

Linear operators $\mathcal{L}_S$, $\mathcal{L}_X$, $\mathcal{L}_{SX}$ are defined by their action on a function ψ:

$$\mathcal{L}_S\psi = -(r - q)\frac{d}{dS}\,(S\,\psi) + \frac{1}{2}\frac{d^2}{dS^2}\left(f(t, X)\sigma\,(t, S)^2\,S^2\,\psi\right) \qquad (12.13a)$$

$$\mathcal{L}_X\psi = k\frac{d}{dX}(X\,\psi) + \frac{1}{2}\frac{d^2\psi}{dX^2} \qquad\qquad (12.13b)$$

$$\mathcal{L}_{SX}\psi = \frac{d^2}{dSdX}\left(\rho\sqrt{f(t, X)}\sigma\,(t, S)\,S\,\psi\right) \qquad\qquad (12.13c)$$

$\mathcal{L}_S$, $\mathcal{L}_{SX}$ involve the local volatility function $\sigma(t, S)$, thus (12.12) has to be solved self-consistently with (12.11). The idea of calibrating $\sigma\,(t, S)$ via a forward PDE-based algorithm was first proposed by Alex Lipton in [70].

Finite-difference algorithm

(12.12) is usually solved with a finite-difference algorithm. We assume that the reader has some familiarity with the numerical solution of parabolic equations – see [83] for an introduction. X and S – or more typically $\ln S$ – are discretized on a two-dimensional grid (S_i, X_j); $i = 0 \ldots n_S - 1$, $j = 0 \ldots n_X - 1$, with uniform spacings δS and δX: $S_{i+1} - S_i = \delta S$ and $X_{i+1} - X_i = \delta X$. Density φ is replaced with a vector of dimension $n_S n_X$: $\varphi_{i+n_S j} = \varphi\,(S_i, X_j)$.

Derivatives are replaced by their centered finite-difference approximations:

$$\frac{df}{dx} \simeq \frac{f_{i+1} - f_{i-1}}{2\delta x}, \quad \frac{d^2 f}{dx^2} \simeq \frac{f_{i+1} + f_{i-1} - 2f_i}{\delta x^2} \qquad (12.14a)$$

$$\frac{d^2 f}{dxdy} \simeq \frac{f_{i+1,j+1} - f_{i-1,j+1} - f_{i+1,j-1} + f_{i-1,j-1}}{4\delta x\delta y} \qquad (12.14b)$$

whose errors are of order two in $\delta x, \delta y$. Action of operators $\mathcal{L}_S$, $\mathcal{L}_X$, $\mathcal{L}_{SX}$ on φ in the right-hand side of (12.12) thus becomes a matrix/vector multiplication.

In what follows, we use the same notation for operators $\mathcal{L}_S$, $\mathcal{L}_X$, $\mathcal{L}_{SX}$ or their discretized version, and likewise for φ.[2] Let us first assume for simplicity that

---

[2] Matrices $\mathcal{L}_S$, $\mathcal{L}_X$ are easy to invert as they are block-diagonal. The inversion of $\mathcal{L}_S$, for example, consists in the independent inversion of n_X sub-matrices, each of dimension n_S. Moreover, because derivatives at point S_i are approximated using only values for S_{i-1}, S_i, S_{i+1}, these sub-matrices are tridiagonal, thus the computational cost of each inversion is linear in n_S.
The total cost of inverting $\mathcal{L}_S$ is thus of order $n_S n_X$ – and likewise for $\mathcal{L}_X$. Compare this to the cost of inverting $\mathcal{L}_{SX}$, which is proportional to $(n_S n_X)^3$. Multiplication by $\mathcal{L}_{SX}$, on the other hand, is achieved at a cost proportional to $n_S n_X$.

$\mathcal{L}_S$, $\mathcal{L}_X$, $\mathcal{L}_{SX}$ do not depend explicitly on time. The formal solution of (12.12) over $[t, t + \delta t]$ reads:

$$\varphi(t + \delta t) = e^{(\mathcal{L}_S + \mathcal{L}_X + \mathcal{L}_{SX})\delta t} \varphi(t)$$

where the exponential of operator $\mathcal{O}$ is defined by: $e^{\mathcal{O}}\varphi = \Sigma_0^\infty \frac{\mathcal{O}^n}{n!}\varphi$.

Vanishing correlation

Assume that $\rho = 0$ so that $\mathcal{L}_{SX}$ vanishes.

Several numerical schemes exist that approximate $e^{(\mathcal{L}_S + \mathcal{L}_X + \mathcal{L}_{SX})\delta t}$ up to second order in δt. The most popular is the Peaceman-Rachford (PR) algorithm, which consists in the following sequence and makes use of the intermediate vector φ^*:

$$\left(1 - \frac{\delta t}{2}\mathcal{L}_X\right)\varphi^* = \left(1 + \frac{\delta t}{2}\mathcal{L}_S\right)\varphi(t) \tag{12.15a}$$

$$\left(1 - \frac{\delta t}{2}\mathcal{L}_S\right)\varphi(t + \delta t) = \left(1 + \frac{\delta t}{2}\mathcal{L}_X\right)\varphi^* \tag{12.15b}$$

Each step involves the inversion of a block-diagonal matrix (this is the so-called implicit sub-step), which as mentioned above, is computationally economical and a multiplication of a block-diagonal matrix on a vector (the so called explicit sub-step), which is computationally frugal as well. Expressing $\varphi(t + \delta t)$ directly in terms of $\varphi(t)$:

$$\varphi(t + \delta t) = \mathcal{U}_{t,t+\delta t}\varphi(t)$$
$$\mathcal{U}_{t,t+\delta t} = \left(1 - \frac{\delta t}{2}\mathcal{L}_S\right)^{-1}\left(1 + \frac{\delta t}{2}\mathcal{L}_X\right)\left(1 - \frac{\delta t}{2}\mathcal{L}_X\right)^{-1}\left(1 + \frac{\delta t}{2}\mathcal{L}_S\right)$$

One can check by hand that $\mathcal{U}_{t,t+\delta t}$ approximates $e^{(\mathcal{L}_S + \mathcal{L}_X)\delta t}$ up to order two in δt.

$$\mathcal{U}_{t,t+\delta t} = 1 + \delta t(\mathcal{L}_S + \mathcal{L}_X) + \frac{\delta t^2}{2}(\mathcal{L}_S + \mathcal{L}_X)^2 + o(\delta t^2)$$

Note that no assumption is made regarding the commutation of $\mathcal{L}_S$, $\mathcal{L}_X$.

Let us introduce operators/matrices $\mathcal{E}_S$ (standing for explicit) and $\mathcal{I}_S$ (standing for implicit):

$$\mathcal{E}_S = \left(1 + \frac{\delta t}{2}\mathcal{L}_S\right) \qquad \mathcal{I}_S = \left(1 - \frac{\delta t}{2}\mathcal{L}_S\right)^{-1}$$

and likewise for $\mathcal{E}_X$ and $\mathcal{I}_X$. With these notations the PR algorithm simply reads:

$$\varphi(t + \delta t) = \mathcal{I}_S\mathcal{E}_X\mathcal{I}_X\mathcal{E}_S\,\varphi(t) \tag{12.16}$$

$\mathcal{E}$ and $\mathcal{I}$ commute; moreover:

$$\mathcal{E}\mathcal{I} = \mathcal{I}\mathcal{E} = 2\mathcal{I} - 1 \tag{12.17}$$

thus there exist different equivalent implementations of the PR algorithm, for example:

$$\varphi(t + \delta t) = \mathcal{I}_S(2\mathcal{I}_X - 1)\mathcal{E}_S\,\varphi(t)$$

which is implemented through the following sequence:

$$\varphi^* = \mathcal{E}_S \varphi(t)$$
$$\mathcal{I}_X^{-1} \varphi^{**} = \varphi^*$$
$$\mathcal{I}_S^{-1} \varphi(t + \delta t) = (2\varphi^{**} - \varphi^*)$$

Non-vanishing correlation

If $\rho \neq 0$, $\mathcal{L}_{SX}$ does not vanish. An algorithm that is correct up to order two in δt is the so-called predictor-corrector algorithm, which consists in two successive iterations of the PR algorithm, where $\mathcal{L}_{SX}$ is always treated explicitly – see footnote 2 on page 458. This is the well-known Craig-Sneyd algorithm.

The predictor step reads:

$$\mathcal{I}_X^{-1} \varphi^{**} = \mathcal{E}_S \varphi(t) + \frac{\delta t}{2} \mathcal{L}_{SX} \varphi(t) \qquad (12.18a)$$

$$\mathcal{I}_S^{-1} \varphi^*(t + \delta t) = \mathcal{E}_X \varphi^{**} + \frac{\delta t}{2} \mathcal{L}_{SX} \varphi(t) \qquad (12.18b)$$

and generates $\varphi^*(t + \delta t)$. The corrector step is similar, except $\mathcal{L}_{SX}$ is applied to the average of $\varphi(t)$ and $\varphi^*(t + \delta t)$. Define $\overline{\varphi}$ as:

$$\overline{\varphi} = \frac{1}{2}(\varphi(t) + \varphi^*(t + \delta t))$$

The corrector step reads:

$$\mathcal{I}_X^{-1} \varphi^{**} = \mathcal{E}_S \varphi(t) + \frac{\delta t}{2} \mathcal{L}_{SX} \overline{\varphi} \qquad (12.19a)$$

$$\mathcal{I}_S^{-1} \varphi(t + \delta t) = \mathcal{E}_X \varphi^{**} + \frac{\delta t}{2} \mathcal{L}_{SX} \overline{\varphi} \qquad (12.19b)$$

We use the same notation φ^{**} for the intermediate results in (12.18) and (12.19) – they are different vectors. The predictor step (12.18) reads:

$$\varphi^*(t + \delta t) = \mathcal{I}_S \mathcal{I}_X \left[\mathcal{E}_X \mathcal{E}_S + \delta t \mathcal{L}_{SX} \right] \varphi(t) \qquad (12.20)$$

where we have used (12.17). The full scheme is compactly expressed through:

$$\varphi(t + \delta t) = \mathcal{U}_{t,t+\delta t} \, \varphi(t)$$
$$\mathcal{U}_{t,t+\delta t} = \mathcal{I}_S \mathcal{I}_X \left(\mathcal{E}_X \mathcal{E}_S + \delta t \mathcal{L}_{SX} \left(\frac{1}{2} + \frac{1}{2} \mathcal{I}_S \mathcal{I}_X \left(\mathcal{E}_X \mathcal{E}_S + \delta t \mathcal{L}_{SX} \right) \right) \right) \quad (12.21)$$

While the predictor step is of order one in δt, $\mathcal{U}_{t,t+\delta t}$ is correct up to second order in δt. The reader is invited to check that indeed:

$$\mathcal{U}_{t,t+\delta t} = 1 + \delta t (\mathcal{L}_S + \mathcal{L}_X + \mathcal{L}_{SX}) + \frac{\delta t^2}{2} (\mathcal{L}_S + \mathcal{L}_X + \mathcal{L}_{SX})^2 + o(\delta t^2)$$

Again, there exist different corrector/predictor sequences implementing (12.21).

When $\mathcal{L}_S$, $\mathcal{L}_X$, $\mathcal{L}_{SX}$ explicitly depend on t – which is the case in practice, if only because $\sigma(t, S)$ enters $\mathcal{L}_S$ and $\mathcal{L}_{SX}$, and also because of the presence of $f(t, X)$ – then in (12.18) and (12.19) $\mathcal{L}_S$ and $\mathcal{L}_{SX}$ have to be evaluated at time $t + \frac{\delta t}{2}$ to preserve the order-two accuracy of our numerical scheme. Thus, over each interval $[t, t + \delta t]$ the local volatility is needed – and is determined – in $t + \frac{\delta t}{2}$.

Implementation

In practice, $\ln S$, rather than S, is used – we still use S in the discussion as there is no difference in implementation. The local volatility $\sigma(t, S)$ is discretized on the same spot grid as φ. The algorithm is started with the initial condition $\varphi(t = 0, S, X) = \delta(S - S_0)\,\delta(X - X_0)$, where S_0 is the initial spot value and we take $X_{t=0} = 0$. In our discretized grid this translates into the following initial condition $\varphi_{i \neq i_0, j \neq j_0} = 0$, $\varphi_{i_0, j_0} = \frac{1}{\delta S}\frac{1}{\delta X}$ where i_0, j_0 are the indexes for the initial values of S, X: $S_{i_0} = S_{t=0}$ and $X_{j_0} = 0$. $\sigma\left(\frac{\delta t}{2}, S\right)$ is simply initialized as: $\sigma\left(\frac{\delta t}{2}, S\right) \equiv \hat{\sigma}\left(K = S, T = \frac{\delta t}{2}\right) / \left(\zeta_0^0 f(0, 0)\right)$.

Time is discretized with a step δt. Application of the finite difference algorithm generates the density φ at times $t_k = k\delta t$.

Assume we have the density φ at time t and the local volatility at time $t - \frac{\delta t}{2}$. Generation of $\varphi(t + \delta t)$ and $\sigma\left(t + \frac{\delta t}{2}, S\right)$ involves the following steps:

- Run the predictor-corrector scheme, where the local volatility $\sigma\left(t + \frac{\delta t}{2}, S\right)$ in $\mathcal{L}_S$, $\mathcal{L}_{SX}$ is taken equal to that determined in the previous step: $\sigma\left(t - \frac{\delta t}{2}, S\right)$. This generates $\varphi(t + \delta t)$.

- Compute $\sigma(t + \delta t, S)$ using (12.11) applied at time $t + \delta t$, using $\varphi(t + \delta t)$, then average the values at t and $\tau + \delta t$:

$$\sigma\left(t + \frac{\delta t}{2}, S\right)^2 = \frac{1}{2}\left(\sigma(t, S)^2 + \sigma(t + \delta t, S)^2\right) \tag{12.22}$$

- Run the predictor-corrector scheme over $[t, t + \delta t]$ again, this time using this final value for $\sigma\left(t + \frac{\delta t}{2}, S\right)$ in $\mathcal{L}_S$, $\mathcal{L}_{SX}$. This generates our final estimate for $\varphi(t + \delta t)$. This step guarantees that our scheme is overall of order two in time – in practice, though, this does not seem necessary.

Some additional points are worthy of note:

- The width of the grids in S (or $\ln S$) and X is defined by choosing a percentile ε and setting $S_{\min}$, $S_{\max}$ and $X_{\min}$, $X_{\max}$ so that $p(S_t \leq S_{\min}) \leq \varepsilon$, $p(S_t \geq S_{\max}) \leq \varepsilon$ and $p(X_t \leq X_{\min}) \leq \varepsilon$, $p(X_t \geq X_{\max}) \leq \varepsilon$ for all $t \in [0, T]$. Typically, setting $S_{\min}$, $S_{\max}$ and $X_{\min}$, $X_{\max}$ so that $p(S_T \leq S_{\min}) = p(S_T \geq S_{\max}) = \varepsilon$ and $p(X_T \leq X_{\min}) = p(X_T \geq X_{\max}) = \varepsilon$ for the furthest maturity of interest T is adequate. Finding $X_{\min}$, $X_{\max}$ is easy as X_T is Gaussian, thus its cumulative density is known in closed form. $S_{\min}$, $S_{\max}$ are also easily found, since $p(S_T \leq S_{\min})$, $p(S_T \geq S_{\max})$ are undiscounted prices of European digital options of maturity T, thus can be read off the market smile directly – see equation 1.24, page 21.

- Boundary conditions for φ need to be specified for $S = S_0, S_{n_S}$ and $X = X_0, X_{n_X}$. Non-trivial boundary conditions for the density are not easy to derive. Typically, one takes wide grids in S (or $\ln S$) and X and imposes that φ vanishes on the edges of the grid. $\varphi(t)$ is a density so should integrate to one for all t; this is not guaranteed by the algorithm above. Once $\varphi(t + \delta t)$ is determined, one typically rescales it so that it integrates (numerically) to one. We refer the reader to Appendix A for an implementation that ensures that φ integrates to one and such that boundary conditions are automatically taken care of.

- For very small/large values of S, the density is small, thus the denominator in the right-hand side of (12.11) is small and subject to numerical noise. It is preferable to extrapolate $\sigma(t, S)$ starting from values of S for which the denominator in (12.11) is still appreciable.

12.2.5 Particle method

The PDE technique outlined in the previous section can in practice only be used for one-factor stochastic volatility models. For models with more than one stochastic volatility factor, the PDE is of dimension three or higher. It is then best to calibrate $\sigma(t, S)$ using the particle method, a Monte Carlo algorithm first published by Pierre Henry-Labordère and Julien Guyon in [52].

The particle algorithm is general, does not depend on the dimensionality of the process driving ζ_t and can also be used to calibrate the local volatility function in a hybrid local-stochastic volatility/stochastic interest rate model, or to calibrate the local correlation of a cross-FX rate or of a basket of equity underlyings to an index smile.

As the particle method is documented in Pierre Henry-Labordère and Julien Guyon's book [53], we only sketch it. The particle method is a Monte Carlo algorithm based on simultaneous simulation of interacting paths.

Time is discretized and we set up a grid of spot values S^* for which the local volatility function will be determined.

- Draw N paths for the pair (S_t, ζ_t) – we denote them by (S_t^k, ζ_t^k), $k = 1 \dots N$, starting from $S_{t=0}^k = S_0$, $\zeta_{t=0}^k = \zeta_0$. Each pair (S_t^k, ζ_t^k) obeys the model's SDE and the Brownian motions driving different pairs are all independent.

- Assume that the local volatility function is known at t_i. Use it to propagate the particles until t_{i+1}. At t_{i+1} use the empirical density defined by:

$$\varphi_{em}(t_{i+1}, S, \zeta) = \frac{1}{N} \Sigma_k \delta(S - S_{t_{i+1}}^k) \delta(\zeta - \zeta_{t_{i+1}}^k) \qquad (12.23)$$

to evaluate the conditional expectation $E[\zeta_t | S_t = S]$ for spot values S^*. Rather than straight Dirac peaks on S one uses in (12.23) a smoother kernel ϕ:

$$\varphi_{em}(t_{i+1}, S, \zeta) = \frac{1}{N} \Sigma_k \phi(S - S_{t_{i+1}}^k) \delta(\zeta - \zeta_{t_{i+1}}^k)$$

Efficient operation of the particle algorithm depends in fact on a proper choice of ϕ. Using expression (12.9) the local volatility at time t_{i+1} is thus given by:

$$\sigma(t_{i+1}, S^*)^2 = \sigma_{\text{Mkt}}(t_{i+1}, S^*)^2 \frac{\Sigma_k \phi\left(S^* - S_{t_{i+1}}^k\right)}{\Sigma_k \zeta_{t_{i+1}}^k \phi\left(S^* - S_{t_{i+1}}^k\right)}$$

Interpolate/extrapolate local volatilities calculated on spot values S^* of the grid to obtain $\sigma(t_{i+1}, S)$. Store this t_{i+1} slice of the local volatility function and simulate the particles until t_{i+2}.

Calibration of the local volatility function and pricing can be performed in one single simulation, using the same paths.

12.3 Usable models

We're now fully equipped for calibrating a mixed model and pricing exotic options. Before we do this, we need to address a concern expressed in Section 12.2.2: are the resulting numbers prices?

The reason for using local-stochastic volatility models is that they are calibrated, by construction, to the vanilla smile. When we build them what we are really trying to do is build a market model for spot and vanilla options.

In such a market model, the hedge instruments are the spot and vanilla options. The price of a derivative is a function of the values of these hedge instruments, and the P&L of a delta-hedged, vega-hedged position is the sum of cross-gamma contributions involving second-order moments of variations of the spot and implied volatilities, accompanied by matching thetas defined by a break-even covariance matrix.

Do mixed models indeed supply a genuine gamma/theta breakdown of the carry P&L? Provided they do, what are the implied break-even volatilities of volatilities and the break-even spot/volatility and volatility/volatility correlations?

We have already carried out such an analysis for the local volatility model, in Section 2.7, page 66. We ask the reader to read that portion of Chapter 2.

As will be made clear shortly, unlike the local volatility model, most mixed models are *not* market models for spot and implied volatilities. Only particular types of mixed models, which we now characterize, give rise to a genuine theta/gamma breakdown of the carry P&L.

12.3.1 Carry P&L

Consider the general case of a mixed model and denote by $P^M(t, x)$ the price of a derivative. x is the vector of inputs: x_1 is the spot price, x_2 the local volatility function, $x_3 \cdots x_n$ are state variables of the underlying stochastic volatility model. For example, if the underlying stochastic volatility model is the two-factor forward variance model, the dynamics of the mixed model is given by (12.2) and P^M reads:

$$P^M(t, x) \equiv P^M(t, S, \sigma, \zeta^u)$$

where the ζ^u make up a curve.

Consider instead a local-stochastic volatility model built on the Heston model:

$$\begin{cases} dS_t = (r - q)S_t dt + \sigma(t, S_t)\sqrt{V_t}S_t dW_t^S \\ dV_t = -k(V_t - V^0)dt + \nu\sqrt{V_t}dW_t^V \end{cases}$$

Here:

$$P^M(t, x) \equiv P^M(t, S, \sigma, V)$$

$P^M(t, x)$ takes as input the local volatility function. Consider now the pricing function $P(t, \widehat{x})$ which takes the set of implied volatilities as an input, rather than the local volatility function; $\widehat{x}_1$ is the spot price, $\widehat{x}_2$ the set of implied volatilities $\widehat{\sigma}_{KT}, \widehat{x}_3 \cdots \widehat{x}_n$ are again the state variables of the underlying stochastic volatility model.

$P(t, \widehat{x})$ is the pricing function we use in trading applications as it takes as inputs market observables, in addition to the state variables of the underlying stochastic volatility model. For the two-factor model:

$$P^M(t, x) \equiv P^M(t, S, \sigma, \zeta^u)$$
$$P(t, \widehat{x}) \equiv P(t, S, \widehat{\sigma}_{KT}, \zeta^u)$$

The reason why we explicitly include the local volatility function σ as an argument of P^M is that σ is not frozen. Indeed, $P(t, S, \widehat{\sigma}_{KT}, \zeta^u)$ implicitly involves recalibration of σ whenever the arguments of P change; as we risk-manage our exotic option using the pricing function P, σ will change and our carry P&L accounts for these changes as well.

In the mixed model, for a set local volatility function, implied volatilities are a *function* of time, spot value, local volatility function and other state variables: $\widehat{x} \equiv \widehat{x}(t, x)$ and we have:

$$P^M(t, x) = P(t, \widehat{x}(t, x))$$

For example, if we use the two-factor model as the underlying stochastic volatility model, we have:

$$\widehat{\sigma}_{KT} \equiv \Sigma_{KT}^M(t, S, \sigma, \zeta^u)$$

and the following relationship between P^M and P:

$$P^M(t, S, \sigma, \zeta^u) = P(t, S, \Sigma_{KT}^M(t, S, \sigma, \zeta^u), \zeta^u)$$

In a trading context, P&L accounting is done with P and involves derivatives of P with respect to $t, S, \widehat{\sigma}_{KT}$. The pricing equation (12.3), however, involves P^M and its derivatives.

Let us thus change variables from (t, x) to $(t, \widehat{x})$:

$$(t, x) \rightarrow (t, \widehat{x}(t, x))$$

The pricing equation (12.3), page 455, of the mixed model – with a set local volatility function – reads:

$$\frac{dP^M}{dt} + \left(\Sigma_k \mu_k \frac{d}{dx_k} + \frac{1}{2} \Sigma_{kl} a_{kl} \frac{d^2}{dx_k dx_l} \right) P^M = 0 \qquad (12.24)$$

where we assume zero interest rates without loss of generality – otherwise consider that P is the undiscounted price.

Switching now to variables $\widehat{x}$, the pricing equation reads:

$$\frac{dP}{dt} + \left(\Sigma_i \widehat{\mu}_i \frac{d}{d\widehat{x}_i} + \frac{1}{2} \Sigma_{ij} \widehat{a}_{ij} \frac{d^2}{d\widehat{x}_i d\widehat{x}_j} \right) P = 0 \qquad (12.25)$$

with:

$$\begin{cases} \widehat{\mu}_i = \dfrac{d\widehat{x}_i}{dt} + \Sigma_k \mu_k \dfrac{d\widehat{x}_i}{dx_k} + \dfrac{1}{2} \Sigma_{kl} a_{kl} \dfrac{d^2 \widehat{x}_i}{dx_k dx_l} \\[2mm] \widehat{a}_{ij} = \Sigma_{kl} a_{kl} \dfrac{d\widehat{x}_i}{dx_k} \dfrac{d\widehat{x}_j}{dx_l} \end{cases} \qquad (12.26)$$

$\widehat{\mu}_i$ is the drift of $\widehat{x}_i$ and $\widehat{a}_{ij}$ is the covariance matrix of $\widehat{x}_i$ and $\widehat{x}_j$ – *as generated by the mixed model with a fixed local volatility function.*

Derivatives $\frac{d\widehat{x}_i}{dx_k}$ are calculated keeping the local volatility function constant; in the two-factor model they involve derivatives $\frac{d\Sigma_{KT}^M}{dt}, \frac{d\Sigma_{KT}^M}{dS}, \frac{d\Sigma_{KT}^M}{d\zeta^u}$.

While the differential operator in (12.24) does not involve derivatives with respect to the local volatility function, the operator in (12.25) does involve derivatives with respect to implied volatilities.

12.3.2 P&L of a hedged position

Consider the P&L of a short option position – unhedged for now – during δt:

$$P\&L = -P(t + \delta t, \widehat{x} + \delta \widehat{x}) + P(t, \widehat{x})$$

Remember that as $t, S, \widehat{\sigma}_{KT}$ move by $\delta t, \delta S$ and $\delta \widehat{\sigma}_{KT}$, the local volatility function of our mixed model is recalibrated.

Expand at order two in $\delta\widehat{x}$ and one in δt, and use (12.25) to express $\frac{dP}{dt}$ in terms of derivatives with respect to the $\widehat{x}_i$:

$$
\begin{aligned}
P\&L &= -\frac{dP}{dt}\delta t - \Sigma_i \frac{dP}{d\widehat{x}_i}\delta\widehat{x}_i - \frac{1}{2}\Sigma_{ij}\frac{d^2 P}{d\widehat{x}_i d\widehat{x}_j}\delta\widehat{x}_i\delta\widehat{x}_j \\
&= -\Sigma_i \frac{dP}{d\widehat{x}_i}\left(\delta\widehat{x}_i - \widehat{\mu}_i\delta t\right) - \frac{1}{2}\Sigma_{ij}\frac{d^2 P}{d\widehat{x}_i d\widehat{x}_j}\left(\delta\widehat{x}_i\delta\widehat{x}_j - \widehat{a}_{ij}\delta t\right)
\end{aligned}
$$

Among components of $\widehat{x}$ we now make a distinction between those that correspond to market observables – $S, \widehat{\sigma}_{KT}$ – which we denote by O_i, and those corresponding to state variables of the underlying stochastic volatility model, which we denote by λ_k: $\widehat{x} \equiv (O, \lambda)$.

$$
\begin{aligned}
P\&L = &- \Sigma_i \frac{dP}{dO_i}\left(\delta O_i - \widehat{\mu}_i\delta t\right) - \frac{1}{2}\Sigma_{ij}\frac{d^2 P}{dO_i dO_j}\left(\delta O_i\delta O_j - \widehat{a}_{ij}\delta t\right) \\
&- \Sigma_k \frac{dP}{d\lambda_k}\left(\delta\lambda_k - \widehat{\mu}_k\delta t\right) \\
&- \frac{1}{2}\Sigma_{kl}\frac{d^2 P}{d\lambda_k d\lambda_l}\left(\delta\lambda_k\delta\lambda_l - \widehat{a}_{kl}\delta t\right) - \Sigma_{ik}\frac{d^2 P}{dO_i d\lambda_k}\left(\delta O_i\delta\lambda_k - \widehat{a}_{ik}\delta t\right)
\end{aligned}
$$

$$(12.27)$$

Consider now a delta and vega-hedged position, so that order-one contributions in δO_i vanish. Because equation (12.27) for the P&L during δt obviously also holds for (a) the underlying itself, (b) vanilla options, cancelling the δO_i terms also cancels the $\widehat{\mu}_i\delta t$ contributions. We denote by P_{H} the value of the delta-hedged, vega-hedged, position.

Using implied volatilities or option prices

Parameters O_i reflect the values of hedge instruments. For the spot we use the spot value itself, but for vanilla options we may use straight option prices, or their implied volatilities, or yet a different parametrization. For the sake of the present discussion, we treat the spot separately from vanilla options. The value of the hedged position is:

$$
P_{\mathrm{H}} = P - \Delta_S S - \sum_i \Delta_i f_i(t, S, O_i)
$$

where the sum runs over vanilla options used as hedges and f_i is the value of a vanilla option as a function of parameter O_i. The delta is $\Delta_S = \frac{dP}{dS}\big|_{\lambda,O}$. The vega hedge ratios Δ_i are given by: $\Delta_i = \frac{dP}{dO_i}\left(\frac{df_i}{dO_i}\right)^{-1}$.

- If O_i is an implied volatility $\widehat{\sigma}_{KT}$ then f_i is the value of a delta-hedged vanilla option in the Black-Scholes model and P_{H} reads:

$$
P_{\mathrm{H}} = P - \frac{dP}{dS}\bigg|_{\lambda,\widehat{\sigma}_{KT}} S - \sum_{KT}\Delta_{KT}\left(P_{KT} - \frac{dP_{KT}^{BS}}{dS}S\right) \qquad (12.28)
$$

with $\Delta_{KT} = \left.\dfrac{dP}{d\hat{\sigma}_{KT}}\right|_{\lambda,S} \left(\dfrac{dP^{BS}_{KT}}{d\hat{\sigma}_{KT}}\right)^{-1}$.

- If instead O_i is the vanilla option price P_{KT}, $\Delta_{KT} = \left.\dfrac{dP}{dP_{KT}}\right|_{\lambda,S}$ and P_{H} reads:

$$P_{\mathrm{H}} = P - \left.\frac{dP}{dS}\right|_{\lambda,P_{KT}} S - \sum_{KT}\Delta_{KT}P_{KT} \qquad (12.29)$$

There is no inconsistency in these two expressions of P_{H}. Obviously, the composition of the hedge portfolio cannot depend on how we decide to represent option prices, either using straight option prices or Black-Scholes implied volatilities: canceling at order one (a) the sensitivity to S and vanilla option prices, or (b) the sensitivity to S and Black-Scholes implied volatilities, is equivalent.

Thus, Δ_{KT} in (12.28) and (12.29) are identical, and so are the deltas in both portfolios:

$$\left.\frac{dP}{dS}\right|_{\lambda,\hat{\sigma}_{KT}} - \sum_{KT}\Delta_{KT}\frac{dP^{BS}_{KT}}{dS} = \left.\frac{dP}{dS}\right|_{\lambda,P_{KT}}$$

We refer the reader to a similar discussion of market-model and sticky-strike deltas in the local volatility model in Section 2.7, page 66.

Splitting the P&L of a hedged position
The P&L of the hedged position reads:

$$P\&L_{\mathrm{H}} = -\frac{1}{2}\Sigma_{ij}\frac{d^2 P_{\mathrm{H}}}{dO_i dO_j}(\delta O_i \delta O_j - \hat{a}_{ij}\delta t) \qquad (12.30\mathrm{a})$$

$$-\Sigma_k \frac{dP_{\mathrm{H}}}{d\lambda_k}(\delta\lambda_k - \hat{\mu}_k\delta t) \qquad (12.30\mathrm{b})$$

$$-\frac{1}{2}\Sigma_{kl}\frac{d^2 P_{\mathrm{H}}}{d\lambda_k d\lambda_l}(\delta\lambda_k\delta\lambda_l - \hat{a}_{kl}\delta t) - \Sigma_{ik}\frac{d^2 P_{\mathrm{H}}}{dO_i d\lambda_k}(\delta O_i\delta\lambda_k - \hat{a}_{ik}\delta t)$$

$$(12.30\mathrm{c})$$

- By construction, the mixed model is calibrated to the market values of hedge instruments $\forall t$, thus we have: $\frac{dO_i}{d\lambda_k} = 0$, $\forall i, \forall k$. This implies that $\frac{dP_{\mathrm{H}}}{d\lambda_k} = \frac{dP}{d\lambda_k}$, $\forall k$ and $\frac{d^2 P_{\mathrm{H}}}{d\lambda_k d\lambda_l} = \frac{d^2 P}{d\lambda_k d\lambda_l}$ as well as $\frac{d^2 P_{\mathrm{H}}}{dO_i d\lambda_k} = \frac{d^2 P}{dO_i d\lambda_k}$: all sensitivities of the hedged position involving λ_k are those of the unhedged position. $P\&L_{\mathrm{H}}$ can be rewritten as:

$$P\&L_{\mathrm{H}} = -\frac{1}{2}\Sigma_{ij}\frac{d^2 P_{\mathrm{H}}}{dO_i dO_j}(\delta O_i\delta O_j - \hat{a}_{ij}\delta t) \qquad (12.31\mathrm{a})$$

$$-\Sigma_k\frac{dP}{d\lambda_k}(\delta\lambda_k - \hat{\mu}_k\delta t) \qquad (12.31\mathrm{b})$$

$$-\frac{1}{2}\Sigma_{kl}\frac{d^2 P}{d\lambda_k d\lambda_l}(\delta\lambda_k\delta\lambda_l - \hat{a}_{kl}\delta t) - \Sigma_{ik}\frac{d^2 P}{dO_i d\lambda_k}(\delta O_i\delta\lambda_k - \hat{a}_{ik}\delta t)$$

$$(12.31\mathrm{c})$$

- Contribution (12.31a) to $P\&L_{\mathrm{H}}$ is the regular theta/gamma P&L involving second-order moments of the variations of market instruments. The matching deterministic terms $\widehat{a}_{ij}\delta t$ are genuine thetas, as $\widehat{a}$ is a valid (positive) covariance matrix – this is obvious from its definition in (12.26). The break-even covariances $\widehat{a}_{ij}$ are those generated by the model with a fixed local volatility function.

- (12.31b) and (12.31c) are unwanted contributions to the P&L, generated by variations of the λ_k state variables, that have no financial significance: V for the Heston model, the ζ^u for the two-factor model. These terms were absent from P&L expression (2.105), page 69, in the local volatility model.

- Note that the $\delta\lambda_k$ are fully in our control. Imagine setting $\delta\lambda_k = \widehat{\mu}_k\delta t$ so that (12.31b) cancels out. There remain the theta and gamma terms in (12.30c).

The conclusion is that, generally, when using a local-stochastic volatility model, a hedged position will generate spurious P&L leakage that does not correspond to any regular gamma/theta P&L.[3,4]

The contribution from $\delta\lambda_k$ obviously vanishes if (a) $\delta\lambda_k = 0$, (b) $\widehat{\mu}_k = 0$ and $\widehat{a}_{kl} = \widehat{a}_{ik} = 0, \forall i, \forall l$. This is the case if λ_k is frozen – in other words if λ_k is a constant parameter, for example the long-run variance V^0 in the Heston model, or constants $k_1, k_2, \theta, \rho_{12}$ in the two-factor model.

12.3.3   Characterizing usable models

Are there instances of mixed models that lead to regular theta/gamma P&L accounting without P&L leakage? The answer is yes.

All we need is for P to not depend on λ_k:[5]

$$\left.\frac{dP}{d\lambda_k}\right|_{S,\widehat{\sigma}_{KT}} = 0, \ \forall k \tag{12.32}$$

---

[3]Could it be that – for some configurations of t, S, O_i, λ and some payoffs – there exists a positive matrix $\widehat{a}_{ij}^*$ such that all δt contributions to the P&L are absorbed in the theta portion of the theta/gamma P&L (12.31a)? $\widehat{a}_{ij}^*$ would then be such that:

$$\frac{1}{2}\Sigma_{ij}\frac{d^2P_{\mathrm{H}}}{dO_idO_j}\widehat{a}_{ij}^* = \frac{1}{2}\Sigma_{ij}\frac{d^2P_{\mathrm{H}}}{dO_idO_j}\widehat{a}_{ij} + \Sigma_k\frac{dP}{d\lambda_k}\widehat{\mu}_k + \frac{1}{2}\Sigma_{kl}\frac{d^2P}{d\lambda_k d\lambda_l}\widehat{a}_{kl} + \Sigma_{ik}\frac{d^2P}{dO_id\lambda_k}\widehat{a}_{ik}$$

By setting $\delta\lambda_k = 0$, our P&L would simply consist of contribution (12.31a) with effective break-even covariances $\widehat{a}_{ij}^*$. Assuming this could hold for (a) all values of t, S, O, λ, (b) all payoffs, is a very irrealistic assumption. It amounts to hoping that a theta that was engineered to offset one cross-gamma offsets a different cross-gamma.

[4]Given a leaky model, can we size up the P&L leakage? Imagine for example that we use the Heston model as the underlying model. Then $V\sigma^2T\left.\frac{d^2P}{dV^2}\right|_{S,\widehat{\sigma}_{KT}}$ is a (very) rough estimate of the leakage generated by the first piece in (12.31c). To assess the magnitude of the leakage from the second piece, one needs second-order derivatives $\left.\frac{d^2P}{dV d\widehat{\sigma}_{KT}}\right|_{S,\widehat{\sigma}_{KT}}$ and $\left.\frac{d^2P}{dV dS}\right|_{S,\widehat{\sigma}_{KT}}$.

[5]Ex-physicists will be tempted to call this a condition of gauge-invariance.

Our P&L then simply reads:

$$P\&L_{\mathrm{H}} = -\frac{1}{2}\Sigma_{ij}\frac{d^2 P_{\mathrm{H}}}{dO_i dO_j}\left(\delta O_i \delta O_j - \hat{a}_{ij}\delta t\right) \tag{12.33}$$

and we have indeed a market model for spot and vanilla options – see our discussion in Section 1.1 of Chapter 1.

Condition (12.32) happens to be fulfilled for the two-factor model. Indeed, consider pricing equations (12.6) together with (12.7), page 456, for the mixed model. Perform the following transformation on the ζ^u and the local volatility function:

$$\zeta_0^u \rightarrow \varphi^u \zeta_0^u$$

$$\sigma(u, S) \rightarrow \sqrt{\frac{1}{\varphi^u}}\sigma(u, S)$$

where φ^u are arbitrary constants. One can see from (12.6) and (12.7) that this leaves the process for S_t and its instantaneous volatility unchanged, thus:

$$\frac{\delta}{\delta\zeta^u}P(t, S, \hat{\sigma}_{KT}, \zeta^u) = 0,\ \forall u$$

This also holds in the version of the two-factor model that generates volatility-of-volatility smile: f in (12.7) is then the sum of two exponentials – see Section 7.7.1, page 263.

- Condition (12.32) does not hold if we use the Heston model as the underlying model since:

$$\frac{d}{dV}P(t, S, \hat{\sigma}_{KT}, V) \neq 0$$

- It does not hold in the Bloomberg model either, defined by the following SDEs – see [45]:

$$\begin{cases} dS_t = (r - q)S_t dt + \sigma(t, S_t)\lambda_t S_t dW_t^S \\ d\lambda_t = -k(\lambda_t - \theta)dt + \xi(t)\lambda_t dW_t^\lambda \end{cases} \tag{12.34}$$

where $\xi(t)$ is a deterministic function of t.

- Condition (12.32) does hold if the instantaneous volatility of the underlying stochastic volatility model is lognormal, as in the SABR model:

$$\begin{cases} dS_t = (r - q)S_t dt + \sigma(t, S_t)\lambda_t S_t dW_t^S \\ d\lambda_t = \nu\lambda_t dW_t^\lambda \end{cases} \tag{12.35}$$

Discussion

We are aiming for a market model that takes as inputs the spot and implied volatilities. In case we make option prices dependent on additional non-financial state variables,[6] it is not surprising that there is P&L leakage, as the model allocates part of its theta as compensation – on average – for second-order contributions from these extra state variables as well.

$\frac{dP}{dV} \neq 0$ is a signal that prices depend on more state variables than hedge instruments, hence the inconsistency in P&L accounting and the fact that the model is unsuitable for trading purposes.

This dependence is of a different nature than the dependence on model parameters, such as V^0 in the Heston model, or k_1, k_2, θ, ρ_{12} in the two-factor model. Model parameters do not generate any P&L leakage, and only give rise to discrete P&Ls when they are changed and the option position is remarked with new parameter values.

In the case of the Heston model used as underlying stochastic volatility model, the "solution" to the P&L leakage – if it can be called a solution – is to include one additional hedge instrument, to which V is then calibrated – say a barrier or forward-start option. We then have as many hedge instruments as there are state variables in our model, and the carry P&L of a hedged option position is again of the genuine gamma/theta form, provided we dynamically trade this additional instrument.

The interesting aspect of models in the admissible class is that the underlying stochastic volatility model *does* affect the dynamics of spot and option prices; yet these additional degrees of freedom do not require any additional hedge instruments, and only impact the covariance structure of hedge instruments.[7]

The carry P&L is of the regular gamma/theta form and involves hedge instruments only.

12.4 Dynamics of implied volatilities

Having derived expression (12.33) of the P&L in an admissible model, we now need to size up covariances $\widehat{a}^*_{ij}$ to assess the suitability of our model's break-even levels.

---

[6]Consider that V is, for example, the temperature in the Luxembourg garden.

[7]It is of course always possible to express a model using more processes than necessary, but this is just cosmetic. Take the Black-Scholes model and write the driving Brownian motion as the sum of two different Brownian motions. The resulting model is still Black-Scholes with exactly the same dynamics and option prices as the original model.

Option prices in the admissible class *are* impacted by the dynamics of the underlying stochastic volatility model.

The $\widehat{a}_{ij}$ are model-generated covariances for spot and implied volatilities for *fixed* strikes and maturities. With regard to the task of calculating realized covariances, however, working with *floating* strikes corresponding to a set moneyness – for example ATMF – and *relative* maturities, is much more convenient.

In what follows, we derive approximations of model-generated variances and covariances of spot and ATMF volatilities. Spot/volatility covariances can equivalently be quantified using the SSR.

In practice, bid/offer costs on options are such that vega hedging is performed less frequently than delta hedging. The resulting carry P&L is different than (12.33) – see the discussion in Section 9.11.3, page 383.

We use the two-factor model as underlying stochastic volatility model and derive expressions of the SSR and the volatility of the ATMF volatility at lowest order, both in:

- the local volatility function – as in Section 2.5

- the volatility of volatility of the underlying two-factor model – as in the expansion of Section 8.2

12.4.1   Components of the ATMF skew

Consider SDE (12.6) for S_t in the mixed two-factor model. Since the two-factor model is in the admissible class, we have $\frac{dP}{d\zeta^\tau} = 0 \ \forall\tau$, thus we can take $\zeta^\tau = 1, \forall\tau$. The SDE for S_t is:

$$dS_t = (r-q)S_t dt + \sigma(t,S_t)\sqrt{f(t,X_t^1,X_t^2)}S_t dW_t^S \qquad (12.36)$$

where f is defined in (12.7).

We now perform an expansion by writing the local volatility function as:

$$\sigma(t,S) = \overline{\sigma}(t) + \delta\sigma(t,S)$$

with $\delta\sigma(t,S)$ given by:

$$\delta\sigma(t,S) = \alpha(t)x, \ x = \ln\frac{S}{F_t}$$

This is the form we use in (2.44), page 46, where we derive the approximate expression of the ATMF skew in the local volatility model. With respect to Section 2.4.5, here we perform the expansion in $\delta\sigma$ around $\overline{\sigma}(t)$ rather than a constant volatility σ_0.

We now turn to f and expand it at order one in ν. For $\nu = 0$, $f = 1$; at order one in volatility of volatility ν,

$$\sqrt{f(t,X_t^1,X_t^2)} = 1 + \frac{\nu}{2}g(t,X_t^1,X_t^2)$$

SDE (12.36) now reads:

$$dS_t = (r - q)S_t dt + \left(\overline{\sigma}(t) + \delta\sigma(t, S_t)\right)\left(1 + \frac{\nu}{2}g(t, X_t^1, X_t^2)\right)S_t dW_t^S \quad (12.37)$$

Write S_t as:

$$S_t = S_t^0 + \delta S_t^{LV} + \delta S_t^{SV} \quad (12.38)$$

where δS_t^{LV} and δS_t^{SV} are, respectively, the order-one corrections in $\delta\sigma$ and in ν. Inserting this expression in SDE (12.37) and equating terms linear in $\delta\sigma$ and ν yields the following SDEs for S_t^0, δS_t^{LV}, δS_t^{SV}:

$$\begin{cases} dS_t^0 = (r - q)S_t^0 dt + \overline{\sigma}(t)\, S_t^0 dW_t^S & S_{t=0}^0 = S_0 \\ d\delta S_t^{LV} = (r - q)\delta S_t^{LV} dt + \delta\sigma(t, S_t^0)S_t^0 dW_t^S & \delta S_{t=0}^{LV} = 0 \\ d\delta S_t^{SV} = (r - q)\delta S_t^{SV} dt + \overline{\sigma}(t)\dfrac{\nu}{2}g S_t^0 dW_t^S & \delta S_{t=0}^{SV} = 0 \end{cases}$$

Expansion (12.38) translates into an expansion of option prices:

$$P = P_0 + \delta P_{LV} + \delta P_{SV} \quad (12.39)$$

where P_0 is the price generated by S_t^0, that is the Black-Scholes price with time-deterministic volatility $\overline{\sigma}(t)$, that is with implied volatilities $\hat{\sigma}_\tau$ defined by:

$$\hat{\sigma}_\tau = \sqrt{\frac{1}{\tau}\int_0^\tau \overline{\sigma}_u^2 du}$$

(12.39) in turn translates into the following expansion of implied volatilities and of the ATMF skew $\mathcal{S}_T = \left.\frac{d\hat{\sigma}_{KT}}{d\ln K}\right|_{F_T}$:

$$\hat{\sigma}_{KT} = \hat{\sigma}_T + \delta\hat{\sigma}_{KT}^{LV} + \delta\hat{\sigma}_{KT}^{SV} \quad (12.40a)$$

$$\mathcal{S}_T = \mathcal{S}_T^{LV} + \mathcal{S}_T^{SV} \quad (12.40b)$$

where the two order-one contributions to $\hat{\sigma}_{KT}$ have been calculated already: $\delta\hat{\sigma}_{KT}^{LV}$ in Chapter 2 – see equation (2.40), page 44; $\delta\hat{\sigma}_{KT}^{SV}$ in Chapter 8 – see equation (8.20), page 314.

- $\mathcal{S}_T^{LV}$ is generated by δS_t^{LV}, which, according to its SDE, corresponds to the perturbation at order one of the Black-Scholes model with deterministic volatility $\overline{\sigma}(t)$ by a local volatility function $\delta\sigma(t, S)$. We can use the skew-averaging expression (2.60), page 51, and get:

$$\mathcal{S}_T^{LV} = \frac{1}{T}\int_0^T \frac{\hat{\sigma}_t^2 t}{\hat{\sigma}_T^2 T}\frac{\overline{\sigma}(t)}{\hat{\sigma}_T}\alpha(t)dt \quad (12.41)$$

If we expand around a constant volatility σ_0, rather than a deterministic volatility $\overline{\sigma}(t)$ we get – replacing $\sigma(t)$, $\widehat{\sigma}_t^2$ and $\widehat{\sigma}_T$ with σ_0 – expression (2.48), page 46:

$$\mathcal{S}_T^{\text{LV}} = \frac{1}{T} \int_0^T \frac{t}{T} \alpha(t) dt \qquad (12.42)$$

- Likewise, $\delta\mathcal{S}_t^{\text{SV}}$ is generated by the perturbation at order one in volatility of volatility in a two-factor forward variance model with an initial variance curve given by:

$$\xi_0^\tau = \overline{\sigma}^2(\tau)$$

We can readily recycle expression (8.54), page 329, of the ATMF skew in the two-factor model:

$$\mathcal{S}_T^{\text{SV}} = \frac{\nu\alpha\theta}{\widehat{\sigma}_T^3 T^2} \int_0^T dt\, \overline{\sigma}(t) \int_t^T du\, \overline{\sigma}^2(u) \left[(1-\theta)\rho_{SX^1} e^{-k_1(u-t)} + \theta\rho_{SX^2} e^{-k_2(u-t)} \right] \qquad (12.43)$$

where $\widehat{\sigma}_T = \sqrt{\frac{1}{T} \int_0^T \xi_0^t dt} = \sqrt{\frac{1}{T} \int_0^T \overline{\sigma}^2(t) dt}$. Expanding around a constant volatility σ_0 leads to the simpler expression (8.55), page 330:

$$\mathcal{S}_T^{\text{SV}} = \nu\alpha\theta \left[(1-\theta)\,\rho_{SX^1} \frac{k_1 T - (1 - e^{-k_1 T})}{(k_1 T)^2} + \theta\rho_{SX^2} \frac{k_2 T - (1 - e^{-k_2 T})}{(k_2 T)^2} \right] \qquad (12.44)$$

Summing both contributions, the ATMF skew in the mixed model at order one both in the local volatility component and in volatility of volatility is thus:

$$\mathcal{S}_T = \frac{1}{T} \int_0^T \frac{\widehat{\sigma}_t^2 t}{\widehat{\sigma}_T^2 T} \frac{\overline{\sigma}(t)}{\widehat{\sigma}_T} \alpha(t) dt \qquad (12.45)$$
$$+ \frac{\nu\alpha\theta}{\widehat{\sigma}_T^3 T^2} \int_0^T dt\overline{\sigma}(t) \int_t^T du\overline{\sigma}^2(u) \left[(1-\theta)\rho_{SX^1} e^{-k_1(u-t)} + \theta\rho_{SX^2} e^{-k_2(u-t)} \right]$$

which, when taking $\overline{\sigma}(t)$ constant, equal to σ_0, simplifies to:

$$\mathcal{S}_T = \frac{1}{T} \int_0^T \frac{t}{T} \alpha(t) dt$$
$$+ \nu\alpha\theta \left[(1-\theta)\,\rho_{SX^1} \frac{k_1 T - (1 - e^{-k_1 T})}{(k_1 T)^2} + \theta\rho_{SX^2} \frac{k_2 T - (1 - e^{-k_2 T})}{(k_2 T)^2} \right]$$

Note that the latter expression of $\mathcal{S}_T$ does not depend on the value of σ_0, the constant volatility around which the order-one expansion is performed.

Expression (12.45) of $\mathcal{S}_T$ is, in itself, useless, as our model is calibrated to market, thus $\mathcal{S}_T$ can be read off the market smile. The expression of $\mathcal{S}_T^{\text{SV}}$, though, will come in handy in what follows.

How should $\overline{\sigma}(t)$ be chosen? We can simply take it constant, equal to the ATMF volatility of the maturity T of interest.

For smiles with a marked term structure, it is preferable to calibrate $\overline{\sigma}(t)$ to the term structure of ATMF volatilities, which are readily read off the market smile, or VS volatilities.

In what follows, we make the former choice. $\widehat{\sigma}_T$ denotes the ATMF volatility of maturity T: $\widehat{\sigma}_T = \widehat{\sigma}_{F_T T}$.

12.4.2   Dynamics of ATMF volatilities

We now focus on the ATMF volatility $\widehat{\sigma}_T$.

From (12.40) the variation during dt of $\widehat{\sigma}_T$, at order one in $\alpha(t)$ and ν consists of two pieces:

$$d\widehat{\sigma}_T = d\delta\widehat{\sigma}_T^{\mathrm{LV}} + d\delta\widehat{\sigma}_T^{\mathrm{SV}}$$

For the sake of calculating covariances, we are only interested in the diffusive portion of $d\widehat{\sigma}_T$ – not in its drift. Thus, in what follows, only the diffusive contributions appear in the expressions of $d\delta\widehat{\sigma}_T^{\mathrm{LV}}$ and $d\delta\widehat{\sigma}_T^{\mathrm{SV}}$.

- $d\delta\widehat{\sigma}_T^{\mathrm{LV}}$ is generated by the local volatility component of the mixed model. From (2.82), page 57, we have:

$$d\widehat{\sigma}_T^{\mathrm{LV}} = \mathcal{R}_T^{\mathrm{LV}} \mathcal{S}_T^{\mathrm{LV}} d\ln S$$

where $\mathcal{R}_T^{\mathrm{LV}}$ is the SSR of the local volatility component, given by expression (2.65), page 52:

$$\mathcal{R}_T^{\mathrm{LV}} = 1 + \frac{1}{T}\int_0^T \frac{\overline{\sigma}^2(t)}{\widehat{\sigma}_t \widehat{\sigma}_T} \frac{\mathcal{S}_t^{\mathrm{LV}}}{\mathcal{S}_T^{\mathrm{LV}}} dt \qquad (12.46)$$

Thus:

$$d\widehat{\sigma}_T^{\mathrm{LV}} = \left(\mathcal{S}_T^{\mathrm{LV}} + \frac{1}{T}\int_0^T \frac{\overline{\sigma}^2(t)}{\widehat{\sigma}_t \widehat{\sigma}_T} \mathcal{S}_t^{\mathrm{LV}} dt\right) d\ln S$$

- $d\delta\widehat{\sigma}_{KT}^{\mathrm{SV}}$ is given by expression (7.36), page 227:

$$d\delta\widehat{\sigma}_T^{\mathrm{SV}} = \nu\alpha_\theta\widehat{\sigma}_T\left((1-\theta)A_1 dW^1 + \theta A_2 dW^2\right)$$

with A_i given by:

$$A_i = \frac{\int_0^T \xi_0^\tau e^{-k_i\tau} d\tau}{\int_0^T \xi_0^\tau d\tau} = \frac{\int_0^T \overline{\sigma}^2(\tau)e^{-k_i\tau}d\tau}{\int_0^T \overline{\sigma}^2(\tau)d\tau} \qquad (12.47)$$

$\delta\widehat{\sigma}_{KT}^{\mathrm{SV}}$ can be calculated from knowledge of the term-structure of ATMF volatilities – to which $\overline{\sigma}(t)$ is calibrated – and the parameters of the two-factor model.

$d\delta\widehat{\sigma}_T^{\mathrm{LV}}$ on the other hand depends on $\mathcal{S}_t^{\mathrm{LV}}$, that is the ATMF skew generated by the local volatility component of our model, which we do not know explicitly. This is readily taken care of by using (12.40): $\mathcal{S}_t^{\mathrm{LV}} = \mathcal{S}_t - \mathcal{S}_t^{\mathrm{SV}}$.

- Bringing now everything together:

$$d\widehat{\sigma}_T = \left((\mathcal{S}_T - \mathcal{S}_T^{\mathrm{SV}}) + \frac{1}{T} \int_0^T \frac{\overline{\sigma}^2(t)}{\widehat{\sigma}_t \widehat{\sigma}_T} (\mathcal{S}_t - \mathcal{S}_t^{\mathrm{SV}}) dt \right) d\ln S \qquad (12.48)$$
$$+ \nu\alpha_\theta \widehat{\sigma}_T \left((1-\theta) A_1 dW^1 + \theta A_2 dW^2 \right)$$

where $\mathcal{S}_t^{\mathrm{SV}}$ is given by (12.43).

- If instead we expand around a constant volatility, equal to the ATMF volatility of the maturity T of interest: $\sigma_0 = \widehat{\sigma}_T$, (12.48) simplifies to:

$$d\widehat{\sigma}_T = \left((\mathcal{S}_T - \mathcal{S}_T^{\mathrm{SV}}) + \frac{1}{T} \int_0^T (\mathcal{S}_t - \mathcal{S}_t^{\mathrm{SV}}) dt \right) d\ln S \qquad (12.49)$$
$$+ \nu\alpha_\theta \widehat{\sigma}_T \left((1-\theta) \frac{1 - e^{-k_1 T}}{k_1 T} dW^1 + \theta \frac{1 - e^{-k_2 T}}{k_2 T} dW^2 \right)$$

where $\mathcal{S}_T^{\mathrm{SV}}$ is given by (12.44).

12.4.2.1 SSR

Recall the definition of the SSR:

$$\mathcal{R}_T = \frac{1}{\mathcal{S}_T} \frac{\langle d\widehat{\sigma}_T d\ln S \rangle}{\langle (d\ln S)^2 \rangle} \qquad (12.50)$$

Writing $d\widehat{\sigma}_T$ as $d\widehat{\sigma}_T^{\mathrm{LV}} + d\widehat{\sigma}_T^{\mathrm{SV}}$ we have:

$$\mathcal{R}_T = \frac{\mathcal{R}_T^{\mathrm{LV}} \mathcal{S}_T^{\mathrm{LV}} + \mathcal{R}_T^{\mathrm{SV}} \mathcal{S}_T^{\mathrm{SV}}}{\mathcal{S}_T^{\mathrm{LV}} + \mathcal{S}_T^{\mathrm{SV}}} \qquad (12.51)$$

where $\mathcal{R}_T^{\mathrm{LV}}$ is the SSR generated by the local volatility component, at order one in $\delta\sigma$ and $\mathcal{R}_T^{\mathrm{SV}}$ the SSR generated by the stochastic volatility component, at order one in ν.

Using formula (12.46) for the SSR in the local volatility model and the fact that $\mathcal{S}_t^{\mathrm{LV}} = \mathcal{S}_t - \mathcal{S}_t^{\mathrm{SV}}$:

$$\mathcal{R}_T^{\mathrm{LV}} \mathcal{S}_T^{\mathrm{LV}} = \mathcal{R}_T^{\mathrm{LV}}(\mathrm{Mkt}) \mathcal{S}_T - \mathcal{R}_T^{\mathrm{LV}}(\mathrm{SV}) \mathcal{S}_T^{\mathrm{SV}}$$

where $\mathcal{R}_T^{\mathrm{LV}}(\mathrm{Mkt})$ (resp. $\mathcal{R}_T^{\mathrm{LV}}(\mathrm{SV})$) are the SSRs of the local volatility model calibrated to the market smile (resp. to the smile generated by the stochastic volatility component), that is given by (12.46) with $\mathcal{S}_t^{\mathrm{LV}}$ replaced with $\mathcal{S}_t$ (resp. with $\mathcal{S}_t^{\mathrm{SV}}$).

Inserting this in (12.51) yields:

$$\mathcal{R}_T = \frac{\mathcal{R}_T^{\mathrm{LV}}(\mathrm{Mkt}) \mathcal{S}_T - \mathcal{R}_T^{\mathrm{LV}}(\mathrm{SV}) \mathcal{S}_T^{\mathrm{SV}} + \mathcal{R}_T^{\mathrm{SV}} \mathcal{S}_T^{\mathrm{SV}}}{\mathcal{S}_T}$$

which supplies our final expression for the SSR of the mixed model:

$$\mathcal{R}_T = \mathcal{R}_T^{\mathrm{LV}}(\mathrm{Mkt}) + \frac{\mathcal{S}_T^{\mathrm{SV}}}{\mathcal{S}_T} \left[\mathcal{R}_T^{\mathrm{SV}} - \mathcal{R}_T^{\mathrm{LV}}(\mathrm{SV}) \right] \qquad (12.52)$$

Let us recall the expressions of $\mathcal{R}_T^{\text{LV}}(\text{Mkt}), \mathcal{R}_T^{\text{LV}}(\text{SV}), \mathcal{R}_T^{\text{SV}}$ for the case of the two-factor model.

No term structure

$$\mathcal{R}_T^{\text{LV}}(\text{Mkt}) = 1 + \frac{1}{T}\int_0^T \frac{S_t}{S_T}\,dt \qquad (12.53)$$

$$\mathcal{R}_T^{\text{LV}}(\text{SV}) = 1 + \frac{1}{T}\int_0^T \frac{S_t^{\text{SV}}}{S_T^{\text{SV}}}\,dt \qquad (12.54)$$

$$\mathcal{S}_T^{\text{SV}} = \nu\alpha_\theta \left[(1-\theta)\,\rho_{SX^1}\frac{k_1 T - (1 - e^{-k_1 T})}{(k_1 T)^2} + \theta\rho_{SX^2}\frac{k_2 T - (1 - e^{-k_2 T})}{(k_2 T)^2}\right]$$

$\mathcal{R}_T^{\text{SV}}$ is given by expression (9.21), page 364:

$$\mathcal{R}_T^{\text{SV}} = \frac{(1-\theta)\,\rho_{SX^1}\frac{1-e^{-k_1 T}}{k_1 T} + \theta\rho_{SX^2}\frac{1-e^{-k_2 T}}{k_2 T}}{(1-\theta)\,\rho_{SX^1}\frac{k_1 T-(1-e^{-k_1 T})}{(k_1 T)^2} + \theta\rho_{SX^2}\frac{k_2 T-(1-e^{-k_2 T})}{(k_2 T)^2}} \qquad (12.55)$$

Using the term structure of ATMF volatilities

$$\mathcal{R}_T^{\text{LV}}(\text{Mkt}) = 1 + \frac{1}{T}\int_0^T \frac{\overline{\sigma}^2(t)}{\widehat{\sigma}_t\widehat{\sigma}_T}\frac{S_t}{S_T}\,dt$$

$$\mathcal{R}_T^{\text{LV}}(\text{SV}) = 1 + \frac{1}{T}\int_0^T \frac{\overline{\sigma}^2(t)}{\widehat{\sigma}_t\widehat{\sigma}_T}\frac{S_t^{\text{SV}}}{S_T^{\text{SV}}}\,dt$$

$$\mathcal{S}_T^{\text{SV}} = \frac{\nu\alpha_\theta}{\widehat{\sigma}_T^3 T^2}\int_0^T dt\,\overline{\sigma}(t)\int_t^T du\,\overline{\sigma}^2(u)\left[(1-\theta)\rho_{SX^1}e^{-k_1(u-t)} + \theta\rho_{SX^2}e^{-k_2(u-t)}\right]$$

$\mathcal{R}_T^{\text{SV}}$ is given by expression (9.19), page 364, with $\xi_0^t = \overline{\sigma}^2(t)$:

$$\mathcal{R}_T^{\text{SV}} = \frac{1}{\overline{\sigma}(0)}\frac{\widehat{\sigma}_T^2 T\int_0^T \overline{\sigma}^2(t)\left[(1-\theta)\rho_{SX^1}e^{-k_1 t} + \theta\rho_{SX^2}e^{-k_2 t}\right]dt}{\int_0^T dt\overline{\sigma}(t)\left(\int_t^T \overline{\sigma}^2(u)\left[(1-\theta)\rho_{SX^1}e^{-k_1(u-t)} + \theta\rho_{SX^2}e^{-k_2(u-t)}\right]du\right)}$$

As observed in Section 9.4, page 359, in the discussion on the bounds of the SSR for a stochastic volatility model, $\mathcal{R}_T$ is very sensitive to the short end of the term structure of ATMF volatilities, that is $\overline{\sigma}(0)$.

A sanity check

- If the market smile happens to be that generated by the underlying stochastic volatility model of our mixed model, so that the local volatility component is a constant, then $\mathcal{S}_T = \mathcal{S}_T^{\text{SV}}$, $\mathcal{R}_T^{\text{LV}}(\text{SV}) = \mathcal{R}_T^{\text{LV}}(\text{Mkt})$. We recover $\mathcal{R}_T = \mathcal{R}_T^{\text{SV}}$.

- If volatility of volatility is switched off so that we are really using a local volatility model, $\mathcal{S}_T^{\text{SV}} = 0$. We recover $\mathcal{R}_T = \mathcal{R}_T^{\text{LV}}(\text{Mkt})$.

To gain accuracy on $\mathcal{R}_T$, as given by (12.52), it is preferable to calculate $\mathcal{S}_T^{\text{SV}}$ numerically. This ensures that, were the input smile generated by the underlying stochastic volatility model, we would have $\frac{\mathcal{S}_T^{\text{SV}}}{\mathcal{S}_T} = 1$ in (12.52) thus would exactly get back the SSR of the stochastic volatility model. Computing $\mathcal{S}_T^{\text{SV}}$ numerically can be done very efficiently in a Monte Carlo simulation; see the examples below.

12.4.2.2 Volatilities of volatilities

Using (12.49) or (12.48), the instantaneous volatility of $\hat{\sigma}_T$ is readily evaluated, as well as its covariance with S_t, hence all volatility/volatility correlations and spot/volatility correlations, at order one in $\alpha(t)$ and in ν.

12.4.3 Numerical evaluation of the SSR and volatilities of volatilities

How do we calculate numerically the exact values of SSR and volatilities of volatilities?

They can be evaluated in a Monte Carlo simulation of the mixed model, without any recalibration, as the gamma/theta break-even levels are those generated by the model *with a fixed local volatility function* – see Section 12.3.2.

Our derivation is similar to that in the pure two-factor model, in Section 9.8, page 368.

In the mixed two-factor model, with a fixed local volatility function, the ATMF volatility $\hat{\sigma}_T$ is a function of S, X^1, X^2:

$$\hat{\sigma}_T \equiv \hat{\sigma}_{F_T T}\left(\ln S, X^1, X^2\right)$$

Expanding at first order in $d\ln S, dX^1, dX^2$:

$$d\hat{\sigma}_T = \frac{d\hat{\sigma}_T}{d\ln S}d\ln S + \frac{d\hat{\sigma}_T}{dX^1}dX^1 + \frac{d\hat{\sigma}_T}{dX^2}dX^2 \qquad (12.56)$$

From the definition of the SSR in (12.50) and using that

$$E[(d\ln S_t)^2] = \sigma_0^2 dt, \quad E[d\ln S_t dX_t^1] = \rho_{SX^1}\sigma_0 dt, \quad E[d\ln S_t dX_t^2] = \rho_{SX^2}\sigma_0 dt$$

we get:

$$\mathcal{R}_T = \frac{1}{\mathcal{S}_T}\frac{E[d\ln S\, d\hat{\sigma}_T]}{E[(d\ln S)^2]} = \frac{1}{\mathcal{S}_T}\frac{1}{\sigma_0}\left(\frac{d\hat{\sigma}_T}{d\ln S}\sigma_0 + \frac{d\hat{\sigma}_T}{dX^1}\rho_{SX^1} + \frac{d\hat{\sigma}_T}{dX^2}\rho_{SX^2}\right)$$

where σ_0 is the instantaneous volatility:

$$\sigma_0 = \sigma(0, S_0)$$

Thus $\mathcal{R}_T$ can be simply evaluated numerically by computing $\hat{\sigma}_T$ with one repricing:

$$\mathcal{R}_T \simeq \frac{1}{S_T}\frac{1}{\sigma_0}\frac{\hat{\sigma}_T\left(\ln S_0 + \varepsilon\sigma_0, X_0^1 + \varepsilon\rho_{SX^1}, X_0^2 + \varepsilon\rho_{SX^2}\right) - \hat{\sigma}_T\left(\ln S_0, X_0^1, X_0^2\right)}{\varepsilon}$$

where ε is a small offset. Typically we take $X_0^2 = X_0^1 = 0$.

As for volatilities of volatilities, $\mathrm{vol}(\hat{\sigma}_{F_T T})$ is obtained by squaring (12.56) and taking its expectation. We have:

$$E[d\ln S^2] = \sigma_0^2 dt \qquad E[(dX^1)^2] = E[(dX^2)^2] = dt$$
$$E[d\ln SdX^1] = \rho_{SX^1}\sigma_0 dt \qquad E[d\ln SdX^2] = \rho_{SX^2}\sigma_0 dt$$

We need $\frac{d\hat{\sigma}_{F_T T}}{d\ln S}$, $\frac{d\hat{\sigma}_{F_T T}}{dX}$, $\frac{d\hat{\sigma}_{F_T T}}{dY}$. Each derivative is obtained with one Monte Carlo simulation (or two when using centered differences). Numerical evaluation of a volatility of volatility thus requires three repricings (or six).

While the order-one approximations in the previous section only apply to ATMF (or VS) volatilities, one can of course numerically compute volatilities of volatilities of arbitrary strikes.

12.5   Numerical examples

We now test our approximations for SSR, volatility of volatility and spot/volatility correlation in a mixed model whose underlying stochastic volatility model is the two-factor model.

We use as market smile a smile generated by the two-factor model, rather than a real market smile, so that the case of pure stochastic volatility is attainable in the mixed model. The parameters we use generate a typical index smile – say, of the Euro Stoxx 50 index; our conclusions hold for general market smiles as well.

The "market smile" is generated with flat VS volatilities equal to 20% and the parameters in Table 12.1, using the mixing-solution technique of Section A.1, Chapter 8.

ρ_{12} is taken equal to zero. Parameters ν, θ, k_1, k_2 are chosen so that volatilities of VS volatilities best match the benchmark form (7.40):[8]

$$\nu_T^B(t) = \sigma_0\left(\frac{T_0}{T-t}\right)^\alpha$$

[8]See page 228 for a discussion of the parametrization of the two-factor model.

ν	θ	k_1	k_2	ρ_{12}	ρ_{SX^1}	ρ_{SX^2}
310%	0.139	8.59	0.47	0%	−54.0%	−62.3%

Table 12.1: Parameters of the two-factor model used for generating the "market smile".

with $\alpha = 0.6$, $\tau_0 = 3$ months and the (lognormal) volatility of the VS volatility for a 3-month VS volatility is $\sigma_0 = 125\%$. Correlations ρ_{SX^1} and ρ_{SX^2} are chosen so as to generate an ATMF skew that approximately decays like $\frac{1}{\sqrt{T}}$, a typical scaling of index smiles. We use zero rate and repo for simplicity.

The ATMF skew of the "market smile" as well as the SSR of the pure two-factor stochastic volatility model are shown in Figure 12.1. In the expansion that produces the approximate formulas of Section 12.4.1 and 12.4.2, $\hat{\sigma}_T$ has been taken equal to the VS volatility for maturity T, here 20%.

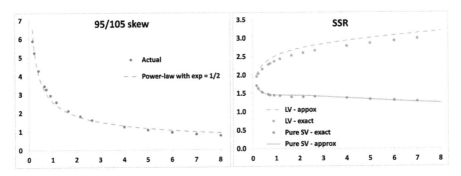

Figure 12.1: Left: term structure of the ATMF skew of the "market smile" expressed as the difference of implied volatilities of the 95% and 105% strikes in volatility points as a function of maturity (years) together with a power-law fit $\frac{1}{T^{\gamma}}$ with $\gamma = \frac{1}{2}$. Right: SSR of (a) the two-factor model, (b) the local volatility model calibrated on the smile of the former, as a function of maturity (years), either calculated in a Monte Carlo simulation (exact) or using, respectively, approximate formulas (12.55) and (12.53) (approx).

Notice how approximate values for (a) the SSR of the two-factor model and (b) the SSR of the local volatility model agree with actual values.

Both values of the SSR start from the value of 2 for very short maturities. Since the ATMF skew decays approximately like $\frac{1}{\sqrt{T}}$, that is with an exponent $\frac{1}{2}$, we expect that, for long maturities, approximately:

- the SSR of the stochastic volatility model tends to $2 - \frac{1}{2} = 1.5$

- the SSR of the local volatility model tends to the value of $\frac{2 - \frac{1}{2}}{1 - \frac{1}{2}} = 3$

This is indeed observed in Figure 12.1.[9] Volatilities of ATMF volatilities and spot/ATMF volatility correlations in both models appear in Figure 12.2.

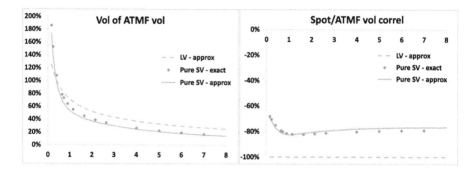

Figure 12.2: Volatilities of ATMF volatilities in the two-factor model and in the local volatility model calibrated on the same smile (left), and spot/ATMF volatility correlation (right) as a function of maturity (years).

Unsurprisingly, spot/volatility correlations in the local volatility model are equal to -1. The approximate values derived from expression (12.48) for $d\widehat{\sigma}_T$ agree well with "exact" values calculated in a Monte Carlo simulation.

Halving spot/volatility correlations – Figure 12.3
Using the same "market smile", we still use the two-factor model as underlying stochastic volatility model, but halve the values of ρ_{SX^1} and ρ_{SX^2}. Roughly half of the skew now needs to be generated by the local volatility component and we expect the SSR of the mixed model to lie in between the two curves in the right-hand graph of Figure 12.1. Figure 12.3 – where those two curves are shown for reference – shows that it is indeed the case, and that formula (12.52) works well.

The "exact" values in figure 12.3 is obtained in a Monte Carlo simulation, with the local volatility component calibrated using the particle method.

Halving volatilities of volatilities – Figure 12.4
Rather than halving ρ_{SX^1} and ρ_{SX^2} we now halve ν. Again, about half of the ATMF skew now needs to be generated by the local volatility component. The SSR, volatilities of ATMF volatilities and spot/ATMF volatility correlation appear in Figure 12.4. Again, the curves for the case of a pure stochastic volatility model and local volatility model, graphed in Figure 12.2, are shown for reference.

---

[9]We refer the reader to Section 9.5, page 361, for a discussion of the relationship of the long-maturity limit of the SSR to the decay of the ATMF skew in time-homogeneous stochastic volatility models, and to Section 2.5.4, page 56, for a discussion of the corresponding relationship in the context of the local volatility model.

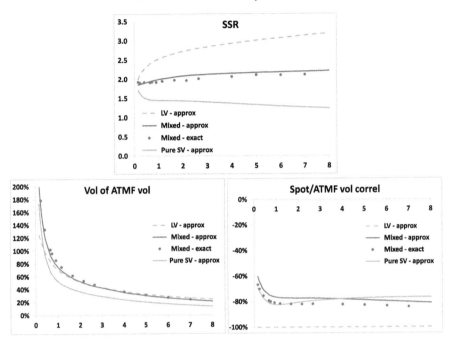

Figure 12.3: Top: SSR of the mixed model, compared to that generated by (a) the local volatility model, (b) the two-factor model used to generate the "market smile", as a function of maturity (years). In the mixed model, ρ_{SX^1} and ρ_{SX^2} are halved. Bottom: volatilities of ATMF volatilities (left) and spot/ATMF volatility correlations (right) in (a) the two-factor model, (b) the local volatility model calibrated to the smile of the former, (c) the mixed model.

Raising volatilities of volatilities – Figure 12.5

What if we increase ν so that the underlying stochastic volatility model generates a steeper smile than the input smile, with the effect that the local volatility component now generates positive skew?

We could use as market smile that produced by the two-factor model with parameters in Table 12.1 and raise ν – say by 50%. With such high level of volatility of volatility, however, calibration by the particle method does not function well anymore.[10]

To circumvent this difficulty, we generate a different "market smile" using parameters in Table 12.1, but with ν halved – $\nu = 155\%$ – and then use $\nu = 232.5\%$ in the mixed model. Numerical results are reported in Figure 12.5.

---

[10]There is indeed no mathematical guarantee that, given a non-arbitrageable market smile and an underlying stochastic volatility model, there exists a local volatility function such that the mixed model recovers the market smile. Deterioration of the quality of calibration is typically observed for (very) large levels of volatility of volatility.

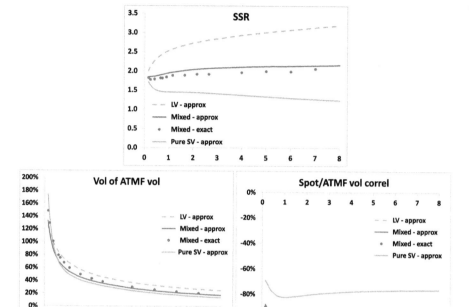

Figure 12.4: Top: SSR of the mixed model, compared to that generated by (a) the local volatility model, (b) the two-factor model used to generate the "market smile", as a function of maturity (years). In the mixed model ν is halved.
Bottom: volatilities of ATMF volatilities (left) and spot/ATMF volatility correlations (right) in (a) the two-factor model, (b) the local volatility model calibrated to the smile of the former, (c) the mixed model.

While qualitatively correct, our expansion at order one in (a) volatility of volatility, (b) local volatility is, in this situation, less accurate, especially for the SSR and the spot/ATMF correlation.

Observe that, by parametrizing the underlying stochastic volatility model so that the ATMF skew it generates is stronger than in the "market smile", the SSR is lower than that of the stochastic volatility model used for generating the "market smile". For sufficiently long maturities, $\mathcal{R}_T \leq 1$.

12.6   Discussion

Graphs in figures 12.3, 12.4, 12.5 confirm that our objective in developing local-stochastic volatility models was reached . Mixed models – in the admissible class

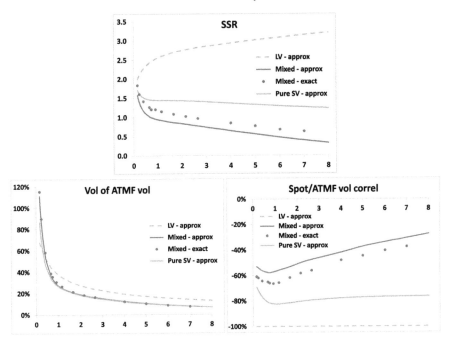

Figure 12.5: Top: SSR of the mixed model, compared to that generated by (a) the local volatility model, (b) the two-factor model used to generate the "market smile", as a function of maturity (years). In the mixed model ν is multiplied by 1.5.
Bottom: volatilities of ATMF volatilities (left) and spot/ATMF volatility correlations (right) in (a) the two-factor model, (b) the local volatility model calibrated to the smile of the former, (c) the mixed model.

– do afford some flexibility as to the SSR, volatility of volatility and spot/volatility correlation.

While our examples seem to show that levels of volatility of volatility in the mixed model vary minimally in figures 12.3 and 12.4, this is due to our choice of parameters: volatilities of volatilities in the two-factor model used to generate the "market smile" and the local volatility model are very similar, in the first place.

This is apparent in the left-hand graph of Figure 12.2. Had we chosen vanishing values of ρ_{SX^1} and ρ_{SX^2} to generate the "market smile", the latter would have been U-shaped with the effect that volatilities of volatilities in the local volatility model would have been almost vanishing.

For example, start from parameters in Table 12.1 and reduce correlations by a factor of 4: $\rho_{SX^1} = -13.5\%$ and $\rho_{SX^2} = -15.6\%$. Figure 12.6 shows volatilities of ATMF volatilities in the two-factor model and in the local volatility model – it is apparent that the difference between volatility-of-volatility levels is larger than in Figure 12.2.

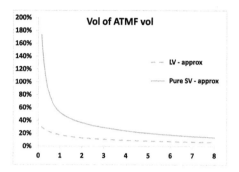

Figure 12.6: Volatilities of ATMF volatilities in the two-factor model and in the local volatility model calibrated on the same smile of the latter, as a function of maturity (years). The parameters of the two-factor model are those of Table 12.1, page 479, except $\rho_{SX^1} = -13.5\%$ and $\rho_{SX^2} = -15.6\%$.

How can the different SSRs in figures 12.1, 12.3, 12.4 be reconciled with the fact that the corresponding vanilla smiles are all identical? The same can be asked of the SSRs in Figure 9.9, page 380.

Recall expression (8.32), page 319, relating the ATMF skew to the weighted average of instantaneous covariances of S and the square of the ATMF volatility for the residual maturity $\widehat{\sigma}_{F_TT}$:

$$\mathcal{S}_T = \frac{1}{2\widehat{\sigma}_T^3(0)T} \int_0^T \frac{T-t}{T} \frac{\left\langle d\ln S_t \, d\widehat{\sigma}_{F_TT}^2(t) \right\rangle}{dt} dt \qquad (12.57)$$

This general expression is correct at order one in volatility of volatility and holds as long as $\left\langle d\ln S_t \, d\widehat{\sigma}_{F_TT}^2(t) \right\rangle$ does not depend on S_t.

In our analysis above, $\widehat{\sigma}_{F_TT}^2$ is the sum of three pieces. From (12.40):

$$\widehat{\sigma}_{F_TT} = \widehat{\sigma}_T + \delta\widehat{\sigma}_{F_TT}^{LV} + \delta\widehat{\sigma}_{F_TT}^{SV}$$

where $\widehat{\sigma}_T$ is the order-zero contribution, which is static, and $\delta\widehat{\sigma}_{F_TT}^{LV}$ and $\delta\widehat{\sigma}_{F_TT}^{SV}$ are respectively the contribution of the local volatility and stochastic volatility components, at order one in $\alpha(t)$ and ν. At order one in $\alpha(t)$ and ν, we can rewrite (12.57) as:

$$\mathcal{S}_T = \frac{1}{\widehat{\sigma}_T^3(0)T} \int_0^T \frac{T-t}{T} \widehat{\sigma}_T(t) \frac{\left\langle d\ln S_t \, d\widehat{\sigma}_{F_TT}(t) \right\rangle}{dt} dt \qquad (12.58)$$

where $\widehat{\sigma}_T(t) = \sqrt{\frac{1}{T-t}\int_t^T \sigma^2(u)du}$. The local volatility function is given by:

$$\sigma(t,S) = \overline{\sigma}(t) + \alpha(t)\ln\frac{S}{F_t}$$

Its contribution to the covariance in (12.58) is:

$$
\left\langle d\ln S_t \, d\delta\widehat{\sigma}^{LV}_{F_T T}(t) \right\rangle = \left\langle \frac{d\delta\widehat{\sigma}^{LV}_{F_T T}}{d\ln S}(d\ln S_t)^2 \right\rangle
$$

$$
= \left(\frac{1}{T-t} \int_t^T \frac{\overline{\sigma}(u)}{\widehat{\sigma}_T(u)} \alpha(u) du \right) \overline{\sigma}^2(t) dt
$$

where we have made use of expression (2.60c), page 51, for $\frac{d\delta\widehat{\sigma}^{LV}_{F_T T}}{d\ln S}$. Thus $\left\langle d\ln S_t \, d\delta\widehat{\sigma}^{LV}_{F_T T}(t) \right\rangle$ does not depend on S_t. From (12.49) $d\delta\widehat{\sigma}^{SV}_{F_T T}$ is given by:

$$
d\delta\widehat{\sigma}^{SV}_{F_T T} = \nu\alpha_\theta\widehat{\sigma}_T\left((1-\theta)A_1 dW^1 + \theta A_2 dW^2 \right)
$$

thus $\left\langle d\ln S_t \, d\delta\widehat{\sigma}^{SV}_{F_T T}(t) \right\rangle$ does not depend on S_t either.

The conclusion is that expression (12.57) for $\mathcal{S}_T$ holds in our mixed model, at order one in $\alpha(t)$ and ν.

Consider two models generating the same smile – say the two-factor stochastic volatility model and a local volatility model calibrated on the smile of the former. From (12.57), the ATMF skew of the smile used for calibration sets the value of the integrated spot/volatility covariance, for all T.

Time-homogeneous models

If the model were time-homogeneous, that is if $\left\langle d\ln S_t \, d\widehat{\sigma}^2_T(t) \right\rangle$ were a function of $T-t$, knowledge of $\mathcal{S}_T$ for all T would determine $\left\langle d\ln S_t \, d\widehat{\sigma}^2_T(t) \right\rangle$ and in particular its value for $t = 0$: the instantaneous covariance of $\ln S$ with the ATMF volatility of maturity T.

Two time-homogeneous models calibrated to the same smile would generate identical SSRs – provided the assumptions needed for (12.57) hold – and also the same future ATMF skews.

The reason why the local volatility model generates a different SSR than a time-homogeneous stochastic volatility model is due to the non-homogeneity of the former. The larger SSRs generated by the local volatility model in Figure 9.9 point to the fact that, in the integral in (12.57), the contribution from covariances for t near 0 to the integral in (12.57) is larger in the local volatility model than in the two-factor model.

12.6.1 Future smiles in mixed models

The implication of the larger SSRs in the local volatility model is that future spot/volatility covariances are lower in the local volatility model; thus accounting for weaker future skews. Mixed models parametrized such that their SSR is lower than that of the local volatility model will thus produce larger future skews. Because, as seen in Section 2.6, page 63, future skews generated by the local volatility component quickly die off, future skews for far-away forward dates are predominantly generated by the stochastic volatility component of the mixed model.

We illustrate these properties using the example of foward-start call spreads struck at $K_{lo} = 95\%$ and $K_{hi} = 105\%$ with maturity 3 months. The payoff of each forward-start option is

$$\left(\frac{S_{T+\Delta}}{S_T} - K_{lo}\right)^+ - \left(\frac{S_{T+\Delta}}{S_T} - K_{hi}\right)^+$$

where $\Delta = 3$ months and the forward-start dates T are quarterly dates, from $T = 0$ to $T = 4$ years and 9 months. We thus have 20 such options.

We take vanishing rate and repo, use as "market smile" the smile generated by the two-factor model, parametrized as in Table 12.1, page 479, and a term structure of VS volatilities flat at 20%.

We price our 20 forward-start options in four different models.[11]

I the two-factor model thus parametrized.

II a mixed model calibrated to the "market smile", with the underlying two-factor model having a value of ν halved – this is the model that was used to produce Figure 12.4.

III the local volatility model calibrated to the "market smile".

IV the two-factor model parametrized as in I, but with ν halved. This is the stochastic volatility portion of model II.

For each foward-start date T, we also price the forward-start ATM option that pays $\left(\frac{S_{T+\Delta}}{S_T} - 1\right)^+$ and imply a Black-Scholes volatility. These implied volatilities appear in Figure 12.7.

Observe that for $T = 0$, our forward option is simply a vanilla option of maturity 3 months, thus has identical prices in models I, II, III as they generate the same vanilla smile. Thus the three curves collapse onto the same value for the first option.

For further-away forward-start dates, implied volatilities – hence prices – are lower in model I. This can be traced mainly to the higher volatility of volatility in model I. We refer the reader to the discussion of forward-start options in Section 3.1.6 of Chapter 3, page 111. For a forward-start ATM call option, adjustment δP_1 is negative and more so when volatility of volatility is higher.

We now turn to the foward call spreads. From the price of each call spread we imply a forward skew $\mathcal{S}$ by equating the model price to a Black-Scholes price calculated using implied volatilities $\widehat{\sigma}_{ATM} + \mathcal{S}\ln(K_{lo})$ for strike K_{lo}, and $\widehat{\sigma}_{ATM} + \mathcal{S}\ln(K_{hi})$ for strike K_{hi}. $\widehat{\sigma}_{ATM}$ is the ATM implied volatility of the corresponding

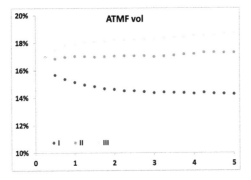

Figure 12.7: Implied volatilities of forward-start ATM options of maturity as a function of their maturities $T + \Delta$.

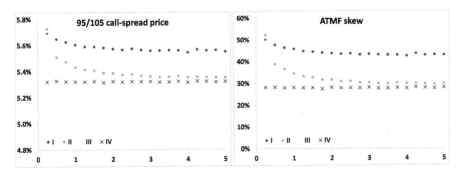

Figure 12.8: Prices (left) and implied skews (right, multiplied by -1) of forward-start 95/105 call spreads as a function of their maturities $T + \Delta$.

ATM call option. Prices and forward skews (multiplied by -1) are shown in Figure 12.8.

- First note the similarity of both graphs in Figure 12.8: the price of a narrow call spread centered on the money directly reflect the ATM skew.

- Then observe how the forward skew in mixed model II is lower than that of the pure stochastic volatility model I, while the forward skew in local volatility model IV is lower still. This confirms what we mentioned above, based on formula (12.58) for the SSR: higher SSRs translate into lower forward skews.

- Finally observe how prices – and implied skews – in model II tend towards those of model IV, for far-away forward dates. Model IV is in fact the stochastic volatility component of model II. This confirms that, indeed, the local volatility

---

[11]I thank Pierre Henry-Labordère for generating these results.

component of mixed model II hardly generates any forward skew far into the future.

Thus parameters of the underlying two-factor model can be chosen to control the forward skew of the mixed model. Long-dated cliquets hardly have any sensitivity to the local volatility component.

12.7   Conclusion

- Provided condition (12.32) holds, local-stochastic volatility models are legitimate models that can be used in trading applications. They are market models for the spot and vanilla options that possess a Markovian representation in terms of the spot price and the state variables of the underlying stochastic volatility model.

  Unlike the local volatility model – also a market model for the same instruments – local-stochastic volatility models afford some control on spot/volatility and volatility/volatility break-even levels, while maintaining calibration to the vanilla smile. This is achieved by appropriately choosing the parameters of the underlying stochastic volatility model.

- The approximate expressions for $d\widehat{\sigma}_T$ in Section 12.4.1 and the resulting approximate formulas for SSR, volatilities of volatilities and spot/volatility correlations allow for a quick assessment of the model's break-even levels for a given market smile, and how the latter vary if the market smile changes – which of course is bound to happen as we risk-manage a derivative position through time.

- In practice, unlike delta hedging, vega hedging is typically not performed on a daily basis, because of larger bid/offer costs. See the discussion in Section 9.11.3, page 383, regarding the carry P&L of a delta-hedged, vega-hedged position when vega and delta rehedging frequencies differ.

- A lower SSR than in the local volatility model also translates into stronger future skews. Future smiles in mixed models are mostly generated by the stochastic volatility component. Thus, mixed models can be parametrized so as to achieve given levels of future skews – which are related to *future* break-even levels for the spot/volatility covariance – rather than choosing parameters to achieve desired *present* break-even levels.

- Local-stochastic volatility models can be used to price and risk-manage path-dependent payoffs that involve spot observations.

  Unlike the forward variances models of Chapter 7, they are not well-suited to options involving VIX observations, as forward variances – let alone VIX futures – are not directly accessible in the model.

Whenever one needs to (a) have a handle on future skews, (b) have explicit access to VIX futures or forward variances, the discrete forward variance models of Chapter 7 should be employed.

Appendix A – alternative schemes for the PDE method

We present here a technique for deriving schemes for the forward equation (12.12) for the density from schemes for the backward equation for prices – this idea was first proposed by Jesper Andreasen and Brian Huge – see [1]. We use here the notations of Section 12.2.4. While the density $\varphi(t, S, X) = E[\delta(S - S_t)\delta(X - X_t)]$ obeys the forward equation (12.12), the undiscounted price $P(t, S, X)$ of a European option satisfies the usual backward equation:

$$\frac{dP}{dt} = -LP \tag{12.59a}$$

$$L = L_S + L_X + L_{SX} \tag{12.59b}$$

with terminal condition $P(t = T, S, X) = g(S)$ where T is the option's maturity and g is the option's payoff. Operators L_S, L_X, L_{SX} read:

$$L_S = (r - q) S \frac{d}{dS} + \frac{1}{2} f(t, X)\sigma(t, S)^2 S^2 \frac{d^2}{dS^2}$$

$$L_X = -kX \frac{d}{dX} + \frac{1}{2} \frac{d^2}{dX^2}$$

$$L_{SX} = \rho\sqrt{f(t, X)}\sigma(t, S) S \frac{d^2}{dSdX}$$

Assume we have φ and P at time t. Then the price at $t = 0$, $p_0 = P(t = 0, S_0, X_0)$ can be written as:

$$p_0 = \iint \varphi(t, S, X) P(t, S, X) \, dSdX \tag{12.60}$$

Since this holds for any $t \in [0, T]$, the derivative with respect to t of the right-hand side vanishes:

$$\iint \left(\frac{d\varphi}{dt} P + \varphi \frac{dP}{dt}\right) dSdX = 0 \tag{12.61}$$

(12.61) is a consistency condition relating the forward PDE (12.12) for φ and the backward PDE (12.59) for P.

Given operator L for the backward equation, the forward operator $\mathcal{L}$ is such that:

$$\iint (\mathcal{L}\varphi)\, P\, dS dX \;=\; \iint \varphi\, (LP)\; dS dX \tag{12.62}$$

(12.62) can be considered a definition of $\mathcal{L}$.[12]

Consider now a discretization of φ and P on the same (S, X) grid with spacings δS, δX. As we did in Section 12.2.4 we now use the notation φ and P to also denote their discretized version – φ and P are then vectors of size $n_S n_X$; similarly, L, L will also designate matrices, generated by replacing differential operators with their discretized versions in (12.14).

Consider a discrete time step δt and assume we have a scheme – i.e. a matrix $U_{t,t+\delta t}$ – for the backward evolution of P over $[t, t + \delta t]$:

$$P(t) \;=\; U_{t,t+\delta t}\, P(t + \delta t) \tag{12.63}$$

In the discretized version of our problem, integrals are converted into sums, and equation (12.60) giving the price at $t = 0$ – a scalar – translates into:

$$p_0 \;=\; \mathcal{N}\varphi(t)^{\mathsf{T}} P(t)$$

where $\varphi(t)^{\mathsf{T}}$ denotes the transpose of vector $\varphi(t)$ and $\mathcal{N}$ is a normalization constant: $\mathcal{N} = \delta S \delta X$. Now use (12.63):

$$\begin{aligned} p_0 &= \mathcal{N}\, \varphi(t)^{\mathsf{T}} U_{t,t+\delta t}\, P(t + \delta t) \\ &= \mathcal{N}\, \left(U_{t,t+\delta t}^{\mathsf{T}} \varphi(t) \right)^{\mathsf{T}} P(t + \delta t) \end{aligned} \tag{12.64}$$

Identifying the right-hand side of (12.64) with $\mathcal{N}\, \varphi(t + \delta t)^{\mathsf{T}} P(t + \delta t)$ yields the following relationship between $\varphi(t)$ and $\varphi(t + \delta t)$:

$$\varphi(t + \delta t) \;=\; U_{t,t+\delta t}^{\mathsf{T}}\, \varphi(t) \tag{12.65}$$

The upshot is that we can obtain a numerical scheme for the forward equation for φ by simply taking the transpose of a scheme for the backward equation for prices – the boundary conditions are automatically taken care of.

Imagine the terminal payoff $g(S)$ is a constant g – then $P(t, S, X) = g\ \forall t, S, X$. The condition that our numerical scheme for P complies with this requirement reads:

$$U_{t,t+\delta t}\mathbf{1} \;=\; \mathbf{1} \tag{12.66}$$

where vector $\mathbf{1}$ has its $n_S n_X$ components all equal to 1. $U_{t,t+\delta t} = e^{\delta t(L_S + L_X + L_{SX})}$. The matrices representing L_S, L_X and L_{SX}, must thus be such that when acting on vector $\mathbf{1}$, whose $n_S n_X$ components are all equal to 1, the resulting vector vanishes – equivalently the components of each line of L_S, L_X, L_{SX} add

---

[12]Starting from the left-hand side of (12.62) and integrating by parts to generate the right-hand equivalent, one can check that one indeed obtains $\mathcal{L}_S, \mathcal{L}_X, \mathcal{L}_{SX}$ in (12.13).

up to zero. Consider for example the operator $\frac{d^2}{dS^2}$. Owing to (12.14) the non-zero matrix elements of line i of its discretized version are given for $i > 0$ by:

$$\frac{d^2}{dS^2}_{i,i-1} = \frac{1}{\delta S^2}, \quad \frac{d^2}{dS^2}_{i,i} = -\frac{2}{\delta S^2}, \quad \frac{d^2}{dS^2}_{i,i+1} = \frac{1}{\delta S^2}$$

and sum up to zero. For $i = 0$ and $i = n_S - 1$ we can enforce the typical boundary condition $\frac{d^2P}{dS^2} = 0$ – which is natural for vanilla payoffs, whose asymptotic profiles are affine, which translates into all elements of the first line $(i = 0)$ and last line $(i = n_S - 1)$ vanishing. The same conditions we impose in the X directions. With regard to L_{SX} conditions on edges and corners have to be such that each line sums up to zero.

Multiplying each side of (12.65) on the left by 1^T yields:

$$1^\mathsf{T}\varphi(t + \delta t) = 1^\mathsf{T}U_{t,t+\delta t}^\mathsf{T}\,\varphi(t) = 1^\mathsf{T}\varphi(t) \qquad (12.67)$$

$1^\mathsf{T}\varphi(t)$ is the sum of the elements of $\varphi(t)$ – up to the normalizing constant N, the integral $\iint \varphi(t, S, X)dSdX$ evaluated numerically. Thus, using $U_{t,t+\delta t}^\mathsf{T}$ as a scheme for $\varphi(t)$ ensures that the numerical integral of the probability density $\varphi(t, S, X)$ is conserved in the forward scheme: starting with an initial density that integrates to one, further densities do so as well.

Besides taking care of boundary conditions, this is another attractive feature of using the transpose of the backward scheme. Note though, that generally it will not be possible to ensure that elements of $\varphi(t)$ are all positive.

The final recipe is simple: choose a scheme for the backward equation and transpose it.

Vanishing correlation
The backward equation is identical to the forward equation, except the initial vector is $P(t + \delta t)$ and the final one is $P(t)$; we can recycle the schemes used in Section 12.2.4.
If ρ vanishes the Peaceman-Rachford algorithm, expressed by (12.16) for the forward equation, reads – for P:

$$P(t) = I_S E_X I_X E_S P(t + \delta t) \qquad (12.68)$$

where E_S, I_S are given by

$$E_S = 1 + \frac{\delta t}{2}L_S \qquad I_S = \left(1 - \frac{\delta t}{2}L_S\right)^{-1}$$

and likewise for E_X, I_X. To get the numerical algorithm for φ, simply transpose (12.68):

$$\varphi(t + \delta t) = E_S^\mathsf{T}I_X^\mathsf{T}E_X^\mathsf{T}I_S^\mathsf{T}\varphi(t)$$

which is implemented, for example, through the following sequence:

$$\left(1 - \frac{\delta t}{2}L_S^\intercal\right)\varphi^* = \varphi(t)$$

$$\left(1 - \frac{\delta t}{2}L_X^\intercal\right)\varphi^{**} = \left(1 + \frac{\delta t}{2}L_X^\intercal\right)\varphi^*$$

$$\varphi(t + \delta t) = \left(1 + \frac{\delta t}{2}L_S^\intercal\right)\varphi^{**}$$

It entails the same number of multiplications/inversions of tridiagonal matrices as in (12.15).

Non-vanishing correlation

Take the predictor-corrector scheme (12.21), express it using backward, rather than forward, operators, and transpose it:

$$\varphi(t + \delta t) = \left(E_S^\intercal E_X^\intercal + \frac{\delta t}{2}L_{SX}^\intercal + \frac{\delta t}{2}\left(E_S^\intercal E_X^\intercal + \delta t L_{SX}^\intercal\right)I_X^\intercal I_S^\intercal L_{SX}^\intercal\right)I_X^\intercal I_S^\intercal \varphi(t) \tag{12.69}$$

This can be implemented through the following sequence consisting of a "predictor" step:

$$\left(1 - \frac{\delta t}{2}L_S^\intercal\right)\varphi^0 = \varphi(t)$$

$$\left(1 - \frac{\delta t}{2}L_X^\intercal\right)\varphi^* = \varphi^0$$

$$\tilde{\varphi} = \delta t L_{SX}^\intercal \varphi^*$$

a "corrector" step:

$$\left(1 - \frac{\delta t}{2}L_S^\intercal\right)\varphi^1 = \tilde{\varphi}$$

$$\left(1 - \frac{\delta t}{2}L_X^\intercal\right)\varphi^2 = \varphi^1$$

$$\overline{\varphi} = \left(1 + \frac{\delta t}{2}L_S^\intercal\right)\left(1 + \frac{\delta t}{2}L_X^\intercal\right)\varphi^2 + \delta t L_{SX}^\intercal \varphi^2$$

and a final step:

$$\varphi(t + \delta t) = \left(1 + \frac{\delta t}{2}L_S^\intercal\right)\left(1 + \frac{\delta t}{2}L_X^\intercal\right)\varphi^* + \frac{1}{2}\left(\tilde{\varphi} + \overline{\varphi}\right)$$

Again, (12.69) can be implemented in multiple manners.

Chapter's digest

12.2 Pricing equation and calibration

▶ In local-stochastic volatility models, the instantaneous volatility of the underlying is written as the product of a local volatility component and a stochastic volatility component, generated by an underlying stochastic volatility model:
$\sigma_t = \sqrt{\zeta_t^t}\,\sigma(t, S_t)$.

▶ The local volatility component is calibrated so that the vanilla smile is recovered. Two methods can be used: (a) a PDE-based method based on the solution of the forward equation for the density, practically applicable to the one-factor case only, (b) the particle method, a general Monte Carlo technique that can be used regardless of the dimensionality of the underlying stochastic volatility model.

🐦 🐦 🐦 🐦 🐦

12.3 Usable models

▶ Local-stochastic volatility models are calibrated to the spot value and the vanilla smile. Are they market models for these assets – that is can they be used for trading purposes? Can the P&L of a delta-hedged, vega-hedged position be written as the sum of gamma/theta contributions involving all hedging instruments, with well-defined and payoff-independent break-even levels?

▶ The condition for a local-stochastic volatility model to be usable is that prices should not depend on the state variables of the underlying stochastic volatility model. Otherwise, spurious contributions to the P&L appear, that have no financial meaningfulness, causing P&L leakage. This condition is not met for most models proposed in the literature.

▶ The two-factor model satisfies the admissibility condition. The Heston model does not; neither does the Bloomberg model. A model whose underlying stochastic volatility model has a lognormal instantaneous volatility meets the admissibility criterion.

🐦 🐦 🐦 🐦 🐦

12.4 Dynamics of implied volatilities

▶ Models belonging to the admissible class possess well-defined break-even gamma/theta levels for the spot and vanilla implied volatilities. The latter can be computed in a Monte Carlo simulation of the model, but it is useful to have approximate values for volatilities of volatilities, spot/volatility correlations, and SSR, for the sake of sizing up model-generated break-even levels, and choosing parameters of the underlying stochastic volatility model.

▶ We derive approximate levels of volatilities of ATMF volatilities, spot/ATMF volatility correlations, and SSR in an expansion at order one in the volatility of volatility of the underlying stochastic volatility model, and in the local volatility component. We obtain expressions that only involve the market smile and parameters of the underlying stochastic volatility model. The (calibrated) local volatility does not appear explicitly.

 za za za za za

12.6 Discussion

▶ We run numerical tests using as market smile that generated by the underlying stochastic volatility model, so that the case of pure stochastic volatility can be spanned.

▶ The accuracies of volatilities of ATMF volatilities, spot/ATMF volatility correlations, and SSR obtained in the order-one expansion are sufficient for choosing the parameters of the underlying two-factor model so as to generate the desired level of SSR.

▶ By selecting the level of volatility of volatility of the underlying stochastic volatility model, it is possible to cover the range of SSRs from that of the pure stochastic volatility to that of the pure local volatility, and also to explore SSRs outside of this range.

▶ A local volatility model calibrated to the smile of a stochastic volatility model produces higher SSRs. This is consistent with the fact that future smiles are weaker in the local volatility model. Future smiles in local-stochastic volatility models are mostly generated by the stochastic volatility component of the model.

Epilogue

Bornons ici cette carrière.
Les longs ouvrages me font peur.
Loin d'épuiser une matière,
On n'en doit prendre que la fleur.

La Fontaine, *Fables, VI*

Et pour ceux qui joignent le bon sens avec l'étude, lesquels seuls je souhaite
pour mes juges [...]

Descartes, *Discours de la méthode*

Quant au mouvement en lui-même, je vous le déclare avec humilité, nous
sommes impuissants à le définir.

Balzac, *La Peau de chagrin*

– This excursion, Reader, ends here. I hope the journey has been pleasant and
instructive.

We attempt to build market models for hedging instruments – the underlying
along with vanilla options, or a subset of them, or variance swaps, ot yet other
convex payoffs – such that (a) the P&L of a hedged option position is of the typical
gamma/theta form, with payoff-independent break-even levels, (b) these break-even
levels are – at least partially – in our control.

We could consider higher-order contributions but, in practice, managing the risk
of all second-order greeks at a book level is a formidable task already.

Vanilla options are not quite independent instruments, however; they are related
to one another and to the underlying itself through their terminal condition. At
order one in volatility of volatility, for example, the SSR of homogeneous stochastic
volatility models is related to the decay of the AMTF skew.

Characterizing the restrictions that the initial configuration of vanilla implied
volatilities places on their future evolution is an unsettled issue, that may be more
easily addressed by modeling other convex payoffs, for example the power payoffs
of Chapter 4.

Even achieving objective (a) is not as straightforward, as highlighted by our
analysis of mixed models in Chapter 12. Expressing the instantaneous volatility as
the product of local and stochastic volatility components, a seemingly reasonable
and innocuous ansatz, generally leads to non-functional models, for lack of a proper
breakdown of the carry P&L.

This serves as a reminder that derivatives modeling does not start with the
assumption of a process for S_t, but with clearly articulated modeling objectives

relating to observable quantities, regardless of how these objectives are achieved mathematically.

A commonplace statement one has been hearing at recent quant conferences is that the age of modeling is now over, that quantitative finance has become but a tedious form of accounting, the preserve of a new order of adjusters – xxA-quants: CVA-quants, FVA-quants, KVA-quants, etc.

Focusing on derivatives' risks, on their proper modeling and the meaningfulness of model-generated prices – the things being adjusted – is the symptom of an old-fashioned, outdated, mindset.

I respectfully dissent. Product risks and modeling choices are still begging for a proper understanding; forty years after Black-Scholes, work on the next generation of models has just started.

What about calibration, a deceptive notion we should strive to abolish? No one says that they "calibrate" the spot value in the Black-Scholes model: a model should naturally take as inputs the market values of instruments used as hedges. Then, *parameters* should be chosen so as to generate the desired break-even levels for the carry P&L.

Should they be calibrated to market prices? This is meaningful only if the difference between their calibrated and realized values can be materialized as the P&L of an actual trading strategy, a rare occurrence.

Therefore, Reader, you will do well to resist the compulsion of calibration and the addictive psychological reward that comes with it.

With these last words of encouragement, Reader, I bid you farewell.

Bibliography

[1] Andreasen, J., Huge B.: *Random grids*, Risk Magazine, July, pp. 66–71, 2011.

[2] Andersen, L., Andreasen, J.: *Jump-Diffusion Processes: Volatility Smile Fitting and Numerical Methods for Pricing*, available at SSRN: http://ssrn.com/abstract=171438.

[3] Avellaneda, M., Levy, A., Paras, A.: *Pricing and hedging derivative securities in markets with uncertain volatilities*, Applied Mathematical Finance 2(2), pp. 73–88, 1995.

[4] Avellaneda, M., Paras, A.:*Pricing and hedging derivative securities in markets with uncertain volatilities: the Lagrangian uncertain volatility model*, Applied Mathematical Finance, 3, pp. 21–52, 1996.

[5] Backus, D., Foresi, S., Wu, L.: *Accounting for biases in Black-Scholes*, available at SSRN: http://ssrn.com/abstract=585623, 1997.

[6] Berestycki, H., Busca, J., Florent, I.: *Asymptotics and calibration of local volatility models*, Quantitative Finance, 2(1), pp. 61–69, 2002.

[7] Berestycki, H., Busca, J., Florent, I.: *Computing the implied volatility in stochastic volatility models*, Communications on Pure and Applied Mathematics, Vol. LVII, pp. 1352–1373, 2004.

[8] Bergomi, L.: *Smile dynamics*, Risk Magazine, September, pp. 117–123, 2004. Also available at SSRN: http://ssrn.com/abstract=1493294.

[9] Bergomi, L.: *Smile dynamics II*, Risk Magazine, October, pp. 67–73, 2005. Also available at SSRN: http://ssrn.com/abstract=1493302.

[10] Bergomi, L.: *Smile dynamics III*, Risk Magazine, October, pp. 90–96, 2008. Also available at SSRN: http://ssrn.com/abstract=1493308.

[11] Bergomi, L.: *Smile dynamics IV*, Risk Magazine, December, pp. 94–100, 2009. Also available at SSRN: http://ssrn.com/abstract=1520443.

[12] Bergomi, L.: *Correlations in asynchronous markets*, Risk Magazine, November, pp. 76–82, 2010. Also available at SSRN: http://ssrn.com/abstract=1635866.

[13] Bergomi, L., Guyon, J.: *Stochastic volatility's orderly smiles*, Risk Magazine, May, pp. 60–66, 2012. Also available at SSRN: http://ssrn.com/abstract=1967470.

[14] Bick, A.: *Quadratic-variation-based dynamic strategies*, Management Science, 41(4), pp. 722–732, 1995.

[15] Björk, T., Blix, M., Landén, C.: *On finite dimensional realizations for the term structure of futures prices*, International Journal of Theoretical and Applied Finance, 9(03), pp. 281–314, 2006.

[16] Bos, M., Vandermark, S.: *Finessing fixed dividends*, Risk Magazine, September, pp. 157–158, 2002.

[17] Bouchaud, J. Ph., Potters, M., Sestovic, D.: *Hedge your Monte Carlo*, Risk Magazine, March, pp. 133–136, 2001. Also available at SSRN: http://ssrn.com/abstract=238868.

[18] Breeden, D., Litzenberger, R.: *Prices of state contingent claims implicit in option prices*, Journal of Business, 51, pp. 621–651, 1978.

[19] Buehler, H.: *Volatility and dividends – volatility modeling with cash dividends and simple credit risk*, available at SSRN: http://ssrn.com/abstract=1141877.

[20] Buehler, H.: *Consistent variance curve models*, Finance and Stochastics, 10(2), pp. 178–203, 2006. Also available at SSRN: http://ssrn.com/abstract=687258.

[21] Carmona, R., Nadtochiy, S.: *Local volatility dynamic models*, Finance and Stochastics (13), pp. 1–48, 2009.

[22] Carr, P., Chou, A.: *Breaking barriers*, Risk Magazine, September, pp. 139–145, 1997.

[23] Carr, P., Madan, D.: *Determining volatility surfaces and option values from an implied volatility smile*, Quantitative Analysis in Financial Markets, Vol II, M. Avellaneda, ed, pp. 163–191, 1998.

[24] Carr, P., Lewis, K.: *Corridor variance swaps*, Risk Magazine, February, pp. 67–72, 2004.

[25] Carr, P., Madan, D.: *Towards a theory of volatility trading*, Volatility, Risk Publications, Robert Jarrow, ed., pp. 417–427, 2002.

[26] Carr, P., Geman, H., Madan, D., Yor, M.: *Stochastic volatility for Lévy processes*, Mathematical Finance 13(3), pp. 345–382, 2003.

[27] Carr, P., Lee, R.: *Robust replication of volatility derivatives*, available at: http://www.math.uchicago.edu/~rl/rrvd.pdf, 2009.

[28] Carr, P., Lee, R. *Hedging variance options on continuous semimartingales*, Finance and Stochastics, 14 (2), pp. 179–207, 2010.

[29] Castagna, A., Mercurio, F.: *The vanna-volga method for implied volatilities*, Risk Magazine, January, pp. 106–111, 2007.

[30] Cheyette, O.: *Markov representation of the Heath-Jarrow-Morton model*, available at SSRN: http://ssrn.com/abstract=6073.

[31] Cherny, A., Dupire, B.: *On certain distributions associated with the range of martingales*, Optimality and Risk – Modern Trends in Mathematical Finance, pp. 29-38, 2010.

[32] Chriss, N., Morokoff, W.: *Market risk of variance swaps*, Risk Magazine, October, pp. 55–59, 1999.

[33] Cont, R., Tankov, P.: *Financial modeling with jump processes*, Chapman & Hall / CRC Press, Financial Mathematics Series, 2003.

[34] Cox, A., Wang, J.: *Optimal robust bounds for variance options*, http://arxiv.org/abs/1308.4363, 2013.

[35] Davis, M. H., Panas, V. G., Zariphopoulou, T.: *European option pricing with transaction costs*, SIAM Journal on Control and Optimization, 31(2), pp. 470-493, 1993.

[36] De Marco, S., Henry-Labordère, P.: *Linking vanillas and VIX options: a constrained martingale optimal transport problem*, available at SSRN: http://ssrn.com/abstract=2354898.

[37] Derman, E., Kani, I.: *Riding on a smile*, Risk Magazine, February, pp. 32–39, 1994.

[38] Duanmu, Z.: *Rational pricing of options on realized volatility – the Black Scholes way*, Global Derivatives conference, Madrid, 2004.

[39] Derman, E., Kani, I.: *Stochastic implied trees: arbitrage pricing with stochastic term and strike structure of volatility*, International Journal of Theoretical and Applied Finance, 1(01), pp. 61–110, 1998.

[40] Dupire, B.: *Pricing with a smile*, Risk Magazine, January, pp. 18–20, 1994.

[41] Dupire, B.: *Functional Ito calculus and volatility hedge*, Global Derivatives conference, Paris, 2009.

[42] Dupire, B.: *Arbitrage bounds for volatility derivatives as free boundary problem*, available at http://www.math.kth.se/pde_finance/presentations/Bruno.pdf, 2005.

[43] Durrleman, V.: *From implied to spot volatilities*, Finance and Stochastics, 14 (4), pp. 157–177, 2010.

[44] Durrleman, V., El Karoui, N.: *Coupling smiles*, Quantitative Finance, 8(6), pp. 573–590, 2008. Also available at SSRN: http://ssrn.com/abstract=1005332.

[45] Fisher, T., Tataru, G.: *Non-parametric stochastic local volatility modeling*, Global Derivatives conference, Paris, 2010.

[46] Fukasawa, M.: *The normalizing transformation of the implied volatility smile*, Mathematical Finance, 22(4), pp. 753–762, 2012. Preprint version available at http://arxiv.org/abs/1008.5055, 2010.

[47] Galichon, A., Henry-Labordère, P., Touzi, N.: *A stochastic control approach to no-arbitrage bounds given marginals, with an application to lookback options*, Annals of Applied Probability, 24(1), pp. 312–336, 2014.

[48] Gatheral, J.: *The volatility surface: a practitioner's guide*, Wiley Finance, 2006.

[49] Gatheral, J., Jacquier, A.: *Arbitrage-free SVI volatility surfaces*, Quantitative Finance, 14(1), pp. 59–71, 2014. Also available at SSRN: http://ssrn.com/abstract=2033323.

[50] Green, R.C., Jarrow, R.A.: *Spanning and completeness in markets with contingent claims*, Journal of Economic Theory, 41, pp. 202–210, 1987.

[51] Guyon, J., Henry-Labordère, P.: *From spot volatilities to implied volatilities*, Risk Magazine, June, pp. 79–84, 2011. Also available at SSRN: http://ssrn.com/abstract=1663878.

[52] Guyon, J., Henry-Labordère, P.: *Being particular about calibration*, Risk Magazine, January, pp. 92–97, 2012.

[53] Guyon, J., Henry-Labordère, P.: *Nonlinear option pricing*, Chapman & Hall/CRC Press, Financial Mathematics Series, 2013.

[54] Gyöngi, I.: *Mimicking the one-dimensional marginal distributions of processes having an Ito differential*, Probability Theory and Related Fields, 71, pp. 501–516, 1986.

[55] Hagan, P.S., Kumar, D., Lesniewski, A.S., Woodward, D.E.: *Managing smile risk*, Wilmott Magazine, September, pp. 84–108, 2002.

[56] Henry-Labordère, P.: *Analysis, geometry, and modeling in finance: advanced methods in option pricing*, Chapman & Hall/CRC Press, Financial Mathematics Series, 2008.

[57] Henry-Labordère, P.: *Automated option pricing: numerical methods*, available at SSRN: http://ssrn.com/abstract=1968344, 2011.

[58] Henry-Labordère, P.: *Calibration of local stochastic volatility models to market smiles: a Monte-Carlo approach*, Risk Magazine, September, pp. 113–117, 2009. Also available at SSRN: http://ssrn.com/abstract=1493306.

[59] Henry-Labordère, P.: *Vega decomposition of exotics on vanillas: a Monte-Carlo approach*, available at SSRN: http://ssrn.com/abstract=2229990.

[60] Heston, S.: *A closed-form solution for options with stochastic volatility with applications to bond and currency options*, Review of Financial Studies, 6(2), pp. 327–343, 1993.

[61] Higham, N. J.: *Computing the nearest correlation matrix – a problem from finance*, IMA Journal of Numerical Analysis, 22(3), pp. 329–343, 2002.

[62] Hobson, D.: *The Skorokhod embedding problem and model-independent bounds for option prices*, Paris-Princeton Lectures on Mathematical Finance, pp. 267–318, 2010.

[63] Hobson, D., Klimmek, M.: *Model-independent hedging strategies for variance swaps*, Finance and Stochastics, 16(4), pp. 611–649, 2012.

[64] Hunt, P., Kennedy, J., Pelsser, A.: *Markov-functional interest rate models*, Finance and Stochastics, 4(4), pp. 391–408, 2000.

[65] Lee, R.: *Weighted variance swap*, Encyclopedia of Quantitative Finance, 2010. Also available at http://math.uchicago.edu/~rl/EQF_weightedvarianceswap.pdf.

[66] Lee, R.: *Implied and local volatilities under stochastic volatility*, International Journal of Theoretical and Applied Finance, 4(01), pp. 45–89, 2001.

[67] Lee, R.: *The moment formula for implied volatility at extreme strikes*, Mathematical Finance, 14(3), pp. 469–480, 2004.

[68] Leland, H. E.: *Option pricing and replication with transaction costs*, The Journal of Finance, 40(5), pp. 1283–1301, 1985.

[69] Lewis, A.: *Option valuation under stochastic volatility*, Finance Press, 2000.

[70] Lipton, A.: *The vol smile problem*, Risk Magazine, February, pp. 61–65, 2002.

[71] Lyons, T.: *Uncertain volatility and the risk-free synthesis of derivatives*, Applied Mathematical Finance, 2(2), pp. 117–133, 1995.

[72] Marcinkiewicz, J.: *Sur une propriété de la loi de Gauss*, Mathematische Zeitschrift, 44, pp. 612–618, 1939.

[73] Matytsin, A.: *Perturbative analysis of volatility smiles*, Columbia Practitioners Conference on the Mathematics of Finance, New York, 2000.

[74] Nachman, D.: *Spanning and completeness with options*, Review of Financial Studies, 3, 31, pp. 311–328, 1988.

[75] Neuberger, A.: *The log contract*, Journal of Portfolio Management 20, 1994.

[76] Nicolay, D.: *Asymptotic Chaos Expansions in Finance: Theory and Practice*, Springer Finance, Springer Finance Lecture Notes, 2014.

[77] Papanicolaou, A.: *Extreme-strike comparisons and structural bounds for SPX and VIX Options*, available at SSRN: http://ssrn.com/abstract=2532020.

[78] Piterbarg, V.: *Time to smile*, Risk Magazine, May, pp. 71–75, 2005.

[79] Romano, M., Touzi, N.: *Contingent claims and market completeness in a stochastic volatility model*, Mathematical Finance, 7(4), pp. 399–412, 1997.

[80] Schönbucher, P. J.: *A market model for stochastic implied volatility*, Phil. Trans. Roy. Soc., Ser. A 357, pp. 2071–2092, 1999.

[81] Schweizer, M., Wissel, J.: *Term structures of implied volatilities: absence of arbitrage and existence results*, Mathematical Finance, 18(1), pp. 77–114, 2008.

[82] Sepp, A.: *Efficient Numerical PDE Methods to Solve Calibration and Pricing Problems in Local Stochastic Volatility Models*, available at http://kodu.ut.ee/~spartak/.

[83] Strikwerda, J.: *Finite Difference Schemes and Partial Differential Equations*, Society for Industrial and Applied Mathematics, 2nd edition, 2007.

[84] Whalley, A. E., Wilmott P.: *An asymptotic analysis of an optimal hedging model for option pricing with transaction costs*. Mathematical Finance 7(3), pp. 307–324, 1997.

[85] Willard, G.A.: *Calculating prices and sensitivities for path-independent derivative securities in multifactor models.* Journal of Derivatives, 5(1), pp. 45–61, 1997.

[86] Wissel, J.: *Arbitrage-free market models for option prices*, Working Paper Series, Working Paper No. 248, FINRISK, 2007.

Index

For Product Safety Concerns and Information please contact our EU
representative GPSR@taylorandfrancis.com
Taylor & Francis Verlag GmbH, Kaufingerstraße 24, 80331 München, Germany

www.ingramcontent.com/pod-product-compliance
Ingram Content Group UK Ltd.
Pitfield, Milton Keynes, MK11 3LW, UK
UKHW021836240425
457818UK00006B/216